D1045223

A General Theory
of Optimal Algorithms

This is a volume in the
ACM MONOGRAPH SERIES

Editor: THOMAS A. STANDISH, *University of California at Irvine*

A complete list of titles in this series appears at the end of this volume.

A General Theory
of Optimal Algorithms

J. F. TRAUB

Departments of Computer Science
and Mathematics
Columbia University
New York, New York

H. WOŹNIAKOWSKI

Institute of Informatics
University of Warsaw
Warsaw, Poland

1980

ACADEMIC PRESS

A Subsidiary of Harcourt Brace Jovanovich, Publishers

New York   London   Toronto   Sydney   San Francisco

COPYRIGHT © 1980, BY ACADEMIC PRESS, INC.
ALL RIGHTS RESERVED.
NO PART OF THIS PUBLICATION MAY BE REPRODUCED OR
TRANSMITTED IN ANY FORM OR BY ANY MEANS, ELECTRONIC
OR MECHANICAL, INCLUDING PHOTOCOPY, RECORDING, OR ANY
INFORMATION STORAGE AND RETRIEVAL SYSTEM, WITHOUT
PERMISSION IN WRITING FROM THE PUBLISHER.

ACADEMIC PRESS, INC.
111 Fifth Avenue, New York, New York 10003

United Kingdom Edition published by
ACADEMIC PRESS, INC. (LONDON) LTD.
24/28 Oval Road, London NW1 7DX

Library of Congress Cataloging in Publication Data

Traub, Joe Fred, Date
　A general theory of optimal algorithms.

　(ACM monograph series)
　Includes bibliographies and index.
　1.　Approximation theory−−Data processing.
2.　Mathematical optimization−−Data processing.
3.　Computational complexity−−Data processing.
I.　Woźniakowski, H. , joint author.　II.　Title.
III.　Series:　Association for Computing Machinery.
ACM monograph series.
QA297.T68　　　　　511'.4　　　　79−8859
ISBN　0−12−697650−3

AMS (MOS) 1970 Subject Classifications: 68C25, 65J10,
65J15

PRINTED IN THE UNITED STATES OF AMERICA

80 81 82 83　　9 8 7 6 5 4 3 2 1

QA
297
T68

To Pamela McCorduck
and Grażyna Woźniakowska
We believe they are optimal,
but have not yet done a complete search.

10-16-80   EBS   32.40   87

467086

Contents*

PREFACE   *xi*
RECOMMENDED READING   *xiii*

Overview   **1**

PART A   GENERAL INFORMATION MODEL
Introduction   *5*

Chapter 1   **Basic Concepts**
1. Introduction   *8*
2. Diameter and Radius of General Information   *9*
3. Complexity of General Information   *17*
4. The Scope of Analytic Computational Complexity   *22*

Chapter 2   **Theory of Linear Information**
1. Introduction   *25*
2. Cardinality of Linear Information   *27*
3. Index of a Linear Problem   *29*

*Chapter 4 of Part A was written jointly with G. W. Wasilkowski, University of Warsaw. Sections 4–8 of Part B were written jointly with B. Kacewicz, University of Warsaw.

4.   Optimal Linear Information Operators   *33*
5.   Convergence and Minimal Subspaces for a Hilbert Case   *38*
6.   Relations to Gelfand *n*-Widths   *41*
7.   Adaptive Linear Information   *47*

Chapter 3   Linear Algorithms for Linear Problems

1.   Introduction   *51*
2.   Preliminaries   *52*
3.   Linear Optimal Error Algorithms for Linear Functionals   *54*
4.   Linear Algorithms for Linear Operators   *60*
5.   Optimal Linear Algorithms and Linear Kolmogorov *n*-Widths   *64*

Chapter 4   Spline Algorithms for Linear Problems

1.   Introduction   *68*
2.   Deviation   *70*
3.   Splines   *71*
4.   Spline Algorithms   *72*
5.   Hilbert Case   *75*
6.   Non-Hilbert Case   *77*
7.   Summary   *81*

Chapter 5   Complexity for the Linear Case

1.   Introduction   *82*
2.   Complexity of Linear Information   *82*

Chapter 6   Applications for Linear Problems

1.   Introduction   *89*
2.   Linear Functionals   *91*
3.   Interpolation   *99*
4.   Integration   *107*
5.   Approximation   *121*
6.   Linear Partial Differential Equations   *132*
7.   Summary and Open Problems   *146*

Chapter 7   Theory of Nonlinear Information

1.   Introduction   *150*
2.   Cardinality of Nonlinear Information   *151*
3.   Optimal Nonlinear Information   *153*
4.   Relations to Kolmogorov *n*-Widths and Entropy   *157*

Chapter 8   Applications for Nonlinear Problems

1.   Introduction   *163*
2.   Optimal Nonadaptive Linear Information for Nonlinear Scalar Equations   *165*
3.   Adaptive Linear Information for Nonlinear Scalar Equations   *168*
4.   Nonlinear Information for Nonlinear Scalar Equations   *170*

5.   Multivariate and Abstract Nonlinear Equations   *173*
6.   Nearly Optimal Nonadaptive Linear Information for the Search
     for the Maximum of Unimodal Functions   *176*
7.   Adaptive Linear Information for the Search for the Maximum
     of Unimodal Functions   *179*

Chapter 9   **Complexity Hierarchy**

1.   Introduction   *185*
2.   Basic Definitions   *185*
3.   Complexity Hierarchy for Class Ψ_f^{non} *187*
4.   Complexity Hierarchy for Class Ψ_f^a *189*
5.   Complexity Hierarchy for Class Ψ_L^{non} *190*
6.   Complexity Hierarchy for Class Ψ_L^a *191*
7.   Complexity Hierarchy for Class Ψ_{NON} *192*
8.   Complexity Hierarchy for a Fixed Problem S *192*
9.   Summary   *193*

Chapter 10   **Different Models of Analytic Complexity**

1.   Introduction   *195*
2.   Average Case Model   *195*
3.   Relative Model   *196*
4.   Perturbed Model   *198*
5.   Asymptotic Model   *199*

Bibliography   **203**

Glossary   **218**

PART B   ITERATIVE INFORMATION MODEL

1.   Introduction   *223*
2.   Diameter and Order of Information   *227*
3.   Complexity of General Information   *234*
4.   Cardinality of Linear Information   *239*
5.   When Is the Class of Iterative Algorithms Empty?   *242*
6.   Index of the Problem S *245*
7.   The mth Order Iterations   *249*
8.   Complexity Index for Iterative Linear Information   *253*
9.   Information Operator with Memory   *257*
10.   Extensions and Open Problems   *264*
11.   Comparison of Results from General and Iterative Information Models   *268*
      Appendix   *269*

Bibliography   **272**

Glossary   **275**

PART C BRIEF HISTORY
AND ANNOTATED BIBLIOGRAPHY

1.   Introduction   *277*
2.   Brief History   *278*
3.   Annotated Bibliography   *281*

AUTHOR INDEX   *331*
SUBJECT INDEX   *335*

Preface

The purpose of this monograph is to create a general framework for the study of optimal algorithms for problems that are solved approximately. For generality the setting is abstract, but we present many applications to practical problems and provide examples to illustrate concepts and major theorems.

The work presented here is motivated by research in many fields. Influential have been questions, concepts, and results from complexity theory, algorithmic analysis, applied mathematics and numerical analysis, the mathematical theory of approximation (particularly the work on n-widths in the sense of Gelfand and Kolmogorov), applied approximation theory (particularly the theory of splines), as well as earlier work on optimal algorithms. But many of the questions we ask (see Overview) are new. We present a different view of algorithms and complexity and must request the reader's indulgence because new concepts and questions require some new vocabulary.

Because we believe our readers come from diverse backgrounds and have varied interests, we provide (see Recommended Reading) eight suggested tracks through the book.

The monograph marks the coming together of two streams of research. One stream, which involved work on optimal algorithms for problems such as the search for the maximum, integration, and approximation, had its inception with the work of Kiefer, Sard, and Nikolskij around 1950. These pioneer researchers, and those who followed them, generally worked, with only a few significant exceptions, on rather specific problems. Only rarely was the complexity of the optimal algorithm included. The second stream, which studied the solution of nonlinear equations, began with Traub in 1961. See Part C for a history and an extensive annotated bibliography.

In two long reports (General Theory of Optimal Error Algorithms and Analytic Complexity, Parts A and B) we showed that both streams could be united in one general framework. This monograph includes extended and improved material from these two reports.

Often a long monograph summarizes knowledge in a field. This monograph, however, may be viewed as a report on *work in progress*. We provide a foundation for a scientific field that is rapidly changing. Therefore we list numerous conjectures and open problems as well as alternative models which need to be explored. There are many more questions which we have chosen not to list.

We want to acknowledge many debts. B. Kacewicz and G. W. Wasilkowski coauthored some parts of this book as indicated in the table of contents. In addition, they carefully read the manuscript and suggested valuable improvements. H. T. Kung and K. Sikorski checked portions of the manuscript and A. G. Sukharev provided valuable references to the Soviet literature. A. Bojańczyk, A. Kiełbasiński, and A. Werschulz commented on an earlier version of this manuscript. D. Josephson, our superb technical typist, prepared the entire manuscript and was always patient about making just one more change. N. K. Brassfield was invaluable in helping us with proofreading and index preparation.

Much of the research was done in the splendid research environments of the Computer Science Department at Carnegie-Mellon University and of the Institute of Informatics at the University of Warsaw. We are indebted to the hospitality of the University of California at Berkeley, where we spent the academic year 1978–1979 while completing the research and writing of the manuscript. We are indebted to M. Blum and R. Karp for arranging this visit and for stimulating conversations. Finally, we are pleased to thank the National Science Foundation (Grant MCS-7823676) and the Office of Naval Research (Contract N00014-76-C-0370) for supporting the research reported here.

Recommended Reading

We recommend different portions of the book to readers with various interests. To get some feel for the material, we suggest that all readers should be familiar with the Preface, Overview, Introductions to Parts A–C and the Introductions to each of the 10 chapters in Part A. For readers with particular interests, we make the following recommendations.

RESEARCHERS INTERESTED IN OPEN PROBLEMS   Our theory suggests many new problems. Conjectures, questions, and open problems are scattered throughout this book. See, in particular, Part A: Chapter 6, Section 7; Chapter 8, Sections 3 and 7; Chapter 10, Sections 2 and 5; Part B: Section 10.

RESEARCHERS INTERESTED IN THE LITERATURE ON HISTORY   Bibliographical references and historical material may be found in many places. Part C is devoted entirely to a history and an annotated bibliography. Bibliographies may be found at the end of Parts A and B. See also Part A: Chapter 1, Section 2; Chapter 4, Section 1; Chapter 6, Sections 3–6; Chapter 8, Section 1; Part B: Section 2.

THEORETICAL COMPUTER SCIENTISTS   [those interested in algorithms and complexity but not specifically in analytic complexity] Part A: Chapter 1, Sections 2, 3; Chapters 5, 9, 10; Part B: Sections 2, 8, 11.

MATHEMATICIANS   [material of greatest mathematical interest] Part A: Chapters 2 and 7; Part B: Sections 4–7.

MATHEMATICAL THEORY OF APPROXIMATION   Part A: Chapter 2, Section 6; Chapter 3, Section 5; Chapter 7, Section 4.

APPLIED APPROXIMATION THEORY   Part A: Chapters 4 and 6; Part C.

NUMERICAL ANALYSTS AND APPLIED MATHEMATICIANS   Part A: Chapters 3, 4, 6, 8–10; Part C.

SCIENTISTS AND ENGINEERS   Appropriate sections of the applications chapters (Part A: Chapters 6 and 8), depending on individual interests.

A General Theory
of Optimal Algorithms

*Although this may seem a paradox, all exact science
is dominated by the idea of approximation.*

B. Russell

Overview

We provide the mathematical foundations of a general theory of optimal algorithms for problems which are solved approximately. This theory is called analytic computational complexity. See Part A, Chapter 1, Section 4 for a discussion of the nature of this field.

In this Overview we have to use words such as problem, information, and optimal algorithm without definition. They are defined rigorously and in great generality later.

We pose new kinds of questions and provide at least partial answers to many of them. Some of these questions are listed later in this Overview. Because of the richness of our domain, we must leave many open problems which we hope will be settled later.

We cover topics ranging from those of concern to theoretical computer scientists and mathematicians to those also of practical interest. We give two examples.

Problem complexity is a measure of the intrinsic difficulty of obtaining the solution to a problem no matter how that solution is obtained. We shall define it (see Part A, Chapter 1, Sections 3 and 4) as the complexity of the "optimal complexity algorithm" for solving the problem. (Problem complexity can only be specified with respect to a model of computation and a class of "permissible" information operators; we ignore such specification here.) The determination of problem complexity is difficult and deep; it has been completely solved for very few problems. When the complexity of a number of problems is at least approximately determined, we can hierarchically arrange problems according to their difficulty. Several partial problem hierarchies are summarized in Part A, Chapter 9. We believe that the determination of problem complexities and problem hierarchies should be one of the central problems of theoretical computer science and mathematics.

A practical application of our work is the rationalization of the synthesis of algorithms. Traditionally, algorithms are derived by ad hoc criteria. We shall find that algorithms obtained by commonly used criteria may have no relation to optimal algorithms (see Part A, Chapter 6, Remark 4.1) and that algorithms using commonly used information may pay an arbitrarily high penalty compared with the optimal algorithm using optimal information (see Part A, Chapter 6, Section 6).

A central issue in computer science is the selection of the best algorithm for solving a problem. Selection of the best algorithm is a multivariate optimization problem with dimensions including time complexity, space complexity, simplicity of program implementation, robustness, and stability. In this monograph we deal only with time complexity although conclusions concerning space complexity could also be easily obtained by changing the set of primitives in the model of computation. In later work we shall investigate tradeoffs in various dimensions for important problems. First we must understand how to optimize in the important time complexity dimension.

The analysis needed to characterize and construct an optimal algorithm (optimal in any of the senses of this book) for a particular problem can be a difficult mathematical problem and will, in any case, require substantial analysis. This may, however, be viewed as a precomputing cost.

We have not stated the domain or scope of this monograph. We must defer a more precise specification until Part A, Chapter 1, Section 4 after certain ideas have been introduced. Here we limit ourselves to a vague description.

We study problems which either cannot be solved exactly with finite complexity or problems which we choose to solve only approximately for reasons of efficiency. To the first class belong "most" problems of mathematics, science, and engineering. See Part A, Chapters 6 and 8 for many applications. Prominent exceptions are combinatorial and certain algebraic problems. Examples of the second class are the approximate solution of large sparse linear systems and of certain hard (for example, NP-complete) combinatorial optimization problems. Indeed, our model includes algebraic complexity as a special case. See Part A, Chapter 1, Examples 3.2 and 3.3, and Section 4.

The generality and power of the theory stems from the central role of information. Adversary arguments based on the information used by an algorithm lead to lower bound theorems. While the notion of information leads to generality it also permits remarkable simplicity. Numerous earlier papers obtain an optimal algorithm under various technical assumptions on the class of algorithms and the class of problem elements. These assumptions are often not verifiable. Our results depend only on the information used and broad verifiable properties of a class of problem elements. They are independent of the structure of an algorithm; they depend only on the information the algorithm uses.

Our theory uses two mathematical models; for convenience we refer to them here as models α and β. In what follows we limit our description to Part A; a similar dichotomy exists for Part B (see Introduction to Part B).

In model α the basic optimality concepts of "optimal error algorithm" and "optimal information" are defined independently of a model of computation. Model β consists of model α as well as concepts related to computation; the basic optimality concept here is "optimal complexity algorithm" for a problem. Much of Part A is devoted to model α for a number of reasons. Negative results can often be proven using model α. Since these results are independent of a model of computation, they are all the stronger. Furthermore, we can use powerful existing mathematical techniques to determine the optimal error algorithm. Indeed, the optimal error algorithm using optimal information must be determined before we can get good bounds on the optimal complexity algorithm.

We mentioned the difficulty of determining problem complexity. We must almost always settle for upper and lower bounds. This is not surprising. Even in algebraic complexity, which is included as a special case in our framework, it is rare to know the exact problem complexity. However, we often have extremely tight bounds on problem complexity. For example, if the optimal error algorithm is a *linear algorithm*, then the bounds are always very tight. (See Part A, Chapters 5 and 6.) For examples of tight complexity bounds when there is no linear optimal algorithm, see Part A, Chapter 8 and Part B, Section 8.

We list 20 of the general questions to be studied. These 20 questions are completely or partially answered, with the exception of two questions for which we conjecture the answer.

1. What is a lower bound on the error of any algorithm for solving a problem using given information?

2. In general is there an algorithm which gets arbitrarily close to this lower bound?

3. When is the information strong enough to solve a problem to within a given accuracy?

4. What is the optimal information for solving a problem?

5. What is the minimal number of linear functionals to solve a problem to within a given accuracy?

6. For linear problems is there always a linear algorithm with optimal error? If not, is there always a linear algorithm whose error is within a constant factor of the optimal error? In a Hilbert space setting what is the value of this constant?

7.   Given a specific problem, how do we characterize and construct an optimal algorithm for its solution?

8.   Given a problem, what are tight upper and lower bounds on its complexity?

9.   Can it be established that one problem is intrinsically harder than another?

10.   Do there exist linear problems with arbitrary complexity? In particular, do there exist linear problems which are arbitrarily hard?

11.   What can be said in general about the dependence of complexity upon the regularity of the class of "Problem elements"?

12.   Are adaptive algorithms more powerful than nonadaptive algorithms for solving linear problems?

13.   Are adaptive algorithms more powerful than nonadaptive algorithms for solving nonlinear problems?

14.   What is the power of linear information operators?

15.   What is the power of nonlinear information operators?

16.   What is the error of the best algorithm using optimal adaptive linear information for computing a simple zero of a nonlinear scalar function?

17.   What is the error of the best algorithm using optimal adaptive linear information for searching for the maximum of a unimodal function (generalized Kiefer problem)?

The following questions deal with iterative information and iterative algorithms:

18.   What is the maximal order of any algorithm using given information?

19.   What is the class of all problems which can be solved by iteration using linear information?

20.   What is the minimal number of linear functionals to iteratively solve a system of nonlinear equations in N dimensions?

We describe the overall structure of this monograph. It is divided into three parts. Parts A and B treat, respectively, a general information model and an iterative information model. Part C consists of a short history and an annotated bibliography of well over 300 items.

We end this Overview by describing our system for referring to material within the text. Theorems, equations, remarks, etc. are separately numbered for each section. A reference to material within the same chapter does not name the chapter. A reference to material within the same part but different chapter names the chapter. A reference to material in a different part names the part and chapter.

PART A

GENERAL INFORMATION MODEL

INTRODUCTION

Central to our theory of optimal algorithms is the notion of information. Part A is devoted to the study of "general information," while Part B treats "iterative information."

We develop a theory of optimal algorithms using given information or optimally chosen information. Two types of optimal algorithms will be of interest, "optimal error algorithms" and "optimal complexity algorithms." In Chapter 6, we apply this theory to a wide range of linear problems, and in Chapter 8 to a number of nonlinear problems.

A large portion of Part A (Chapters 2–6) is devoted to the theory of linear problems using linear information and to the applications of this theory. This emphasis is due to a number of factors.

1. Our results in the linear theory are very strong. In some cases, the questions we pose can be completely answered.

2. Many important applications are specified by linear problems and linear information.

3. The *class* of nonlinear information is too powerful (see Chapter 7) in the general information setting.

We broadly summarize the contents of Part A. See the introductions to each chapter for more detailed summaries.

CHAPTER 1 The basic concepts are formalized. An adversary principle leads to sharp lower bounds on algorithm error. Our model of computation is defined. The basic notions of optimal error algorithm and optimal complexity algorithm are introduced. The scope of analytic complexity and of this book are discussed.

CHAPTER 2 Chapters 2–6 treat linear problems and linear information. In Chapter 2, the general theory of linear information is developed. We vary the information operator and study the optimal information for solving a problem. In a Hilbert space setting, we give a complete solution to the problem of optimal information. We study adaptive information and prove the surprising result that adaptive information is not more powerful than nonadaptive information for a linear problem. Relations between Gelfand n-widths and nth minimal diameters are determined.

CHAPTER 3 We study *linear* algorithms for linear problems; such algorithms must have good time and space complexity. Very tight lower and upper complexity bounds are established for any problem with a linear optimal error algorithm. An especially constructed example establishes the existence of a linear problem for which there is no linear optimal error algorithm; no such example is known for "naturally" occurring problems. We use Smolyak's theorem to establish that algorithms optimal in the sense of Sard and Nikolskij are optimal error algorithms. Relations between optimal linear algorithms and the linear Kolmogorov n-widths are established.

CHAPTER 4 Generally, our analysis is worst case over all the problem elements f in a class $\mathfrak{I}_0$. Here we study algorithms for which the *local* error is almost as small as possible for *every* f from $\mathfrak{I}_0$. An algorithm which enjoys this property has small deviation. Do there exist linear algorithms with small deviation which are optimal or nearly optimal error algorithms? We introduce spline algorithms and show they permit us to answer this question.

CHAPTER 5 We specify our model of computation for the linear case. We show that there exist linear problems with essentially arbitrary complexity. This implies the existence of arbitrarily hard linear problems and that there are no "gaps" in the complexity function.

CHAPTER 6 We apply our theory to the solution of many different linear problems. Included are the approximation of a linear functional, interpolation, integration, approximation, and the solution of linear partial differential equations. Results are obtained for various spaces of problem elements. We obtain optimal error and nearly optimal complexity algorithms, optimal information operators, and very tight bounds on problem complexity for many applications. In some cases, our new algorithms are faster than commonly used algorithms by an unbounded factor.

CHAPTER 7   The theory of *nonlinear* information is developed. We show that in the general information setting the *class* of nonlinear information operators is too powerful to be of interest. Relations are established between Kolmogorov n-widths and the errors of n-dimensional algorithms as well as between ε-entropy and the cardinality of information operators with radius of information less than ε.

CHAPTER 8   We apply our general theory to the solution of some nonlinear problems. We show that adaptive linear information is *exponentially* better than nonadaptive linear information for the nonlinear problems considered in this chapter. We prove that in the search for the maximum of a unimodal function, "Kiefer information" is not optimal in the class of adaptive linear information using n linear functionals. We propose a conjecture on the error of the optimal algorithm using optimal linear information.

CHAPTER 9   One of the central problems of analytic complexity is to establish the intrinsic complexity of a problem. We present a partial hierarchy for the problems considered in Chapters 6 and 8.

CHAPTER 10   Earlier chapters have been devoted to a worst case model of analytic complexity. In this chapter, we briefly discuss four other models: average case, relative, perturbed, and asymptotic.

Chapter 1

Basic Concepts

1. INTRODUCTION

In this chapter, formal definitions of the basic concepts used for the remainder of Part A are presented. Fundamental are the concepts of solution operator S, information operator $\mathfrak{N}$, and algorithm φ.

Our focus on the information used by an algorithm permits a vast simplification of the theory of optimal algorithms. It leads us to a powerful adversary principle from which we obtain a lower bound on *any* algorithm for solving S using the information $\mathfrak{N}$. We show that the radius of information $r(\mathfrak{N},S)$ is the best possible lower bound. The problem S cannot be solved to within ε if $r(\mathfrak{N},S) \geq \varepsilon$. This negative result is independent of the model of computation used.

As mentioned in the Overview, we deal with two models, α and β. The negative result mentioned above belongs to model α. Also, in model α, we introduce the idea of an "optimal error algorithm" as the algorithm whose error is the infimum of the errors of all algorithms for S using information $\mathfrak{N}$. We introduce the concepts of central algorithm and interpolatory algorithm. A central algorithm is always an optimal error algorithm. It enjoys an even stronger optimality property (see Remark 2.2 and Chapter 4). The error of an interpolatory algorithm is within a factor of at most two from the error of an optimal error algorithm. Central and interpolatory algorithms are useful in practice as well as in the general theory. See Chapters 3 and 4 for theoretical aspects and Chapters 6 and 8 for applications.

To discuss complexity, we must have a model of computation. Our model consists of a set of primitives and the notion of permissible information operator and permissible algorithm. In model β, we introduce the idea of an "optimal complexity algorithm" as the algorithm whose complexity is the infimum of the complexity of all algorithms for S using information $\mathfrak{N}$ with error less than ε. In general, the complexity of an optimal error algorithm need not be close to the complexity of an optimal complexity algorithm. (See Remark 3.2.) However, for certain classes of algorithms (see Chapters 5 and 6 for linear algorithms) or for certain nonlinear problems (see Chapter 8), there is a close connection between the two types of optimality.

Our setting is so general that algebraic and combinatorial complexity are included as special cases. See Examples 3.2 and 3.3 as well as the discussion preceding these examples. See also Section 4 for another instance of how algebraic complexity fits into our setting.

We summarize the results of this chapter. In Section 2, basic concepts such as solution operator, information operator, ε-approximation, radius and diameter of information, algorithm, interpolatory algorithm, central algorithm, and optimal error algorithm are introduced. An adversary principle leads (Theorem 2.1) to the conclusion that the radius of information $\mathfrak{N}$ for the problem S is a lower bound on the error of any algorithm for S using the information $\mathfrak{N}$. It is possible to find an ε-approximation for all problem elements if and only if the radius of information is less than ε.

Our model of computation is presented in Section 3. Fundamental are the concepts of primitive operations, permissible information operators, and permissible algorithms. The concept of optimal complexity algorithm is introduced and Remark 3.2 illustrates the difference between optimal error algorithm and optimal complexity algorithm. Discussion and examples at the end of the section consider the relation with algebraic complexity.

The concepts introduced in Sections 2 and 3 permit us to define the scope of analytic complexity in the concluding section.

2.   DIAMETER AND RADIUS OF GENERAL INFORMATION

Let $\mathfrak{I}_0$ be a subset of a linear space $\mathfrak{I}_1$ over the real or complex field. Consider a linear or nonlinear operator S such that

$$(2.1) \qquad\qquad S: \mathfrak{I}_0 \to \mathfrak{I}_2,$$

where $\mathfrak{I}_2$ is a linear normed space over the real or complex field. Let $\varepsilon > 0$ be a given number. Our problem is to find an *ε-approximation* $x = x(f)$, $x \in \mathfrak{I}_2$, to $\alpha = S(f)$, i.e.,

$$(2.2) \qquad\qquad \|x - \alpha\| < \varepsilon$$

for all $f \in \mathfrak{I}_0$. We shall call S the *solution operator*, f a *problem element*, and α a *solution element*. We shall often refer to S and its domain $\mathfrak{I}_0$ as the *problem S*.

To find an ε-approximation, we must know something about the problem elements. Let

$$(2.3) \qquad\qquad \mathfrak{N} : D_{\mathfrak{N}} \to \mathfrak{I}_3$$

be an *information operator* (not necessarily linear), where $\mathfrak{I}_0 \subset D_{\mathfrak{N}} \subset \mathfrak{I}_1$ and $\mathfrak{I}_3$ is a given space. $\mathfrak{N}(f)$ is called the *information of f*. For most problems, the information operator $\mathfrak{N}$ is not one-to-one and $\mathfrak{N}(f)$ does not uniquely define the solution element $\alpha = S(f)$. Thus, there may exist many different problem elements $f \in \mathfrak{I}_0$ with the same information.

Let $f \in \mathfrak{I}_0$. Let

$$(2.4) \qquad\qquad V(f) = \{ \tilde{f} : \mathfrak{N}(\tilde{f}) = \mathfrak{N}(f) \text{ and } \tilde{f} \in \mathfrak{I}_0 \}$$

be the preimage set of $y = \mathfrak{N}(f)$ in $\mathfrak{I}_0$, $V(f) = \mathfrak{N}^{-1}(y) \cap \mathfrak{I}_0$. Note that $V(f)$ is not empty since $f \in V(f)$ for every $f \in \mathfrak{I}_0$. Furthermore, let

$$(2.5) \qquad\qquad U(f) = \{ S(\tilde{f}) : \tilde{f} \in V(f) \}$$

be the set of all solutions $S(\tilde{f})$ of problem elements $\tilde{f}$ which share the same information as f, $U(f) = S(\mathfrak{N}^{-1}(y) \cap \mathfrak{I}_0)$. Then knowing only $\mathfrak{N}(f)$, it is impossible to recognize which solution element $\alpha = S(f)$ or $\tilde{\alpha} = S(\tilde{f})$ is being actually approximated for all $\tilde{f} \in V(f)$. This *adversary principle* can be schematized as shown in Figure 1.

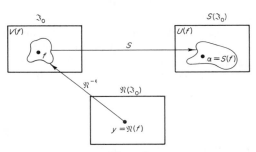

Figure 1

As we shall see, the diameter and the radius of the set $U(f)$ play essential roles. These concepts are defined as follows.

Recall that for a set A, $A \subset \mathfrak{I}_2$,

$$(2.6) \qquad\qquad \operatorname{diam}(A) = \sup_{a_1, a_2 \in A} \| a_1 - a_2 \|$$

is called the *diameter of A* and

$$(2.7) \qquad\qquad \operatorname{rad}(A) = \inf_{a \in \mathfrak{I}_2} \sup_{a_1 \in A} \| a - a_1 \|$$

is called the *radius of A*. Roughly speaking, rad(A) is the minimal radius of a "ball" which contains A. If there exists c, $c \in \mathfrak{I}_2$, such that

$$(2.8) \qquad \sup_{a_1 \in A} \|c - a_1\| = \text{rad}(A),$$

then c is a *center of A*. Note that c can be an element outside A and need not be unique.

Definition 2.1 We shall say $d(\mathfrak{N},S)$ is the *diameter of information* $\mathfrak{N}$ *for the problem S* iff

$$(2.9) \qquad d(\mathfrak{N},S) = \sup_{f \in \mathfrak{I}_0} \text{diam}(U(f)) \quad \left(= \sup_{f \in \mathfrak{I}_0} \sup_{\tilde{f} \in V(f)} \|S(\tilde{f}) - S(f)\| \right).$$

We shall say $r(\mathfrak{N},S)$ is the *radius of information* $\mathfrak{N}$ *for the problem S* iff

$$(2.10) \qquad r(\mathfrak{N},S) = \sup_{f \in \mathfrak{I}_0} \text{rad}(U(f)) \quad \left(= \sup_{f \in \mathfrak{I}_0} \inf_{a \in \mathfrak{I}_2} \sup_{\tilde{f} \in V(f)} \|a - S(\tilde{f})\| \right). \quad \blacksquare$$

It is obvious that

$$(2.11) \qquad r(\mathfrak{N},S) \le d(\mathfrak{N},S) \le 2r(\mathfrak{N},S).$$

For many $\mathfrak{N}$ and S, it is much easier to compute the diameter $d(\mathfrak{N},S)$ than the radius $r(\mathfrak{N},S)$.

REMARK 2.1 Suppose that for every $f \in \mathfrak{I}_0$ there exists $c = c(f) \in \mathfrak{I}_2$ such that $U = U(f)$ is symmetric with respect to c, i.e., $h + c \in U$ implies $-h + c \in U$. Then c is a center of U and

$$(2.12) \qquad d(\mathfrak{N},S) = 2r(\mathfrak{N},S).$$

Indeed, for some $a \in \mathfrak{I}_2$, assume $\sup_{u \in U}\|a - u\| < \sup_{u \in U}\|c - u\|$. Take $x \in U$ such that $\|a - u\| < \|c - x\|$, $\forall u \in U$. Let $x = c + h$. Then $c - h \in U$ and

$$2\|h\| \le \|a - (c + h)\| + \|a - (c - h)\| < 2\|h\|,$$

which is a contradiction. Thus, c is a center of U.

Assume without loss of generality that rad(U) $< +\infty$. Let $h = u - c$, where $u \in U$, $\|c - u\| \ge \text{rad}(U) - \delta$ for $\delta > 0$. Define $u_1 = c + h$, $u_2 = c - h$. Then $u_i \in U$ and $\|u_1 - u_2\| = 2\|c - u\| \ge 2(\text{rad}(U) - \delta)$. This proves that diam($U$) $= 2\,\text{rad}(U)$ and implies (2.12).

An example of a problem S and an information operator $\mathfrak{N}$ for which $d(\mathfrak{N},S) < 2r(\mathfrak{N},S)$ may be found in Micchelli and Rivlin [77, p. 9]. $\blacksquare$

We shall show that the radius $r(\mathfrak{N},S)$ is a lower bound on the error of any algorithm for solving $\alpha = S(f)$. By an *algorithm*, we mean an operator $\varphi : \mathfrak{N}(\mathfrak{I}_0) \to \mathfrak{I}_2$. (See also the definition of "permissible algorithm" in Section 3.) Let $\Phi(\mathfrak{N},S)$ be the class of *all* algorithms. Since $\varphi(\mathfrak{N}(\tilde{f})) = \varphi(\mathfrak{N}(f))$ for all $\tilde{f} \in V(f)$, φ has to approximate any element of the set $U(f)$. This is shown in Figure 2.

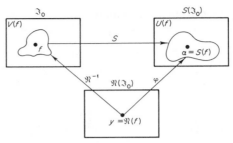

Figure 2

Definition 2.2   We shall say $e(\varphi)$ is the *error of algorithm* φ iff

$$(2.13) \qquad\qquad e(\varphi) = \sup_{f \in \mathfrak{I}_0} \|\varphi(\mathfrak{N}(f)) - S(f)\|. \quad \blacksquare$$

Note that (2.13) can be rewritten as

$$(2.14) \quad e(\varphi) = \sup_{f \in \mathfrak{I}_0} \sup_{\tilde{f} \in V(f)} \|\varphi(\mathfrak{N}(f)) - S(\tilde{f})\| = \sup_{f \in \mathfrak{I}_0} \sup_{\tilde{\alpha} \in U(f)} \|\varphi(\mathfrak{N}(f)) - \tilde{\alpha}\|.$$

It is intuitively obvious that the radius $r(\mathfrak{N},S)$ is a lower bound on the error of any algorithm. A formal proof is provided by

Theorem 2.1   For any algorithm φ, $\varphi \in \Phi(\mathfrak{N},S)$,

$$(2.15) \qquad\qquad\qquad e(\varphi) \geq r(\mathfrak{N},S). \quad \blacksquare$$

PROOF   Let $f \in \mathfrak{I}_0$. Then due to (2.7) and (2.14), we get

$$\mathrm{rad}(U(f)) \leq \sup_{\tilde{\alpha} \in U(f)} \|\varphi(\mathfrak{N}(f)) - \tilde{\alpha}\| \leq e(\varphi).$$

Thus $r(\mathfrak{N},S) = \sup_{f \in \mathfrak{I}_0} \mathrm{rad}(U(f)) \leq e(\varphi)$ which proves (2.15). $\blacksquare$

This generalizes Theorem 4 in Micchelli and Rivlin [77], where S and $\mathfrak{N}$ are assumed to be linear.

We define "interpolatory algorithms" and show they are within a factor of 2 of the radius $r(\mathfrak{N},S)$.

Definition 2.3   An algorithm φ^{I}, $\varphi^{\mathrm{I}} \in \Phi(\mathfrak{N},S)$, is an *interpolatory algorithm for the problem S with the information $\mathfrak{N}$* (briefly an interpolatory algorithm) iff

$$(2.16) \qquad\qquad\qquad \varphi^{\mathrm{I}}(\mathfrak{N}(f)) = S(\tilde{f})$$

for some $\tilde{f} \in V(f)$. $\blacksquare$

This means that knowing the information $\mathfrak{N}(f)$, one finds a problem element $\tilde{f}$ (which always exists) which has the same information as f, $\tilde{f} \in V(f)$, and $\tilde{\alpha} = S(\tilde{f})$ is proposed as an approximation to $\alpha = S(f)$. In practice, $\tilde{f}$ is chosen to be "simpler" than f. Note that $\varphi^{\mathrm{I}}(\mathfrak{N}(f)) \in U(f)$. In some cases, an assumption how to choose a unique $\tilde{f}$ is added. Examples of interpolatory algorithms are known for such problems as nonlinear equations, approximation, interpolation, and integration.

Theorem 2.2   For any interpolatory algorithm φ^I, $\varphi^I \in \Phi(\mathfrak{N},S)$,

$$(2.17) \qquad\qquad e(\varphi^I) \leq d(\mathfrak{N},S) \leq 2r(\mathfrak{N},S). \qquad \blacksquare$$

PROOF   Take any $f \in \mathfrak{I}_0$. Then

$$\left\| \varphi^I(\mathfrak{N}(f)) - S(f) \right\| \leq \left\| S(\tilde{f}) - S(f) \right\| \leq d(\mathfrak{N},S)$$

since $\tilde{f} \in V(f)$. Taking the supremum with respect to f, we get (2.17).   ∎

We seek "optimal error algorithms" which are defined as follows.

Definition 2.4   We shall say $e(\mathfrak{N},S)$ is the *optimal error* iff

$$(2.18) \qquad\qquad e(\mathfrak{N},S) = \inf_{\varphi \in \Phi(\mathfrak{N},S)} e(\varphi).$$

We shall say φ^{oe}, $\varphi^{oe} \in \Phi(\mathfrak{N},S)$, is an *optimal error algorithm for the problem S with the information* $\mathfrak{N}$ (briefly an optimal error algorithm) iff

$$(2.19) \qquad\qquad e(\varphi^{oe}) = e(\mathfrak{N},S). \qquad \blacksquare$$

Combining Theorems 2.1 and 2.2, we see that any interpolatory algorithm is nearly an optimal error algorithm.

Corollary 2.1   For any interpolatory algorithm φ^I, $\varphi^I \in \Phi(\mathfrak{N},S)$, with the convention $0/0 = 1$,

$$(2.20) \qquad\qquad 1 \leq \frac{e(\varphi^I)}{e(\mathfrak{N},S)} \leq 2. \qquad \blacksquare$$

We now prove that the optimal error $e(\mathfrak{N},S)$ is equal to the radius $r(\mathfrak{N},S)$.

Theorem 2.3

$$(2.21) \qquad\qquad e(\mathfrak{N},S) = r(\mathfrak{N},S). \qquad \blacksquare$$

PROOF   Let $\delta > 0$ be an arbitrary number. Define an algorithm φ_δ as follows. Let

$$(2.22) \qquad\qquad \varphi_\delta(\mathfrak{N}(f)) = c_\delta(f),$$

where $\left\| c_\delta(f) - \tilde{\alpha} \right\| \leq \mathrm{rad}(U(f)) + \delta$ for all $\tilde{\alpha} \in U(f)$. Thus, $c_\delta(f)$ is almost a center of $U(f)$. Then

$$e(\mathfrak{N},S) \leq e(\varphi_\delta) = \sup_{f \in \mathfrak{I}_0} \left\| S(f) - \varphi_\delta(\mathfrak{N}(f)) \right\| = \sup_{f \in \mathfrak{I}_0} \sup_{\tilde{\alpha} \in U(f)} \left\| \tilde{\alpha} - c_\delta(f) \right\|$$

$$\leq \sup_{f \in \mathfrak{I}_0} \mathrm{rad}(U(f)) + \delta = r(\mathfrak{N},S) + \delta.$$

Since δ is arbitrary, $e(\mathfrak{N},S) \leq r(\mathfrak{N},S)$. Due to Theorem 2.1, we know that $e(\mathfrak{N},S) \geq r(\mathfrak{N},S)$ which proves (2.21).   ∎

See Micchelli and Rivlin [77], where a similar result is established for $\mathfrak{N}$ and S linear. Theorem 2.3 motivates using a center $c(f)$, if it exists, as an approximation to $\alpha = S(f)$. Suppose that $U(f)$ has a center for any $f \in \mathfrak{J}_0$.

Definition 2.5   An algorithm φ^c, $\varphi^c \in \Phi(\mathfrak{N},S)$ is a *central algorithm for the problem S with the information* $\mathfrak{N}$ (briefly a central algorithm) iff

$$(2.23) \qquad \varphi^c(\mathfrak{N}(f)) = c(f),$$

where $c(f)$ is a center of $U(f)$. ∎

Theorem 2.4   Any central algorithm is an optimal error algorithm, i.e.,

$$(2.24) \qquad e(\varphi^c) = r(\mathfrak{N},S).$$ ∎

PROOF   Note that

$$e(\varphi^c) = \sup_{f \in \mathfrak{J}_0} \left\| S(f) - \varphi^c(\mathfrak{N}(f)) \right\| = \sup_{f \in \mathfrak{J}_0} \sup_{\alpha \in U(f)} \left\| \tilde{\alpha} - c(f) \right\|$$

$$= \sup_{f \in \mathfrak{J}_0} \operatorname{rad}(U(f)) = r(\mathfrak{N},S).$$ ∎

As we shall see in Chapters 3 and 4, an interpolatory algorithm may turn out to be an optimal error algorithm.

REMARK 2.2   The central algorithms φ^c enjoy an even stronger property than optimality; namely, they produce a best possible approximation to $S(f)$ for every $f \in \mathfrak{J}_0$. More precisely, for $\varphi \in \Phi(\mathfrak{N},S)$ and $f \in \mathfrak{J}_0$, consider the *local error*

$$e(\varphi,f) = \sup_{\tilde{f} \in V(f)} \left\| \varphi(\mathfrak{N}(f)) - S(\tilde{f}) \right\|.$$

For fixed f we want to minimize $e(\varphi,f)$. It is clear that

$$\inf_{\varphi \in \Phi(\mathfrak{N},S)} e(\varphi,f) = \operatorname{rad}(U(f)) = e(\varphi^c,f) \qquad \forall f \in \mathfrak{J}_0.$$

Thus, the central algorithms minimize the local error for every problem element. ∎

Recall that we wish to find an ε-approximation to $\alpha = S(f)$ for all $f \in \mathfrak{J}_0$, i.e., to find $x(f)$ such that $\|x(f) - \alpha\| < \varepsilon$. Due to Theorem 2.3, we get

Corollary 2.2   It is possible to find an ε-approximation to $\alpha = S(f)$ for all $f \in \mathfrak{J}_0$ iff

$$(2.25) \qquad r(\mathfrak{N},S) < \varepsilon.$$ ∎

We wish to stress that an information operator $\mathfrak{N}$ has to be defined in such a way that $\mathfrak{N}(f)$ is "computable" for every $f \in \mathfrak{J}_0$. This rules out many operators as "permissible" information operators. For instance, let $\mathfrak{N}(f) = f$ be the identity operator I. Then, since I is one-to-one, $r(I,S) = 0$ for any solution operator S. However, $\mathfrak{N}(f) = f$ is "computable" iff f can be represented by a finite-dimensional vector, i.e., $\mathfrak{I}_1$ is a finite-dimensional space. As a second example, consider $\mathfrak{N}(f) = S(f)$. Then $r(S,S) = 0$, but for most problems $S(f)$ is not "computable" and $\mathfrak{N} = S$ is not a "permissible" information operator. See Section 3 and Chapter 5 for a precise definition of our model of computation. Examples of computable operators will be found in Chapters 6 and 8.

Example 2.1 To illustrate the preceding concepts, we consider the following problem. Let $\mathfrak{I}_1 = C^n[0,1]$ be the class of n times continuously differentiable functions of one variable, $n \geq 1$. Let $\mathfrak{I}_2 = C^0[0,1]$. Define

$$(2.26) \qquad\qquad S(f) = f;$$

that is, $S = I$. Note that (2.26) is a formulation of the approximation problem. Let

$$f \in \mathfrak{I}_0 = \left\{ f : f \in \mathfrak{I}_1 \text{ and } \max_{0 \leq t \leq 1} \left| \frac{1}{n!} f^{(n)}(t) \right| \leq 1 \right\}.$$

Consider the information operator $\mathfrak{N}$ given by

$$(2.27) \qquad\qquad \mathfrak{N}(f) = [f(t_1), f(t_2), \ldots, f(t_n)]$$

for some distinct points $t_i \in [0,1]$. This means that we want to approximate f from $\mathfrak{I}_0$ knowing only the values of f at n points. Let $\omega(t) = \prod_{i=1}^n (t - t_i)$. Then $\tilde{f} \in V(f)$ implies

$$\tilde{f}(t) = f(t) + g(t)\omega(t),$$

where g is the nth divided difference of $\tilde{f} - f$ and $\|g\| = \max_{0 \leq t \leq 1} |g(t)| \leq 2$. It is easy to show that $d(\mathfrak{N},S) = 2r(\mathfrak{N},S)$ and

$$(2.28) \qquad\qquad r(\mathfrak{N},S) = \|\omega\| \geq 2/4^n.$$

Furthermore, (2.28) holds with equality for information $\mathfrak{N}$ of the form (2.27) with

$$t_i = t_i^* = \left(1 + \cos\left(\frac{\pi}{2n} + \frac{i-1}{n}\pi \right) \right) \Big/ 2$$

which are the zeros of the Chebyshev polynomial $T_n(2t - 1)$.

   If $\varepsilon > 2/4^n$, we can find an ε-approximation to $\alpha = S(f) = f$ for all $f \in \mathfrak{I}_0$ using the information operator $\mathfrak{N}$ with $t_i = t_i^*$. For $\varepsilon \leq 2/4^n$, the information operator $\mathfrak{N}$ of the form (2.27) does not supply enough information to find ε-approximations for *any* choice of points t_i. ∎

   REMARK 2.3 Finding the solution elements exactly corresponds formally to $\varepsilon = 0$. In this case we do not need to assume that the space $\mathfrak{I}_1$ is linear or the space $\mathfrak{I}_2$ is linear and normed. It is easy to verify that we can find the exact solution iff the set $U(f)$ consists of exactly one element for every f from $\mathfrak{I}_0$. ∎

We conclude this section with a historical note. In the annotated bibliography of Part C, the reader can find a list of relevant papers as well as a short history of analytic complexity which references such pioneering work as the papers of Sard [49], Nikolskij [50], Kiefer [53], Golomb and Weinberger [59], and Traub [61, 64].

Here we confine ourselves to a partial list of papers in which the ideas presented in this section have been explicitly (or more typically, implicitly) used for a *particular* problem or a class of problems.

General linear problems (mostly approximation of linear functionals) were considered by, among others, Ahlberg and Nilson [66], Babuška and Sobolev [65], Bakhvalov [68, 70, 71a], Gal and Micchelli [78], Golomb and Weinberger [59], Grebennikov and Morozov [77], Grebennikov [78], Ivanov [75, 77], Knauff and Kress [74, 76], Larkin [70], Mangasarian and Schumaker [73], Mansfield [71, 72], Marchuk and Osipenko [75], Meyers and Sard [50a,b], Meinquet [67], Melkman and Micchelli [77], Micchelli [75], Micchelli and Pinkus [77], Micchelli and Rivlin [77], Nielson [73], Osipenko [76], Reinsch [74], Richter-Dyn [71b], Rice [73, 76], Ritter [70], Sard [49, 67, 73], Schoenberg [64a], Schultz [74], Smolyak [65], Weinberger [61, 72] and Winograd [76].

Some ideas presented here have been also used for nonlinear problems. See Booth [67, 69], Chernousko [68], Eichhorn [68], Gross and Johnson [59], Hyafil [77], Kiefer [57], Majstrovskij [72], Micchelli and Miranker [75], Sonnevend [77], Sukharev [76], and Todd [76], who considered the solution of nonlinear equations. The search for the maximum for a class of unimodal functions was considered by Kiefer [53], who initiated research in that area. His followers include Adamski, Korytowski, and Mitkowski [77], Aphanasjev [74], Aphanasjev and Novikov [77], Avriel and Wilde [66], Beamer and Wilde [69, 70, 71], Chernousko [70a,b], Danilin [71], Eichhorn [68], Fine [66], Gal [72], Ganshin [76, 77], Ivanov [72a], Johnson [56], Judin and Nemirovsky [76a,b, 77], Karp and Miranker [68], Kiefer [57], Krolak [66, 68], Krolak and Cooper [63], Kuzovkin and Tikhomirov [67], Levin [65], Mockus [72], Newman [65], Piavsky [72], Sonnevend [77], Strongin [78], Sukharev [71, 72, 75], Tarassova [78], Wilde [64], Zaliznyak and Ligun [78], and Zhilinskas [75].

For the solution of nonlinear equations, the information operators usually depend on a current approximation to a solution, $\mathfrak{N} = \mathfrak{N}(f, x)$. The theory of such information operators (which are called iterative information operators) is developed in Part B. Brent, Winograd, and Wolfe [73], Kacewicz [75, 76a,b], Kung [76], Kung and Traub [74, 76], Meersman [76a,b], Traub [61, 64], Traub and Woźniakowski [76a], and Woźniakowski [72, 74, 75, 76] considered the optimal iterative solution of scalar or abstract nonlinear equations.

See also Werschulz [77a,b] who dealt with "discretization" information operators and considered the maximal order of numerical integration and differentiation.

Even though it is not mentioned, most of these papers are at least implicitly based on the *adversary principle*, i.e., to find a problem element $\tilde{f}$ which shares the same information as f and the distance between $S(f)$ and $S(\tilde{f})$ is an inherent error of any algorithm. The general significance of the adversary principle was pointed out by Winograd [76] and Woźniakowski [75]. Winograd [76] introduced a very general "fooling" technique and showed its importance for a number of problems. Some of these ideas were already used in Brent, Winograd, and Wolfe [73], where the optimality of nonstationary one-point iterations with memory for the solution of scalar nonlinear equations is established for a special information operator. For the solution of nonlinear operator equations, Woźniakowski [75] introduced the basic concept of order of information which is primarily based on the adversary principle. The idea of order of information has been used by a number of authors to establish the optimality of many stationary iterations for solving nonlinear equations.

3.   COMPLEXITY OF GENERAL INFORMATION

We present our model of computation which consists of a set of primitive operations, permissible information operators, and permissible algorithms. In what follows, we shall use the words cost and complexity interchangeably. Context will make it clear whether we mean algorithm complexity or problem complexity.

Model of Computation

(i)   We assume that the computations are performed on a random access machine. See Aho, Hopcroft, and Ullman [74, Chapter 1]. Let p be a *primitive operation*. Examples of primitive operations include arithmetic operations, the evaluation of a square root or of an integral. Let comp(p) be the complexity of p; comp(p) must be finite. Suppose that P *is a given collection of primitives*. The choice of P and comp(p), $p \in P$, are arbitrary and can depend on the particular problem being solved.

(ii)   Let $\mathfrak{N}$ be an information operator. We say that $\mathfrak{N}$ is a *permissible information operator with respect to* P if there exists a program using a finite number of primitive operations from P which computes $\mathfrak{N}(f)$ for all $f \in \mathfrak{J}_0$. Let comp($\mathfrak{N}(f)$) denote the *information complexity* of computing $\mathfrak{N}(f)$. We assume that if $\mathfrak{N}(f)$ requires the evaluation of primitives $p_1, p_2, \ldots, p_k$, then comp($\mathfrak{N}(f)$) $= \sum_{i=1}^{k}$ comp(p_i).

(iii)   Let φ be an algorithm which uses the permissible information $\mathfrak{N}$. To evaluate $\varphi(\mathfrak{N}(f))$, we

(a)   compute $y = \mathfrak{N}(f)$,
(b)   compute $\varphi(y)$.

The complexity of computing y is given by (ii). We say that φ is a *permissible algorithm with respect to P* if there exists a program using a finite number of primitive operations from P which computes $\varphi(y)$ for all $y = \mathfrak{N}(f)$, $f \in \mathfrak{I}_0$. Let $\text{comp}(\varphi(y))$ be the *combinatory complexity* of computing $\varphi(y)$. We assume that if $\varphi(y)$ requires the evaluation of primitives $q_1, q_2, \ldots, q_j$, then $\text{comp}(\varphi(y)) = \sum_{i=1}^{j} \text{comp}(q_i)$. ∎

REMARK 3.1   Let $\mathfrak{N}$ be a permissible information operator. This means that $\mathfrak{N}(f)$ can be computed from the set of primitives P. Often there exist many different algorithms for computing $\mathfrak{N}(f)$ and the optimal computation of $\mathfrak{N}(f)$ can be treated as a subproblem. However, we assume that an algorithm (possibly not optimal) for the computation of $\mathfrak{N}(f)$ is defined by a "user." ∎

Example 3.1   Suppose we wish to approximate $S(f) = \int_0^1 f(t)\,dt$, where f is an absolutely continuous scalar function and $\int_0^1 (f'(t))^2\,dt \leq 1$. Define two sets of primitives, $P_1 = \{\text{the evaluation of an integral}\}$ and $P_2 = \{\text{arithmetic operations, the evaluation of a function}\}$. Note that $\mathfrak{N}(f) = S(f)$ is permissible with respect to P_1 and not permissible with respect to P_2. Of course, $r(S,S) = 0$. However, S is a primitive only in P_1. An example of a permissible information operator for P_2 is $\mathfrak{N}(f) = [f(t_1), f(t_2), \ldots, f(t_n)]$ for equally spaced $t_i \in [0,1]$. It is shown in Section 4 of Chapter 6 that $r(\mathfrak{N},S) = O(1/n)$. ∎

We showed in Section 2 that a necessary and sufficient condition for finding an ε-approximation to $\alpha = S(f)$ is $r(\mathfrak{N},S) < \varepsilon$. If $r(\mathfrak{N},S) \geq \varepsilon$, then the information operator $\mathfrak{N}$ does not supply sufficient information to solve the problem. We say that the problem S with an information operator $\mathfrak{N}$ is *ε-noncomputable* if $r(\mathfrak{N},S) \geq \varepsilon$. If $\mathfrak{N}$ is permissible, $r(\mathfrak{N},S) < \varepsilon$, and there exists a permissible algorithm φ such that $e(\varphi) < \varepsilon$, then the problem S with $\mathfrak{N}$ is called *ε-computable with respect to P*.

Suppose then that $r(\mathfrak{N},S) < \varepsilon$ for a permissible $\mathfrak{N}$ and assume that $\Phi(\varepsilon)$, the class of all permissible algorithms for which $e(\varphi) < \varepsilon$, is nonempty. We want to derive lower and upper bounds on the complexity of finding an ε-approximation using any $\varphi \in \Phi(\varepsilon)$.

Since the set of primitives P is fixed, we do not mention the dependence of complexity on P. Let $\varphi \in \Phi(\varepsilon)$. Then the complexity of an algorithm φ is defined by

$$(3.1) \qquad \text{comp}(\varphi) = \sup_{f \in \mathfrak{I}_0} (\text{comp}(\mathfrak{N}(f)) + \text{comp}(\varphi(\mathfrak{N}(f)))).$$

We define optimal complexity algorithm (Definition 3.1) and complexity of a problem (see Definition 3.2).

Definition 3.1   We shall say $\text{comp}(\mathfrak{N},S,\varepsilon)$ is the *ε-complexity of the problem S with the information $\mathfrak{N}$* (briefly the ε-complexity of S with $\mathfrak{N}$) iff

$$(3.2) \qquad \text{comp}(\mathfrak{N},S,\varepsilon) = \begin{cases} \inf_{\varphi \in \Phi(\varepsilon)} \text{comp}(\varphi) & \text{if } r(\mathfrak{N},S) < \varepsilon \text{ and } \Phi(\varepsilon) \neq \varnothing, \\ +\infty, & \text{otherwise.} \end{cases}$$

We shall say φ^{oc}, $\varphi^{oc} \in \Phi(\varepsilon)$ is an *optimal complexity algorithm for the problem S with the information* $\mathfrak{N}$ (briefly an optimal complexity algorithm for S with $\mathfrak{N}$) iff

(3.3) $$\text{comp}(\varphi^{oc}) = \text{comp}(\mathfrak{N},S,\varepsilon). \quad \blacksquare$$

REMARK 3.2 We stress the difference between an optimal error algorithm and an optimal complexity algorithm. The optimal error algorithm minimizes the error but not necessarily the complexity, since its combinatory complexity may be high. As an example, consider the solution of a system of linear equations $A\mathbf{x} = \mathbf{b}$, where A is an $n \times n$ nonsingular matrix and $\mathbf{b}$ is an $n \times 1$ vector. Then $f = [A,\mathbf{b}]$ and

$$S(f) = A^{-1}\mathbf{b}.$$

Note that, in our setting, the solution of a system of *linear* equations is a *nonlinear* problem since S is a nonlinear solution operator.

Every exact algorithm φ for this problem uses the information operator $\mathfrak{N}(f) = f$ and is an optimal error algorithm since $e(\varphi) = r(\mathfrak{N},S) = 0$. Thus, all classical numerical algorithms such as Gauss elimination, Householder or Gram–Schmidt orthogonalization methods are optimal error algorithms. They are not, however, optimal complexity algorithms (at least for large n) since their combinatory complexities are proportional to n^3. It is known that the ε-complexity of this problem with $\mathfrak{N}(f) = f$ is at most proportional to n^β, where the best β known, due to Pan [78], is about 2.79. The actual value of the ε-complexity is still unknown. (Note added in proof: Schönhage has reduced the value of β to about 2.55.)

It also may happen that an optimal error algorithm is not permissible and therefore it is not an optimal complexity algorithm. As an example, define

$$S(f) = \sqrt{f}, \qquad f \in \mathfrak{I}_0 = [1,2],$$
$$\mathfrak{N}(f) = f,$$

and let the set of primitives P be the set of four arithmetic operations $\{+, -, *, /\}$. Note that $\mathfrak{N}$ is permissible and $r(\mathfrak{N},S) = 0$. The unique optimal error algorithm $\varphi(\mathfrak{N}(f)) = \sqrt{f}$ is, however, not permissible, since the square root operation does not belong to P.

On the other hand, there exists an algorithm which is permissible and whose error is less than ε; namely, let

$$x_0 = 1,$$
$$x_{i+1} = (x_i + f/x_i)/2, \qquad i = 0, 1, \ldots, k-1,$$

where $k = \lceil \log \log(1/\varepsilon)/\log 1/(\sqrt{2}-1) \rceil$. Thus, x_k is generated by Newton–Heron iteration. Define the algorithm

$$\varphi(\mathfrak{N}(f)) = x_k.$$

It is easy to check that $e(\varphi) < \varepsilon$. The algorithm φ is permissible and assuming that every arithmetic operation costs unity, its combinatory complexity is $3k$. Thus, the ε-complexity of this problem with $\mathfrak{N}(f) = f$ is bounded by $3k$. The actual value of the ε-complexity is unknown. ∎

Let

$$(3.4) \qquad \mathrm{comp}(\mathfrak{N}) = \sup_{f \in \mathfrak{I}_0} \mathrm{comp}(\mathfrak{N}(f))$$

be the *information complexity of* $\mathfrak{N}$. Suppose that the combinatory complexity of every algorithm φ, $\varphi \in \Phi(\varepsilon)$, for $\mathfrak{N}(f)$ such that $\mathrm{comp}(\mathfrak{N}(f)) = \mathrm{comp}(\mathfrak{N})$ is bounded below by $m(\mathfrak{N})$. More precisely, let

$$(3.5) \qquad m(\mathfrak{N}) = \inf_{\varphi \in \Phi(\varepsilon)} \sup_{f: \mathrm{comp}(\mathfrak{N}(f)) = \mathrm{comp}(\mathfrak{N})} \mathrm{comp}(\varphi(\mathfrak{N}(f))).$$

In general, $m(\mathfrak{N})$ depends on the total number of "independent pieces" of information $\mathfrak{N}$. See Section 2 of Chapter 2 and Chapter 5, where the "cardinality" of information $\mathfrak{N}$ is introduced and its influence on the combinatory complexity of φ is shown. For some linear problems, as we shall see in Chapter 3, it is possible to find optimal error algorithms whose combinatory complexity is proportional to the "cardinality" of $\mathfrak{N}$.

From (3.4) and (3.5), we get

$$(3.6) \qquad \mathrm{comp}(\mathfrak{N},S,\varepsilon) \geq \mathrm{comp}(\mathfrak{N}) + m(\mathfrak{N}).$$

Furthermore, if there exists $\varphi \in \Phi(\varepsilon)$ such that $\mathrm{comp}(\varphi(\mathfrak{N}(f))) \ll \mathrm{comp}(\mathfrak{N})$ for all $f \in \mathfrak{I}_0$, then

$$(3.7) \qquad \mathrm{comp}(\mathfrak{N},S,\varepsilon) \cong \mathrm{comp}(\varphi) \cong \mathrm{comp}(\mathfrak{N});$$

i.e., φ is essentially an optimal complexity algorithm for S with $\mathfrak{N}$. Equations (3.6) and (3.7) motivate our interest in $\mathrm{comp}(\mathfrak{N})$.

Suppose that ε tends to zero. Then if $r(\mathfrak{N},S) > 0$, the fixed information $\mathfrak{N}$ is weak for the problem S for sufficiently small ε. However, in many cases, we can choose a permissible information operator $\mathfrak{N} = \mathfrak{N}(\varepsilon)$ such that $r(\mathfrak{N}(\varepsilon),S) < \varepsilon$ and solve our problem using the information $\mathfrak{N}(\varepsilon)$.

Let Ψ be a class of permissible information operators. Let

$$(3.8) \qquad r(\Psi,S) = \inf_{\mathfrak{N} \in \Psi} r(\mathfrak{N},S).$$

If $r(\Psi,S) = 0$, then we can solve the problem S for *any* ε using a suitably chosen information operator from Ψ. If $r(\Psi,S) > 0$, then the class Ψ does not supply enough information to solve the problem S for $\varepsilon \leq r(\Psi,S)$.

Definition 3.2 We shall say $\mathrm{comp}(\Psi,S,\varepsilon)$ is the ε-*complexity of the problem S in the class* Ψ (briefly the ε-complexity of S in Ψ) iff

$$(3.9) \qquad \mathrm{comp}(\Psi,S,\varepsilon) = \inf_{\mathfrak{N} \in \Psi} \mathrm{comp}(\mathfrak{N},S,\varepsilon).$$

We shall refer to the ε-complexity of the problem S in the class Ψ as the *problem complexity*.

We shall say φ^{oc} is an *optimal complexity algorithm for the problem S in the class Ψ* (briefly an optimal complexity algorithm for S in Ψ) iff

(3.10) $$\text{comp}(\varphi^{oc}) = \text{comp}(\Psi,S,\varepsilon). \quad \blacksquare$$

REMARK 3.3   Note that although we like to think of problem complexity as an intrinsic property of the problem, it is defined with respect to a model of computation and a class of information operators. $\blacksquare$

Note that $\text{comp}(\Psi,S,\varepsilon)$ is a nonincreasing function of ε. If $\varepsilon \leq r(\Psi,S)$ then (3.2) yields that $\text{comp}(\Psi,S,\varepsilon) = +\infty$. We shall prove in Chapter 5 that $\text{comp}(\Psi,S,\varepsilon)$ can be an essentially arbitrary nonincreasing function of ε.

Our setting is sufficiently general that it includes problems for which information operators do not play a role. Examples are combinatorial problems and such problems of linear algebra as matrix mutiplication and the direct solution of linear systems. For such problems the information operator is the identity operator $\mathfrak{N}(f) = f$, where f belongs to a finite-dimensional space. Furthermore, the information complexity $\text{comp}(I) = 0$ since there is no cost in computing $\mathfrak{N}(f) = f$. Note that $r(I,S) = 0$ for any problem S because $\mathfrak{N} = I$ is one-to-one. Thus, we can define $\Psi = \{I\}$. Typically, we seek the exact solution $\alpha = S(f)$; thus, $\varepsilon = 0$. The complexity is given by

$$\text{comp}(I,S,0) = \inf_{\varphi\,:\,e(\varphi)=0} \text{comp}(\varphi(f)).$$

Therefore, in algebraic and combinatorial complexity, we seek an algorithm which finds $\alpha = S(f)$ and has minimal combinatory complexity.

Example 3.2 *Matrix Multiplication*   Let $f = [A,B]$, where A and B are $n \times n$ matrices. Then the matrix multiplication problem may be formulated as $S(f) = A * B$. Let $\mathfrak{N}(f) = f = [A,B]$. This means that all coefficients of A and B are known and we seek an algorithm with minimal combinatory complexity which yields the matrix $x(f) = A * B$. If the cost of each arithmetic operation is taken as unity, then

$$c_1 n^2 \leq \text{comp}(I,S,0) \leq c_2 n^\beta,$$

for some positive constants c_1 and c_2, where the best β known, due to Pan [78], is about 2.79. The actual value of $\text{comp}(I,S,0)$ is unknown. $\blacksquare$

Example 3.3 *Sorting*   Let $f = [f_1,f_2,\ldots,f_n]$, where $f_i \in D$ and D is an ordered set. Define

$$S(f) = [f_{i_1},f_{i_2},\ldots,f_{i_n}],$$

where $f_{i_1} \leq f_{i_2} \leq \cdots \leq f_{i_n}$ and $i_1,\ldots,i_n$ is a permutation of $1,\ldots,n$. Then the sorting problem may be formulated as $\alpha = S(f)$. Let $\mathfrak{N}(f) = f$ and $\varepsilon = 0$. (As we mentioned in Remark 2.3, if $\varepsilon = 0$ the spaces need not be linear or normed.) We seek an algorithm which finds $S(f)$ with minimal cost, where the cost is taken as the number of comparisons. The complexity satisfies

$$\text{comp}(I,S,0) = (n\log_2 n)(1 + o(1)). \quad \blacksquare$$

Recently there has been an interest in finding ε-approximations to the solutions of algebraic and combinatory problems. For some problems, the complexity $\text{comp}(I,S,\varepsilon)$ is significantly smaller for positive ε than $\text{comp}(I,S,0)$. Examples may be found in Garey and Johnson [76].

Example 3.4   *Polynomial Zero*   Let $f = [a_0,a_1,\ldots,a_n]$, where the a_i are the coefficients of an nth degree polynomial, $P(x)$. Let $P(\alpha) = 0$, $\alpha = S(f)$. For algorithms which require knowledge of all the a_i, $\mathfrak{N}_1(f) = f$. On the other hand, there exist iterative algorithms requiring only that P and P' can be evaluated at any point and $\mathfrak{N}_2(f) = [x,P(x),P'(x)]$. $\blacksquare$

4. THE SCOPE OF ANALYTIC COMPUTATIONAL COMPLEXITY

We have reached a point where we can discuss the scope of analytic complexity more precisely. See particularly Definition 4.3 and the remarks which follow the definition.

Fix a set of primitives P and a class of permissible information operators Ψ.

Definition 4.1   We shall say a problem S is a *finite-complexity problem in* Ψ (briefly finite-complexity problem) iff there exists a permissible algorithm using an $\mathfrak{N}$ from Ψ which solves it exactly ($\varepsilon = 0$) with finite complexity. A problem S is an *infinite-complexity problem in* Ψ (briefly infinite-complexity problem) iff it is not a finite-complexity problem. $\blacksquare$

REMARK 4.1   Whether a problem is finite-complexity or infinite-complexity can only be determined with respect to P and Ψ. See the square root example in Remark 3.2. $\blacksquare$

With a "reasonable" set of primitives "most" problems of mathematics, science, and engineering are infinite-complexity problems. Major exceptions are combinatorial and certain algebraic problems.

Even though finite-complexity problems can, by definition, be solved with finite cost, that cost may be very high. We may therefore elect to solve such a problem approximately to lower the complexity. Important examples are provided by the approximate solution of NP-complete problems and the iterative solution of large sparse linear systems.

Since computation can only deal with objects generated with finite cost, it consists of approximately computed infinite-complexity problems and exactly or approximately computed finite-complexity problems.

It is convenient to distinguish between two types of information operators.

Definition 4.2 If an information operator $\mathfrak{N}$ is one-to-one in $\mathfrak{I}_0$, we say $\mathfrak{N}$ is *complete* (in $\mathfrak{I}_0$). If $\mathfrak{N}$ is many-to-one in $\mathfrak{I}_0$, we say $\mathfrak{N}$ is *partial* (in $\mathfrak{I}_0$). ∎

Algorithms for infinite-complexity problems often, but not always, use partial information. Algorithms for the approximate solution of finite-complexity problems sometimes use partial information. Our adversary principle (see Section 2) is trivial (with respect to the radius of information) unless $\mathfrak{N}$ is partial.

Example 4.1 As in Remark 3.2, let $S(f) = \sqrt{f}$, $f \in \mathfrak{I}_0 = [1,2]$, $\mathfrak{N}(f) = f$. Let the set of primitives be $\{ +, -, *, / \}$. Then S is an infinite-complexity problem. Define the permissible algorithm

$$x_0 = 1,$$
$$x_{i+1} = (x_i + f/x_i)/2, \qquad i = 0, 1, \ldots, k-1.$$

The algorithm uses complete information. ∎

Consider any permissible algorithm φ which uses information $y = \mathfrak{N}(f)$. As observed in Section 3, to evaluate φ, we

(a) compute $y = \mathfrak{N}(f)$,
(b) compute $\varphi(y)$.

The cost of the computation of $\varphi(y)$ for a worst y is called the combinatory complexity of the algorithm φ.

Note that the problem of minimizing the combinatory complexity may always be viewed as a finite-complexity problem with complete information. Thus, this problem of algebraic complexity is always a subproblem in our setting. See Brent and Kung [78, Section 6] and Trojan [79] for interesting results in algebraic complexity arising from analytic complexity. We observed in Examples 3.2 and 3.3 and the discussion preceding Example 3.2 that algebraic complexity can be formally included in our setting. The problem of minimizing combinatory complexity is a nontrivial instance of how algebraic complexity fits into our framework.

We are now ready to specify the scope of our subject.

Definition 4.3 *Analytic computational complexity* is the study of optimal algorithms for problems which are solved approximately. Analytic complexity may require the solution of problems in algebraic complexity as a subproblem. Often the information operator is partial. ∎

REMARK 4.2 Examples of optimal algorithms are optimal error algorithms and optimal complexity algorithms in Part A and maximal order algorithms and minimal complexity index algorithms in Part B. ∎

REMARK 4.3 The complexity of the optimal complexity algorithm in a class of information operators is the problem complexity. (See Definition 3.2 and Remark 3.3.) Problem complexity is one of the basic issues in analytic complexity. ∎

REMARK 4.4   The definition of optimal algorithm is model dependent. Among the models of interest are worst case, average case, and asymptotic case. (See Chapter 10.) This book treats a worst case model. Thus, by problem complexity we mean "worst case problem complexity."   ∎

This monograph is primarily devoted to infinite-complexity problems. Research on approximate solution of finite-complexity problems will be reported elsewhere.

We exhibit some relations between Definitions 4.1 and 4.2 in

REMARK 4.5   If S is a finite-complexity problem in Ψ, then there exists an $\mathfrak{N} \in \Psi$ such that $U(f)$ is a singleton for all $f \in \mathfrak{I}_0$. If, in addition, S is one-to-one, then $\mathfrak{N}$ is complete.

If S is an infinite-complexity problem in Ψ, then one of the following two cases holds:

(i)   For all $\mathfrak{N} \in \Psi$, there exists $f \in \mathfrak{I}_0$ such that $U(f)$ is not a singleton. Every $\mathfrak{N}$ from Ψ is partial.

(ii).   There exists an $\mathfrak{N} \in \Psi$ such that $U(f)$ is a singleton for all $f \in \mathfrak{I}_0$ and for all permissible φ, $e(\varphi) > 0$. (See Remark 3.2 for an example.) Then if S is one-to-one, $\mathfrak{N}$ is complete.   ∎

Chapter 2

Theory of Linear Information

1. INTRODUCTION

In the next five chapters, we shall assume that the solution operator S and the information operator $\mathfrak{N}$ are both linear.

Many important problems are linear and therefore covered by the theory of this chapter. Examples are integration, interpolation, approximation, and linear partial differential equations (see Chapter 6). Note, however, that the solution of linear algebraic equations is not a linear problem.

The linearity of $\mathfrak{N}$ is not as strong an assumption as might be supposed (see Chapter 7).

In Chapter 1, a problem is specified by a solution operator S and a set of problem elements $\mathfrak{J}_0$. In this chapter, we add to the specification a linear restriction operator T which restricts the set of problem elements f in the set $\mathfrak{J}_0$. It is often used to impose a regularity condition. It would be difficult to overemphasize the importance of T in our work. Many examples of its use may be found in the applications of Chapter 6.

We introduce the concept of the cardinality of $\mathfrak{N}$, card($\mathfrak{N}$), which is a measure of "how much information" there is in $\mathfrak{N}$. Card($\mathfrak{N}$) may be viewed as the number n of linearly independent linear functionals comprising $\mathfrak{N}$.

We vary $\mathfrak{N}$ and ask for the most "relevant" information operator for solving a problem. More precisely, we seek $\mathfrak{N}$ which minimizes the diameter of information among all information operators with cardinality at most n. We call this best $\mathfrak{N}$ the nth optimal information. In a Hilbert space setting, we give a complete solution to the problem of optimal information.

Returning to the general case, we show there exist linear problems so that for every ε, *no matter how large*, one cannot find an ε-approximation using *any* finite number of linear functionals. This negative result is independent of any model of computation.

In Sections 2–6, we consider *nonadaptive* information which consists of n independently given linear functionals. In Section 7, we introduce *adaptive* information, where previously computed functionals may be used to determine the next functional. Adaptive information is widely used in practice and it is known to be more powerful than nonadaptive information for some *nonlinear* problems. We shall prove the surprising result that adaptive information is not more powerful than nonadaptive information for *linear* problems.

In several portions of this monograph, we shall show relations between concepts in mathematical approximation theory and analytic complexity theory. This permits us to use deep results from approximation theory and conversely analytic complexity may be used to solve pure approximation theory problems. In Section 6, we determine relations between Gelfand n-widths and nth minimal diameters.

We summarize the major results of this chapter. In Section 2, the cardinality card($\mathfrak{N}$) of a linear information operator is defined, and we show (Lemma 2.2) that information operators with finite cardinality equal to n can be represented by n linearly independent linear functionals. In Section 3, we consider problems specified by a linear solution operator S and a linear restriction operator T. We define index(S,T) and show (Theorem 3.2) that if card($\mathfrak{N}$) < index(S,T) the solution cannot be approximated to within ε even for arbitrarily large ε. In particular, if index(S,T) $= \infty$, the problem cannot be solved by any information operator with finite cardinality.

In the next section, we study what is the most relevant information for solving a problem. The nth minimal diameter of information $d(n,S,T)$ is the diameter if the best information of cardinality at most n is used. Theorem 4.1 shows that $d(n,S,T)$ is completely determined by the operator ST^{-1}. A problem is ε-noncomputable if $d(S,T) = \lim_{n \to \infty} d(n,S,T) \geq 2\varepsilon$ and is convergent if $d(S,T) = 0$. We show (Lemma 4.5) that $d(S,T)$ can be any number.

In Section 5, we show that in a Hilbert space setting the problem (S,T) is convergent iff ST^{-1} is a compact operator. In a Hilbert space, the problem of most relevant information of cardinality n is completely solved (Theorem 5.3).

In Section 6, we find relations between nth minimal diameters and Gelfand n-widths (Theorem 6.1). This is used to find optimal or nearly optimal information operators.

In the concluding section, we define adaptive linear information and prove (Theorem 7.1) that for linear problems adaptive information is not more powerful than nonadaptive information.

2. CARDINALITY OF LINEAR INFORMATION

Let $\mathfrak{N}:\mathfrak{I}_1 \to \mathfrak{I}_3$ be a linear information operator where $\mathfrak{I}_3$ is a linear space. Let $\ker \mathfrak{N} = \{f:\mathfrak{N}(f) = 0\}$ be the kernel of $\mathfrak{N}$. We shall prove in Section 3 that the dependence of the diameter of information on $\mathfrak{N}$ is only through the kernel of $\mathfrak{N}$. This suggests we should not distinguish between two information operators with the same kernel.

Let $\mathfrak{N}_1:\mathfrak{I}_1 \to \mathfrak{I}_3$ and $\mathfrak{N}_2:\mathfrak{I}_1 \to \mathfrak{I}'_3$ be two linear information operators, where $\mathfrak{I}'_3$ is a linear space not necessarily equal to $\mathfrak{I}_3$.

Definition 2.1 We shall say $\mathfrak{N}_1$ is *contained in* $\mathfrak{N}_2$ (briefly $\mathfrak{N}_1 \subset \mathfrak{N}_2$) iff $\ker \mathfrak{N}_2 \subset \ker \mathfrak{N}_1$. We shall say $\mathfrak{N}_1$ is *equivalent to* $\mathfrak{N}_2$ (briefly $\mathfrak{N}_1 \asymp \mathfrak{N}_2$) iff $\ker \mathfrak{N}_1 = \ker \mathfrak{N}_2$. ∎

Note that "$\asymp$" is an equivalence relation.

We want to show that $\mathfrak{N}_1 \subset \mathfrak{N}_2$ can be characterized by the rank of a certain matrix. We first briefly recall some facts on linear spaces. See, for instance, Edwards [65]. Let A be a linear subspace of $\mathfrak{I}_1$. Then there exists a linear subspace $A^\perp$ of $\mathfrak{I}_1$ such that

$$(2.1) \qquad \mathfrak{I}_1 \overset{\mathrm{df}}{=} A \oplus A^\perp.$$

In general, $A^\perp$ is not uniquely defined. However, if $\mathfrak{I}_1$ is a Hilbert space and A is closed, then there exists a unique orthogonal $A^\perp$ to A such that (2.1) holds. In either case, $A^\perp$ is isomorphic to the quotient space $\mathfrak{I}_1/A$ and

$$(2.2) \qquad \operatorname{codim} A \overset{\mathrm{df}}{=} \dim A^\perp = \dim \mathfrak{I}_1/A.$$

The space $A^\perp$ is called an *algebraic complement* of A in the space $\mathfrak{I}_1$.

Let $L_1, L_2, \ldots, L_m$ be linearly independent linear functions. By

$$(2.3) \qquad \mathfrak{N}_1 = [L_1, L_2, \ldots, L_m]^t,$$

we mean $\mathfrak{N}_1(f) = [L_1(f), L_2(f), \ldots, L_m(f)]^t \in \mathbb{C}^m$, where t denotes the transpose of a vector and $f \in \mathfrak{I}_1$.

Lemma 2.1 Let $\mathfrak{N}_1 = [L_1, L_2, \ldots, L_n]^t$ and $\mathfrak{N}_2 = [L_{n+1}, L_{n+2}, \ldots, L_{n+k}]^t$ be information operators.

(i) $\mathfrak{N}_1 \subset \mathfrak{N}_2$ iff $k \geq n$ and there exists an $n \times k$ matrix M such that

$$\mathfrak{N}_1 = M\mathfrak{N}_2 \qquad \text{and} \qquad \operatorname{rank} M = n.$$

(ii) Let $k = n$. Then $\mathfrak{N}_1 \subset \mathfrak{N}_2$ iff $\mathfrak{N}_1 \asymp \mathfrak{N}_2$. ∎

PROOF (i) Suppose that $\ker \mathfrak{N}_2 \subset \ker \mathfrak{N}_1$. Let $\mathfrak{I}_1 = \ker \mathfrak{N}_2 \oplus (\ker \mathfrak{N}_2)^\perp$ and $(\ker \mathfrak{N}_2)^\perp = \operatorname{lin}(\xi_1, \xi_2, \ldots, \xi_k)$, where $L_{n+j}(\xi_i) = \delta_{ij}$ and δ_{ij} denotes the Kronecker delta. Then $f = f_0 + \sum_{i=1}^k L_{n+i}(f)\xi_i$, where $f_0 \in \ker \mathfrak{N}_2$ and $f \in \mathfrak{I}_1$.

Since $f_0 \in \ker \mathfrak{N}_1$, we get

$$(2.4) \qquad L_j(f) = \sum_{i=1}^{k} L_{n+i}(f)L_j(\xi_i) \qquad \text{for} \quad j = 1, 2, \ldots, n.$$

This yields $\mathfrak{N}_1 = M\mathfrak{N}_2$ with $M = (L_j(\xi_i))$. Let $(\ker \mathfrak{N}_1)^\perp = \text{lin}(\eta_1, \eta_2, \ldots, \eta_n)$, where $L_j(\eta_i) = \delta_{ij}$. Set $f = \eta_i$ in (2.4) for $i = 1, 2, \ldots, n$. Then

$$I = M[\mathfrak{N}_2(\eta_1), \mathfrak{N}_2(\eta_2), \ldots, \mathfrak{N}_2(\eta_n)],$$

where I is the $n \times n$ identity matrix. This implies that rank $M = n$ and completes this part of the proof. Suppose now that $\mathfrak{N}_1 = M\mathfrak{N}_2$. Then $h \in \ker \mathfrak{N}_2$ implies $\mathfrak{N}_1(h) = M\mathfrak{N}_2(h) = 0$ which yields $\mathfrak{N}_1 \subset \mathfrak{N}_2$.

(ii) Suppose that $\mathfrak{N}_1 \subset \mathfrak{N}_2$. Due to the first part of Lemma 2.1, we get $\mathfrak{N}_1 = M\mathfrak{N}_2$, where the $n \times n$ matrix M is nonsingular. Then $\mathfrak{N}_2 = M^{-1}\mathfrak{N}_1$ which implies $\mathfrak{N}_2 \subset \mathfrak{N}_1$ and $\mathfrak{N}_1 \asymp \mathfrak{N}_2$. The second part of (ii) is trivial. Hence Lemma 2.1 is proven. ∎

We now show that any information operator $\mathfrak{N}$, where $n = \text{codim} \ker \mathfrak{N}$ is finite, may be represented by n linearly independent linear functionals.

Lemma 2.2   Let $\mathfrak{N}$ be an information operator and $n = \text{codim} \ker \mathfrak{N} < +\infty$. Then there exist linearly independent linear functionals $L_1, L_2, \ldots, L_n$ such that

$$\mathfrak{N} \asymp \mathfrak{N}_1 \qquad \text{where} \quad \mathfrak{N}_1 = [L_1, L_2, \ldots, L_n]^t. \quad ∎$$

PROOF   Let $(\ker \mathfrak{N})^\perp = \text{lin}(\xi_1, \xi_2, \ldots, \xi_n)$. Every element f has a unique representation $f = f_0 + \sum_{i=1}^{n} L_i(f)\xi_i$, where $f_0 \in \ker \mathfrak{N}$ and $L_1, L_2, \ldots, L_n$ are linearly independent linear functionals. Since $\ker \mathfrak{N} = \{f : L_i(f) = 0, \ i = 1, 2, \ldots, n\} = \ker \mathfrak{N}_1$, we get $\mathfrak{N} \asymp \mathfrak{N}_1$. ∎

Lemma 2.2 states that an information operator with finite $n = \text{codim} \ker \mathfrak{N}$ is equivalent to an information operator generated by n linearly independent linear functionals. Observe that to know $\mathfrak{N}_1(f)$ one has to evaluate n linear functionals. This suggests the following definition of the cardinality of $\mathfrak{N}$.

Definition 2.2   We shall say that card$(\mathfrak{N})$ is the *cardinality of the information* $\mathfrak{N}$ iff

$$(2.5) \qquad \text{card}(\mathfrak{N}) = \text{codim} \ker \mathfrak{N} \quad (= \dim \mathfrak{N}(\mathfrak{I}_1)). \quad ∎$$

We shall prove in Section 3 that unless the cardinality of the information $\mathfrak{N}$ is sufficiently large, the diameter of information is infinity and the problem cannot be solved with this information.

To illustrate the concept of cardinality, we consider two examples.

Example 2.1   Let $\mathfrak{N} = [L_1, L_2, \ldots, L_n]^t$. From the preceding considerations easily follows that card$(\mathfrak{N}) \leq n$ and card$(\mathfrak{N}) = n$ iff $L_1, L_2, \ldots, L_n$ are linearly independent. ∎

Example 2.2 Let $f: D \subset \mathbb{C}^m \to \mathbb{C}^m$ be a k-times differentiable function. Let

$$\mathfrak{N}(f) = f^{(k)}(x) \qquad \text{for} \quad x \in D.$$

Note that $f(x) = [f_1(x), f_2(x), \ldots, f_m(x)]^t$, where $f_j: D \to \mathbb{C}^1$ is a scalar function. The $f_j^{(k)}(x)$ can be represented by $\binom{m+k-1}{k}$ linearly independent functionals of the form $L(f) = \partial^k f / \partial x_1^{p_1} \cdots \partial x_m^{p_m}$, where $x = [x_1, x_2, \ldots, x_m]^t$ and $p_i \geq 0$, $p_1 + p_2 + \cdots + p_m = k$. This yields

$$\text{card}(\mathfrak{N}) = m \binom{m+k-1}{k}.$$

Let $\mathfrak{N}(f) = [f(x), f'(x), \ldots, f^{(n-1)}(x)]$. Then it is called standard information and

$$\text{card}(\mathfrak{N}) = \sum_{k=0}^{n-1} m \binom{m+k-1}{k} = m \binom{m+n-1}{n-1}.$$

This shows the dependence of cardinality on the dimension of the space $\mathbb{C}^m$. ∎

We end this section by showing that there is a one-to-one correspondence between information operators and subspaces of $\mathfrak{I}_1$.

Let $\mathfrak{N}$ be an information operator with $\text{card}(\mathfrak{N}) = n$. Then $A(\mathfrak{N}) = \ker \mathfrak{N}$ has codimension equal to n. Furthermore, $\mathfrak{N}_1 \asymp \mathfrak{N}_2$ implies $A(\mathfrak{N}_1) = A(\mathfrak{N}_2)$. We now show that the converse statement is also true.

Lemma 2.3 Let A be an arbitrary linear subspace of $\mathfrak{I}_1$ such that codim $A = n$. Then there exists a unique (up to the equivalence relation) information operator $\mathfrak{N}$ with $\text{card}(\mathfrak{N}) = n$ such that $A = \ker \mathfrak{N}$. ∎

PROOF Let $\mathfrak{I}_1 = A \oplus A^\perp$ where $A^\perp = \text{lin}(\xi_1, \xi_2, \ldots, \xi_n)$. Then $f = f_0 + \sum_{i=1}^n L_i(f)\xi_i$, where $f_0 \in A$ and $L_i(\xi_j) = \delta_{ij}$. Define

$$\mathfrak{N} = [L_1, L_2, \ldots, L_n]^t.$$

Since $L_1, L_2, \ldots, L_n$ are linearly independent, $\text{card}(\mathfrak{N}) = n$ and $\ker \mathfrak{N} = A$. To prove the uniqueness, observe that if $A = \ker \mathfrak{N}_1 = \ker \mathfrak{N}_2$ then $\mathfrak{N}_1 \asymp \mathfrak{N}_2$. This completes the proof. ∎

3. INDEX OF A LINEAR PROBLEM

We consider in this section linear information operators for the solution element $\alpha = S(f)$, where $S: \mathfrak{I}_1 \to \mathfrak{I}_2$ is a linear operator. Furthermore, we assume that $\mathfrak{I}_0$ is defined as

(3.1) $$\mathfrak{I}_0 = \{f \in \mathfrak{I}_1 : \|Tf\| \leq 1\},$$

where $T: \mathfrak{I}_1 \to \mathfrak{I}_4 = T(\mathfrak{I}_1)$ is a linear operator and $\mathfrak{I}_4$ is a linear normed space over the real or complex field. We shall call T the *restriction operator*. This means we want to find an ε-approximation to the solution $\alpha = S(f)$ for all f such that $\|Tf\| \le 1$.

To stress the dependence on T, we shall replace S by (S,T) in all basic definitions. For instance, we shall refer to the *problem* (S,T), the *diameter* $d(\mathfrak{N},S,T)$, etc., where $\mathfrak{N}$ is a linear information operator.

Without loss of generality, we choose a bound $\|Tf\| \le 1$ instead of $\|Tf\| \le c$ for a positive constant c. Indeed, let $T_1 = c^{-1}T$. Then $\|T_1 f\| \le 1$ is equivalent to $\|Tf\| \le c$. It is easy to observe that $d(\mathfrak{N},S,T) = cd(\mathfrak{N},S,T_1)$ and all estimates on complexity are linear in c.

We now show that the dependence of $d(\mathfrak{N},S,T)$ on $\mathfrak{N}$ is only through the kernel of $\mathfrak{N}$.

Lemma 3.1

$$(3.2) \qquad d(\mathfrak{N},S,T) = 2 \sup_{h \in V(0)} \|Sh\|,$$

where $V(0) = \ker \mathfrak{N} \cap \mathfrak{I}_0$. (See (2.4) of Chapter 1.) ∎

PROOF Set $c = 2 \sup_{h \in V(0)} \|Sh\|$. Let $f \in \mathfrak{I}_0$ and $\tilde{f} \in V(f)$. Then $h = \frac{1}{2}(\tilde{f} - f) \in \ker \mathfrak{N}$ and $\|Th\| \le 1$. This yields

$$\|S\tilde{f} - Sf\| = 2\|Sh\| \le c.$$

Taking the supremum with respect to f and $\tilde{f}$, we get $d(\mathfrak{N},S,T) \le c$. To prove the reverse inequality, let $h \in V(0)$. Set $f = h$ and $\tilde{f} = -h$. Then $\tilde{f} \in V(f) = V(0)$ and

$$2\|Sh\| = \|S\tilde{f} - Sf\| \le d(\mathfrak{N},S,T).$$

Thus, $c \le d(\mathfrak{N},S,T)$ which completes the proof. ∎

From Lemma 3.1 we immediately get the following corollary.

Corollary 3.1 If $\mathfrak{N}_1 \subset \mathfrak{N}_2$, then $d(\mathfrak{N}_2,S,T) \le d(\mathfrak{N}_1,S,T)$. If $\mathfrak{N}_1 \asymp \mathfrak{N}_2$, then $d(\mathfrak{N}_2,S,T) = d(\mathfrak{N}_1,S,T)$. ∎

REMARK 3.1 To prove Lemma 3.1 and Corollary 3.1, we used only two properties of $\mathfrak{I}_0$; namely, that $\mathfrak{I}_0$ is convex, i.e., $f, g \in \mathfrak{I}_0$ implies $tf + (1 - t)g \in \mathfrak{I}_0$, $\forall t \in [0,1]$, and that $\mathfrak{I}_0$ is balanced, i.e., $f \in \mathfrak{I}_0$ implies $-f \in \mathfrak{I}_0$. In fact, it is easy to see that

$$d(\mathfrak{N},S) = 2 \sup_{h \in \ker \mathfrak{N} \cap \mathfrak{I}_0} \|Sh\|,$$

$$\mathfrak{N}_1 \subset \mathfrak{N}_2 \quad \text{implies} \quad d(\mathfrak{N}_1,S) \le d(\mathfrak{N}_2,S) \quad \text{and} \quad r(\mathfrak{N}_1,S) \le r(\mathfrak{N}_2,S),$$
$$\mathfrak{N}_1 \asymp \mathfrak{N}_2 \quad \text{implies} \quad d(\mathfrak{N}_1,S) = d(\mathfrak{N}_2,S) \quad \text{and} \quad r(\mathfrak{N}_1,S) = r(\mathfrak{N}_2,S)$$

for any linear operators S and $\mathfrak{N}$, and for any convex balanced set $\mathfrak{I}_0$. See Chapter 7, where this problem is considered for the general nonlinear case. ∎

In Section 2 of Chapter 1, we showed that the radius $r(\mathfrak{N},S,T)$ is the intrinsic error of the information $\mathfrak{N}$ and the problem (S,T). Due to Lemma 3.1 we get

$$c/2 \le r(\mathfrak{N},S,T) \le c,$$

where $c = 2 \sup_{h \in V(0)} \|Sh\|$. We now show when $r(\mathfrak{N},S,T) = c/2$.

Lemma 3.2 If for any $f \in \mathfrak{I}_0$ there exists $h_0 \in \ker \mathfrak{N}$ such that $Th_0 = Tf$, then

$$r(\mathfrak{N},S,T) = \sup_{h \in V(0)} \|Sh\|. \quad \blacksquare$$

PROOF Let $a = S(f - h_0)$. Then for any $\tilde{f} = f + h \in V(f)$, $h \in \ker \mathfrak{N}$, we get $\|a - S\tilde{f}\| = \|Sz\|$, where $z = h_0 + h \in \ker \mathfrak{N}$. Since $Th_0 = Tf$,

$$\|Tz\| = \|T(z - h_0) + Tf\| = \|T\tilde{f}\| \le 1.$$

From (2.10) of Chapter 1 and Lemma 3.1, we get

$$r(\mathfrak{N},S,T) \le \sup_{f \in \mathfrak{I}_0} \sup_{\tilde{f} \in V(f)} \|a - S\tilde{f}\| \le \sup_{z \in V(0)} \|Sz\| = \tfrac{1}{2} d(\mathfrak{N},S,T).$$

Since $r(\mathfrak{N},S,T) \ge \tfrac{1}{2} d(\mathfrak{N},S,T)$ for any $\mathfrak{N}$, S, and T, Lemma 3.2 is proven. $\blacksquare$

We want to examine when the diameter $d(\mathfrak{N},S,T)$ is equal to infinity. (Of course, $d(\mathfrak{N},S,T) = +\infty$ implies $r(\mathfrak{N},S,T) = +\infty$.) We begin with

Theorem 3.1 If $\ker \mathfrak{N} \cap \ker T \not\subset \ker S$, then $d(\mathfrak{N},S,T) = +\infty$. $\blacksquare$

PROOF Let $h \in \ker \mathfrak{N} \cap \ker T$ and $h \notin \ker S$. Then $T(ch) = 0$, $\mathfrak{N}(ch) = 0$ for any constant c. Further, $\|S(ch)\| = |c| \, \|Sh\| \to +\infty$ with $|c| \to +\infty$. Due to Lemma 3.1, we get $d(\mathfrak{N},S,T) = +\infty$. $\blacksquare$

Theorem 3.1 states that $\ker \mathfrak{N} \cap \ker T$ *has to* be contained in $\ker S$ for $d(\mathfrak{N},S,T)$ to be finite. We prove that $\ker \mathfrak{N} \cap \ker T \subset \ker S$ implies that the cardinality of $\mathfrak{N}$ is at least as large as the "problem index." Let

(3.3)
$$\ker T = (\ker T \cap \ker S) \oplus A(T,S),$$
$$A(T,S) = \lin(\xi_1^*, \xi_2^*, \ldots, \xi_{n^*}^*),$$

where $A(T,S)$ is an algebraic complement of $\ker T \cap \ker S$ in the space $\ker T$ and $\xi_1^*, \ldots, \xi_{n^*}^*$ form a basis of $A(T,S)$, $n^* = n^*(T,S) \le +\infty$.

Definition 3.1 We shall say that $\operatorname{index}(S,T) = \dim A(T,S)$ is the *index of the problem* (S,T). We shall sometimes write $\operatorname{index}(S,T) = n^*$. $\blacksquare$

Note that

$$\operatorname{index}(S,T) = \dim(\ker T) - \dim(\ker T \cap \ker S)$$

whenever $\dim(\ker T \cap \ker S)$ is finite. We are ready to prove the main result of this section.

Theorem 3.2   If $\text{card}(\mathfrak{N}) < \text{index}(S,T)$, then $d(\mathfrak{N},S,T) = +\infty$. ∎

PROOF   We show that $\ker \mathfrak{N} \cap \ker T \not\subset \ker S$. Define $f = \sum_{i=1}^{n^*} c_i \xi_i^* \in A(T,S)$.
We want to find a nonzero vector $(c_1, c_2, \ldots, c_{n^*})$ such that $f \in \ker \mathfrak{N}$. From
Lemma 2.2 it follows that there exists an information operator $\mathfrak{N}_1 =$
$[L_1, L_2, \ldots, L_m]^t$, where $m = \text{card}(\mathfrak{N}) < \text{index}(S,T)$ such that $\ker \mathfrak{N}_1 = \ker \mathfrak{N}$.
Thus, $f \in \ker \mathfrak{N}$ iff $L_j(f) = \sum_{i=1}^{n^*} c_i L_j(\xi_i^*) = 0$ for $j = 1, 2, \ldots, m$. Hence, we get
a homogeneous system of m linear equations in n^* unknowns. Since $m < n^*$,
there exists a nonzero vector $(c_1, c_2, \ldots, c_{n^*})$ which is a solution of the system.
Thus, $0 \neq f \in A(T,S) \cap \ker \mathfrak{N}$. This means that a nonzero f belongs to
$\ker \mathfrak{N} \cap \ker T$ and $f \notin \ker S$. Due to Theorem 3.1, we get $d(\mathfrak{N},S,T) = +\infty$. ∎

Theorem 3.2 states that every information operator with cardinality less
than the index of S and T does not supply enough information to solve the
problem. For $\text{index}(S,T) = +\infty$, we get the following corollary.

Corollary 3.2   If $\text{index}(S,T) = +\infty$, then the problem (S,T) *cannot* be solved
by *any* information operator with *finite* cardinality. ∎

We illustrate the preceding results for some restriction operators T. We
begin with $T = 0$.

Lemma 3.3   (No restriction operator)   Let $T = 0$. Then

(i) $d(\mathfrak{N},S,0)$ is either zero or infinity. More precisely, $\ker \mathfrak{N} \not\subset \ker S$ implies
$d(\mathfrak{N},S,0) = +\infty$, $\ker \mathfrak{N} \subset \ker S$ implies $d(\mathfrak{N},S,0) = 0$.
(ii) $\text{index}(S,0) = \dim(\ker S)^\perp$ is finite iff S is a finite-dimensional operator,
i.e., $\dim S(\mathfrak{I}_1) < +\infty$. ∎

PROOF   Since $T = 0$, $\mathfrak{I}_0 = \mathfrak{I}_1$ and $\ker T = \mathfrak{I}_1$. If $\ker \mathfrak{N} \cap \ker T = \ker \mathfrak{N} \not\subset$
$\ker S$, then $d(\mathfrak{N},S,0) = +\infty$ due to Theorem 3.1. If $\ker \mathfrak{N} \subset \ker S$, then $Sh = 0$
for all $h \in V(0) = \ker \mathfrak{N} \subset \ker S$. Thus $d(\mathfrak{N},S,0) = 0$ by Lemma 3.1. This proves (i).

From (3.3), we get $A(0,S) = (\ker S)^\perp$ and $\text{index}(S,0) = \dim(\ker S)^\perp$. It is well
known that $\dim(\ker S)^\perp$ is finite iff S is a finite-dimensional operator. This proves
Lemma 3.3. ∎

As our second illustration, consider $T = D^k$, $k \geq 0$, i.e., $Tf = f^{(k)}$ for a scalar
function f. If S is a one-to-one operator, then

$$A(T,S) = \ker T = \{f : f^{(k)} \equiv 0\}$$

and $\text{index}(S,D^k) = \dim(\ker T) = k$. Hence, we have to compute k linear func-
tionals to assure that $\text{card}(\mathfrak{N}) \geq \text{index}(S,D^k)$ and $\ker \mathfrak{N} \cap \ker T = \ker S = \{0\}$.

REMARK 3.2   We show that modulo a technical detail, the assumption that
$\mathfrak{I}_0$ is generated by a restriction operator and that $\mathfrak{I}_0$ is convex, balanced, and
absorbing are equivalent. By absorbing, we mean that for any $f \in \mathfrak{I}_1$ there
exists a positive constant c such that $cf \in \mathfrak{I}_0$.

To show the equivalence in one direction, assume that $\mathfrak{I}_0 = \{f \in \mathfrak{I}_1 : \|Tf\| \leq 1\}$
is generated by a restriction operator T. Then it is obvious that $\mathfrak{I}_0$ is convex,
balanced, and absorbing.

Now assume that $\mathfrak{I}_0$ is convex, balanced, and absorbing. We show that there exist a linear normed space $\mathfrak{I}_4$ and a linear operator $T:\mathfrak{I}_1 \to \mathfrak{I}_4$ such that

$$(3.4) \qquad\qquad \underline{\mathfrak{I}}_0 \subset \mathfrak{I}_0 \subset \overline{\mathfrak{I}}_0,$$

where $\underline{\mathfrak{I}}_0 = \{f \in \mathfrak{I}_1 : \|Tf\| < 1\}$ and $\overline{\mathfrak{I}}_0 = \{f : \|Tf\| \le 1\}$.

To prove (3.4), recall the definition of the Minkowski functional (gauge) $q:\mathfrak{I}_1 \to \mathbb{R}$,

$$q(f) = \inf\{c^{-1} : cf \in \mathfrak{I}_0, c > 0\}.$$

It is known, see for instance Wilansky [78], that q is a seminorm and

$$(3.5) \qquad\qquad \{f \in \mathfrak{I}_1 : q(f) < 1\} \subset \mathfrak{I}_0 \subset \{f \in \mathfrak{I}_1 : q(f) \le 1\}.$$

Let

$$A = \{f \in \mathfrak{I}_0 : q(f) = 0\}.$$

If $f, g \in A$, then for any constants c_1 and c_2 we have

$$q(c_1 f + c_2 g) \le |c_1| q(f) + |c_2| q(g) = 0$$

which shows that $c_1 f + c_2 g \in A$. Thus, A is a linear subspace of $\mathfrak{I}_1$. Let $\mathfrak{I}_1 = A \oplus A^\perp$. Then $f = f_1 + f_2$, where $f_1 \in A$ and $f_2 \in A^\perp$. Define

$$\mathfrak{I}_4 = A^\perp \qquad \text{and} \qquad \|f_2\| = q(f_2).$$

Since $q(f_2) = 0$ implies $f_2 \in A$, then $f_2 = 0$ and the seminorm q becomes a norm on $\mathfrak{I}_4$. Thus, $\mathfrak{I}_4$ is a linear normed space. Define a linear operator

$$Tf = f_2.$$

Since $\|f_2\| = q(f_2) = q(f)$, we have

$$\underline{\mathfrak{I}}_0 = \{f \in \mathfrak{I}_1 : q(f) < 1\}, \qquad \overline{\mathfrak{I}}_0 = \{f \in \mathfrak{I}_1 : q(f) \le 1\}.$$

Hence (3.5) yields (3.4).

Note that it can happen that $\mathfrak{I}_0$ is neither $\underline{\mathfrak{I}}_0$ nor $\overline{\mathfrak{I}}_0$. It is, however, obvious that the solution operators $\underline{S} = S|_{\underline{\mathfrak{I}}_0}$ and $\overline{S} = S|_{\overline{\mathfrak{I}}_0}$ have the same radii and diameters of information, i.e.,

$$r(\mathfrak{N},\underline{S}) = r(\mathfrak{N},S) = r(\mathfrak{N},\overline{S}),$$
$$d(\mathfrak{N},\underline{S}) = d(\mathfrak{N},S) = d(\mathfrak{N},\overline{S}) \qquad \forall \mathfrak{N}.$$

This shows that the problems $\underline{S}$, S, and $\overline{S}$ are only insignificantly different. ∎

4. OPTIMAL LINEAR INFORMATION OPERATORS

Assume throughout the rest of this chapter that $n^* = \text{index}(S,T) < +\infty$. We construct an information operator $\mathfrak{N}^*$ with $\text{card}(\mathfrak{N}^*) = \text{index}(S,T)$ such that $\ker \mathfrak{N}^* \cap \ker T \subset \ker S$. Recall that $A(T,S)$ is defined by (3.3) and $A(T,S) = \text{lin}(\xi_1^*, \xi_2^*, \ldots, \xi_{n^*}^*)$. Let $\mathfrak{I}_1 = A(T,S) \oplus A(T,S)^\perp$, where $A(T,S)^\perp = (\ker T \cap \ker S) \oplus (\ker T)^\perp$. Then $f = \sum_{i=1}^{n^*} L_i^*(f)\xi_i^* + f_1$, where $f_1 \in A(T,S)^\perp$ and $L_i^*(\xi_j^*) = \delta_{ij}$.

Lemma 4.1   Let

(4.1) $\mathfrak{N}^* = [L_1^*, L_2^*, \ldots, L_{n^*}^*]^t$ $(\ker \mathfrak{N}^* = A(T,S)^\perp).$

Then $\ker \mathfrak{N}^* \cap \ker T \subset \ker S$.   ∎

PROOF   Let $f \in \ker T$. Then $f = f_0 + \sum_{i=1}^{n^*} L_i^*(f)\xi_i^*$, where $f_0 \in \ker T \cap \ker S$. If $f \in \ker \mathfrak{N}^*$, then $L_i^*(f) = 0$ for $i = 1, 2, \ldots, n^*$ and $f = f_0 \in \ker S$. This proves Lemma 4.1.   ∎

To simplify further considerations and to assure that $\ker \mathfrak{N} \cap \ker T \subset \ker S$, we shall consider throughout this section only information operators $\mathfrak{N}$ such that $\mathfrak{N}^* \subset \mathfrak{N}$. (This means $\ker \mathfrak{N} \subset \ker \mathfrak{N}^*$ and $\ker \mathfrak{N} \cap \ker T \subset \ker \mathfrak{N}^* \cap \ker T \subset \ker S$ due to Lemma 4.1.)

We show the diameter $d(\mathfrak{N},S,T)$ can be computed in terms of the inverse operator T^{-1} which is defined as follows. (T is not one-to-one in general.) Recall that $\mathfrak{I}_4 = T(\mathfrak{I}_1)$ and let

(4.2) $\mathfrak{I}_1 = \ker T \oplus (\ker T)^\perp.$

Thus, $f = f_1 + f_2$, where $f_1 \in \ker T$ and $f_2 \in (\ker T)^\perp$. Define a linear operator $T^{-1} : \mathfrak{I}_4 \to \mathfrak{I}_1$ such that

(4.3) $T^{-1}z = f_2,$

where $z = Tf$.

We check that T^{-1} is well defined. Let $z = Tf = Tg$, where $f, g \in \mathfrak{I}_1$. Since $T(f - g) = 0$, $f - g = (f_1 - g_1) + (f_2 - g_2) \in \ker T$ which yields $f_2 = g_2$. This shows that $T^{-1}z$ does not depend on a particular choice of pre-image of z. Hence T^{-1} is well defined. Note that T^{-1} depends on a particular choice of $(\ker T)^\perp$ in (4.2).

As an example, observe that $T = 0$ implies $\mathfrak{I}_4 = \{0\}$ and $0^{-1} = 0$.

Let $K : \mathfrak{I}_4 \to \mathfrak{I}_2$ be a linear operator and let B be a linear subspace of $\mathfrak{I}_4$. Denote

(4.4) $\|K\|_B \stackrel{\mathrm{df}}{=} \sup_{\|z\| \le 1,\, z \in B} \|Kz\|.$

We are ready to prove

Lemma 4.2   Let $\mathfrak{N}^* \subset \mathfrak{N}$. Then

(4.5) $d(\mathfrak{N},S,T) = 2\|ST^{-1}\|_{T(\ker \mathfrak{N})}.$ ∎

PROOF   Lemma 3.1 states that $d(\mathfrak{N},S,T) = 2 \sup \|Sh\|$, where $\mathfrak{N}h = 0$ and $\|Th\| \le 1$. Let $h = h_1 + h_2$ due to (4.2). Since $h \in \ker \mathfrak{N} \subset \ker \mathfrak{N}^* = A(T,S)^\perp$, then (3.3) yields $h_1 \in \ker T \cap \ker S$. Then $Sh = Sh_2$ and $Th = Th_2$. Let $z = Th \in T(\ker \mathfrak{N})$. Observe that $T^{-1}z = h_2$ and $ST^{-1}z = Sh_2$. This proves (4.5).   ∎

Lemma 4.2 states that the diameter $d(\mathfrak{N},S,T)$ is equal to twice the norm of the linear operator $K = ST^{-1}$ in a certain linear subspace $B = T(\ker \mathfrak{N})$. This suggests the following problem. For a fixed integer n, find the most relevant information operator $\mathfrak{N}$, $\mathrm{card}(\mathfrak{N}) \leq n$, that is, the operator which minimizes $d(\mathfrak{N},S,T)$ among all information operators with cardinality $\leq n$. This is equivalent, as we shall prove, to finding a linear subspace B with codim $B \leq n - n^*$ which minimizes $\|K\|_B$ among all linear subspaces of codimension $\leq n - n^*$.

To formalize this problem, let Ψ_n be the class of *all* linear information operators $\mathfrak{N}$ such that $\mathfrak{N}^* \subset \mathfrak{N}$, $\mathrm{card}(\mathfrak{N}) \leq n$, where $n \geq \mathrm{index}(S,T)$.

Definition 4.1   We shall say $d(n,S,T)$ is the nth *minimal diameter of information* iff

(4.6)
$$d(n,S,T) = \begin{cases} \inf\limits_{\mathfrak{N} \in \Psi_n} d(\mathfrak{N},S,T) & \text{if } n \geq \mathrm{index}(S,T), \\ +\infty & \text{if } n < \mathrm{index}(S,T). \end{cases}$$

We shall say $\mathfrak{N}_n^{\mathrm{oi}}$ is an nth *optimal information* iff

(4.7)
$$d(n,S,T) = d(\mathfrak{N}_n^{\mathrm{oi}},S,T), \qquad \mathfrak{N}_n^{\mathrm{oi}} \in \Psi_n. \quad \blacksquare$$

We define $d(n,S,T) = +\infty$ for $n < \mathrm{index}(S,T)$, since for any $\mathfrak{N}$ with cardinality less than $\mathrm{index}(S,T)$, $d(\mathfrak{N},S,T) = +\infty$. See Section 6, where the relations between $d(n,S,T)$ and the Gelfand n-widths are studied. We illustrate Definition 4.1 by the following example.

Example 4.1   Let dim $\mathfrak{I}_1 = +\infty$ and let $T = I$ be the identity operator. Define $S = cI$ for a positive constant c. Then

(4.8)
$$d(n,cI,I) = 2c \qquad \forall n.$$

Indeed, let $\mathfrak{N} \in \Psi_n$. Then ker $\mathfrak{N} \neq \{0\}$ and $d(\mathfrak{N},S,T) = 2\|cI\|_{\mathrm{ker}\,\mathfrak{N}} = 2c$. Note that $\varphi(\mathfrak{N}(f)) \equiv 0$ is an optimal error algorithm, since $e(\varphi) = c = \tfrac{1}{2}d(\mathfrak{N},S,T) = r(\mathfrak{N},S,T)$. This means that no matter how many linear functionals are computed, the zero of the space $\mathfrak{I}_1$ is the best approximation to the solution $Sf = cf$ for some f such that $\|f\| \leq 1$. See Schultz [74] for related material.

However, for the identity information operator $\mathfrak{N}(f) \equiv f$, we get ker $\mathfrak{N} = \{0\}$ and $d(I,cI,I) = 0$. Note that $\mathrm{card}(I) = +\infty$. This shows that $d(n,S,T)$ can be a discontinuous function of n at infinity. $\blacksquare$

From Example 4.1, we get the following corollary.

Corollary 4.1   For every ε (no matter how large), there exists a linear problem (S,T) with finite index for which one cannot find an ε-approximation using any finite number of linear functionals. $\blacksquare$

We show that the nth minimal diameter and the nth optimal information are fully determined by the operator $K = ST^{-1}$. Let

(4.9)
$$b(m,K) = 2 \inf_{B \subset \mathfrak{I}_4,\ \mathrm{codim}\ B \leq m} \|K\|_B$$

be the mth *minimal norm of the linear operator* K.

Suppose there exists a sequence $\{B_m\}$, $m \geq 0$, such that

(4.10) $b(m,K) = 2\|K\|_{B_m}$ and $\text{codim } B_m \leq m$.

Let $\Im_4 = B_m \oplus B_m^\perp$ and

(4.11) $g = g_0 + \sum_{i=1}^{k} L_{i,m}(g)\eta_{i,m}$,

where $g_0 \in B_m$ and $B_m^\perp = \text{lin}(\eta_{1,m}, \eta_{2,m}, \ldots, \eta_{k,m})$ with $k = k(m) = \text{codim } B_m \leq m$. We shall call B_m an mth *minimal subspace of the linear operator K*.

Recall that $L_1^*, L_2^*, \ldots, L_{n^*}^*$ form $\mathfrak{N}^*$. See (3.3) and (4.1). Define

(4.12) $\mathfrak{N}_n = [L_1^*, L_2^*, \ldots, L_{n^*}^*, L_{1,n-n^*}T, \ldots, L_{k(n-n^*),n-n^*}T]^t$.

We are ready to prove the main result of this section.

Theorem 4.1   The information $\mathfrak{N}_n$ defined by (4.12) is the nth optimal information and

(4.13)
$$d(\mathfrak{N}_n,S,T) = d(n,S,T) = b(n - n^*,K),$$
$$K = ST^{-1}, \qquad n^* = \text{index}(S,T). \quad \blacksquare$$

PROOF   To prove Theorem 4.1, we need two lemmas.

Lemma 4.3   Let B be any linear subspace of $\Im_4$ with $\text{codim } B = k < +\infty$. Then there exists a unique (up to the equivalence relation) information $\mathfrak{N}$ such that

   (i) $\mathfrak{N}^* \subset \mathfrak{N}$,
   (ii) $T(\ker \mathfrak{N}) = B$,
   (iii) $\text{card}(\mathfrak{N}) = k + n^*$. $\quad \blacksquare$

PROOF OF LEMMA 4.3   Let $\Im_4 = B \oplus B^\perp$ and $B^\perp = \text{lin}(\eta_1, \eta_2, \ldots, \eta_k)$. Thus, for every $g \in \Im_4$, we have $g = g_0 + \sum_{i=1}^{k} L_i(g)\eta_i$, where $g_0 \in B$ and $L_i(\eta_j) = \delta_{ij}$. Define

(4.14) $\mathfrak{N} = [L_1^*, L_2^*, \ldots, L_{n^*}^*, L_1 T, L_2 T, \ldots, L_k T]^t$.

Then $\ker \mathfrak{N} = \{f : L_i^*(f) = 0, L_j(Tf) = 0, i = 1, 2, \ldots, n^*, j = 1, 2, \ldots, k\} \subset \ker \mathfrak{N}^*$. This proves (i). Let $h \in \ker \mathfrak{N}$. Then $L_i(Th) = 0$ for $i = 1, 2, \ldots, k$ and $Th \in B$. Hence $T(\ker \mathfrak{N}) \subset B$. Now let g be an arbitrary element of B, i.e., $L_i(g) = 0$ for $i = 1, 2, \ldots, k$. Since $g \in \Im_4$, there exists $f \in \Im_1$ such that $g = Tf$. Decompose $f = f_1 + f_2$, where $f_1 \in \ker T$ and $f_2 \in (\ker T)^\perp$. See (4.2). Then $g = Tf = Tf_2$. Since $L_i^*(f_2) = 0$ for $i = 1, 2, \ldots, n^*$ and $L_i(Tf_2) = 0$ for $i = 1, 2, \ldots, k$, we get $f_2 \in \ker \mathfrak{N}$ and $g = Tf_2 \in T(\ker \mathfrak{N})$. This yields $T(\ker \mathfrak{N}) = B$ which proves (ii).

To prove that $\text{card}(\mathfrak{N}) = k + n^*$, we show that $L_1^*, \ldots, L_{n^*}^*, L_1 T, \ldots, L_k T$ are linearly independent. (See Example 2.1.) Assume that

$$\left(\sum_{i=1}^{n^*} c_i L_i^* + \sum_{i=1}^{k} d_i L_i T \right) f = 0 \qquad \forall f \in \mathfrak{I}_1.$$

Set $f = \xi_i^*$, where $\xi_1^*, \xi_2^*, \ldots, \xi_{n^*}^*$ form a basis of $A(T,S)$. (See (3.3).) Then $T\xi_i^* = 0$ and $L_j^*(\xi_i^*) = \delta_{ij}$. This yields $c_i = 0$ for $i = 1, 2, \ldots, n^*$. Now let $\eta_i = Tf_i$ and set $f = f_i$. Since $L_j(Tf_i) = L_j(\eta_i) = \delta_{ij}$, we get $d_i = 0$ for $i = 1, 2, \ldots, k$. This proves that $\text{card}(\mathfrak{N}) = k + n^*$.

We now show the uniqueness of $\mathfrak{N}$. Suppose that an information operator $\mathfrak{N}_1 = [\tilde{L}_1, \tilde{L}_2, \ldots, \tilde{L}_{k+n^*}]^t$ satisfies (i)–(iii). Thus, $\ker \mathfrak{N}_1 \subset \ker \mathfrak{N}^*$ means that $h \in \ker \mathfrak{N}_1$ implies $L_i^*(h) = 0$ for $i = 1, 2, \ldots, n^*$. Next, $T(\ker \mathfrak{N}_1) = B$ means that $h \in \ker \mathfrak{N}_1$ implies $Th \in B$, i.e., $L_i(Th) = 0$, $i = 1, 2, \ldots, k$. Thus, $h \in \ker \mathfrak{N}$ and $\ker \mathfrak{N}_1 \subset \ker \mathfrak{N}$. Since $\text{card}(\mathfrak{N}_1) = \text{card}(\mathfrak{N})$, from Lemma 2.1 we get $\mathfrak{N}_1 \asymp \mathfrak{N}$. This completes the proof of Lemma 4.3. ∎

Due to the uniqueness of $\mathfrak{N}$, we shall write $\mathfrak{N} = \mathfrak{N}(T,B)$. Note that $\mathfrak{N}_n$ defined by (4.12) is equal to $\mathfrak{N}_n(T, B_{n-n^*})$, where B_{n-n^*} is the $(n - n^*)$th minimal subspace of K.

Lemma 4.4 Let $\text{card}(\mathfrak{N}) = n$ and $\mathfrak{N}^* \subset \mathfrak{N}$. Then $\text{codim } T(\ker \mathfrak{N}) \leq n - n^*$.
∎

PROOF Let $B = T(\ker \mathfrak{N})$ and let $k = \text{codim } B$. From Lemma 4.3, $\mathfrak{N}_1 = \mathfrak{N}_1(T,B)$ has the properties $\mathfrak{N}^* \subset \mathfrak{N}_1$ and $\text{card}(\mathfrak{N}_1) = k + n^*$. Repeating a part of the proof of Lemma 4.3, it is easy to show that $\ker \mathfrak{N} \subset \ker \mathfrak{N}_1$. From Lemma 2.1, we get $\mathfrak{N}_1 = M\mathfrak{N}$, where the $(k + n^*) \times n$ matrix M has rank $k + n^*$. This is possible only if $k \leq n - n^*$ which completes the proof. ∎

We proceed to prove Theorem 4.1. From Lemma 4.3, we know

$$d(\mathfrak{N}_n, S, T) = 2\|K\|_{B_{n-n^*}} = b(n - n^*, K).$$

Let $\mathfrak{N}$ be any information operator from Ψ_n. From Lemma 4.4, we get $\text{codim } T(\ker \mathfrak{N}) \leq n - n^*$ and

$$d(\mathfrak{N}, S, T) = 2\|K\|_{T(\ker \mathfrak{N})} \geq b(n - n^*, K) = d(\mathfrak{N}_n, S, T).$$

This proves that $\mathfrak{N}_n$ is the nth optimal information and $d(\mathfrak{N}, S, T) = d(n, S, T) = b(n - n^*, K)$. This completes the proof of Theorem 4.1. ∎

If $d(n, S, T) \geq 2\varepsilon$, then it is impossible to find an ε-approximation no matter which information operator $\mathfrak{N}$ with $\text{card}(\mathfrak{N}) \leq n$ is used. In this case, we have to increase n and possibly find such $m > n$ that $d(m, S, T) < 2\varepsilon$. This motivates our interest in the dependence of the nth optimal diameter $d(n, S, T)$ on n. Note that $d(n, S, T)$ is a nonincreasing function of n.

Definition 4.2 We shall say $d(S,T)$ is the *diameter of problem error* in the class of information of finite cardinality iff

$$(4.15) \qquad\qquad d(S,T) = \lim_{n \to \infty} d(n,S,T).$$

We shall say that the *problem* (S,T) is *strongly noncomputable* if $d(S,T) = +\infty$, is *ε-noncomputable* if $d(S,T) \geq 2\varepsilon > 0$, and is *convergent* if $d(S,T) = 0$. ∎

We now show that the diameter of problem error $d(S,T)$ can be any number. This shows that for any ε there exist linear problems which are ε-noncomputable.

Lemma 4.5 Let $\delta \in [0, +\infty]$. Then there exists a linear problem (S,T) such that

$$(4.16) \qquad\qquad d(S,T) = \delta. ∎$$

PROOF Let $\delta = +\infty$. Define $T = 0$ and let S be a one-to-one operator. From Lemma 3.3, we get $\mathrm{index}(S,0) = +\infty$ for infinite-dimensional $\mathfrak{I}_1$. Thus, by Theorem 3.2, $d(n,S,0) = +\infty$ for any finite n and $d(S,T) = +\infty = \delta$. Now let $\delta \in [0, +\infty)$. From Example 4.1, we get $d(n,(\delta/2)I,I) = \delta$ for any n. Thus, $d(S,T) = \delta$ which completes the proof. ∎

In the next section, we show when the problem (S,T) is convergent and how to find the nth minimal information.

5. CONVERGENCE AND MINIMAL SUBSPACES
FOR A HILBERT CASE

In this section, we will find relations between the diameter of problem error $d(S,T)$ and the operator $K = ST^{-1} : \mathfrak{I}_4 \to \mathfrak{I}_2$.

Theorem 5.1 Let $\{K_n\}$ be an arbitrary sequence of finite-dimensional linear operators $K_n : \mathfrak{I}_4 \to \mathfrak{I}_2$, $\dim(K_n(\mathfrak{I}_4)) < +\infty$. Then

$$(5.1) \qquad\qquad d(S,T) \leq 2 \inf_n \|K - K_n\|. ∎$$

PROOF Define $B_p = \ker K_n$, where $p = p(n) = \mathrm{codim}\ \ker K_n = \dim(K_n(\mathfrak{I}_4)) < +\infty$. From Lemma 4.3, we know there exists a unique information operator $\mathfrak{N}_{p+n^*} = \mathfrak{N}_{p+n^*}(T,B_p)$ such that $\mathfrak{N}^* \subset \mathfrak{N}_{p+n^*}$ and $\mathrm{card}(\mathfrak{N}_{p+n^*}) = p + n^*$. From Lemma 4.2, we get $d(\mathfrak{N}_{p+n^*},S,T) = 2\|K\|_{B_p}$. Since $Kg = (K - K_n)g$ for any $g \in B_p$, we have $\|Kg\| \leq \|K - K_n\|\ \|g\|$. Hence, $\|K\|_{B_p} \leq \|K - K_n\|$. Finally,

$$d(S,T) \leq d(\mathfrak{N}_{p+n^*},S,T) \leq 2\|K - K_n\|$$

which proves (5.1). ∎

From Theorem 5.1, it follows that if K can be uniformly approximated by $\{K_n\}$, $\lim_n \|K - K_n\| = 0$, then the problem (S,T) is convergent. If $\mathfrak{I}_2$ is a Hilbert space, this holds iff K is compact.

We now show that sometimes $d(S,T)$ can be bounded from below by $2\rho^{-1}$ $\inf_n \|K - K_n\|$ for suitably chosen ρ and $\{K_n\}$. In order to do this, we define $\rho = \rho(\mathfrak{I}_4)$ as follows. Let A be an arbitrary closed subspace of $\mathfrak{I}_4$ of finite codimension. Suppose there exists an algebraic complement $A^\perp$ of A, dim $A^\perp < +\infty$, such that for every $g = g_1 + g_2$, $g_1 \in A$, and $g_2 \in A^\perp$ we have

$$(5.2) \qquad \|g_1\| \le \rho \|g\|.$$

Note that if $\mathfrak{I}_4$ is a Hilbert space, then $A^\perp$ can be the orthogonal complement of A and $\|g_1\| = \sqrt{\|g\|^2 - \|g_2\|^2}$ which implies $\rho(\mathfrak{I}_4) = 1$. Let $Pg = g_1$, $\forall g \in \mathfrak{I}_4$. Then P is a continuous projector and (5.2) states that $\|P\| \le \rho$. Thus, the existence of a finite ρ means that every continuous projector with range of finite codimension has norm bounded by ρ.

We are ready to prove

Theorem 5.2 Let K be continuous and let $\rho = \rho(\mathfrak{I}_4)$ be finite. Then there exists a sequence of finite-dimensional continuous linear operators $\{K_n\}$ such that

$$(5.3) \qquad 2 \inf_n \|K - K_n\| \le \rho \, d(S,T). \quad \blacksquare$$

PROOF Let $\{\mathfrak{N}_n\}$ be a sequence of information operators such that $\operatorname{card}(\mathfrak{N}_n) \le n$ and $\inf_n d(\mathfrak{N}_n,S,T) = d(S,T)$. Due to Lemmas 4.2 and 4.4, we get $d(\mathfrak{N}_n,S,T) = 2\|K\|_{B_n}$, $B_n = T(\ker \mathfrak{N}_n)$, and codim $B_n \le n - n^*$. Since K is continuous, $\|K\|_{B_n} = \|K\|_{\bar{B}_n}$. Let $\mathfrak{I}_4 = \bar{B}_n \oplus \bar{B}_n^\perp$. Thus, $g = g_1 + g_2$, where $g_1 \in \bar{B}_n$, $g_2 \in \bar{B}_n^\perp$, and $\|g_1\| \le \rho \|g\|$, since codim $\bar{B}_n \le n - n^*$. Define

$$K_n g = K g_2.$$

Then K_n is a continuous linear operator from $\mathfrak{I}_4$ to $\mathfrak{I}_2$ and dim $K_n(\mathfrak{I}_4) = $ dim $K(\bar{B}_n^\perp) \le$ dim $\bar{B}_n^\perp \le n - n^* < +\infty$. Furthermore,

$$\|(K - K_n)g\| = \|K g_1\| \le \|K\|_{B_n} \|g_1\| \le \rho \|K\|_{B_n} \|g\|.$$

This proves that $\|K - K_n\| \le \rho \|K\|_{B_n} = \rho \, d(\mathfrak{N}_n,S,T)/2$ and $2 \inf_n \|K - K_n\| \le \rho \, d(S,T)$. Hence (5.3) is proven. $\blacksquare$

From Theorems 5.1 and 5.2, we get the following corollary.

Corollary 5.1 Let $K : \mathfrak{I}_4 \to \mathfrak{I}_2$ be continuous. Let $\mathfrak{I}_2$ and $\mathfrak{I}_4$ be Hilbert spaces. Then the problem (S,T) is convergent iff K is compact. $\blacksquare$

Corollary 5.1 states necessary and sufficient conditions for the problem (S,T) to be convergent. Note that in many cases $K = ST^{-1}$ is not compact. This holds, for instance, for $S = T = I$ and infinite-dimensional $\mathfrak{I}_4$.

We show how to find minimal subspaces of K assuming that K is compact and $K(\mathfrak{I}_4) \subset \mathfrak{I}_4$ for a Hilbert space $\mathfrak{I}_4$. Let K^* be an adjoint operator of K. Define a self-adjoint compact operator

$$(5.4) \qquad K_1 \overset{\mathrm{df}}{=} K^* K : \mathfrak{I}_4 \to \mathfrak{I}_4.$$

Decompose $\mathfrak{I}_4 = \ker K_1 \oplus (\ker K_1)^\perp$, where the orthogonal complement $(\ker K_1)^\perp$ is spanned by eigenvectors of K_1, i.e.,

(5.5) $(\ker K_1)^\perp = \lin(\xi_1, \xi_2, \ldots, \xi_r)$, $r \leq +\infty$, $K_1 \xi_i = \lambda_i \xi_i$,

where $\lambda_i > 0$ and $\lambda_1 \geq \lambda_2 \geq \cdots$, $(\xi_i, \xi_j) = \delta_{ij}$. If r is finite, we formally put $\lambda_i = 0$ and $\xi_i = 0$ for $i \geq r + 1$. Due to compactness of K_1, $\lim_i \lambda_i = 0$. Every element of $f \in \mathfrak{I}_4$ has the unique decomposition $f = f_0 + \sum_{i=1}^\infty (f, \xi_i) \xi_i$, where $f_0 \in \ker K_1$. Define

(5.6) $B_{n-n^*} = \ker K_1 \oplus \lin(\xi_{n-n^*+1}, \ldots, \xi_r)$, $n \geq n^*$,

and an information operator

(5.7) $\mathfrak{N}_n(f) = [L_1^*(f), \ldots, L_{n^*}^*(f), (Tf, \xi_1), \ldots, (Tf, \xi_{n-n^*})]^t$,

where $L_1^*, \ldots, L_{n^*}^*$ are given by (4.1). We are ready to prove

Theorem 5.3   Let $\mathfrak{I}_4$ be a Hilbert space and let $K = ST^{-1}$ be a compact operator such that $K(\mathfrak{I}_4) \subset \mathfrak{I}_4$.

The information operator $\mathfrak{N}_n$ defined by (5.7) is the nth optimal information, B_{n-n^*} defined by (5.6) is the $(n - n^*)$th minimal subspace of K and

(5.8) $d(\mathfrak{N}_n, S, T) = d(n, S, T) = b(n - n^*, K) = 2\sqrt{\lambda_{n-n^*+1}}$.   ∎

PROOF   We first show that B_k is the kth minimal subspace, $k = n - n^*$. Let $f \in B_k$. Then $f = f_0 + \sum_{i=k+1}^\infty (f, \xi_i) \xi_i$, where $f_0 \in \ker K_1$ and $\|Kf\|^2 = (Kf, Kf) = (K_1 f, f) = \sum_{i=k+1}^\infty |(f, \xi_i)|^2 \lambda_i \leq \lambda_{k+1} \|f\|^2$. Since this bound is sharp, we get $\|K\|_{B_k} = \sqrt{\lambda_{k+1}}$. Now let B be any linear subspace such that codim $B \leq k$. Then $\mathfrak{I}_4 = B \oplus B^\perp$ and $B^\perp = \lin(\eta_1, \eta_2, \ldots, \eta_m)$, where $m = $ codim $B \leq k$. Furthermore, $f = f_0 + \sum_{i=1}^m L_i(f) \eta_i$ for certain linear functionals $L_1, L_2, \ldots, L_m$ and $f_0 \in B$. Thus, $f \in B$ iff $L_i(f) = 0$ for $i = 1, 2, \ldots, m$. Let $f = \sum_{i=1}^{k+1} c_i \xi_i$. Then $L_i(f) = 0$ for $i = 1, 2, \ldots, m$ is equivalent to $Mc = 0$, where $M = (L_i(\xi_j))$ is the $m \times (k+1)$ matrix and $c = [c_1, c_2, \ldots, c_{k+1}]^t$. Since $m < k + 1$, there always exists a nonzero solution c and therefore a nonzero $f = \sum_{i=1}^{k+1} c_i \xi_i$ which belongs to B. Then

$$\|Kf\|^2 = (K_1 f, f) = \sum_{i=1}^{k+1} |c_i|^2 \lambda_i \geq \lambda_{k+1} \|f\|^2$$

which yields $\|K\|_B \geq \sqrt{\lambda_{k+1}} = \|K\|_{B_k}$. This proves that B_k is the kth minimal subspace and $b(n - n^*, K) = 2\sqrt{\lambda_{n-n^*+1}}$.

Note that $B_k^\perp = \lin(\xi_1, \xi_2, \ldots, \xi_k)$ and $L_{ik}(f) = (f, \xi_i)$ for $i = 1, 2, \ldots, k$. (See (4.11).) Thus the information operator $\mathfrak{N}_n$ defined by (5.7) is identical with (4.12). From Theorem 4.1, we get that $\mathfrak{N}_n$ is the nth optimal information and

$$d(\mathfrak{N}_n, S, T) = d(n, S, T) = b(n - n^*, k) = 2\sqrt{\lambda_{n-n^*+1}}.$$

This completes the proof.   ∎

The information $\mathfrak{N}_n$ supplies the best possible information on the problem (S,T) in the class Ψ_n. Note that the evaluation of (Tf, ξ_i) means that we compute the ith component of Tf in the eigenvalue decomposition of K_1.

6. RELATIONS TO GELFAND n-WIDTHS

This is the first of several sections in which we show that some basic results in approximation theory can be helpful for finding or estimating the basic quantities of the theory of analytic complexity. In this section, we show a relation between the nth minimal diameters and the Gelfand n-widths. In Section 5 of Chapter 3, we prove that the minimal errors of linear algorithms are related to the linear Kolmogorov n-widths. Finally, in Section 4 of Chapter 7, we show that the Kolmogorov n-widths are related to the minimal errors of n-dimensional algorithms and also that the cardinalities of linear information operators with radii less than ε are related to the ε-entropy of the range of a solution operator.

These relations are mathematically interesting. In addition, they are also of great practical interest since many deep or difficult-to-prove results in approximation theory can be used to establish optimality of information operators and/or optimality of algorithms.

Conversely, the results on optimality of information operators and algorithms may sometimes be useful for solving pure approximation problems.

We begin with relations between the nth minimal diameters and the Gelfand n-widths. In Sections 3–5, we assumed that the domain $\mathfrak{I}_0$ of a linear solution operator S is of the form (3.1) generated by a restriction operator T. (See Remark 3.2.) Here we replace (3.1) by the assumption that $\mathfrak{I}_0$ is a balanced convex subset of $\mathfrak{I}_1$. Then as we mentioned in Remark 3.1,

$$(6.1) \qquad d(\mathfrak{N},S) = 2 \sup_{h \in \ker \mathfrak{N} \cap \mathfrak{I}_0} \|Sh\|$$

for any linear information operator $\mathfrak{N}$.

We need to generalize the concept of the nth minimal diameter and nth optimal information introduced in (4.6). In Section 4, we defined Ψ_n as the class of all linear information operators $\mathfrak{N}$ such that $\mathfrak{N}^* \subset \mathfrak{N}$ and $\text{card}(\mathfrak{N}) \leq n$ where $n \geq \text{index}(S,T)$. Here we extend the definition of Ψ_n assuming that this is the class of all linear information operators with $\text{card}(\mathfrak{N}) \leq n$.

Definition 6.1 We shall say $d(n) = d(n,S,\mathfrak{I}_0)$ is the nth *minimal diameter of information* iff

$$(6.2) \qquad d(n) = \inf_{\mathfrak{N} \in \Psi_n} d(\mathfrak{N},S) \quad \left(= 2 \inf_{\mathfrak{N} \in \Psi_n} \sup_{h \in \ker \mathfrak{N} \cap \mathfrak{I}_0} \|Sh\| \right).$$

We shall say $\mathfrak{N}_n^{oi}$ is an nth *optimal information* iff

$$(6.3) \qquad d(n) = d(\mathfrak{N}_n^{oi}, S), \qquad \mathfrak{N}_n^{oi} \in \Psi_n. \quad \blacksquare$$

Note that if $\mathfrak{I}_0$ is of the form (3.1) and $\mathfrak{N}^*$ is contained in all information operators, then Definition 6.1 coincides with Definition 4.1 and $d(n) = d(n,S,T)$.

We show that the nth minimal diameters $d(n)$ are closely related to the Gelfand n-widths which are defined as follows. Let X be a balanced subset of a linear normed space $\mathfrak{I}_2$. The Gelfand n-width of X is defined as

$$(6.4) \qquad d^n(X,\mathfrak{I}_2) = \inf_{A^n} \sup_{x \in X \cap A^n} \|x\|,$$

where A^n is a subspace of $\mathfrak{I}_2$, $\mathrm{codim}(A^n) \le n$. Thus $A^n = \{x \in \mathfrak{I}_2 : R_i(x) = 0,$ $i = 1,2,\ldots,n\}$ for some linear functionals $R_1, R_2, \ldots, R_n$. Roughly speaking, the Gelfand n-width is the maximal norm of x from X subject to n properly chosen linear constraints. (See Tikhomirov [65].) Let

$$(6.5) \qquad d^n = d^n(S(\mathfrak{I}_0),\mathfrak{I}_2)$$

be the Gelfand n-width of the range of the solution operator in $\mathfrak{I}_2$. To show the relations between $d(n)$ and d^n, we proceed as follows. Let $\mathfrak{I}_1 = \ker S \oplus \ker S^\perp$, i.e., every f from $\mathfrak{I}_1$ has a unique decomposition $f = f_1 + f_2$, where $f_1 \in \ker S$ and $f_2 \in \ker S^\perp$. Suppose for a moment that $\mathfrak{I}_0$ is absorbing and define

$$(6.6) \qquad q = q(S,\mathfrak{I}_0) = \inf_{f \in \mathfrak{I}_0} \sup_{cf_2 \in \mathfrak{I}_0} c.$$

Note that q is well defined. Furthermore, q is either infinity or no greater than one. Indeed, if $\ker S^\perp \subset \mathfrak{I}_0$, then obviously $q = +\infty$. We show that $\ker S^\perp \not\subset \mathfrak{I}_0$ implies $q \le 1$. Suppose then that $f_2 \in \ker S^\perp$, $f_2 \notin \mathfrak{I}_0$, and $\sup\{c : cf_2 \in \mathfrak{I}_0\} > 1$. Then there exists a constant c, $c > 1$, such that $cf_2 \in \mathfrak{I}_0$. Since $\mathfrak{I}_0$ is balanced and convex, $-cf_2 \in \mathfrak{I}_0$ and $[tc - (1-t)c]f_2 \in \mathfrak{I}_0, \forall t \in [0,1]$. Setting $t = (1+c)/(2c)$, we get $f_2 \in \mathfrak{I}_0$ which is a contradiction.

As an example, observe that $q = 1$ if $\ker S^\perp \not\subset \mathfrak{I}_0$ and $f \in \mathfrak{I}_0$ implies $f_2 \in \mathfrak{I}_0$. The last implication holds if, for instance, S is one-to-one or $\mathfrak{I}_1$ is a Hilbert space, $\ker S^\perp$ is the orthogonal complement of the closed kernel of S, and $\mathfrak{I}_0$ is the unit ball in $\mathfrak{I}_1$.

We are ready to prove a theorem which shows that the sequence of nth minimal diameters $\{d(n)\}$ behaves essentially as the sequence of Gelfand n-widths $\{d^n\}$.

Theorem 6.1   Let $S : \mathfrak{I}_1 \to \mathfrak{I}_2$ be a linear operator. If $\mathfrak{I}_0$ is balanced and convex, then

$$(6.7) \qquad d(n) \le 2d^n \qquad \forall n.$$

If, additionally, $\mathfrak{I}_0$ is absorbing, then

$(6.8) \quad q \le 1 \qquad$ implies $\quad d(n) \ge 2qd^n \qquad \forall n,$

$(6.9) \quad q = +\infty \quad$ and $\quad \dim \mathfrak{I}_2 \le n$ imply $\quad d(n) = d^n = 0,$

$(6.10) \quad q = +\infty \quad$ and $\quad \dim \mathfrak{I}_2 > n$ imply $\quad d(n) = d^n = +\infty.$ ∎

PROOF We first prove that $d(n) \leq 2d^n$. Let A be a linear subspace of $\mathfrak{I}_2$ such that codim $A = k \leq n$. Then $A = \{x \in \mathfrak{I}_2 : R_i(x) = 0, i = 1, 2, \ldots, k\}$ for some linearly independent linear functionals $R_1, R_2, \ldots, R_k$. Define the information operator $\mathfrak{N} = [R_1 S, R_2 S, \ldots, R_k S]^t$. Then ker $\mathfrak{N} = \{f \in \mathfrak{I}_1 : Sf \in A\}$ and $S(\ker \mathfrak{N}) \subset A$. Due to (6.1),

$$\tfrac{1}{2} d(\mathfrak{N}, S) = \sup_{h \in \ker \mathfrak{N} \cap \mathfrak{I}_0} \|Sh\| \leq \sup_{x \in A \cap S(\mathfrak{I}_0)} \|x\|.$$

Taking the infimum with respect to A, we get $\tfrac{1}{2} d(n) \leq d^n$.

We now prove (6.8). If $q = 0$, (6.8) is trivial. Suppose, then, $0 < q \leq 1$ and choose $\delta \in (0, q)$. Let $\mathfrak{N} = [L_1, L_2, \ldots, L_n]^t$ be an arbitrary information operator. Define

$$R_i(x) = L_i(f_2), \qquad i = 1, 2, \ldots, n,$$

where $x = Sf + x_2$, $Sf \in S(\mathfrak{I}_1)$, $x_2 \in (\mathfrak{I}_1)^{\perp}$, and $f = f_1 + f_2, f_1 \in \ker S, f_2 \in \ker S^{\perp}$. The linear functional R_i is well defined since $x = y$ with $y = Sg + y_2$ implies $Sf = Sg$ and $f - g \in \ker S$. This means that $f_2 = g_2$ and $R_i(x) = R_i(y)$. Let $A = \{x \in \mathfrak{I}_2 : R_i(x) = 0, i = 1, 2, \ldots, n\}$. Then codim $A \leq n$ and

$$(6.11) \quad d^n \leq \sup_{x \in A \cap S(\mathfrak{I}_0)} \|x\|$$

$$\leq \sup\{\|Sh_2\| : Sh_2 \in S(\mathfrak{I}_0), h_2 \in \ker S^{\perp}, L_i(h_2) = 0, i = 1, 2, \ldots, n\}.$$

Since $Sh_2 \in S(\mathfrak{I}_0)$, there exists $f \in \mathfrak{I}_0$ such that $Sf = Sh_2$. This yields $f_2 = h_2$. From (6.6), it follows that there exists a constant $c = c(h_2) \geq q - \delta > 0$ such that $ch_2 \in \mathfrak{I}_0$. Thus, from (6.11), we get

$$d^n \leq \sup\{c^{-1} \|Sch_2\| : ch_2 \in \mathfrak{I}_0 \text{ and } L_i(ch_2) = 0, i = 1, 2, \ldots, n\}$$

$$\leq \frac{1}{q - \delta} \sup_{h \in \ker \mathfrak{N} \cap \mathfrak{I}_0} \|Sh\| = \frac{1}{2(q - \delta)} d(\mathfrak{N}, S).$$

Since δ is arbitrary, $d(\mathfrak{N}, S) \geq 2qd^n$. Taking the infimum with respect to $\mathfrak{N}$, we get $d(n) \geq 2qd^n$.

Assume now that $q = +\infty$ and dim $\mathfrak{I}_2 \leq n$. From (6.4) with $A^n = \{0\}$, it follows that $d^n = 0$. From (6.7), $d(n) = 0$. Let dim $\mathfrak{I}_2 > n$. Then there exist linearly independent elements $Sf_1, Sf_2, \ldots, Sf_{n+1}$, where $f_i \in \ker S^{\perp}$. Thus, $f_1, f_2, \ldots, f_{n+1}$ are also linearly independent. Let $\mathfrak{N} = [L_1, L_2, \ldots, L_n]^t$ be any information operator. Define $f = \sum_{i=1}^{n+1} c_i f_i$, where $c_1, c_2, \ldots, c_{n+1}$ is a nonzero solution of n linear homogeneous equations $L_j(f) = \sum_{i=1}^{n+1} c_i L_j(f_i) = 0$, $j = 1, 2, \ldots, n$. Thus, $f \in \ker \mathfrak{N} \cap \ker S^{\perp}$. Since $q = +\infty$, then $\ker S^{\perp} \subset \mathfrak{I}_0$ which yields that $cf \in \mathfrak{I}_0$ for every c and $\|Scf\| = |c| \|Sf\| \to +\infty$ with $|c| \to +\infty$. This proves that $d(\mathfrak{N}, S) = +\infty$ for every $\mathfrak{N}$. Hence, $d(n) = +\infty$ and (6.7) yields $d^n = +\infty$. This completes the proof. ∎

The following example show that the inequality in (6.7) can be strict.

Example 6.1   Let $\mathfrak{I}_1 = \mathbb{R}^3$ and $\mathfrak{I}_0 = \{f = [f_1,f_2,f_3]^t : |f_1| \le 1, |f_2 - f_1| \le a,$ $|f_3 - f_1| \le a\}$, where a is a constant. Let $Sf = [0,f_2,f_3]^t$ and $\mathfrak{I}_2 = S(\mathfrak{I}_1) = \{[0,f_2,f_3]^t : f_i \in \mathbb{R}, \ i = 2,3\}$ with the L_∞ norm. Then $\ker S = \{[f_1,0,0,]^t : f_1 \in \mathbb{R}\}$ and $\ker S^\perp = \mathfrak{I}_2$. Note that $f \in \mathfrak{I}_0$ implies $|f_2| \le a + 1$ and $|f_3| \le a + 1$ which easily yields

$$q(S,\mathfrak{I}_0) = a/(a + 1).$$

To find the Gelfand 1-width, observe that

$$d^1 = \inf_L \sup\{\|Sf\| : Sf \in S(\mathfrak{I}_0) \text{ and } L(Sf) = 0\},$$

where L is a linear functional and $S(\mathfrak{I}_0) = \{[0,f_2,f_3]^t : \exists f_1, |f_1| \le 1 \text{ such that }$ $|f_2 - f_1| \le a, |f_3 - f_1| \le a\}$. Since $L([0,f_2,f_3]^t) = f_2 c_2 + f_3 c_3$ for some constants c_2 and c_3, we get

$$d^1 = \inf_{c_2,c_3} \sup\{\max(|f_2|,|f_3|) : [0,f_2,f_3]^t \in S(\mathfrak{I}_0)$$

$$\text{and } f_2 c_2 + f_3 c_3 = 0\} = a + 1.$$

To find $d(1)$, observe that the information operator $\mathfrak{N}(f) = f_2 + f_3 - f_1$ has cardinality one and the algorithm $\varphi(\mathfrak{N}(f)) = [0,\mathfrak{N}(f),\mathfrak{N}(f)]^t$ satisfies

$$\|Sf - \varphi(\mathfrak{N}(f))\| = \|[0,f_1 - f_3, f_1 - f_2]^t\| \le a \qquad \forall f \in \mathfrak{I}_0.$$

Thus, $d(1) \le d(\mathfrak{N},S) \le 2e(\varphi) \le 2a$. From (6.8), we get $d(1) \ge (2a/(a + 1))(a + 1) = 2a$ which yields

$$d(1) = 2qd^1 = 2a.$$

Note that $2d^1/d(1) = (a + 1)/a$ can be arbitrarily large for small a. Of course, $d(n) = d^n = 0, \forall n \ge 2$. ∎

REMARK 6.1   For $q = 1$, Theorem 6.1 states

$$d(n) = 2d^n.$$

This establishes a "two-way" relation between the theory of approximation and the theory of analytic complexity. This means that if d^n is known from approximation theory, then we know $d(n)$, and conversely if $d(n)$ is known from analytic complexity, we know d^n. ∎

Theorem 6.1 allows us to find optimal or nearly optimal information operators. We remind the reader that A^n is called an nth *extremal subspace of* $X = S(\mathfrak{I}_0)$ *in the sense of Gelfand* if

(6.12) $$\sup_{x \in X \cap A^n} \|x\| = d^n(X,\mathfrak{I}_2), \qquad \text{codim } A^n \le n.$$

That is, if the infimum in (6.4) is attained for A^n, then A^n is an nth extremal subspace. Let $A^n = \{x \in \mathfrak{I}_2 : R_i(x) = 0, i = 1,2,\ldots,k\}$, where $R_1, R_2, \ldots, R_k$ are linearly independent linear functionals and $k = \text{codim } A^n \leq n$. Define the information operator

(6.13) $$\mathfrak{N} = [R_1S, R_2S, \ldots, R_kS]^t.$$

From the proof of Theorem 6.1, we get

$$d(\mathfrak{N}, S) \leq 2d^n \leq d(n)/q.$$

This yields the following corollary.

Corollary 6.1 Under the hypotheses of Theorem 6.1, with $q = 1$, the information operator defined by (6.13) is an nth optimal information and

$$d(\mathfrak{N}, S) = d(n) = 2d^n. \quad \blacksquare$$

The Gelfand widths and extremal subspaces for sets of practical interest may be found in several papers. See among others Micchelli and Pinkus [77] and Tikhomirov [69, 76]. Due to Corollary 6.1, we can therefore establish optimality of information operators for many solution operators. See Chapter 6, where the results of this section are applied for several linear problems.

If S is a finite-dimensional linear operator, $k = \dim S(\mathfrak{I}_1)$, then $d(n) = d^n = 0$ for $n \geq k$. However, for most such cases, nth optimal information operators are not permissible. For instance, if $Sf = \int_0^1 f(t) \, dt$, for a smooth scalar function f, then we can usually compute *only* the values of f and its derivatives. Although Theorem 6.1 does not seem to be applicable, we now show how it can be used for such problems.

For a linear solution operator S, let $\Psi = \Psi(S)$ be a class of permissible information operators with cardinality at most n. We are interested in finding an *optimal information* $\mathfrak{N}^o$ in Ψ, i.e.,

(6.14) $$d(\mathfrak{N}^o, S) = \inf_{\mathfrak{N} \in \Psi} d(\mathfrak{N}, S), \qquad \mathfrak{N}^o \in \Psi.$$

Note that if Ψ is not the class of all information operators with cardinality at most n, then $\mathfrak{N}^o$ is not necessarily an nth optimal information operator in the sense of Definition 6.1.

Suppose there exists a linear operator $S_1 : \mathfrak{I}_0 \to \mathfrak{I}_5$, where $\mathfrak{I}_5$ is a linear normed space, such that

(6.15) $$\inf_{\mathfrak{N} \in \Psi} d(\mathfrak{N}, S) = \inf_{\mathfrak{N} \in \Psi} d(\mathfrak{N}, S_1).$$

The significance of (6.15) is that S_1 can be an infinite-dimensional operator (even if S is finite-dimensional) and the Gelfand n-width of $S_1(\mathfrak{I}_0)$ provides a lower bound on $d(\mathfrak{N}, S)$.

Let $A^n = \{x \in \mathfrak{I}_5 : R_i(x) = 0, \; i = 1, 2, \ldots, k\}$ be an nth extremal subspace of $S_1(\mathfrak{I}_0)$ in the sense of Gelfand, where $R_1, R_2, \ldots, R_k$ are linearly independent linear functionals and $k = \text{codim } A'' \leq n$. Define the information operator

(6.16) $\mathfrak{N}^\circ = [R_1 S_1, R_2 S_1, \ldots, R_k S_1]^t.$

Let $q_1 = q(S_1, \mathfrak{I}_0)$ be defined by (6.6) and let $d_1^n = d^n(S_1(\mathfrak{I}_0), S_1(\mathfrak{I}_1))$ be the Gelfand n-width. If $q_1 = 1$, then Corollary 6.1 yields that $\mathfrak{N}^\circ$ is an nth optimal information operator for the problem S_1. From Theorem 6.1, we get the following corollary.

Corollary 6.2   Let $\mathfrak{I}_0$ be balanced convex and absorbing. If (6.15) holds, then

(6.17) $\inf_{\mathfrak{N} \in \Psi} d(\mathfrak{N}, S) \geq 2q_1 d_1^n.$

If $q_1 = 1$ and $\mathfrak{N}^\circ$ defined by (6.16) belongs to Ψ, then $\mathfrak{N}^\circ$ is an optimal information operator in Ψ and

(6.18) $d(\mathfrak{N}^\circ, S) = 2d_1^n.$ ∎

Corollary 6.2 states that optimal information from ψ for the problem S can be sometimes found by an nth optimal information operator for the problem S_1. We illustrate Corollary 6.2 by the following example.

Example 6.2   We now show how an optimal information operator for the integration problem can be obtained by the solution of a corresponding approximation problem.

Let $\mathfrak{I}_1$ be a linear space of scalar functions $f : [-1, 1] \to \mathbb{R}$ which are absolutely continuous and $\|f'\|_\infty < +\infty$. $\mathfrak{I}_0 = \{f : f \text{ is abs. cont. and } \|f'\|_\infty \leq 1\}$. Consider the *integration* operator

$$S(f) = \int_{-1}^1 f(x)\,dx$$

with $\mathfrak{I}_2 = \mathbb{R}$. Since S is a linear functional, $\dim S(\mathfrak{I}_1) = 1$, and $d(n) = d^n = 0$, $\forall n \geq 1$. Let Ψ be the class of information operators of the form

$$\mathfrak{N}(f) = [f(x_1), f(x_2), \ldots, f(x_n)]^t$$

for distinct points $x_i \in [-1, 1]$, $x_1 < x_2 < \cdots < x_n$. Note that

(6.19) $d(\mathfrak{N}, S) = 2 \sup\left\{ \left| \int_{-1}^1 h(x)\,dx \right| : h(x_i) = 0,\ i = 1, \ldots, n,\ h \in \mathfrak{I}_0 \right\}$

Note that the supremum in (6.19) is attained for a perfect spline $\bar{h}$,

$$\bar{h}(x) = \begin{cases} x_1 - x, & -1 \leq x \leq x_1, \\ x - x_i, & x_i \leq x \leq (x_i + x_{i+1})/2, & i = 1, 2, \ldots, n-1, \\ x_{i+1} - x, & (x_i + x_{i+1})/2 \leq x \leq x_{i+1}, & i = 1, 2, \ldots, n-1, \\ x - x_n, & x_n \leq x \leq 1. \end{cases}$$

Furthermore, for every $h \in \mathfrak{I}_0$ such that $h(x_i) = 0$, $i = 0, 1, \ldots, n$, we get

$$-\overline{h}(x) \le h(x) \le \overline{h}(x) \qquad \forall x \in [-1,1].$$

From this, we conclude

(6.20) $d(\mathfrak{N},S) = 2 \sup \left\{ \int_{-1}^{+1} |h(x)| \, dx : h(x_i) = 0, i = 1, \ldots, n, h \in \mathfrak{I}_0 \right\}$

for any n and any distinct points x_i. Define the approximation operator

$$S_1 f = f$$

equipped with the L_1 norm ($\mathfrak{I}_5 = L_1$). Then (6.19) and (6.20) imply

$$d(\mathfrak{N},S) = d(\mathfrak{N},S_1) \qquad \forall \mathfrak{N} \in \Psi.$$

Hence, (6.15) holds. Of course, $q(S_1(\mathfrak{I}_0),L_1) = 1$. It is known that $d^n(S_1(\mathfrak{I}_0),L_1) = 1/n$ and the set $A^n = \{ f : f(z_i) = 0, z_1 = -1 + h, z_i = -1 + 2(i - 1)h, h = 1/n, i \in [2,n] \}$ is an nth extremal subspace of $S_1(\mathfrak{I}_0)$. Thus, Corollary 6.2 yields

$$\inf_{\mathfrak{N} \in \Psi} d(\mathfrak{N},S) = d(\mathfrak{N}^o,S) = 2/n,$$

where $\mathfrak{N}^o(f) = [f(z_1), f(z_2), \ldots, f(z_n)]^t$ is an optimal information operator in Ψ.

In Section 4 of Chapter 6, we show that the integration problem for many sets $\mathfrak{I}_0$ of practical interest can be studied in terms of the approximation problem. See also Korotkov [77], where a related question is studied. ∎

7. ADAPTIVE LINEAR INFORMATION

In Sections 2–6, we deal with linear information operators of the form

(7.1) $\mathfrak{N}(f) = [L_1(f), L_2(f), \ldots, L_n(f)]^t,$

defined by n *independently* given linear functionals $L_1, L_2, \ldots, L_n$. A natural generalization is an *adaptive* linear information operator of the form

(7.2) $\mathfrak{N}^a(f) = [L_1(f), L_2(f; y_1), \ldots, L_n(f; y_1, \ldots, y_{n-1})]^t$

where L_i depends linearly on its first argument and $y_i = L_i(f; y_1, \ldots, y_{i-1})$. This form enables us to use the previously computed functionals to determine the next functional. In contrast to (7.2), we shall call an information operator of the form (7.1) *nonadaptive*.

Adaptive information is widely used in practice in a number of application areas. For some problems, adaptive information is much more effective than nonadaptive information. For instance, consider the *nonlinear* problem of searching for the maximum of a function belonging to the class of unimodal scalar functions. It is known that the error of the Kiefer algorithm based on n adaptively computed function evaluations is exponentially decreasing in n (see Kiefer [53, 57]). We shall show in Section 6 of Chapter 8 that the error of an optimal error algorithm based on n simultaneously computed function evaluations is inversely proportional to n.

We ask whether adaptive information can help for *linear* problems. More precisely, for linear problems, does there exist adaptive information which is more efficient than any nonadaptive information based on the same number of linear functionals?

We shall prove that the answer to this question is negative. We shall show that for any linear problem and any adaptive information based on n linear functionals, there exists a nonadaptive information operator of cardinality at most n with no greater diameter of information. Thus, the much more complicated structure of the class of adaptive information does not supply more knowledge about linear problems than the relatively simple structure of the class of nonadaptive information.

We wish to stress that this result means that for any adaptive information $\mathfrak{N}^a$, there exists a solution element f_0 from $\mathfrak{I}_0$ for which $\mathfrak{N}^a(f_0)$ does not supply more information than $\mathfrak{N}^{non}(f_0)$ for suitably chosen nonadaptive information $\mathfrak{N}^{non}$. Of course, it may happen that for many solution elements f different from f_0, the adaptive information $\mathfrak{N}^a(f)$ is much more efficient than $\mathfrak{N}^{non}(f)$ by making use of special properties of such f. (See Example 7.1.)

In numerical practice, the idea of adaption is also used for information operators whose cardinality may vary and which depend on the particular problem element f being actually approximated. For instance, for the integration problem $\int_0^1 f(t)\,dt$, one often evaluates the function f at the points x_1, $x_2, \ldots$, where $x_j = x_j(f(x_1), \ldots, f(x_{j-1}))$, until the termination criterion

$$\left| \varphi_j(f(x_1), \ldots, f(x_j)) - \varphi_{j+1}(f(x_1), \ldots, f(x_{j+1})) \right| \le \delta$$

is satisfied. Here φ_j is a quadrature formula and δ is a given small number. The total number of function evaluations (i.e., cardinality of information) now depends on a particular f.

This is a different model of analytic complexity which is not studied in this book. See Chapter 10, where we discuss different models of analytic complexity.

To demonstrate the announced results, let $\mathfrak{N}^a$ be an adaptive linear information operator of the form (7.2). Define the nonadaptive information operator

$$\mathfrak{N}^{non}(f) = [L_1(f), L_2(f;0), \ldots, L_n(f;0, \ldots, 0)]^t.$$

Of course, $\operatorname{card}(\mathfrak{N}^{non}) \le n$.

Theorem 7.1   Let S be a linear operator and let $\mathfrak{I}_0$ be balanced and convex. Then

(7.3) $d(\mathfrak{N}^a, S) \ge d(\mathfrak{N}^{non}, S).$ ∎

PROOF By Remark 3.1, we have

$$d(\mathfrak{N}^{non}, S) = 2 \sup_{h \in \ker \mathfrak{N}^{non} \cap \mathfrak{I}_0} \|Sh\|.$$

Let δ be an arbitrary positive number. Choose $h \in \ker \mathfrak{N}^{non} \cap \mathfrak{I}_0$ such that $\|Sh\| \geq \frac{1}{2} d(\mathfrak{N}^{non}, S) - \delta$ if $d(\mathfrak{N}^{non}, S) < +\infty$ and $\|Sh\| \geq \delta$ if $d(\mathfrak{N}^{non}, S) = +\infty$. From (2.9) of Chapter 1, we get

$$d(\mathfrak{N}^a, S) = \sup_{f \in \mathfrak{I}_0} \sup \{\|S\tilde{f} - Sf\| : \tilde{f} \in \mathfrak{I}_0, \ \mathfrak{N}(\tilde{f}) = \mathfrak{N}(f)\}.$$

Define $f = h$ and $\tilde{f} = -h \in \mathfrak{I}_0$. Then $L_1(f) = L_1(\tilde{f}) = 0$, $L_2(f;0) = L_2(\tilde{f};0) = 0, \ldots, L_n(f;0, \ldots, 0) = L_n(\tilde{f};0, \ldots, 0) = 0$. This means $\mathfrak{N}^a(\tilde{f}) = \mathfrak{N}^a(f)$ and

$$d(\mathfrak{N}^a, S) \geq \|S\tilde{f} - Sf\| = 2\|Sh\| \geq \begin{cases} d(\mathfrak{N}^{non}, S) - 2\delta & \text{if} \quad d(\mathfrak{N}^{non}, S) < +\infty, \\ 2\delta & \text{if} \quad d(\mathfrak{N}^{non}, S) = +\infty. \end{cases}$$

Since δ is arbitrary, (7.3) follows. ∎

Let Ψ_n^a denote the class of all adaptive information operators of the form (7.2). Since any nonadaptive information operator with cardinality at most n belongs to Ψ_n^a, Theorem 7.1 yields

Corollary 7.1

$$\inf_{\mathfrak{N}^a \in \Psi_n^a} d(\mathfrak{N}^a, S) = d(n),$$

where $d(n)$ is the nth minimal diameter of information defined in (6.2). ∎

A similar result was established by Bakhvalov [71a] assuming that S is a linear functional and by Gal and Micchelli [78] using a different proof technique. Also, Kiefer [57] proved that adaptive information cannot help for the integration problem defined on a particular nonbalanced class $\mathfrak{I}_0$.

The same result also holds for *some* nonlinear problems. See Sukharev [71] and Zaliznyak and Ligun [78], who considered the search for the maximum of a function from a class $\mathfrak{I}_0$. Sukharev [71] dealt with the class of scalar functions of several variables satisfying a Lipschitz condition, and Zaliznyak and Ligun [78] with the class of functions which is the algebraic sum of a balanced convex compact set and a finite-dimensional linear space.

Theorem 7.1 and Corollary 7.1 state that as far as the diameter of information is concerned, we cannot gain by using adaptive information. This means that there exists $f \in \mathfrak{I}_0$ for which the adaptive information $\mathfrak{N}^a(f)$ is not better than an appropriately chosen nonadaptive $\mathfrak{N}^{non}(f)$. However, for some special $f \in \mathfrak{I}_0$, adaptive information $\mathfrak{N}^a(f)$ can be very useful and much more efficient than nonadaptive $\mathfrak{N}^{non}(f)$. This is shown by the following example.

Example 7.1 Let $\mathfrak{I}_1$ be the class of scalar absolutely continuous functions $f : [-1,1] \rightarrow \mathbb{R}$ such that $f' \in L_\infty$. Define $\mathfrak{I}_0 = \{f \in \mathfrak{I}_1 : \|f'\|_\infty \leq 1\}$. Consider the *approximation* problem $S(f) = f$. From Example (6.2), we can easily conclude that

(7.4) $$\mathfrak{N}_n^{non}(f) = [f(z_1), f(z_2), \ldots, f(z_n)]^t$$

is an nth (nonadaptive) optimal information operator and $d(\mathfrak{N}_n^{non}, I) = d(n) = 2/n$. Suppose that we apply $\mathfrak{N}_n^{non}$ to the function f such that $f \in \mathfrak{I}_0$ and $f(z_k) = z_k$ and $f(z_p) = z_p$ for $k \ll p$. Observe that knowing $f(z_k)$ and $f(z_p)$, we immediately conclude that $f(t) \equiv t$ for $t \in [z_k, z_p]$. Therefore, there is no need to compute $f(z_{k+1}), f(z_{k+2}), \ldots, f(z_{p-1})$. Adaptive information would recognize this favorable case and evaluate f outside of the interval $[z_k, z_p]$. ∎

Chapter 3

Linear Algorithms for Linear Problems

1. INTRODUCTION

An optimal error algorithm may have large combinatory complexity. Indeed, it may happen that the computation of $\varphi(y)$, given $y = \mathfrak{N}(f)$, dominates the cost of computing $\mathfrak{N}(f)$. In this chapter we shall study *linear* algorithms; such algorithms are guaranteed to have small combinatory complexity.

Let $\mathfrak{N}(f) = [L_1(f), \ldots, L_n(f)]^t$. Then we say an algorithm φ is a linear algorithm if

$$(1.1) \qquad \varphi(\mathfrak{N}(f)) = \sum_{i=1}^{n} L_i(f)g_i,$$

where the $g_i = g_i(S, \mathfrak{N}_0, \varphi)$ are elements of $\mathfrak{I}_2$. Since $g_1, \ldots, g_n$ are independent of f, they can be precomputed. Given the g_i, we perform at most n multiplications of elements from $\mathfrak{I}_2$ by a scalar and $n - 1$ additions of elements from $\mathfrak{I}_2$ to compute $\varphi(y)$. The combinatory complexity of a linear algorithm is linear in n and is in general small with respect to the complexity of computing $\mathfrak{N}(f)$. See Chapter 5 for a detailed discussion of the time complexity of linear algorithms. We show (Chapter 5, Lemma 2.2) that a linear optimal error algorithm is a "nearly optimal complexity algorithm." In addition, the space complexity of a linear algorithm is good. In particular, if $S(f)$ is a linear functional, then the space complexity is at most $2n$.

We are therefore very interested in linear algorithms which are optimal or nearly optimal error algorithms.

Does every linear problem have a linear optimal error algorithm? See Section 4 for an example, due to Micchelli [78], of a linear problem for which there is no linear optimal error algorithm. This example is especially constructed; we know of no such example arising in real applications.

Although it is not always possible to find a linear optimal error algorithm, we show how to construct a linear algorithm whose error differs from the radius of information by a factor c depending only on $T(\ker \mathfrak{N})$. In a Hilbert space setting, c is unity.

Smoljak [65] proved that if S is a real linear functional and if $\mathfrak{N}$ is formed by n real linear functionals, then there exists a linear optimal error algorithm. In a classic paper, Sard [49] studied optimal algorithms for integration. His information is the values of f at n *fixed* points. He *assumes* that the algorithms under consideration are linear and integrate all polynomials up to a certain degree exactly. Using Smoljak's theorem, we show that *neither* of Sard's assumptions is needed. See Section 3 for additional discussion of this point.

Nikolskj [50] studied optimal algorithms for integration with *optimally chosen points* at which f is evaluated. He also *assumes* linear algorithms. We prove under rather weak assumptions that optimality in the sense of Nikolskij and optimality in the sense of Sard (with optimally chosen points) coincide.

We present additional relations (see also Chapter 2, Section 6) between mathematical approximation theory and analytic complexity. We define the nth minimal linear error and relate it to the linear Kolmogorov n-width.

We summarize the results of this chapter. Section 2 consists of mathematical preliminaries. In Section 3 we adapt (Theorem 3.1) Bakhvalov's proof of Smoljak's theorem. We use Smoljak's theorem to show (Theorem 3.3) that optimal algorithms in the sense of Sard or Nikolskij are also optimal error algorithms.

In Section 4, Example 4.1 exhibits a linear problem for which there is no linear optimal algorithm. In Theorem 4.1, we exhibit a linear algorithm which is within a factor of at most c of a linear optimal error algorithm. The concluding section relates optimal linear algorithms which use linear information operators with cardinality at most n and the linear Kolmogorov n-width.

2. PRELIMINARIES

In this section, we consider some special cases for which one can find linear algorithms which are not only optimal error algorithms but which are also central and interpolatory.

Suppose first that $r(\mathfrak{N},S) = 0$. Then any interpolatory algorithm φ^I (see Section 2 of Chapter 1) has error $e(\varphi^I) = 0$. Thus, $\varphi^I(\mathfrak{N}(f)) = Sf$ is linear and central. This yields

Corollary 2.1 If $r(\mathfrak{N},S) = 0$, then there exists a linear central interpolatory algorithm. ∎

Note that $r(\mathfrak{N},S) = 0$ for any linear S iff ker $\mathfrak{N} \cap \mathfrak{I}_0 = \{0\}$. This holds, for instance, if card$(\mathfrak{N}) = \dim \mathfrak{I}_1$, since then ker $\mathfrak{N} = \{0\}$. If card$(\mathfrak{N}) = \dim \mathfrak{I}_1 = n$, then there exist $f_1, f_2, \ldots, f_n$ such that $\mathfrak{I}_1 = \lin(f_1, f_2, \ldots, f_n)$ and $L_i(f_j) = \delta_{ij}$. Then for every $f \in \mathfrak{I}_1$, we have $f = \sum_{i=1}^n L_i(f)f_i$. Define the algorithm

$$(2.1) \qquad \varphi(\mathfrak{N}(f)) = \sum_{i=1}^{n} L_i(f)g_i, \qquad g_i = Sf_i.$$

Since $\varphi(\mathfrak{N}(f)) = Sf$, we get

Corollary 2.2 If card$(\mathfrak{N}) = \dim \mathfrak{I}_1$, then algorithm φ defined by (2.1) is a linear central interpolatory algorithm and $e(\varphi) = 0$. ∎

We now deal with a case for which $r(\mathfrak{N},S)$ is not necessarily equal to zero.

Lemma 2.1 Let $\mathfrak{I}_0 = \{f \in \mathfrak{I}_1 : \|Tf\| \leq 1\}$. Suppose there exist elements $f_1, f_2, \ldots, f_n$ from $\mathfrak{I}_1$ such that

$$(2.2) \qquad L_i(f_j) = \delta_{ij} \qquad \text{and} \qquad T(f_j) = 0 \qquad \text{for} \quad i, j = 1, 2, \ldots, n.$$

Then $\varphi(\mathfrak{N}(f)) = \sum_{i=1}^n L_i(f)g_i$, $g_i = S(f_i)$, is a linear central interpolatory algorithm and

$$(2.3) \qquad e(\varphi) = r(\mathfrak{N},S,T) = \tfrac{1}{2}d(\mathfrak{N},S,T) = \sup_{h \in \ker \mathfrak{N}} \|Sh\|/\|Th\|. \qquad ∎$$

PROOF Let $\tilde{f}_0 = \sum_{j=1}^n L_j(f)f_j$. Then $\tilde{f}_0 \in \ker T$ and $L_i(\tilde{f}_0) = \sum_{j=1}^n L_j(f)L_i(f_j) = L_i(f)$. This proves that φ is linear and interpolatory. To prove centrality of φ, recall that $U = U(f)$, defined by (2.5) of Chapter 1, is the set of all solution elements $S\tilde{f}$ of problem elements $\tilde{f}$ from $\mathfrak{I}_0$ which share the same information as f. Observe that U is symmetric with respect to $S\tilde{f}_0$ for every $f \in \mathfrak{I}_0$. Indeed, $S\tilde{f}_0 + Sh \in U$ yields $h \in \ker \mathfrak{N}$ and $\|Th\| \leq 1$. Then $S\tilde{f}_0 - Sh$ also belongs to U. From Remark 2.1 of Chapter 1, we conclude that $\varphi(\mathfrak{N}(f)) = S\tilde{f}_0$ is central and (2.3) holds due to (2.12) of Chapter 1 and (3.2) of Chapter 2. ∎

Note that (2.2) means that $\dim \ker T \geq$ card$(\mathfrak{N})$ and the linear functionals $L_1, L_2, \ldots, L_n$ are linearly independent on ker T. Lemma 2.2 has many interesting applications especially for certain classical problems. This is shown in several sections of Chapter 6.

We end this section with a lemma which will be needed in the next section to establish a relation between optimal error algorithms and optimal algorithms in the sense of Sard.

Lemma 2.2 Let $\mathfrak{I}_0 = \{f \in \mathfrak{I}_1 : \|Tf\| \leq 1\}$. If φ is a homogeneous algorithm with $e(\varphi) < +\infty$, then

$$(2.4) \qquad \varphi(\mathfrak{N}(f)) = Sf \qquad \forall f \in \ker T. \qquad ∎$$

PROOF   Let f be an arbitrary element from ker T. Then $cf \in \mathfrak{I}_0$ for any c and

$$\|\varphi(\mathfrak{N}(cf)) - S(cf)\| = |c| \, \|\varphi(\mathfrak{N}(f)) - Sf\| \le e(\varphi) < +\infty.$$

This implies $\varphi(\mathfrak{N}(f)) = Sf$ and proves (2.4).   ∎

Lemma 2.2 states that any homogeneous algorithm with finite error is exact for elements from ker T. As an example, assume $Sf = \int_a^b f(t)\,dt$, $Tf = f^{(k)}$, and $\mathfrak{N}f = [f(t_1), f(t_2), \ldots, f(t_n)]^t$. Then for $n \ge k$, $r(\mathfrak{N}, S, T) < +\infty$ and any homogeneous algorithm with finite error is exact for all polynomials of degree at most $k - 1$.

3. LINEAR OPTIMAL ERROR ALGORITHMS
FOR LINEAR FUNCTIONALS

In this section, we assume that S is a real linear functional, i.e., $\mathfrak{I}_2 = \mathbb{R}$, and $\mathfrak{N} = [L_1, L_2, \ldots, L_n]^t$ is formed by real linear functionals $L_1, L_2, \ldots, L_n$. We prove a theorem of Smolyak which guarantees the existence of a linear optimal error algorithm. Using this theorem, we show that the optimal algorithms in the sense of Sard or Nikolskij are optimal error algorithms.

We begin with the following lemma.

Lemma 3.1   If S is a real linear functional and $\mathfrak{I}_0$ is balanced and convex, then

(3.1) $r(\mathfrak{N}, S) = \tfrac{1}{2} d(\mathfrak{N}, S) = \sup_{h \in \ker \mathfrak{N} \cap \mathfrak{I}_0} Sh.$ ∎

PROOF   Since S is a real linear functional and $\mathfrak{I}_0$ is convex, then the set $U(f)$ is an interval. Thus, $\mathrm{rad}(U(f)) = \mathrm{diam}(U(f))/2$, $\forall f \in \mathfrak{I}_0$. This yields $r(\mathfrak{N}, S) = d(\mathfrak{N}, S)/2$. By Remark 3.1 of Chapter 2, we get $r(\mathfrak{N}, S) = \mathrm{rad}(U(0)) = \sup\{|Sh| : h \in \ker \mathfrak{N} \cap \mathfrak{I}_0\}$. Since $U(0)$ is balanced, we can omit the modulus in $|Sh|$. This proves (3.1).   ∎

Let

(3.2) $r_i(x) = \sup\{Sf : f \in \mathfrak{I}_0, \, L_i(f) = x, \, L_j(f) = 0 \text{ for } j \ne i\}.$

We are ready to prove the following theorem due to Smolyak.

Theorem 3.1   (Smolyak [65])   Let S be a real linear functional defined on a balanced convex set $\mathfrak{I}_0$ and $\mathfrak{N} = [L_1, \ldots, L_n]^t$ be a linear real information operator. Then

(i)   there exists a linear optimal error algorithm,
(ii)   if $r_i'(0)$ exists for $i = 1, 2, \ldots, n$, then $\varphi(\mathfrak{N}(f)) = \sum_{i=1}^n L_i(f) r_i'(0)$ is a unique linear optimal error algorithm.   ∎

PROOF   We adapt the Bakhvalov [71a] proof of this theorem. Consider first $r(\mathfrak{N},S) = +\infty$. Then every algorithm φ is an optimal error algorithm since $e(\varphi) = r(\mathfrak{N},S) = +\infty$. Next, if $r(\mathfrak{N},S) = 0$, then Corollary 2.1 yields (i). Thus, without loss of generality, we can assume that $r = r(\mathfrak{N},S) \in (0,+\infty)$. Assume first that $L_1, L_2, \ldots, L_n$ are linearly independent on $\mathfrak{I}_0$, i.e.,

$$\sum_{j=1}^{n} c_j L_j(f) = 0 \qquad \forall f \in \mathfrak{I}_0 \qquad \text{implies} \quad c_1 = c_2 = \cdots = c_n = 0.$$

Let

$$Y = \{(S(f), L_1(f), \ldots, L_n(f)): f \in \mathfrak{I}_0\} \subset \mathbb{R}^{n+1}.$$

The set Y is balanced and convex due to the assumptions on $\mathfrak{I}_0$. It is known from the theory of convex sets that for every point p on the boundary of a convex set X there exists a support hyperplane of X passing through p. This means there exist real numbers $c_0, c_1, \ldots, c_n$ such that the hyperplane

$$c_0(y_0 - r) + \sum_{j=1}^{n} c_j y_j = 0$$

passes through the boundary point $(r, 0, \ldots, 0)$ and is supporting to the set Y, i.e.,

(3.3) $$c_0(S(f) - r) + \sum_{j=1}^{n} c_j L_j(f) \le 0 \qquad \forall f \in \mathfrak{I}_0.$$

Suppose that $c_0 = 0$. Since $\mathfrak{I}_0$ is balanced, then $\sum_{j=1}^{n} c_j L_j(f) = 0$, $\forall f \in \mathfrak{I}_0$, which contradicts our assumption that $L_1, L_2, \ldots, L_n$ are linearly independent. Hence $c_0 \ne 0$. Since Y is balanced, then the hyperplane

$$c_0(y_0 + r) + \sum_{j=1}^{n} c_j y_j = 0$$

passes through the boundary point $(-r, 0, \ldots, 0)$ and is supporting to the set Y, i.e.,

(3.4) $$c_0(S(f) + r) + \sum_{j=1}^{n} c_j L_j(f) \ge 0 \qquad \forall f \in \mathfrak{I}_0.$$

Define $q_j = -c_j/c_0$. Then (3.3) and (3.4) yield

$$\left| S(f) - \sum_{j=1}^{n} L_j(f) q_j \right| \le r = r(\mathfrak{N},S).$$

This proves that the linear algorithm $\varphi(\mathfrak{N}(f)) = \sum_{j=1}^{n} L_j(f) q_j$ is an optimal error algorithm. This completes this part of the proof.

Assume now that $L_1, L_2, \ldots, L_n$ are linearly dependent on $\mathfrak{I}_0$. Without loss of generality, we can assume that there exists an integer k, $k < n$, such that

$L_1, L_2, \ldots, L_k$ are linearly independent on $\mathfrak{I}_0$ and

$$L_{k+i}(f) = \sum_{j=1}^{k} c_{ij} L_j(f) \qquad \forall f \in \mathfrak{I}_0,$$

for some constants c_{ij}, $i = 1, 2, \ldots, n - k$. Define the information operator $\mathfrak{N}_1(f) = [L_1(f), \ldots, L_k(f)]^t$. Then ker $\mathfrak{N} =$ ker $\mathfrak{N}_1$ and $r(\mathfrak{N}, S) = r(\mathfrak{N}_1, S)$. Let φ_1 be a linear optimal error algorithm for $\mathfrak{N}_1$, $\varphi_1(\mathfrak{N}_1(f)) = \sum_{j=1}^{k} L_j(f) q_j$. Then $e(\varphi_1) = r(\mathfrak{N}_1, S) = r(\mathfrak{N}, S)$ which shows that φ is also a linear optimal error algorithm for $\mathfrak{N}$. Hence (i) is proven.

To prove (ii), let $\varphi(\mathfrak{N}(f)) = \sum_{j=1}^{n} L_j(f) q_j$ be an optimal error algorithm. Choose $f \in \mathfrak{I}_0$ such that $L_i(f) = x$, $L_j(f) = 0$ for $j \neq i$ and for sufficiently small x. Then

(3.5) $$|S(f) - xq_i| \leq e(\varphi) = r.$$

Note that $r = r_i(0)$ which implies that r is finite. Since $S(f)$ can be arbitrarily close to $r_i(x)$ (see (3.2)), we get

$$r_i(x) - r_i(0) \leq xq_i.$$

Dividing by x, we have

$$\frac{r_i(|x|) - r_i(0)}{|x|} \leq q_i \leq \frac{r_i(-|x|) - r_i(0)}{-|x|}.$$

Since $r_i'(0)$ exists, letting x tend to zero, we conclude $q_i = r_i'(0)$. This shows that $\varphi(\mathfrak{N}(f)) = \sum_{j=1}^{n} L_j(f) r_j'(0)$ is a unique linear optimal error algorithm. ∎

Theorem 3.1 was generalized for a perturbed information operator $\mathfrak{N}$, i.e., instead of $\mathfrak{N}(f)$, we know z such that $\|\mathfrak{N}(f) - z\| < \delta$ for a given δ, by Marchuk and Osipenko [75], for the complex case, i.e., S, $L_1, L_2, \ldots, L_n$ are complex linear functionals, by Osipenko [76], and for perturbed information operators with arbitrary cardinality, i.e., card($\mathfrak{N}$) can be infinity, by Micchelli and Rivlin [77].

As an application of Theorem 3.1, we show that the optimal algorithms in the sense of Sard or Nikolskij are optimal error algorithms.

Sard [49] considers approximation of the integration problem $Sf = \int_a^b f(t) \, dt$ for a class of scalar functions $f: [a,b] \to \mathbb{R}$. The information is the values of f at n fixed points $t_1, t_2, \ldots, t_n$. For a fixed nonnegative integer r, $r \leq n$, let $\Phi = \Phi(n,r)$ be a class of algorithms φ such that φ uses the information $\mathfrak{N}(f) = [f(t_1), f(t_2), \ldots, f(t_n)]^t$ and

(i) φ is linear, that is, $\varphi(\mathfrak{N}(f)) = \sum_{i=1}^{n} f(t_i) k_i$ for some $k_i = k_i(\varphi)$, $i = 1, 2, \ldots, n$,

(ii) φ is exact for the class Π_{r-1} of polynomials of degree at most $r - 1$, i.e.,

$$\varphi(\mathfrak{N}(f)) = Sf \qquad \forall f \in \Pi_{r-1}.$$

Then assuming that $f^{(r-1)}$ is absolutely continuous, he concludes that there exists a function k such that

(3.6)
$$Sf - \varphi(\mathfrak{N}(f)) = \int_a^b f^{(r)}(t)k(t)\,dt.$$

In fact, it is known that

$$k(t) = \frac{1}{r!}(b - t)^r - \sum_{i=1}^n \frac{(t_i - t)_+^{r-1}}{(r - 1)!} k_i,$$

where $t_+ = \max(t,0)$. Sard defines φ, $\varphi \in \Phi(n,r)$ as a *best* quadrature formula if

(3.7)
$$\int_a^b k^2(t)\,dt$$

is minimized with respect to all possible k_i. We shall refer to a Sard best algorithm as an optimal algorithm in the sense of Sard.

Optimality of algorithms in the sense of Sard seems to be restrictive. It is not clear a priori why an algorithm has to be linear or to be exact for the class of polynomials of degree at most $r - 1$. One might hope that permitting nonlinear algorithms which are not necessarily exact for polynomials, it is possible to find an algorithm whose error is less than the error of an optimal algorithm in the sense of Sard.

We now show that this is not the case. Using Smolyak's theorem, we prove that an optimal error algorithm in the unrestricted class of algorithms belongs to the Sard class $\Phi(n,r)$ and is also an optimal algorithm in the sense of Sard.

In order to do this, define $\mathfrak{I}_1 = W_2^r[a,b]$ as the space of scalar functions $f:[a,b] \to \mathbb{R}$ whose $(r-1)$th derivative is absolutely continuous and $f^{(r)}$ belongs to L_2. Let

$$\mathfrak{I}_0 = \{f \in \mathfrak{I}_1 : \|Tf\|_2 \le 1\}, \qquad Tf = f^{(r)}.$$

Since $\mathfrak{I}_0$ is convex and balanced, the theorem of Smolyak implies that there exists a linear optimal error algorithm φ^{oe} which uses the information $\mathfrak{N}(f) = [f(t_1), f(t_2), \ldots, f(t_n)]^t$. In this case, $\varphi^{oe}(\mathfrak{N}(f)) = \int_a^b \sigma(t)\,dt$, where σ is a unique natural spline of degree $2r - 1$, which interpolates f at $t_1, t_2, \ldots, t_n$. (See Chapter 4 and Section 4 of Chapter 6.) Since

$$e(\varphi^{oe}) = r(\mathfrak{N},S,T)$$

$$= \sup\left\{\int_a^b f(t)\,dt : f(t_i) = 0, i = 1, \ldots, n, \|f^{(r)}\|_2 \le 1\right\} < +\infty,$$

Lemma 2.2 guarantees that φ^{oe} is exact for elements from $\ker T$, i.e., for polynomials of degree at most $r - 1$. Thus, φ^{oe} belongs to the Sard class $\Phi(n,r)$ which yields the following theorem.

Theorem 3.2   Neither assumption (i) or (ii) is necessary. That is, an optimal algorithm in the sense of Sard is an optimal error algorithm for the integration problem with $\mathfrak{I}_0 = W_2^r[a,b]$ and the information operator $\mathfrak{N}(f) = [f(t_1), \ldots, f(t_n)]^t$.   ∎

Optimal algorithms in the sense of Sard for approximation of other linear functionals are defined in a similar way. It is possible to show that they also are optimal error algorithms for suitable chosen $\mathfrak{I}_0$.

There are many papers dealing with optimal algorithms in the sense of Sard. See, among others, Karlin [69, 71], Lee [77], Lipow [73], Mangasarian and Schumaker [73], Mansfield [72], Meyers and Sard [50a,b], Ritter [70], Sard [63, 67], and Schoenberg [64a,b, 65, 66, 69, 70].

REMARK 3.1   Sard considered *best* quadrature formulas which use the values of f at n *fixed* points. Schoenberg [69] defined *optimal* quadrature formulas in the sense of Sard as formulas for which points t_i as well as weights k_i are chosen to minimize the value $\int_a^b k^2(t)\, dt$. Since k depends nonlinearly on t_i, this is a much harder problem. As we shall see, this problem is related to the problem posed by Nikolskij [50].   ∎

We now discuss optimal algorithms in the sense of Nikolskij. Nikolskij [50] considers approximation of the integration problem $Sf = \int_a^b f(t)\, dt$ for a class $\mathfrak{I}_0$ of scalar functions. He defines

(3.8) $$E_n(\mathfrak{I}_0; p_i, x_i) = \sup_{f \in \mathfrak{I}_0} \left| \int_a^b f(t)\, dt - \sum_{i=1}^n p_i f(x_i) \right|,$$

(3.9) $$E_n(\mathfrak{I}_0; a, b) = \inf_{p_i, x_i} E_n(\mathfrak{I}_0; p_i, x_i).$$

An algorithm $\varphi(\mathfrak{N}(f)) = \sum_{i=1}^n p_i f(x_i)$ with $\mathfrak{N}(f) = [f(x_1), \ldots, f(x_n)]^t$ is called *optimal in the sense of Nikolskij* if $E_n(\mathfrak{I}_0; p_i, x_i) = E_n(\mathfrak{I}_0; a, b)$. (Sometimes it is mentioned that this problem was posed by Kolmogorov.)

Thus, Nikolskij considers *linear* algorithms with *optimally chosen points* at which f is evaluated. He assumes (i) but not (ii), i.e., his algorithms are not necessarily exact for some polynomials.

Let $\Psi_f(n)$ be the class of information operators $\mathfrak{N}$ of the form $\mathfrak{N}(f) = [f(t_1), f(t_2), \ldots, f(t_n)]^t$, $t_i = t_i(\mathfrak{N})$ for $i = 1, 2, \ldots, n$. If $\mathfrak{I}_0$ is balanced and convex, then Smolyak's theorem guarantees that for any $\mathfrak{N} \in \Psi_f(n)$ there exists a linear optimal error algorithm.

Since

$$\inf_{p_i} E_n(\mathfrak{I}_0; p_i, x_i) = r(\mathfrak{N}, S),$$

where $\mathfrak{N}(f) = [f(x_1), \ldots, f(x_n)]^t$, we can rewrite (3.9) as

(3.10) $$E_n(\mathfrak{I}_0; a, b) = \inf_{\mathfrak{N} \in \Psi_f(n)} r(\mathfrak{N}, S).$$

Thus, the Nikolskij problem is equivalent to the minimization of the radius of information with respect to points x_i. As we mentioned in Section 6 of Chapter 2, this is related to the Gelfand n-width and to the extremal subspaces of the approximation problem in the space L_1.

From this discussion it follows that Nikolskij's assumption on linearity of his algorithms is not restrictive. An optimal algorithm in the sense of Nikolskij is also an optimal error algorithm. Furthermore, if $\mathfrak{I}_0 = \{f : f^{(r-1)}$ is abs. cont. and $\|f^{(r)}\|_2 \le 1\}$, then due to Lemma 2.2 optimality in the sense of Nikolskij and optimality in the sense of Sard for optimally chosen points coincide. We summarize this in the following theorem.

Theorem 3.3 If $\mathfrak{I}_0$ is balanced and convex, then an optimal algorithm in the sense of Nikolskij is an optimal error algorithm for the integration problem with the information operator $\mathfrak{N}(f) = [f(t_1), \ldots, f(t_n)]^t$, where the points t_i are chosen to minimize the radius of information.

If $\mathfrak{I}_0 = \{f : f^{(r-1)}$ is abs. cont. and $\|f^{(r)}\|_2 \le 1\}$, then optimality in the sense of Nikolskij and optimality in the sense of Sard with optimally chosen points coincide. ∎

Optimal algorithms in the sense of Nikolskij have been studied by many people including Alhimova [72], Bojanov [76], Kornejčuk [74], Kornejčuk and Lušpaj [69], Krylov [62], Levin, Giršovič, and Arro [76], Ligun [76], Lušpaj [66, 69, 74], Motornyj [73, 74, 76], Nikolskij [58], Šajdaeva [59], and Žensykbaev [76, 77, 78].

So far we have confined ourselves to linear functionals. From Smolyak's theorem, there follows the existence of a linear optimal error algorithm for certain linear *operators* S. Namely, it is enough to assume that the range space $\mathfrak{I}_2$ of the linear solution operator S satisfies:

(3.11) $\mathfrak{I}_2$ is a space of real functions g defined on a set X,

(3.12)   the norm of $\mathfrak{I}_2$ is the sup norm, i.e.,

$$\|g\| = \sup_{x \in X} |g(x)|.$$

Indeed, let $f \in \mathfrak{I}_0$ and denote $g = Sf$. Knowing

$$\mathfrak{N}(f) = [L_1(f), L_2(f), \ldots, L_n(f)]^t,$$

we approximate $g(x)$ for $x \in X$. Due to the theorem of Smolyak, for every $x \in X$ there exists $g_i = g_i(x)$, $i = 1, 2, \ldots, n$, such that the linear algorithm

(3.13) $$\varphi(\mathfrak{N}(f), x) = \sum_{i=1}^{n} L_i(f) g_i(x)$$

is an optimal error algorithm. This means

$$|(Sf)(x) - \varphi(\mathfrak{N}(f), x)| = \sup_{h \in \ker \mathfrak{N} \cap \mathfrak{I}_0} (Sh)(x) \le \sup_{h \in \ker \mathfrak{N} \cap \mathfrak{I}_0} \|Sh\|.$$

Due to (3.12), we have

$$\left\|Sf - \varphi(\mathfrak{N}(f),\cdot)\right\| = r(\mathfrak{N},S)$$

which proves that φ is a linear optimal error algorithm. This is summarized in

Corollary 3.1   If $\mathfrak{I}_2$ satisfies (3.11) and (3.12), there exists a linear optimal error algorithm. ∎

We end this section with a bibliographical remark. The approximation of linear functionals has been studied by many people. We have already mentioned papers dealing with optimal algorithms in the sense of Sard and Nikolskij. Some related results may be also found in Ahlberg and Nilson [66], Bojanov and Chernogorov [77], Chawla and Kaul [73], Gal and Micchelli [78], Golomb and Weinberger [59], Kiefer [57], Knauff and Kress [74, 76], Larkin [70], Mansfield [71], Meinguet [67], Micchelli [75], Nielson [73], Reinsch [74], Richter-Dyn [71b], Secrest [65a], and Weinberger [61].

4.   LINEAR ALGORITHMS FOR LINEAR OPERATORS

In this section, we assume that S is a linear operator and $\mathfrak{I}_0 = \{f : \|Tf\| \leq 1\}$ for a linear restriction operator T. We show that for some S, T, and $\mathfrak{N}$, there exists no linear optimal error algorithm. Next we construct a linear algorithm whose error differs from the radius of information by a factor c which depends only on $T(\ker \mathfrak{N})$. If the range of T lies in a Hilbert space and $T(\ker \mathfrak{N})$ is closed, the factor c is equal to unity. In this case, we get a linear optimal error algorithm which is also central. This algorithm is closely related to spline algorithms which are studied in Chapter 4.

We begin with an example of a linear problem for which there exists no linear optimal error algorithm. This example was communicated to us by Micchelli [78] and is based on Example 1.1 of Melkman and Micchelli [77] and also on the paper by Micchelli and Rivlin [77].

As we shall see, this example is especially constructed and uses a nonstandard norm in the space $\mathfrak{I}_2$. We do not know an example of a linear problem arising in real applications for which there exists no linear optimal error algorithm.

Example 4.1   Let $\mathfrak{I}_1 = \mathbb{R}^3$, and for $f = (f_1, f_2, f_3)$ define $Sf = (f_1, f_2) \in \mathfrak{I}_2$ with the norm $\|g\|_{\mathfrak{I}_2} = \sqrt[4]{\lambda_1 g_1^4 + \lambda_2 g_2^4}$, where $\lambda_1 > \lambda_2 > 0$. Let $T = I : \mathfrak{I}_1 \to \mathfrak{I}_4$ $= \mathbb{R}^3$ with the norm $\|g\|_{\mathfrak{I}_4} = \max(\sqrt{g_1^2 + g_2^2}, |g_3|)$. Finally, set $\mathfrak{N}(f) = f_1 + af_3$, where $0 < a \leq \mu = \sqrt{2\lambda_2/(\lambda_1 + \lambda_2)}/3$. Observe that

(4.1) $$d(\mathfrak{N}, S, T) = 2 \sup_{h \in \ker \mathfrak{N} \cap \mathfrak{I}_0} \|Sh\| = 2\sqrt[4]{\lambda_2}.$$

Indeed, $\sup\{\lambda_1 f_1^4 + \lambda_2 f_2^4 : f_1 + af_3 = 0, f_1^2 + f_2^2 \leq 1, |f_3| \leq 1\} = \max(\lambda_2, \lambda_1 a^4 + \lambda_2(1 - a^2)^2) = \lambda_2$, since $\lambda_1 a^4 + \lambda_2(1 - a^2)^2 \leq \lambda_2$ for $a \leq 3\mu$. This implies (4.1).

Define the nonlinear algorithm

$$\varphi(y) = \begin{cases} (0,0), & |y| \le 2a, \\ (y,0), & |y| > 2a. \end{cases}$$

Consider $e = \|Sf - \varphi(\mathfrak{N}(f))\|$. If $|\mathfrak{N}(f)| \le 2a$, then

$$e^4 \le \sup\{\lambda_1 f_1^4 + \lambda_2 f_2^4 : |f_1 + af_3| \le 2a, f_1^2 + f_2^2 \le 1, |f_3| \le 1\}$$
$$= \max(\lambda_2, \lambda_1(3a)^4 + \lambda_2(1 - (3a)^2)^2) = \lambda_2$$

since $a \le \mu$. For $|\mathfrak{N}(f)| > 2a$, we get $|f_1| > a$, $f_2^2 \le 1 - a^2$, and $e^4 = \lambda_1(af_3)^4 + \lambda_2 f_2^4 \le \lambda_1 a^4 + \lambda_2(1 - a^2)^2 \le \lambda_2$. Thus, $e(\varphi) = \sqrt[4]{\lambda_2}$. This and (4.1) proves that

$$r(\mathfrak{N},S,T) = \sqrt[4]{\lambda_2}.$$

Consider now an arbitrary linear algorithm $\varphi(y) = (c_1 y, c_2 y)$, where c_1 and c_2 are real numbers. Then

$$e^4(\varphi) = \sup_{f \in \mathfrak{I}_0} (\lambda_1(f_1 - c_1(f_1 + af_3))^4 + \lambda_2(f_2 - c_2(f_1 + af_3))^4).$$

Setting $f_1 = 1$, $f_2 = f_3 = 0$ and $f_1 = 0$, $f_2 = 1$, $f_3 = \pm 1$, we get

$$e^4(\varphi) \ge \max(\lambda_1(1 - c_1)^4 + \lambda_2 c_2^4, \lambda_1(c_1 a)^4 + \lambda_2(1 \pm ac_2)^4).$$

If $c_2 \ne 0$ or $c_2 = 0$ and $c_1 \ne 0$, the second term in the max yields $e^4(\varphi) > \lambda_2$. If $c_1 = c_2 = 0$, the first term yields $e^4(\varphi) \ge \lambda_1 > \lambda_2$. Thus, always

$$e(\varphi) > \sqrt[4]{\lambda_2} = r(\mathfrak{N},S,T)$$

which implies that there exists no linear optimal error algorithm for this problem. ∎

Although it is not always possible to find a linear optimal error algorithm, we now construct a linear algorithm whose error differs from the radius of information by a factor c depending only on $T(\ker \mathfrak{N})$.

Without loss of generality, we consider the information operator

(4.2) $$\mathfrak{N} = [L_1, L_2, \ldots, L_n]^t$$

for linearly independent linear functionals. Due to Corollary 2.2, we confine ourselves to $n = \text{card}(\mathfrak{N}) < \dim \mathfrak{I}_1$. In Section 4 of Chapter 2, we showed that unless $\ker \mathfrak{N} \cap \ker T$ is contained in $\ker S$, $d(\mathfrak{N},S,T) = +\infty$. This assumption holds if $\mathfrak{N}^* \subset \mathfrak{N}$, where

(4.3) $$\mathfrak{N}^* = [L_1^*, L_2^*, \ldots, L_{n^*}^*]^t, \qquad L_i^*(\xi_j^*) = \delta_{ij},$$

is defined by (4.1) of Chapter 2 with $n^* = \text{index}(S,T) < +\infty$. Therefore, we assume that $\mathfrak{N}$ defined by (4.2) satisfies

(4.4) $$L_i = L_i^* \qquad \text{for} \quad i = 1, 2, \ldots, n^*, \qquad n \ge n^*.$$

Let

(4.5) $$\Im_4 = T(\ker \Re) \oplus T(\ker \Re)^\perp,$$

where $T(\ker \Re)^\perp = \text{lin}(\eta_1, \eta_2, \ldots, \eta_k)$. From Lemma 4.4 of Chapter 2, we know that $k = \dim T(\ker \Re)^\perp \le n - n^*$. Then for every $g \in \Im_4$ we have

(4.6) $$g = g_0 + \sum_{i=1}^{k} R_i(g)\eta_i,$$

where $g_0 \in T(\ker \Re)$ and $R_1, R_2, \ldots, R_k$ are linearly independent linear functionals such that $R_i(\eta_j) = \delta_{ij}$. Define

(4.7) $$c = \sup_{g \in \Im_4} \frac{\|g_0\|}{\|g\|}.$$

Note that c depends on $T(\ker \Re)$ but is independent of $\Im_1$ and S. Furthermore, $c \ge 1$, and if $T(\ker \Re)$ is closed, then c is finite. If $\Im_4$ is a Hilbert space and $T(\ker \Re)$ is closed, then we can assume that $T(\ker \Re)^\perp$ is the orthogonal complement of $T(\ker \Re)$, $(\eta_i, \eta_j) = \delta_{ij}$, $R_i(g) = (g, \eta_i)$, and $\|g\|^2 = \|g_0\|^2 + \sum_{i=1}^{k} |(g, \eta_i)|^2$, which implies $c = 1$.

Let $\Im_1 = \ker \Re \oplus (\ker \Re)^\perp$. Then $f = f_0 + \sum_{i=1}^{n} L_i(f)\xi_i$, where $f_0 \in \ker \Re$ and $L_i(\xi_j) = \delta_{ij}$ for $i, j = 1, 2, \ldots, n$, $\xi_i = \xi_i^*$ for $i = 1, 2, \ldots, n^*$, where ξ_i^* is defined by (4.3). Note that $Tf = Tf_0 + \sum_{i=n^*+1}^{n} L_i(f)T\xi_i$, since $T\xi_i^* = 0$. Thus

$$k = \dim \text{lin}(T\xi_{n^*+1}, \ldots, T\xi_n).$$

There exist linearly independent elements $\xi_{n^*+1}^*, \ldots, \xi_{n-k}^*$ such that $T\xi_i^* = 0$ and $\xi_i^* \in \text{lin}(\xi_{n^*+1}, \ldots, \xi_n)$, $i = n^* + 1, \ldots, n - k$.

Let $m = n - k$. Since $\Im_4 = T(\Im_1)$, there exist $\xi_{m+1}^*, \ldots, \xi_n^*$ such that $\eta_i = T\xi_{m+1}^*$ for $i = 1, 2, \ldots, k$. Define

(4.8) $$M = (L_i(\xi_j^*)), \qquad i, j = 1, 2, \ldots, n.$$

We show that M is nonsingular. Indeed, let

$$c_1 L_i(\xi_1^*) + \cdots + c_n L_i(\xi_n^*) = 0 \qquad \text{for} \quad i = 1, 2, \ldots, n.$$

Then $\xi = c_1 \xi_1^* + \cdots + c_n \xi_n^* \in \ker \Re$ and $T\xi \in T(\ker \Re)$. Since $T\xi_i^* = 0$ for $i = 1, 2, \ldots, m = n - k$, we get $T\xi = c_{m+1}T\xi_{m+1}^* + \cdots + c_n T\xi_n^* = c_{m+1}\eta_1 + \cdots + c_n\eta_k \in T(\ker \Re)^\perp$. This implies $c_{m+1} = \cdots = c_n = 0$. Hence, $\xi = c_1\xi_1^* + \cdots + c_m\xi_m^* \in (\ker \Re)^\perp$ which yields $c_1 = \cdots = c_m = 0$. This proves that M is nonsingular. Define

(4.9) $$[f_1, f_2, \ldots, f_n]^t = (M^t)^{-1}[\xi_1^*, \xi_2^*, \ldots, \xi_n^*]^t.$$

Note that $L_i(f_j) = \delta_{ij}$ for $i, j = 1, 2, \ldots, m$. We are ready to prove the main result of this section.

Theorem 4.1 Let $n^* = \text{index}(S,T) \leq n = \text{card}(\mathfrak{N}) < +\infty$. Let $c, f_1, f_2, \ldots, f_n$ be defined by (4.7) and (4.9), respectively. Then

$$(4.10) \qquad \varphi(\mathfrak{N}(f)) = \sum_{i=1}^{n} L_i(f)Sf_i$$

is a linear optimal error algorithm to within a factor of at most c, i.e.,

$$(4.11) \qquad r(\mathfrak{N},S,T) \leq e(\varphi) \leq (c/2)d(\mathfrak{N},S,T) \leq cr(\mathfrak{N},S,T). \quad \blacksquare$$

PROOF   Since $Tf = Tf_0 + \sum_{i=1}^{n} L_i(f)T\xi_i$, we get

$$(4.12) \qquad R_j(Tf) = \sum_{i=1}^{n} L_i(f)R_j(T\xi_i) \qquad \text{for} \quad j = 1, 2, \ldots, k.$$

Set $f = \xi_i^*$ for $i = 1, 2, \ldots, n$ in (4.12). Since $T\xi_i^* = 0$ for $i \leq n - k$ and $R_j(T\xi_i^*) = R_j(\eta_{i-m}) = \delta_{i-m,j}$ for $i > m$, we get

$$(4.13) \qquad [0,I] = M_1 M,$$

where 0 is the $k \times (n - k)$ zero matrix, I is the $k \times k$ unit matrix, and $M_1 = (R_i(T\xi_j))$, $i = 1, 2, \ldots, k, j = 1, 2, \ldots, n$.

Since M is nonsingular, we have

$$(4.14) \qquad M_1 = [0,I]M^{-1}.$$

From (4.9) and (4.14), we get

$$[Tf_1, \ldots, Tf_n]^t = (M^t)^{-1}[0, \ldots, 0, \eta_1, \eta_2, \ldots, \eta_k]^t,$$

$$(4.15) \qquad Tf_i = \sum_{j=1}^{k} R_j(T\xi_i)\eta_j.$$

We are ready to prove optimality of $\varphi(\mathfrak{N}(f)) = \sum_{i=1}^{n} L_i(f)Sf_i$. Let $\tilde{f} = \sum_{i=1}^{n} L_i(f)f_i$ for $f \in \mathfrak{I}_0$. Since $L_i(f_j) = \delta_{ij}$, then $\mathfrak{N}(\tilde{f}) = \mathfrak{N}(f)$. Let $h = f - \tilde{f}$, $h \in \ker \mathfrak{N}$. From (4.15), (4.12), and (4.6) we get

$$Th = Tf - \sum_{i=1}^{n} L_i(f)Tf_i = Tf - \sum_{i=1}^{n} L_i(f) \sum_{j=1}^{k} R_j(T\xi_i)\eta_j$$

$$= Tf - \sum_{j=1}^{k} \left(\sum_{i=1}^{n} L_i(f)R_j(T\xi_i) \right)\eta_j = Tf - \sum_{j=1}^{k} R_j(Tf)\eta_j$$

$$= (Tf)_0 \in T(\ker \mathfrak{N}).$$

Then $\|Th\| = \|(Tf)_0\| \leq c\|Tf\| \leq c$, due to (4.7). Thus,

$$\|\varphi(\mathfrak{N}(f)) - Sf\| = \|Sh\| \leq c \sup_{\substack{h \in \ker \mathfrak{N} \\ \|Th\| \leq 1}} \|Sh\| = (c/2)d(\mathfrak{N},S,T) \leq cr(\mathfrak{N},S,T).$$

Since f is an arbitrary element of $\mathfrak{I}_0$, $e(\varphi) \leq (c/2)d(\mathfrak{N},S,T)$ and of course $e(\varphi) \geq r(\mathfrak{N},S,T)$. Hence (4.11) is proven which completes the proof. $\blacksquare$

Theorem 4.1 shows how to construct a linear algorithm φ whose error is within at most a factor of c of the optimal error. Note that $\varphi(\mathfrak{N}(f)) = S\tilde{f}$, where $\mathfrak{N}(\tilde{f}) = \mathfrak{N}(f)$. Since $\tilde{f}$ need not be in $\mathfrak{I}_0$, the algorithm φ is *not*, in general, interpolatory in the sense of Definition 2.3 of Chapter 1.

For $c = 1$, the algorithm φ is an optimal error algorithm. We now show when it is also central.

Theorem 4.2   Let $\mathfrak{I}_4$ be a Hilbert space and $T(\ker \mathfrak{N})$ be closed. Then the algorithm φ defined by (4.10) is a linear central interpolatory algorithm and

(4.16) $e(\varphi) = r(\mathfrak{N},S,T) = \tfrac{1}{2}d(\mathfrak{N},S,T).$ ∎

PROOF   Let $\mathfrak{I}_4 = T(\ker \mathfrak{N}) \oplus T(\ker \mathfrak{N})^{\perp}$ be the orthogonal decomposition of $\mathfrak{I}_4$. For every $f \in \mathfrak{I}_0$, we have $Tf = (Tf)_0 + (Tf)_1$ and $\|Tf\|^2 = \|(Tf)_0\|^2 + \|(Tf)_1\|^2 \leq 1$. We show that $\varphi(\mathfrak{N}(f)) = S\tilde{f}$ is a center of $U(f)$. See (2.5) of Chapter 1. Let $f = \tilde{f} + h$. Then $h \in \ker \mathfrak{N}$. We showed in the proof of Theorem 4.1 that $Th = (Tf)_0$. This implies that $T\tilde{f} = (Tf)_1$ is orthogonal to Th and $\tilde{f} \in \mathfrak{I}_0$. Thus, the algorithm φ is interpolatory. Furthermore, $\mathfrak{N}(\tilde{f} - h) = \mathfrak{N}(f)$ and $\|T(\tilde{f} - h)\|^2 = \|T\tilde{f}\|^2 + \|Th\|^2 = \|Tf\|^2 \leq 1$. This means that $S\tilde{f} - Sh \in U(f)$ which implies that $U(f)$ is symmetric with respect to $S\tilde{f}$. From Remark 2.1 of Chapter 1, we conclude that φ is central and (4.16) holds.   ∎

See Chapter 4, where we discuss spline algorithms and we show that under the hypotheses of Theorem 4.2 the algorithm (4.10) is a spline algorithm.

5. OPTIMAL LINEAR ALGORITHMS AND
LINEAR KOLMOGOROV n-WIDTHS

This is the second section in which we present relations between some basic concepts of approximation theory and the theory of analytic complexity. (See also Section 6 of Chapter 2 and Section 4 of Chapter 7.)

For fixed S and $\mathfrak{I}_0$, we consider optimal linear algorithms for the class of linear information operators with cardinality at most n. It was observed in several papers that the error of a linear algorithm which uses $\mathfrak{N}$ with $\mathrm{card}(\mathfrak{N}) \leq n$ is no less than the Kolmogorov n-width of the solution set $S(\mathfrak{I}_0)$. See among others Babuška and Sobolev [65], Bakhvalov [62a, 68], Chzhan Guan-Tszyuan [62], Golomb [77], Melkman [77], Melkman and Micchelli [77], Micchelli and Pinkus [77], Micchelli and Rivlin [77], and Schultz [74].

We shall strengthen this observation by proving that the errors of such linear algorithms are more closely related to *linear* Kolmogorov n-widths than to Kolmogorov n-widths. Furthermore, we show that the minimal error of such linear algorithms differs by a factor depending only on S and $\mathfrak{I}_0$ from an appropriate linear Kolmogorov n-width.

Let $\Phi_{\mathrm{L}}(n)$ be the class of linear algorithms which use a linear information operator with cardinality at most n, i.e., $\varphi \in \Phi_{\mathrm{L}}(n)$ means there exists a linear information operator $\mathfrak{N} = [L_1, L_2, \ldots, L_n]^t$ such that $\varphi(\mathfrak{N}(f)) = \sum_{i=1}^n L_i(f)g_i$ for some elements $g_1, g_2, \ldots, g_n$ from $\mathfrak{J}_2$. Let

$$(5.1) \qquad \lambda(n) = \lambda(n,S) = \inf_{\varphi \in \Phi_L(n)} e(\varphi)$$

be the nth *minimal linear error* of linear algorithms from the class $\Phi_L(n)$. We remind the reader of the concept of the linear Kolmogorov n-width. Let X be a balanced subset of $\mathfrak{J}_2$. The *linear Kolmogorov n-width* $\lambda_n(X,\mathfrak{J}_2)$ of the set X is defined as

$$(5.2) \qquad \lambda_n(X,\mathfrak{J}_2) = \inf_{A_n} \inf_{A:\, AX \subset A_n} \sup_{x \in X} \|x - Ax\|,$$

where A is a linear operator with the range AX in a linear subspace A_n of dimension $\leq n$. Thus, in (5.2) we approximate the identity operator in X by a linear operator whose range is at most n-dimensional. The linear Kolmogorov n-width (briefly the linear n-width) is the infimum error of such an approximation. (See Tikhomirov [60].) The Gelfand n-width of X is no greater than the linear Kolmogorov n-width, $d^n(X,\mathfrak{J}_2) \leq \lambda_n(X,\mathfrak{J}_2), \forall n$. See Ismagilov [74], where a relation between these two widths is established.

REMARK 5.1   The *Kolmogorov n-width* $d_n(X,\mathfrak{J}_2)$ is defined similarly to the linear Kolmogorov n-width dropping the assumption that A is a linear operator, i.e.,

$$(5.3) \qquad d_n(X,\mathfrak{J}_2) = \inf_{A_n} \sup_{x \in X} \inf_{y \in A_n} \|x - {}_{(}y\|.$$

Of course, $d_n(X,\mathfrak{J}_2) \leq \lambda_n(X,\mathfrak{J}_2)$. It can happen that $d_n(X,\mathfrak{J}_2) < \lambda_n(X,\mathfrak{J}_2)$. However, for many sets X of practical interest, $d_n(X,\mathfrak{J}_2) = \lambda_n(X,\mathfrak{J}_2)$. (See for instance Tikhomirov [76] and Kornejčuk [76].) In Section 4 of Chapter 7, we establish a relation between Kolmogorov n-widths and the minimal error of n-dimensional (not necessarily linear) algorithms.

For any linear algorithm φ from $\Phi_L(n)$, we have $\varphi(\mathfrak{N}(f)) \in A_n = \mathrm{lin}(g_1, g_2, \ldots, g_n)$ and $\|Sf - \varphi(\mathfrak{N}(f))\| \geq \inf_{y \in A_n}\|Sf - y\|$. Thus,

$$(5.4) \qquad e(\varphi) \geq d_n(S(\mathfrak{J}_0),\mathfrak{J}_2) \qquad \forall \varphi \in \Phi_L(n),$$

and, of course, $\lambda(n) \geq d_n(S(\mathfrak{J}_0),\mathfrak{J}_2)$. ∎

Let

$$(5.5) \qquad \lambda_n = \lambda_n(S(\mathfrak{J}_0),\mathfrak{J}_2)$$

be the linear n-width for the solution set $S(\mathfrak{J}_0)$ in $\mathfrak{J}_2$. Recall that $q = q(S,\mathfrak{J}_0)$ is defined by (6.6) of Chapter 2. We are ready to prove an analogous theorem to Theorem 6.1 of Chapter 2 which shows that the sequence of nth minimal linear errors $\{\lambda(n)\}$ behaves essentially as the sequence of linear Kolmogorov n-widths $\{\lambda_n\}$.

Theorem 5.1   Let $S: \mathfrak{J}_1 \to \mathfrak{J}_2$ be a linear operator. If $\mathfrak{J}_0$ is balanced and convex, then

$$(5.6) \qquad\qquad\qquad \lambda(n) \le \lambda_n \qquad \forall n.$$

If additionally $\mathfrak{J}_0$ is absorbing, then

$$(5.7) \qquad\qquad q \le 1 \qquad \text{implies} \quad \lambda(n) \ge q\lambda_n \qquad \forall n,$$

$$(5.8) \qquad q = +\infty \qquad \text{and} \qquad \dim S(\mathfrak{J}_1) \le n \qquad \text{imply} \quad \lambda(n) = \lambda_n = 0,$$

$$(5.9) \quad q = +\infty \qquad \text{and} \qquad \dim S(\mathfrak{J}_1) > n \qquad \text{imply} \quad \lambda(n) = \lambda_n = +\infty. \quad \blacksquare$$

PROOF   We first prove that $\lambda(n) \le \lambda_n$. Let A be a linear operator such that $AS(\mathfrak{J}_0)$ lies in an n-dimensional subspace of $\mathfrak{J}_2$. Then $ASf = \sum_{i=1}^{n} R_i(Sf)\xi_i$, $\forall f \in \mathfrak{J}_0$, for some elements $\xi_1, \xi_2, \ldots, \xi_n$ of $\mathfrak{J}_2$ and some linear functionals $R_1, R_2, \ldots, R_n$. Define the information operator $\mathfrak{N} = [R_1 S, R_2 S, \ldots, R_n S]$ and the algorithm $\varphi(\mathfrak{N}(f)) = ASf$. Then $\varphi \in \Phi_L(n)$ and $e(\varphi) = \sup\{\|x - Ax\| : x \in S(\mathfrak{J}_0)\}$. Taking the infimum with respect to A, we get $\lambda(n) \le \lambda_n$.

To prove (5.7), assume without loss of generality that $q \in (0,1]$. Let $\varphi \in \Phi_L(n)$, i.e., $\varphi(\mathfrak{N}(f)) = \sum_{i=1}^{n} L_i(f)\xi_i$ for some $\mathfrak{N} = [L_1, L_2, \ldots, L_n]^t$. Define the linear operator $A: \mathfrak{J}_2 \to \mathfrak{J}_2$, $Ax = \sum_{i=1}^{n} L_i(f_2)\xi_i$, where $x = Sf + x_2$, $Sf \in S(\mathfrak{J}_1)$, $x_2 \in S(\mathfrak{J}_1)^{\perp}$, and $f = f_1 + f_2$, where $f_1 \in \ker S$, $f_2 \in \ker S^{\perp}$. Note that A is well defined and $AS(\mathfrak{J}_0) \subset \text{lin}(\xi_1, \xi_2, \ldots, \xi_n)$. Consider

$$(5.10) \qquad\qquad Sf - A(Sf) = Sf_2 - \sum_{i=1}^{n} L_i(f_2)\xi_i \qquad \forall f \in \mathfrak{J}_0.$$

Let $\delta \in (0,q)$. From (6.6) of Chapter 2, it follows that for every $f \in \mathfrak{J}_0$ there exists a constant $c = c(f) \ge q - \delta > 0$ such that $cf_2 \in \mathfrak{J}_0$. From (5.10), we get

$$\|Sf - A(Sf)\| = c^{-1}\|S(cf_2) - \varphi(\mathfrak{N}(cf_2))\| \le (q - \delta)^{-1} e(\varphi).$$

Since φ and δ are arbitrary, we get $\lambda_n \le q^{-1}\lambda(n)$ which proves (5.7).

Suppose now that $q = +\infty$. If $S(\mathfrak{J}_1)$ is a k-dimensional subspace, $k \le n$, then of course, $\lambda_n = 0$. From (5.6), we get $\lambda(n) = 0$ which yields (5.8). Assume that $\dim S(\mathfrak{J}_1) > n$ and suppose there exists an algorithm $\varphi \in \Phi_L(n)$ such that $e(\varphi) < +\infty$. Since $q = +\infty$ and $\mathfrak{J}_0$ is balanced and convex, then $f \in \mathfrak{J}_0$ implies that $cf_2 \in \mathfrak{J}_0$ for every c. Then $\|S(cf_2) - \varphi(\mathfrak{N}cf_2)\| = |c| \; \|Sf_2 - \varphi(\mathfrak{N}(f_2))\| \le e(\varphi) < +\infty$ implies $Sf = Sf_2 = \varphi(\mathfrak{N}(f_2))$. Thus, $S(\mathfrak{J}_0) \subset \varphi(\mathfrak{N}(\mathfrak{J}_0))$. But $\mathfrak{J}_0$ is also absorbing, which yields that $\dim \text{lin } S(\mathfrak{J}_0) = \dim S(\mathfrak{J}_1) > n$. This means that $\dim \text{lin } \varphi(\mathfrak{N}(\mathfrak{J}_0)) > n$ which is a contradiction. Thus, $e(\varphi) = +\infty$. This implies $\lambda(n) = +\infty$ and (5.9) follows from (5.6). Hence the proof of Theorem 5.1 is completed. $\blacksquare$

The following example shows that the inequality in (5.6) can be strict.

Example 5.1   Consider the problem described in Example 6.1 of Chapter 2. It is easy to show that $\lambda_1 = a + 1$. Observe that the algorithm $\varphi(y) = [0, y, y]^t$

with $y = \mathfrak{N}(f) = f_2 + f_3 - f_1$ is linear and $e(\varphi) = a$. Since $q = a/(a + 1)$, we conclude from (5.5) that $\lambda(1) = a$. Thus, $\lambda_1/\lambda(1) = (a + 1)/a$ can be arbitrarily large for small a. Of course, $\lambda(n) = \lambda_n = 0, \forall n \geq 2$. ∎

REMARK 5.2   For $q = 1$, Theorem 5.1 states

$$\lambda(n) = \lambda_n.$$

This establishes a "two-way" relation between the theory of approximation and the theory of analytic complexity. This means that if λ_n is known from approximation theory, then we know $\lambda(n)$. Conversely, if $\lambda(n)$ is known from analytic complexity, then we know λ_n. ∎

Theorems 5.1 and 6.1 of Chapter 2 can be used to establish optimality of information operators and linear algorithms. To do this, observe first that

$$\lambda(n) \geq \tfrac{1}{2}d(n),$$

where $d(n)$ is the nth minimal diameter defined by (6.2) of Chapter 2. It is very desirable to have $\lambda(n) = \tfrac{1}{2}d(n)$, since this essentially means that for an nth optimal information operator there exists a linear optimal error algorithm. A precise statement is given by the following corollary.

Corollary 5.1   Suppose there exists $\mathfrak{N} = [L_1, L_2, \ldots, L_n]^t$ and a linear algorithm φ using $\mathfrak{N}$ such that

(5.11) $e(\varphi) = \lambda(n)$.

If

(5.12) $\lambda(n) = \tfrac{1}{2}d(n)$,

then $\mathfrak{N}$ is an nth optimal information and φ is a linear optimal error algorithm. ∎

PROOF   Note that $\lambda(n) = e(\varphi) \geq r(\mathfrak{N},S) \geq \tfrac{1}{2}d(\mathfrak{N},S) \geq \tfrac{1}{2}d(n) = \lambda(n)$. Thus $d(\mathfrak{N},S) = d(n)$ and $e(\varphi) = r(\mathfrak{N},S)$ which proves Corollary 5.1. ∎

To verify whether $\lambda(n)$ is equal to $\tfrac{1}{2}d(n)$, we can use the Gelfand and linear Kolmogorov n-widths. From Theorems 5.1 and 6.1 of Chapter 2, we conclude

Corollary 5.2   Let $q = q(S,\mathfrak{I}_0) = 1$. Then

(5.13) $\lambda(n) = \tfrac{1}{2}d(n)$ iff $\lambda_n = d^n$. ∎

There are many sets $S(\mathfrak{I}_0)$ of practical interest for which $\lambda_n = d^n$. (See Tikhomirov [76].) In Chapter 6 we exhibit several examples.

Chapter 4

Spline Algorithms for Linear Problems

1. INTRODUCTION

This chapter is based primarily on Wasilkowski and Woźniakowski [78]. Optimal error algorithms minimize the error for all problem elements f in a class. For "worst" elements f, i.e., f such that $\|Sf - \varphi(\mathfrak{N}(f))\| = e(\varphi)$, the optimal error algorithms produce the best possible approximations. However, it may happen that an optimal error algorithm does not produce a best possible approximation for "easy" elements f. By an easy f we mean $\|Sf - \varphi(\mathfrak{N}(f))\| \ll e(\varphi)$. For the user who may want to solve the problem for just such an easy f, this is a very undesirable property. Therefore, in this chapter, we study algorithms which are not only optimal (or nearly optimal) but for which the *local* error is almost as small as possible for *every* f from $\mathfrak{I}_0$.

As we mentioned in Section 2 of Chapter 2, the central algorithms have the smallest possible local error for every f. We define the deviation dev(φ) of an algorithm φ as the ratio, for the worst case f, between the local error of φ and the error of a central algorithm. Thus, dev(φ) $\in [1, +\infty]$ and dev(φ) $= 1$ iff φ is a central algorithm. We are interested in algorithms with small deviation.

As always, we want to find an algorithm with small combinatory complexity. We know that the class of linear algorithms enjoys this property. Therefore, we shall be mostly interested in algorithms which are linear and have small deviation.

These twin desiderata of small deviation and linearity suggest the following questions:

(1.1)   Do there exist linear algorithms with small deviation?

(1.2)   Do there exist linear algorithms with small deviation which are optimal error algorithms?

We introduce the concept of a spline algorithm and show that spline algorithms permit us to answer (1.1) and (1.2).

Splines are extensively used in numerical mathematics and in the theory of approximation. There are enormous numbers of papers dealing with many theoretical and practical aspects of splines. Many optimal properties of splines are known. Probably Schoenberg [64a] was the first who realized the close connection between splines and optimal algorithms in the sense of Sard.

Splines were used to establish optimal algorithms (sometimes in the sense of Sard) for many problems. For instance, see Coman [72], Karlin [71], Kornejčuk [74], Kornejčuk and Lušpaj [69], Lee [77], Ligun [76], Lipow [73], Schoenberg [64b, 65, 69, 70], and Secrest [65b], who considered the integration problem; Bojanov [75], Forst [77], Gaffney [77a,b], and Gaffney and Powell [76], who considered the interpolation problem; de Boor [77], Golomb [77], Melkman [77], Micchelli and Pinkus [77], and Micchelli, Rivlin, and Winograd [76], who considered the approximation problem; Ahlberg and Nilson [66], Mangasarian and Schumaker [73], Mansfield [72], Nielson [73], Reinsch [74], Ritter [70], Schoenberg [64a], and Secrest [65a], who considered approximation of linear functionals; Grebennikov and Morozov [77], and Micchelli and Rivlin [77], who considered approximation of linear operators; and Grebennikov [78], who considered approximation of nonlinear operators. Also a classic paper of Golomb and Weinberger [59] dealing with approximation of linear functionals implicitly made use of optimal properties of splines.

In this chapter we unify and generalize many known results and develop new optimality properties of spline algorithms.

We summarize the results of this chapter. In Section 2 we introduce the concept of deviation which plays a fundamental role in this chapter. In Section 3 we recall the definition of splines in a linear normed space. We give a general definition of a spline algorithm in Section 4. Spline algorithms are homogeneous and not necessarily linear. We prove their deviation is no greater than two. Assuming the uniqueness of the spline algorithm we prove that any linear (in fact, even any homogeneous) nonspline algorithm has infinite deviation. This yields the answer to (1.1). Namely, the class of linear algorithms with finite deviation consists only of linear spline algorithms. So, this class is empty iff spline algorithms are nonlinear. In Section 5 we specialize to spline algorithms in a Hilbert case. We show that then the spline algorithm is linear and central.

This means that its deviation is equal to unity and its combinatory complexity is proportional to the cardinality of information. Section 6 returns to the general case. We give necessary and sufficient conditions for a spline algorithm to be central or optimal. This answers (1.2). We illustrate the analysis by several examples which show the sharpness of our lemmas and theorems. In Section 7 we summarize the results of this chapter.

2.   DEVIATION

Let S be a linear solution operator $\mathfrak{I}_0 = \{f : \|Tf\| \leq 1\}$, where T is a linear restriction operator, and let $\mathfrak{N} = [L_1, L_2, \ldots, L_n]^t$ be a linear information operator card$(\mathfrak{N}) = n$. We assume throughout this chapter that the radius $r(\mathfrak{N}, S, T)$ of information is finite. Recall that

$$(2.1) \qquad r(\mathfrak{N}, S, T) = c \sup_{h \in \ker \mathfrak{N}} \|Sh\| / \|Th\|,$$

where $c \in [1, 2]$. Thus, $r(\mathfrak{N}, S, T) < +\infty$ implies

$$(2.2) \qquad \ker \mathfrak{N} \cap \ker T \subset \ker S.$$

Let $\Phi = \Phi(\mathfrak{N}, S, T)$ be the class of all algorithms which use the information operator $\mathfrak{N}$ for the problem (S, T). For $\varphi \in \Phi$ and $f \in \mathfrak{I}_0$ define (as in Remark 2.2 of Chapter 1) the local error $e(\varphi, f)$ as

$$(2.3) \qquad e(\varphi, f) = \sup_{\tilde{f} \in V(f)} \|\varphi(\mathfrak{N}(f)) - S\tilde{f}\|,$$

where $V(f)$ is defined by (2.4) of Chapter 1. Of course, $e(\varphi) = \sup\{e(\varphi, f) : f \in \mathfrak{I}_0\}$. Remark 2.2 of Chapter 1 states that

$$(2.4) \qquad \inf_{\varphi \in \Phi} e(\varphi, f) = \operatorname{rad} U(f) \qquad \forall f \in \mathfrak{I}_0,$$

where $U(f)$ is defined by (2.5) of Chapter 1. Thus, the radius rad $U(f)$ is the sharp lower bound on the local error of an algorithm for every $f \in \mathfrak{I}_0$. Since we want to assure that the local error of an algorithm φ is nearly as small as possible, we compare $e(\varphi, f)$ with rad $U(f)$. This leads us to the concept of deviation defined as follows.

Definition 2.1 We shall say dev(φ) is the *deviation of the algorithm* φ, $\varphi \in \Phi(\mathfrak{N}, S, T)$, with the convention $0/0 = 1$, iff

$$(2.5) \qquad \operatorname{dev}(\varphi) = \sup_{f \in \mathfrak{I}_0} \frac{e(\varphi, f)}{\operatorname{rad} U(f)} \left(= \sup_{f \in \mathfrak{I}_0} \sup_{\tilde{f} \in V(f)} \frac{\|\varphi(\mathfrak{N}(f)) - S\tilde{f}\|}{\operatorname{rad} U(f)} \right). \quad \blacksquare$$

Of course, dev$(\varphi) \geq 1$ for any φ and dev$(\varphi) = 1$ iff φ is a central algorithm. The problem of linear algorithms with small deviation is closely related to spline algorithms which will be introduced in Section 4. Before that we remind the reader of the definition of splines in a linear normed space.

3. SPLINES

We remind the reader of the definition and some basic properties of splines in linear normed spaces and introduce notation we shall use in this chapter. See among others Anselone and Laurent [68], Atteia [65], and Holmes [72].

Let $y \in \mathbb{C}^n$. Define

$$(3.1) \qquad A(y) = \{ f \in \mathfrak{I}_1 : \mathfrak{N}(f) = y \}.$$

Note that for any y, the set $A(y)$ is nonempty since $\mathfrak{N}$ is defined by n linearly independent linear functionals, i.e., $\mathfrak{N}(\mathfrak{I}_1) = \mathbb{C}^n$. (See Section 2.)

Definition 3.1   An element $\sigma(y) \in A(y)$ is called a *spline interpolating y* (briefly σ is a *spline*) iff

$$(3.2) \qquad \|T\sigma(y)\| = \min_{f \in A(y)} \|Tf\|. \quad \blacksquare$$

Let $z \in \mathfrak{I}_4$. Define

$$(3.3) \qquad P(z) = \{ h \in \ker \mathfrak{N} : \|Th - z\| = \mathrm{dist}(T(\ker \mathfrak{N}), z) \}.$$

Thus, every element of $T(P(z))$ is a best approximation of the element z from the set $T(\ker \mathfrak{N})$. It is easy to observe that the concepts of splines and best approximation are closely connected. Namely, the following relations hold:

(3.4)   There exists a spline $\sigma(y)$ interpolating y iff the set $P(Tf)$ is nonempty for some $f \in A(y)$.

(3.5)   An element $\sigma \in A(y)$ is a spline interpolating y iff $f - \sigma \in P(Tf)$ for every $f \in A(y)$.

(3.6)   There exists a unique spline $\sigma(y)$ iff $\ker \mathfrak{N} \cap \ker T = \{0\}$ and $P(Tf)$ is a singleton set (i.e., $P(Tf)$ has exactly one element), $\forall f \in A(y)$.

Splines are homogeneous, i.e., if $\sigma(y)$ is a spline interpolating y, then $c\sigma(y)$ is a spline interpolating cy for any constant $c \in \mathbb{C}$. This means that $\sigma(cy) = c\sigma(y)$ whenever the spline $\sigma(y)$ is uniquely defined.

Suppose that $\mathfrak{I}_4 = T(\mathfrak{I}_1)$ is a Hilbert space. Then $P(Tf)$ is nonempty for any $f \in \mathfrak{I}_1$ iff $T(\ker \mathfrak{N})$ is closed. Furthermore, $\sigma(y)$ is a spline iff $\sigma(y) \in A(y)$ and $(T\sigma(y), Th) = 0, \forall h \in \ker \mathfrak{N}$. A spline σ depends linearly on y, i.e., if splines $\sigma(y_1)$ and $\sigma(y_2)$ interpolate y_1 and y_2 respectively, then $c_1\sigma(y_1) + c_2\sigma(y_2)$ is a spline interpolating $c_1 y_1 + c_2 y_2$ for any constants $c_1, c_2 \in \mathbb{C}$. The splines $\sigma(y)$ are uniquely defined iff $\ker \mathfrak{N} \cap \ker T = \{0\}$.

4. SPLINE ALGORITHMS

In this section we introduce the concept of spline algorithms and prove their optimality properties in the class of homogeneous and interpolatory algorithms. To assure the existence of splines we shall assume throughout the rest of this chapter that $P(Tf)$ is a nonempty set for any $f \in \Im_1$.

Definition 4.1   We shall say φ^s is a *spline algorithm*, $\varphi^s \in \Phi^s$, *for the problem S with the information* $\Re$ (briefly a spline algorithm) iff for any $f \in \Im_0$

(4.1) $\varphi^s(y) = S\sigma(y)$,   where $y = \Re(f)$,

and $\sigma(y)$ is a spline interpolating y.   ∎

Note that a spline algorithm is interpolatory which implies that $e(\varphi^s, f) \leq$ diam $U(f) \leq 2$ rad $U(f)$, and dev$(\varphi) \leq 2$. Since splines are homogeneous, it is obvious that a spline algorithm is also homogeneous.

REMARK 4.1   To compute $\varphi^s(y)$ given y, in general we need to know a spline $\sigma(y)$, i.e., to solve the optimization problem (3.2). The complexity of solving (3.2) can be high. However, if a spline algorithm is linear, i.e., $\varphi^s(\Re(f)) = \sum_{i=1}^{n} L_i(f)g_i$, then we need to compute only elements $g_1, g_2, \ldots, g_n$. Since the elements $g_1, g_2, \ldots, g_n$ are independent of $\Re(f)$, the idea of precomputing can be used. This means that in many cases we have to solve the optimization problem (3.2) only once. For nonlinear spline algorithms, the idea of precomputing cannot in general be used and the combinatory complexity of such algorithms can be very high.   ∎

We establish the optimality properties of spline algorithms in the class $\Phi = \Phi(\Re, S, T)$.

Lemma 4.1   Let φ, $\varphi \in \Phi$, be a homogeneous algorithm and $y \in \Re(\Im_0)$. If $e(\varphi) < +\infty$ and $\sigma(y) \in \ker T$, then

(4.2) $\varphi(y) = S\sigma(y)$.   ∎

PROOF   Since $\sigma(y) \in \ker T$, then $c\sigma(y) \in \Im_0$ for any $c \in \mathbb{C}$ and $\Re(c\sigma(y)) = cy$. Consider $\varphi(cy) - S(c\sigma(y)) = c(\varphi(y) - S\sigma(y))$. Observe that $|c| \, \|\varphi(y) - S\sigma(y)\| \leq e(\varphi) < +\infty, \forall c \in \mathbb{C}$. This implies $\varphi(y) = S\sigma(y)$ which proves (4.2).   ∎

Lemma 4.2   Let φ, $\varphi \in \Phi$, be a homogeneous algorithm which is interpolatory. Then φ is a spline algorithm.   ∎

PROOF   Since φ is interpolatory, then $e(\varphi) \leq 2r(\Re, S, T) < +\infty$. Take an arbitrary y from $\Re(\Im_0)$ and consider a spline $\sigma(y)$. If $\sigma(y) \in \ker T$, then Lemma 4.1 yields $\varphi(y) = S\sigma(y)$. Thus we can assume that $T\sigma(y) \neq 0$. Let $\tilde{y} = y/\|T\sigma(y)\|$ and $\sigma(\tilde{y})$ be a spline which interpolates $\tilde{y}$. Note that $\|T\sigma(\tilde{y})\| = 1$ which implies that $\sigma(\tilde{y}) \in \Im_0$ and $\Re(\sigma(\tilde{y})) = \tilde{y}$. Consider the set $V = V(\sigma(\tilde{y})) = \{f \in \Im_1 : \Re(f) = \tilde{y},$

$\|Tf\| \leq 1\}$. Since $1 = \|T\sigma(\tilde{y})\| \leq \|Tf\|$ for any $f \in V$, every element of V is a spline interpolating $\tilde{y}$.

The algorithm φ is interpolatory. Thus $\varphi(\tilde{y}) = S\sigma(\tilde{y})$ for a spline $\sigma(\tilde{y})$. Since φ is also homogeneous $\varphi(y) = \|T\sigma(y)\|\varphi(\tilde{y}) = \|T\sigma(y)\|S\sigma(\tilde{y}) = S\sigma(y)$, where $\sigma(y)$ is a spline which interpolates y. This proves that φ is a spline algorithm. ∎

We wish to examine when there exists a unique spline algorithm. It is easy to prove the following lemma.

Lemma 4.3 There exists a unique spline algorithm iff $SP(Tf)$ is a singleton set for any $f \in \mathfrak{I}_0$. ∎

PROOF Let $f \in \mathfrak{I}_0$ and $y = \mathfrak{N}(f)$. Consider the splines $\sigma_1(y)$ and $\sigma_2(y)$. From Section 3 we know that $f - \sigma_i(y) = h_i \in P(Tf)$ for $i = 1, 2$. Then

$$S\sigma_1(y) - S\sigma_2(y) = Sh_2 - Sh_1.$$

Thus, $SP(Tf)$ is singleton iff $S\sigma_1(y) = S\sigma_2(y)$, $\forall y \in \mathfrak{N}(\mathfrak{I}_0)$. This proves Lemma 4.3. ∎

We are ready to consider the deviation of homogeneous algorithms belonging to Φ.

Theorem 4.1 Let $SP(Tf)$ be a singleton set for any $f \in \mathfrak{I}_0$. Let φ, $\varphi \in \Phi$, be a homogeneous algorithm. Then

$$(4.3) \qquad \operatorname{dev}(\varphi) = \begin{cases} \leq 2 & \text{if} \quad \varphi \text{ is a spline algorithm,} \\ +\infty & \text{otherwise.} \end{cases} \quad ∎$$

PROOF The inequality $\operatorname{dev}(\varphi) \leq 2$ for a spline algorithm φ was already proven. If $e(\varphi) = +\infty$, then $\operatorname{dev}(\varphi) = +\infty$ since $\operatorname{rad} U(f) \leq r(\mathfrak{N}, S, T) < +\infty$. Thus, without loss of generality we can assume φ is a nonspline algorithm with $e(\varphi) < +\infty$. This means there exists $y \in \mathfrak{N}(\mathfrak{I}_0)$ such that $\varphi(y) \neq S\sigma(y)$, where $\sigma(y)$ is the unique spline interpolating y. Lemma 4.1 guarantees that $T\sigma(y) \neq 0$. As in the proof of Lemma 4.2, define $\tilde{y} = y/\|T\sigma(y)\|$ and consider the singleton set $U = U(\sigma(\tilde{y})) = \{S\sigma(\tilde{y})\}$. Of course, $\operatorname{rad} U = 0$. Since φ is homogeneous,

$$\varphi(\tilde{y}) = \varphi(y)/\|T\sigma(y)\| \neq S\sigma(y)/\|T\sigma(y)\| = S\sigma(\tilde{y}).$$

Thus, $e(\varphi, \sigma(\tilde{y})) \neq 0$ and $\operatorname{dev}(\varphi) \geq e(\varphi, \sigma(\tilde{y}))/\operatorname{rad} U(\sigma(\tilde{y})) = +\infty$. This completes the proof. ∎

The assumption that $SP(Tf)$ is singleton for any $f \in \mathfrak{I}_0$ is essential. To see that consider the following example.

Example 4.1 Let $\mathfrak{I}_1 = \mathfrak{I}_2 = \mathfrak{I}_4 = C[0,1]$ be the class of continuous functions on $[0,1]$ with the sup norm $\|f\| = \max_{0 \leq x \leq 1} |f(x)|$. Let $S = T = I$ be the identity operator and $y = \mathfrak{N}(f) = [f(x_1), f(x_2), \ldots, f(x_n)]^t$ for some distinct points $x_i \in [0,1]$. Thus, we want to recover the function f knowing its n values at x_i and the bound $\|f\| \leq 1$. It is easy to show that

$$\text{rad } U(f) = 1 \qquad \forall f \in \mathfrak{I}_0,$$

and the unique center of $U(f)$ is the zero function. Note that the center of $U(f)$ does not belong to $U(f)$ for $\mathfrak{N}(f) \neq 0$. Furthermore, every function σ, $\sigma \in \mathfrak{I}_1$, which agrees with f at x_i, i.e., $\sigma(x_i) = f(x_i) = y_i$, $i = 1, 2, \ldots, n$, and $\|\sigma\| \leq \max_{1 \leq i \leq n} |f(x_i)| = \|y\|_\infty$ is a spline. Thus, if $f \notin \ker \mathfrak{N}$, i.e., $y \neq 0$, then there exist infinitely many splines and of course

$$SP(Tf) = P(f) = \{h \in \ker \mathfrak{N} : \|h - f\| \leq \|y\|_\infty\}$$

is not a singleton set.

Consider the central linear algorithm $\varphi(y) \equiv 0$. Of course, $\text{dev}(\varphi) = 1$. Since φ is not interpolatory, φ is not a spline algorithm. Furthermore, it can be shown that any interpolatory algorithm φ has the local error $e(\varphi, f) = 2$ for any $f \in \mathfrak{I}_0$ such that $\|\mathfrak{N}(f)\|_\infty = 1$, and $\text{dev}(\varphi) = 2$. ∎

REMARK 4.2   Theorem 4.1 states that among homogeneous algorithms, only the spline algorithm has finite deviation. This provides the answer to question (1.1). When the spline algorithm is nonlinear, the class of linear algorithms with finite deviation is empty. When the spline algorithm is linear, the class of linear algorithms with finite deviation consists of exactly one element; namely, the unique linear spline algorithm. ∎

Therefore, it is important to know when a spline algorithm is linear. Although we defined a spline algorithm φ only for $y \in \mathfrak{N}(\mathfrak{I}_0)$, it is obvious that $\varphi(y) = S\sigma(y)$ where $\sigma(y)$ is a spline interpolating y for $y \in \mathbb{C}^n$, is the needed extension. Assume that $SP(Tf)$ is a singleton set for $f \in \mathfrak{I}_0$. Note that the set $P(z)$ defined by (3.3) is homogeneous. Thus, since $SP(z)$ is a singleton set for any $z \in T(\mathfrak{I}_0)$, $SP(z) = \{a(z)\}$ is also singleton for any $z \in T(\mathfrak{I}_1) = \mathfrak{I}_4$. Therefore, we can define an operator $R : \mathfrak{I}_4 \to \mathfrak{I}_2$ such that

(4.4) $R(z) = a(z).$

The spline algorithm can be represented by R as

(4.5) $S\sigma(y) = Sf - R(Tf) \qquad \forall y = \mathfrak{N}(f).$

Indeed, since $\sigma(y)$ is a spline, then $f - \sigma(y) \in P(Tf)$ for any f such that $\mathfrak{N}(f) = y$. Then $S(f - \sigma(y)) \in SP(Tf) = \{a(Tf)\}$ and $S\sigma(y) = Sf - S(f - \sigma(y)) = Sf - R(Tf)$ which proves (4.5). From (4.5) we immediately get

Lemma 4.4   Let $SP(z)$ be a singleton set for any $z \in \mathfrak{I}_4$. Then the spline algorithm is linear iff R is a linear operator. ∎

Elsewhere we give examples for which a linear spline algorithm exists. We now illustrate Lemma 4.4 by an example of a unique spline algorithm which is nonlinear.

Example 4.2 Let $\mathfrak{I}_1 = \mathfrak{I}_2 = \mathfrak{I}_4$ be the space of polynomials of one variable of degree $\leq n$. Define $\|f\| = \max_{0 \leq t \leq 1} |f(t)|$ and let $S = T = I$. The information operator is given by

$$y = \mathfrak{N}(f) = \left[f'(0), \frac{f''(0)}{2!}, \dots, \frac{f^{(n)}(0)}{n!} \right]^t.$$

Thus, $h \in \ker \mathfrak{N}$ implies $h(t) = \text{const.}$ Then

$$P(f) = \left\{ h(t) \equiv h_0: \sup_{0 \leq t \leq 1} |f(t) - h_0| = \inf_{c \in \mathbb{R}} \sup_{0 \leq t \leq 1} |f(t) - c| \right\},$$

i.e., $h \in P(f)$ iff h is a constant function and h is the best approximation of f among all zeroth-degree polynomials. It is well known that $P(f) = \{a(f)\}$ is singleton and

$$R(f) = a(f) = (\bar{f} + \underline{f})/2,$$

where $\bar{f} = \max_{0 \leq t \leq 1} f(t)$ and $\underline{f} = \min_{0 \leq t \leq 1} f(t)$. Of course, R is nonlinear which means that the unique spline algorithm is also nonlinear. Define $g(y) = \sum_{i=1}^{n} y_i t^i$ where $y_i = f^{(i)}(0)/i!$. From (4.5), we get

$$\varphi^s(y) = S\sigma(y) = f - R(f) = g(y) + f(0) - (\bar{f} + \underline{f})/2$$
$$= g(y) + f(0) - (\bar{g} + 2f(0) + \underline{g})/2 = g(y) - (\bar{g} + \underline{g})/2.$$

It can be shown that the spline algorithm is central and

$$e(\varphi^s, f) = 1 - (\bar{g} - \underline{g})/2 \leq r(\mathfrak{N}, S, T) = 1. \quad \blacksquare$$

5. HILBERT CASE

In this section, we assume that the operator T maps onto a Hilbert space $\mathfrak{I}_4 = T(\mathfrak{I}_1)$ and $T(\ker \mathfrak{N})$ is closed. Observe that the set $SP(Tf)$ is singleton for any $f \in \mathfrak{I}_1$. Indeed, let $\sigma_1(y)$ and $\sigma_2(y)$ be splines which interpolate $y = \mathfrak{N}(f)$. Then $T\sigma_i(y)$ is orthogonal to $T(\ker \mathfrak{N})$. Let $h = \sigma_1(y) - \sigma_2(y)$. Then $h \in \ker \mathfrak{N}$ and $\|T\sigma_1\|^2 = \|T\sigma_2\|^2 = \|T\sigma_1 - Th\|^2 = \|T\sigma_1\|^2 + \|Th\|^2$. This implies that $Th = 0$. Thus, $h \in \ker \mathfrak{N} \cap \ker T$ which due to (2.2) yields $h \in \ker S$, i.e., $S\sigma_1(y) = S\sigma_2(y)$.

The unique spline algorithm may be derived as follows. Let

$$e^i = [0, \dots, 1, \dots, 0]^t \in \mathbb{C}^n$$

denote the ith unit vector, $i = 1, 2, \dots, n$. Find $\sigma_i \in \mathfrak{I}_1$ such that $\mathfrak{N}(\sigma_i) = e^i$ and $T\sigma_i$ is orthogonal to $T(\ker \mathfrak{N})$. Of course, σ_i is a spline interpolating e^i. Then $\sigma(y) = \sum_{i=1}^{n} L_i(f)\sigma_i$ is also orthogonal to $T(\ker \mathfrak{N})$ and therefore $\sigma(y)$ is a spline interpolating $y = [L_1(f), L_2(f), \dots, L_n(f)]^t$. The unique spline algorithm

φ^s is of the form

$$(5.1) \qquad \varphi^s(y) = S\sigma(y) = \sum_{i=1}^{n} L_i(f)S\sigma_i$$

for $y = \mathfrak{N}(f)$. This shows that the spline algorithm is linear.

Optimality of spline algorithms in a Hilbert case was established by a number of people for particular cases. See, for instance, Micchelli and Rivlin [77], where the case $\mathfrak{I}_4 = \mathfrak{I}_1$ and $T = I$ is considered. Applying a proof technique similar to that used by Micchelli and Rivlin [77], we establish

Theorem 5.1   If $\mathfrak{I}_4$ is a Hilbert space and $T(\ker \mathfrak{N})$ is closed, then the spline algorithm φ^s is central and

$$(5.2) \qquad e(\varphi^s, f) = \operatorname{rad} U(f) = \sqrt{1 - \|T\sigma(y)\|^2}\, r(\mathfrak{N}, S, T), \qquad y = \mathfrak{N}(f),$$

where the radius of information is equal to

$$(5.3) \qquad r(\mathfrak{N}, S, T) = \sup_{h \in \ker \mathfrak{N}} \frac{\|Sh\|}{\|Th\|}. \quad \blacksquare$$

PROOF   Let $\tilde{f} \in V(f)$. Then $\tilde{f} = \sigma(y) + h$, where $y = \mathfrak{N}(f)$, $h \in \ker \mathfrak{N}$, and $1 \ge \|T\tilde{f}\|^2 = \|T\sigma(y)\|^2 + \|Th\|^2$. Thus,

$$V(f) = \{\sigma(y) + h : h \in \ker \mathfrak{N} \text{ and } \|Th\|^2 \le 1 - \|T\sigma(y)\|^2\}.$$

We now show that $U(f) = SV(f)$ is symmetric with respect to $S\sigma(y)$. Indeed, let $S\sigma(y) + Sh \in U(f)$. Then $h \in \ker \mathfrak{N}$ and $\|Th\|^2 \le 1 - \|T\sigma(y)\|^2$. The element $S\sigma(y) - Sh$ also belongs to $U(f)$ since $\mathfrak{N}(\sigma(y) - h) = y$ and $\|T(\sigma(y) - h)\|^2 = \|T\sigma(y)\|^2 + \|Th\|^2 \le 1$. From Remark 2.1 of Chapter 1, we conclude that $S\sigma(y)$ is a center of $U(f)$.

Hence, $\varphi^s(y) = S\sigma(y)$ is a central algorithm and

$$\begin{aligned}
e(\varphi^s, f) &= \operatorname{rad} U(f) = \sup\{\|S(\sigma(y) + h) - S\sigma(y)\| : \sigma(y) + h \in V(f)\} \\
&= \sup\{\|Sh\| : h \in \ker \mathfrak{N}, \|Th\| \le \sqrt{1 - \|T\sigma(y)\|^2}\} \\
&= \sqrt{1 - \|T\sigma(y)\|^2}\, \sup\{\|Sh\|/\|Th\| : h \in \ker \mathfrak{N}\}.
\end{aligned}$$

Observe that $r(\mathfrak{N}, S, T) = \sup_{y \in \mathfrak{N}(\mathfrak{I}_0)} \operatorname{rad} U(f) = \operatorname{rad} U(0)$ which completes the proof. $\blacksquare$

REMARK 5.1   Since there exists exactly one linear algorithm which has finite deviation, we conclude that the algorithm defined by (4.10) of Chapter 3 is a spline algorithm and (4.10) of Chapter 3 coincides with (5.1). $\blacksquare$

Theorem 5.1 states that the spline algorithm is central. It is also linear. These are very desirable and useful properties. This affirmatively answers our questions from Section 1 for any linear operators S, T, and $\mathfrak{N}$ (assuming that $\mathfrak{I}_4$ is a Hilbert space and $T(\ker \mathfrak{N})$ is closed). We illustrate Theorem 5.1 by two examples.

Example 5.1 Suppose that $\mathfrak{I}_1 = \mathfrak{I}_4$ is a Hilbert space with an orthonormal basis $\{\xi_i\}_{i=1}^{\infty}$. Let T be an orthogonal operator, i.e., $T^*T = TT^* = I$. Let $f \in \mathfrak{I}_1$, i.e., $f = \sum_{i=1}^{\infty} (f,\xi_i)\xi_i$. Define the information operator $\mathfrak{N}$ as

$$y = \mathfrak{N}(f) = [(f,\xi_1),(f,\xi_2), \ldots ,(f,\xi_n)]^t.$$

Then $\sigma(y) = \sum_{i=1}^{n} (f,\xi_i)\xi_i$ satisfies $\mathfrak{N}(\sigma(y)) = y$ and for any $h \in \ker \mathfrak{N} = \{h : (h,\xi_i) = 0, \ i = 1,2, \ldots ,n\}$ we get $(T\sigma(y),Th) = (\sigma,h) = \sum_{i=1}^{n} (f,\xi_i)(\xi_i,h) = 0$. This shows that a truncated Fourier series $\sigma(y)$ is a spline. Thus, for any linear operator $S : \mathfrak{I}_1 \to \mathfrak{I}_2$, the spline algorithm

$$\varphi(y) = \sum_{i=1}^{n} (f,\xi_i)S\xi_i$$

is central and $r(\mathfrak{N},S,T) = \|S\|_{\ker \mathfrak{N}} = \sup\{\|Sh\| : h \in \ker \mathfrak{N} \text{ and } \|h\| \le 1\}$. ∎

Example 5.2 Let $\mathfrak{I}_1 = W_2^r[0,1]$ be a Sobolev space, i.e., the space of functions for which the $(r-1)$st derivative is absolutely continuous and the rth derivative belongs to $\mathfrak{I}_4 = L^2[0,1]$. Let $T = D^r$, i.e., $Tf = f^{(r)}$, and

$$\mathfrak{N}(f) = [f(x_1), \ldots ,f^{(j_1-1)}(x_1), \ldots ,f(x_k), \ldots ,f^{(j_k-1)}(x_k)]^t$$

for distinct $x_i \in [0,1]$ and $\max_{1 \le i \le k} j_i \le r$. Then the cardinality of $\mathfrak{N}$ is equal to $n = j_1 + j_2 + \cdots + j_k$. Assuming that $n \ge r$, it is well-known that the spline $\sigma(y)$ is the natural spline function of degree $2r-1$ with respect to the knots $x_1, x_2, \ldots , x_k$ with multiplicity $j_1, j_2, \ldots , j_k$ respectively.

Many authors dealt with this information operator with different $j_1, j_2, \ldots , j_k$ for different linear operators S. See the Introduction for the reference list. For instance, if $S = I$ or $S(f) = \int_0^1 f(x)\,dx$, it is known that $r(\mathfrak{N},S,T) = +\infty$ for $n < r$ and $r(\mathfrak{N},S,T)$ is roughly n^{-r} for $n \ge r$. Theorem 5.1 states that for any linear operator S the algorithm obtained through the natural spline is central. See Chapter 6 for numerous examples of optimal spline algorithms. ∎

6. NON-HILBERT CASE

In this section we deal with spline algorithms where $\mathfrak{I}_4$ is not necessarily a Hilbert space. We give necessary and sufficient conditions for a spline algorithm to be a central or optimal error algorithm. We also show examples for which spline algorithms are neither central nor optimal error algorithms. We begin with the centrality of spline algorithms. It is intuitively obvious that a spline algorithm is central iff the centers of $U(f)$ belong to $U(f)$ and enjoy the same homogeneous property. A formal proof is provided by

Lemma 6.1 There exists a central spline algorithm φ^c iff there exists a function $c : \mathfrak{N}(\mathfrak{I}_0) \to \mathfrak{I}_2$ such that

(i) for any $f \in \mathfrak{I}_0$, $c(y) \in U(f)$, and $c(y)$ is a center of $U(f)$ where $y = \mathfrak{N}(f)$.

(ii) c is homogeneous, i.e., $tc(y) \in U(tf)$ and is a center of $U(tf)$, where $y = \mathfrak{N}(f)$, $\forall |t| \leq 1$, $\forall f \in \mathfrak{I}_0$. $\blacksquare$

PROOF   Suppose that φ^c is a central spline algorithm. Define $c(y) = \varphi^c(y) = S\sigma(y)$. Since φ^c is interpolatory, central, and $\sigma(y)$ is homogeneous, then c has the desired properties.

Assume now that c satisfies (i) and (ii) which means $c(y)$ is a homogeneous algorithm which is interpolatory. From Lemma 4.2 it follows that $c(y)$ is a spline algorithm. $\blacksquare$

We now show an example where the unique linear spline algorithm is an optimal error algorithm which is not central.

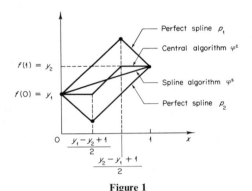

Figure 1

Example 6.1   Let $\mathfrak{I}_1 = \mathfrak{I}_2 = W_\infty^1[0,1]$ be the space of functions which are absolutely continuous and $f' \in L_\infty[0,1]$. Consider $S = I$, $Tf = f'$, and $\mathfrak{N}(f) = [f(0), f(1)]^t$ (see Figure 1). It is easy to verify that there exists a unique spline algorithm

$$\varphi^s(y)(x) = y_1 + (y_2 - y_1)x, \qquad e(\varphi^s) = \tfrac{1}{2} = r(\mathfrak{N}, S, T).$$

But the central algorithm is equal to

$$\varphi^c(y)(x) = \tfrac{1}{2}(p_1(x) + p_2(x)),$$

where p_1 and p_2 are perfect splines,

$$p_1(x) = \begin{cases} x + y_1 & \text{for} \quad x \leq \tfrac{1}{2}(y_2 - y_1 + 1), \\ -x + y_2 + 1 & \text{otherwise,} \end{cases}$$

$$p_2(x) = \begin{cases} -x + y_1 & \text{for} \quad x \leq \tfrac{1}{2}(y_1 - y_2 + 1), \\ x + y_2 - 1 & \text{otherwise,} \end{cases}$$

and $\varphi^s \neq \varphi^c$. $\blacksquare$

We turn to the question when a spline algorithm is an optimal error algorithm. Note that we now deal with nonunique spline algorithms since we do not assume that $SP(Tf)$ is a singleton set.

Lemma 6.2 There exists an optimal error spline algorithm iff there exists a function $\sigma:\mathfrak{N}(\mathfrak{I}_0) \to \mathfrak{I}_2$ such that

   (i) $\sigma(y)$ is a spline interpolating y, $\forall y$,
   (ii) for any $f \in \mathfrak{I}_0$,

$$\|Tf - T\sigma(y)\| > 1 \quad \text{implies} \quad \|Sf - S\sigma(y)\| \le r(\mathfrak{N},S,T),$$

where $y = \mathfrak{N}(f)$. ∎

PROOF   Suppose that $\bar{\varphi}^s$ is an optimal error spline algorithm, $\bar{\varphi}^s(y) = S\bar{\sigma}(y)$, where $\bar{\sigma}(y)$ is a spline interpolating y. Then $\|Sf - S\bar{\sigma}(y)\| \le r(\mathfrak{N},S,T)$ for any $f \in \mathfrak{I}_0$, $y = \mathfrak{N}(f)$. Hence we can put $\sigma(y) = \bar{\sigma}(y)$ which satisfies (i) and (ii).

Assume now that σ satisfies conditions (i) and (ii). Define an algorithm $\varphi(y) = S\sigma(y)$. Due to (i), φ is a spline algorithm. Consider $\|Sf - \varphi(y)\|$ for $y = \mathfrak{N}(f)$. If $\|Tf - T\sigma(y)\| > 1$, then (ii) implies $\|Sf - \varphi(y)\| \le r(\mathfrak{N},S,T)$. If $\|Tf - T\sigma(y)\| \le 1$, then setting $h = f - \sigma(y)$, $h \in \ker \mathfrak{N}$, we have $\|Sf - \varphi(y)\| = \|Sh\|$ and $\|Th\| \le 1$. Thus, $\|Sf - \varphi(y)\| \le \|Sh\|/\|Th\| \le r(\mathfrak{N},S,T)$, due to (2.1). This means that φ is an optimal error spline algorithm, which completes the proof. ∎

Lemma 6.2 states that a spline algorithm $\varphi^s(y) = S\sigma(y)$ is an optimal error algorithm if the elements $h = f - \sigma(y)$ of the large T-norm, $\|Th\| > 1$, are correlated with the operator S in such a way that $\|Sh\|$ is small, i.e., $\|Sh\| \le r(\mathfrak{N},S,T)$. Note that if $\mathfrak{I}_4$ is a Hilbert space, we have a unique spline algorithm for which $\|Th\| = \sqrt{\|Tf\|^2 - \|T\sigma(y)\|^2} \le \|Tf\| \le 1$ and condition (ii) is automatically satisfied. Note that Lemma 6.2 provides the answer to question (1.2). Namely, the class of linear algorithms with finite deviation which are optimal error algorithms is nonempty iff there exists a function σ satisfying (i) and (ii) of Lemma 6.1.

Example 6.1 provides a problem in a non-Hilbert space $\mathfrak{I}_4$ for which condition (ii) holds. We now consider an example for which an unique spline algorithm is not an optimal error algorithm and its deviation is arbitrary close to two.

Example 6.2   Let $\mathfrak{I}_1 = \mathfrak{I}_2 = \mathbb{R}^2 = \{(f_1,f_2):f_i \text{ are real}\}$ with the 1_2-norm $\|f\| = \sqrt{f_1^2 + f_2^2}$. Let $\mathfrak{I}_4 = \mathbb{R}^2$ be equipped with the norm

$$\|f\| = \max(|f_1|,a^{-1}|f_2|),$$

where a parameter $a \in (1, +\infty)$. Define $Sf = f$, $Tf = f$, and $y = \mathfrak{N}(f) = f_1 + f_2$. Thus, knowing the sum of components of f, we want to recover f, where $f \in \mathfrak{I}_0 = \{f:\|Tf\| = \max(|f_1|,a^{-1}|f_2|) \le 1\}$.

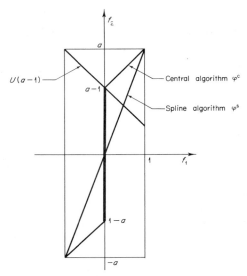

Figure 2

Figure 2 shows the central and spline algorithms. The central algorithm φ^c is equal to

$$\varphi^c(y) = \begin{cases} \left(\dfrac{y + 1 - a}{2}, \dfrac{y - 1 + a}{2} \right) & \text{for} \quad a - 1 \le y \le a + 1, \\[2mm] (0, y) & \text{for} \quad |y| \le a - 1, \\[2mm] \left(\dfrac{y - 1 + a}{2}, \dfrac{y + 1 - a}{2} \right) & \text{for} \quad -1 - a \le y \le -a + 1 \end{cases}$$

and

$$e(\varphi^c, f) = \begin{cases} \sqrt{2} & \text{for} \quad |y| \le a - 1, \\[2mm] \dfrac{\sqrt{2}}{2}(1 + a - |y|) & \text{otherwise.} \end{cases}$$

Hence $r(\mathfrak{N}, S, T) = \sqrt{2}$. The unique spline algorithm φ^s is given by

$$\varphi^s(y) = y \left(\frac{1}{1 + a}, \frac{a}{1 + a} \right).$$

The local error of φ^s is

$$e(\varphi^s, f) = \begin{cases} \dfrac{\sqrt{2}}{1 + a}(y + a + 1) & \text{for} \quad |y| \le a - 1, \\[2mm] \dfrac{\sqrt{2}}{1 + a} a(1 + a - |y|) & \text{otherwise,} \end{cases}$$

and $e(\varphi^s) = e(\varphi^s, a - 1) = (2\sqrt{2}/(1 + a))a = (2a/(1 + a))r(\mathfrak{N},S,T)$. This means that φ^s is *not* an optimal error algorithm. Furthermore,

$$\mathrm{dev}(\varphi^s) = \frac{e(\varphi^s, a - 1)}{\mathrm{rad}\ U(a - 1)} = \frac{2a}{1 + a}.$$

Thus, for large a, $\mathrm{dev}(\varphi^s)$ can be arbitrary close to 2.

We want to stress that for this problem there exists a unique optimal linear algorithm φ^L, $\varphi^L(y) = (0, y)$, which is not an interpolatory algorithm and whose deviation is infinity. ∎

7. SUMMARY

We summarize the results of this chapter. Assuming the uniqueness of the spline algorithm φ^s, the answer to question (1.1) is that the class of linear algorithms with finite deviation is empty if φ^s is nonlinear. This class consists of one element, namely φ^s, if φ^s is linear. The answer to question (1.2) is that the class of linear algorithms with finite deviation which are optimal error algorithms is empty if φ^s is nonlinear or is not an optimal error algorithm. This class consists of one element, namely φ^s, if φ^s is linear and an optimal error algorithm. We give necessary and sufficient conditions for φ^s to be a linear central optimal error algorithm.

If $T(\ker \mathfrak{N})$ is closed in a Hilbert space, there exists a unique spline algorithm φ^s which is central and linear. Due to centrality, it yields the best possible approximation for every f. Due to linearity, its combinatory complexity is small since the idea of precomputing can be used. We stress again that in general the combinatory complexity of a nonlinear spline algorithm is high and that precomputing cannot be used.

Chapter 5

Complexity for the Linear Case

1. INTRODUCTION

We study the complexity of linear problems (S,T) and linear information operators $\mathfrak{N}$. The model of computation is specified for the linear case. (Our general model of computation was defined in Chapter 1, Section 3.)

We observed in the Introduction to Chapter 3 that linear algorithms must enjoy good time and space complexity. Here we quantify this notion by showing that every linear algorithm with error less than ε is a nearly optimal complexity algorithm. We also quantify the relation between the minimal cardinality of $\mathfrak{N}$ such that $r(\mathfrak{N},S,T) < \varepsilon$ and problem complexity.

We propose and answer the following question. What functions can be complexity functions? We establish the surprising result that an arbitrary decreasing real function on the positive reals is "essentially" the complexity of some linear problem. (The precise result is given in Theorem 1.2.) This implies the existence of linear problems with arbitrarily high complexity. Furthermore, in contrast to the theory of recursively computable functions, there are no "gaps" in the complexity functions in our model.

2. COMPLEXITY OF LINEAR INFORMATION

We specify our model of computation for linear problems (S,T) and linear information operators $\mathfrak{N}$ as follows. (For the general case, see Section 3 of Chapter 1.)

Model of Computation for the Linear Case

(i)   Let P be a given collection of primitives. We assume that the addition of two elements of $\Im_2$, $f + g$, and the multiplication of an element of $\Im_2$ by a scalar, cf, are primitive operations which belong to P. We also assume that every linear functional L, $L:\Im_1 \to \mathbb{C}$, is a primitive operation which belongs to P. This implies that any linear information operator $\mathfrak{N} = [L_1, L_2, \ldots, L_n]^t$ of finite cardinality is *permissible*, where L_1, $L_2, \ldots, L_n$ are arbitrary linear functionals.

(ii)   To normalize the measure of the complexity, we assume that the cost of the addition of two elements of $\Im_2$ and the multiplication of an element of $\Im_2$ by a scalar is taken as unity. Note that for a finite-dimensional space $\Im_2$, $m = \dim \Im_2$, unit cost means the cost of m scalar additions or multiplications.

Let comp(L) denote the complexity of evaluating a linear functional L. Let $\mathfrak{N} = [L_1, L_2, \ldots, L_n]^t$ be a linear information operator with linearly independent linear functionals L_1, $L_2, \ldots, L_n$, card($\mathfrak{N}$) $= n$. We assume that $\mathfrak{N}(f)$ is computed by the independent evaluation of $L_1(f), L_2(f), \ldots, L_n(f)$ and the information complexity (see (3.4) of Chapter 1) of $\mathfrak{N}$ is given by

$$\text{comp}(\mathfrak{N}) = \sum_{i=1}^{n} \text{comp}(L_i).$$

If comp(L_i) $\equiv c_1$, then comp($\mathfrak{N}$) $= nc_1$ which shows how the information complexity depends on the cardinality of $\mathfrak{N}$.

(iii)   Let φ be a permissible algorithm which uses $\mathfrak{N}(f)$ and finds an ε-approximation to $\alpha = S(f)$. Let $d(\varphi)$ be the combinatory complexity of φ. For all linear problems of practical interest, φ has to use every $L_i(f)$, $i = 1$, $2, \ldots, n$, at least once and $d(\varphi) \geq n - 1$. We rule out special problems and information operators, assuming that $d(\varphi) \geq n - 1$ for every algorithm under consideration.   ∎

Example 2.1   Let $P = \{$arithmetic operations, the evaluation of linear functionals, the evaluation of an operator, the evaluation of a derivative of an operator$\}$. Let $\mathfrak{N}(f) = \{f(x), f'(x)\}$, where f is a function in a space of dimension m, $m \leq \infty$. Assume $\varphi(\mathfrak{N}(f)) = x - (f'(x))^{-1}f(x)$.

CASE 1 $m < \infty$ Let comp(L) $= c_1$ for every linear functional L. Then comp($f(x)$) $= mc_1$, comp($f'(x)$) $= m^2c_1$. We compute $\varphi(\mathfrak{N}(f))$ by solving the appropriate linear system by Gaussian elimination. Thus, by (ii), $d(\varphi) = O(m^2)$ times unit cost. We conclude

$$\text{comp}(\varphi) = mc_1 + m^2c_1 + d(\varphi).$$

CASE 2 $m = \infty$ We add the solution of a linear system to our set of primitives; let the complexity of this primitive be c_4. Let comp($f(x)$) $= c_2$, comp($f'(x)$) $= c_3$. Then comp(φ) $= c_2 + c_3 + c_4$.   ∎

REMARK 2.1   In (i) of the model of computation we assume that every linear functional is permissible. For some solution operators we are interested in a subclass of permissible information operators. For instance, if $Sf = \int_0^1 f(t)\,dt$ the subclass might be restricted to function evaluations.   ∎

This model of computation is an idealized one since we assume that every linear functional is a primitive operation. However, even in this idealized model, we shall prove (see Theorem 2.2) that the complexity of a linear problem (S,T) as a function of ε can be essentially any decreasing function of ε.

Fix $\mathfrak{N}$, S, and T. Let φ be a linear algorithm, i.e., $\varphi(\mathfrak{N}(f)) = \sum_{i=1}^n L_i(f)g_i$ for certain elements $g_1, g_2, \ldots, g_n$ from $\mathfrak{I}_2$. The elements g_i depend on $\mathfrak{N}$, S, and T, but are independent of f. Therefore, the elements $g_1, g_2, \ldots, g_n$ can be precomputed. The computation of $\varphi(\mathfrak{N}(f))$, given $\mathfrak{N}(f)$, requires at most n multiplications and $n - 1$ additions which are primitive operations with unit cost. Thus, any linear algorithm φ is permissible and its combinatory complexity is at most $2n - 1$. Due to (iii), every linear algorithm is within a factor of 2 of minimal combinatory complexity. Thus,

(2.1) $\mathrm{comp}(\varphi) \leq \mathrm{comp}(\mathfrak{N}) + 2n - 1.$

Recall that $\mathrm{comp}(\mathfrak{N},S,T,\varepsilon)$ is the ε-complexity of the problem (S,T) with the information $\mathfrak{N}$. (See Definition 3.1 of Chapter 1.) If the error of the linear algorithm φ is less than ε, we easily conclude that

(2.2) $\mathrm{comp}(\mathfrak{N},S,T,\varepsilon) = \mathrm{comp}(\mathfrak{N}) + a_1 n - 1,$

where $a_1 \in [1,2]$. From (2.1) and (2.2), it follows that

(2.3) $\mathrm{comp}(\mathfrak{N},S,T,\varepsilon) \leq \mathrm{comp}(\varphi) \leq 2\,\mathrm{comp}(\mathfrak{N},S,T,\varepsilon).$

Thus, the complexity of every linear algorithm with error less than ε is at most twice the ε-complexity of (S,T) with $\mathfrak{N}$. Furthermore,

(2.4)   if $\mathrm{comp}(\mathfrak{N}) \gg n$,   then $\mathrm{comp}(\mathfrak{N},S,T,\varepsilon) \cong \mathrm{comp}(\varphi) \cong \mathrm{comp}(\mathfrak{N}).$

This assumption, which often holds, implies that the complexity of every linear algorithm with error less than ε is essentially equal to the ε-complexity of (S,T) with $\mathfrak{N}$. Equations (2.3) and (2.4) motivate the following definition.

We shall say φ^{noc} is a *nearly optimal complexity algorithm for* (S,T) *with* $\mathfrak{N}$ iff φ^{noc} satisfies (2.3) and (2.4).

Thus, we have proven

Lemma 2.1   Assume there exists a linear algorithm φ using $\mathfrak{N}$ such that $e(\varphi) < \varepsilon$. Then φ is a nearly optimal complexity algorithm for (S,T) with $\mathfrak{N}$.   ∎

For fixed S and T, let Ψ be a class of permissible linear information operators. We are now interested in optimal complexity algorithms for the problem (S,T) in the class Ψ. (See Definition 3.2 of Chapter 1.) As we shall see, this is related to the minimal cardinality of a linear information operator $\mathfrak{N}$ from Ψ such that $r(\mathfrak{N},S,T) < \varepsilon$.

Definition 2.1 We shall say $m(\Psi,S,T,\varepsilon)$ is the ε-*cardinality number of the problem* (S,T) *in the class* Ψ (briefly the ε-cardinality number of (S,T) in Ψ) iff

(2.5) $\qquad m(\Psi,S,T,\varepsilon) = \min\{\mathrm{card}(\mathfrak{N}): \mathfrak{N} \in \Psi, r(\mathfrak{N},S,T) < \varepsilon\}.$ ∎

Note that $m(\Psi,S,T,\varepsilon)$ is a nonincreasing function of ε. Recall that the ε-complexity $\mathrm{comp}(\Psi,S,T,\varepsilon)$ of the problem (S,T) in the class Ψ is defined by (3.9) of Chapter 1. We show how the ε-complexity, $\mathrm{comp}(\Psi,S,T,\varepsilon)$, depends on the ε-cardinality number $m(\Psi,S,T,\varepsilon)$.

Suppose that for every $\mathfrak{N} = [L_1, L_2, \ldots, L_m]^t \in \Psi$, the complexity of evaluating a linear functional L_i is equal to c_1. Thus,

$$\mathrm{comp}(\mathfrak{N}) = nc_1 \qquad \forall \mathfrak{N} \in \Psi, \quad n = \mathrm{card}(\mathfrak{N}).$$

Similarly to (2.3) and (2.4), we shall say an algorithm φ is a *nearly optimal complexity algorithm for* (S,T) *in* Ψ iff

$$\mathrm{comp}(\Psi,S,T,\varepsilon) \leq \mathrm{comp}(\varphi) \leq 2\,\mathrm{comp}(\Psi,S,T,\varepsilon),$$

and if $c_1 \gg 1$, then

$$\mathrm{comp}(\Psi,S,T,\varepsilon) \cong \mathrm{comp}(\varphi).$$

Let $\mathfrak{N}_m$ be an information operator such that $\mathfrak{N}_m \in \Psi$, $r(\mathfrak{N}_m,S) < \varepsilon$, and $\mathrm{card}(\mathfrak{N}_m) = m = m(\Psi,S,T,\varepsilon)$.

Lemma 2.2 Suppose there exists a linear optimal error algorithm φ using $\mathfrak{N}_m$. Then

(i) the ε-complexity of the problem (S,T) in the class Ψ is

$$\mathrm{comp}(\Psi,S,T,\varepsilon) = (c_1 + a_1)m(\Psi,S,T,\varepsilon) - 1,$$

where $a_1 \in [1,2]$,

(ii) φ is a nearly optimal complexity algorithm for (S,T) in Ψ. ∎

PROOF The ε-complexity of the problem (S,T) with the information $\mathfrak{N}_m$ satisfies

$$\mathrm{comp}(\mathfrak{N}_m,S,T,\varepsilon) \leq \mathrm{comp}(\varphi) \leq mc_1 + 2m - 1.$$

Furthermore, for every $\mathfrak{N} \in \Psi$ with $r(\mathfrak{N},S) < \varepsilon$, we have $\mathrm{card}(\mathfrak{N}) \geq m$ and

$$\mathrm{comp}(\mathfrak{N},S,T,\varepsilon) \geq mc_1 + m - 1.$$

Combining these two inequalities, we have

$$\mathrm{comp}(\Psi,S,T,\varepsilon) = (c_1 + a_1)m - 1,$$

where $a_1 \in [1,2]$ which proves (i). It is obvious that $\text{comp}(\varphi) \leq 2\,\text{comp}(\Psi,S,T,\varepsilon)$ and $c_1 \gg 1$ implies that $\text{comp}(\varphi)$ and $\text{comp}(\Psi,S,T,\varepsilon)$ are essentially equal to mc_1. This proves (ii). ∎

Lemma 2.2 exhibits the close relation between the ε-complexity of (S,T) in Ψ and the ε-cardinality number of (S,T) in Ψ. It also shows that a linear optimal error algorithm for the information $\mathfrak{N}_m$ is a nearly optimal complexity algorithm for (S,T) in Ψ. In Chapter 6, we present several examples of practical interest for which Lemma 2.2 is applicable.

Let Ψ_U be the class of *all* linear information operators $\mathfrak{N}$ such that $\text{card}(\mathfrak{N}) < +\infty$. ($\mathfrak{N}$ is permissible due to (i)!) Note that $\bigcup_{n=n^*}^{\infty} \Psi_n \subset \Psi_U$, where Ψ_n (see Section 4 of Chapter 2) is the class of all information operators $\mathfrak{N}$ such that $\mathfrak{N}^* \subset \mathfrak{N}$, $\text{card}(\mathfrak{N}) \leq n$. Furthermore, $\bigcup_{n=0}^{\infty} \Psi_n = \Psi_U$ if $\text{index}(S,T) = 0$. The class Ψ_U contains all information operators of practical interest since every information operator which is to be computed has to have finite cardinality.

We show that $m(\Psi_U,S,T,\varepsilon)$ can be essentially *any* decreasing function of ε. More precisely, assume that ε belongs to the interval $(0,\varepsilon_0]$. Let

$$(2.6) \qquad g:(0,\varepsilon_0] \to \mathbb{R}^+$$

be a decreasing function such that $g(\varepsilon_0) \geq 1$ and $\lim_{\varepsilon \to 0} g(\varepsilon) = +\infty$. For simplicity we shall assume that g is continuous.

Theorem 2.1 For every function g defined by (2.6), there exists a linear problem (S,T) such that

$$(2.7) \qquad g(\varepsilon) - 1 < m(\Psi_U,S,T,\varepsilon) \leq g(\varepsilon) \qquad \forall \varepsilon \in (0,\varepsilon_0].$$

Furthermore, there exists a sequence $\{\varepsilon_i\}$ such that $\varepsilon_i \in (0,\varepsilon_0]$, $\lim_{i \to \infty} \varepsilon_i = 0$, and

$$(2.8) \qquad m(\Psi_U,S,T,\varepsilon_i) = g(\varepsilon_i). \quad ∎$$

PROOF Let $g^{-1}:[g(\varepsilon_0),+\infty) \to \mathbb{R}^+$ be the inverse function of g. Define $\beta_i = \varepsilon_0 + 1$ for $i < g(\varepsilon_0)$ and $\beta_i = g^{-1}(i)$ for $i \geq g(\varepsilon_0)$. Note that $\lim_{i \to \infty} \beta_i = 0$.

Let $\mathfrak{J}_4 = \mathfrak{J}_2 = \text{lin}(\xi_1,\xi_2,\ldots)$ be an infinite-dimensional Hilbert space, where $(\xi_i,\xi_j) = \delta_{ij}$. Define $T = I$ and

$$(2.9) \qquad Sf = \sum_{i=1}^{\infty} \beta_i(f,\xi_i)\xi_i.$$

Thus, S is a self-adjoint and compact operator. Furthermore, $Sf_i = \beta_i \xi_i$ for $i = 1, 2, \ldots$.

Note that $n^* = 0$, where $n^* = \text{index}(S,I)$. From (5.4) of Chapter 2, we get $K_1 = S^2$ and the eigenvalues of K_1 satisfy $\lambda_i = \beta_i^2$ for $i = 1, 2, \ldots$.

Let $m = m(\Psi_U,S,I,\varepsilon)$. This means there exists an information operator $\mathfrak{N}_0$ such that $\text{card}(\mathfrak{N}_0) = m$ and $r(\mathfrak{N}_0,S,T) < \varepsilon$. Moreover, for every $\mathfrak{N}$ such that $\text{card}(\mathfrak{N}) < m$, $r(\mathfrak{N},S,T) \geq \varepsilon$. Due to Theorem 5.3 of Chapter 2 and Theorem 4.2

of Chapter 3, we know that

$$\beta_{m+1} \leq r(\mathfrak{N}_0, S, T) < \varepsilon.$$

Thus, $m + 1 \geq g(\varepsilon_0) + 1$ and $\beta_{m+1} = g^{-1}(m + 1) < \varepsilon$ which yields $m > g(\varepsilon) - 1$. Furthermore, for $\mathfrak{N} = \mathfrak{N}_{m-1}$ defined by (5.7) of Chapter 2, we get $\mathrm{card}(\mathfrak{N}_{m-1}) \leq m$ and $r(\mathfrak{N}_{m-1}, S, I) = \beta^m \geq \varepsilon$. This yields $m \leq g(\varepsilon)$ and proves (2.7).

Let $\varepsilon_i = \beta_{i+1}$ for $i \geq g(\varepsilon_0)$. Then $\varepsilon_i \in (0, \varepsilon_0]$ and $\lim_i \varepsilon_i = 0$. Since $r(\mathfrak{N}, S, I) = \beta_{i+1} = \varepsilon_i$, we get $m(\Psi_U, S, I, \varepsilon) = i + 1 = g(\varepsilon_i)$. This proves (2.8) and completes the proof. ∎

Theorem 2.1 states that $m(\Psi_U, S, T, \varepsilon)$ can be an essentially arbitrary function of ε. From Theorem 2.1, we can conclude that $\mathrm{comp}(\Psi_U, S, T, \varepsilon)$ can depend arbitrarily on ε. To show this, assume for the sake of simplicity that the complexity of evaluating any linear functional L_i is fixed, $\mathrm{comp}(L_i) = c_1$.

Theorem 2.2   For every function g defined by (2.6), there exists a linear problem (S, T) such that

(2.10) $g(\varepsilon)(c_1 + 1) - c_1 - 2 < \mathrm{comp}(\Psi_U, S, T, \varepsilon) \leq g(\varepsilon)(c_1 + 2) - 1$

$$\forall \varepsilon \in (0, \varepsilon_0). \quad \blacksquare$$

PROOF   Consider the problem (S, T) defined in the proof of Theorem 2.1. Thus, the ε-cardinality number $m = m(\Psi_U, S, T, \varepsilon)$ satisfies (2.7) and the information complexity $\mathfrak{N}$ such that $\mathrm{card}(\mathfrak{N}) = m$, $r(\mathfrak{N}, S, T) < \varepsilon$, satisfies

(2.11) $(g(\varepsilon) - 1)c_1 < \mathrm{comp}(\mathfrak{N}) = m(\Psi_U, S, T, \varepsilon)c_1 \leq g(\varepsilon)c_1.$

Since the problem (S, T) is defined in a Hilbert space, Theorem 4.2 of Chapter 3 assures the existence of a linear optimal error algorithm φ, $e(\varphi) = r(\mathfrak{N}, S, T) < \varepsilon$. From Lemma 2.2, we have

(2.12) $m - 1 + \mathrm{comp}(\mathfrak{N}) \leq \mathrm{comp}(\Psi_U, S, T, \varepsilon) \leq 2m - 1 + \mathrm{comp}(\mathfrak{N}).$

From (2.11) and (2.12), we get (2.10). ∎

Theorem 2.2 states that $\mathrm{comp}(\Psi_U, S, T, \varepsilon)$ is roughly the same function of ε as the ε-cardinality number $m(\Psi_U, S, T, \varepsilon)$. Note that the function g can tend to infinity arbitrarily fast as ε tends to zero. This proves

Corollary 2.1   (i)   There exist linear problems with arbitrarily high complexity.

(ii)   There are no "gaps" in the complexity functions. ∎

This may be contrasted with the theory of recursively computable functions in which complexity gaps are known to occur (Borodin [72]).

REMARK 2.2   We assumed that $\mathfrak{N}$ consists of linear functionals which are computed independently and therefore $\mathrm{comp}(\mathfrak{N}) = nc_1$. For some information operators, $\mathfrak{N}(f)$ can be computed faster than nc_1. For instance, assume that $L_i(f) = f(x_i)$ for distinct points x_i, $i = 1, \ldots, n$, where f is a polynomial of degree $n - 1$. Then the complexity of L_i is $O(n)$, but $\mathfrak{N}(f)$ can be computed in $O(n \log^2 n)$. In fact, Theorem 2.2 remains valid under the relaxed assumption that $\text{comp}(\mathfrak{N}(f)) = \omega(n)$, where ω is an increasing function of n with $\lim_{n \to \infty} \omega(n) = +\infty$. ∎

Chapter 6

Applications for Linear Problems

1. INTRODUCTION

We apply the theory developed in Chapters 1–5 to a variety of linear problems. In successive sections, we deal with linear functionals, interpolation, integration, approximation, and linear partial differential equations of parabolic, hyperbolic, and elliptic type. The applications will be set in various function spaces. We have chosen to order the applications according to the power of the mathematical tools used. We are sometimes able to use results of other authors to aid our analysis. We then state such results using our terminology.

Some of our results are counterintuitive. We give two instances.

1.   We show in subsection 4(ii) that for the integration problem defined on the class of periodic smooth functions, the rectangle algorithm is an optimal error algorithm. This algorithm has zero order of exactness. This is an illustration that order of exactness is not related to the optimality of quadrature formulas.

2.   We show that for linear parabolic and elliptic equations studied in subsections 6(ii) and 6(iii), the cost penalty for utilizing the information commonly used in numerical practice, rather than the optimal information, is unbounded.

The results presented here are made possible by the general theory developed in Chapters 1–5 as well as the development of mathematical results in specific application areas over some decades. We expect this to be only the beginning of work on optimal algorithms for various application areas. Some of the future work we expect to see is:

1.   The analysis of the applications investigated here for other norms and problem element classes. Most of the results depend critically on the particular norm used and the class of problem elements; changing either may require a new analysis.

2.   The solution of some of the problems posed in Section 7.

3.   The analysis of "harder" application areas.

4.   Complete analysis of optimal algorithms for important applications. A complete analysis should include time complexity, space complexity, stability considerations, etc. Here we confine ourselves to time complexity.

We define notation which is extensively used in this chapter. Let X be an interval of the real axis $\mathbb{R}$ and let $L_p = L_p(X)$ be the space of functions f for which

$$(1.1) \qquad \|f\|_p = \left(\int_X |f(x)|^p \, dx \right)^{1/p}$$

is finite, $p \in [1, +\infty]$. By q we mean a real number or infinity such that

$$(1.2) \qquad 1/p + 1/q = 1.$$

For a nonnegative integer r, let

$$(1.3) \qquad W_p^r(X) = \{f : f^{(r-1)} \text{ is abs. cont. and } f^{(r)} \in L_p\}.$$

We shall often deal with the linear operator $T = D_p^r : W_p^r(X) \to L_p$ defined as

$$(1.4) \qquad Tf = D_p^r f = f^{(r)}.$$

By W_p^r, we mean

$$(1.5) \qquad W_p^r = \{f : f \in W_p^r(X) \text{ and } \|D_p^r f\|_p \le 1\}.$$

We also consider the spaces of periodic functions. By $\tilde{W}_p^r(X)$ with $X = [a, b]$, we mean

$$(1.6) \qquad \tilde{W}_p^r(X) = \{f : f \text{ is } (b - a)\text{-periodic on } \mathbb{R} \text{ and } f \in W_p^r(X)\}.$$

By $T = \tilde{D}_p^r : \tilde{W}_p^r(X) \to L_p$, we mean the linear operator

$$(1.7) \qquad Tf = \tilde{D}_p^r f = f^{(r)}.$$

Similarly,

$$\tilde{W}_p^r = \{f : f \in \tilde{W}_p^r(X) \text{ and } \|\tilde{D}_p^r f\|_p \le 1\}.$$

By $C = C(X)$, we denote the space of continuous functions f equipped with the sup norm $\|f\| = \sup\{|f(t)| : t \in X\}$, and $\tilde{C} = \tilde{C}(X)$ is the space of periodic continuous functions equipped with the same norm.

By $\lceil x \rceil$, we mean the smallest integer n such that $x \leq n$; $\lfloor x \rfloor$ means the largest integer n such that $n \leq x$. We also use the Θ notation defined as in Knuth [76]. That is, $\Theta(f(\varepsilon))$ denotes the set of all functions $g(\varepsilon)$ such that there exist positive constants c_1, c_2, and ε_0 with

$$c_1 f(\varepsilon) \leq g(\varepsilon) \leq c_2 f(\varepsilon) \qquad \text{for} \quad \varepsilon \in (0, \varepsilon_0].$$

For completeness, we also recall the standard O and o notations. Namely, $O(f(\varepsilon))$ denotes the set of all functions $g(\varepsilon)$ such that there exist positive constants c and ε_0 with

$$|g(\varepsilon)| \leq cf(\varepsilon) \qquad \text{for} \quad \varepsilon \in (0, \varepsilon_0];$$

$o(f(\varepsilon))$ denotes the set of all functions $g(\varepsilon)$ such that

$$\lim_{\varepsilon \to 0} \frac{g(\varepsilon)}{f(\varepsilon)} = 0.$$

2.   LINEAR FUNCTIONALS

In this section, we consider a particular case of approximation of a linear functional

$$(2.1) \qquad\qquad S: W_p^r[-1,1] \to \mathbb{R}.$$

Define $T = D_p^r$ and $\mathfrak{I}_4 = L_p[-1,1]$. Thus,

$$(2.2) \qquad\qquad \mathfrak{I}_0 = W_p^r.$$

We consider information operators of the form

$$(2.3) \qquad \mathfrak{N}_n(f) = [f(x_1), \ldots, f^{(k_1)}(x_1), \ldots, f(x_s), \ldots, f^{(k_s)}(x_s)]^t,$$

where $n = k_1 + k_2 + \cdots + k_s + s$ and $x_1, x_2, \ldots, x_s$ are distinct points from $[-1,1]$. For simplicity, we assume in this section that

$$(2.4) \qquad\qquad n \leq r.$$

The case $n \geq r$ will be considered for several solution operators in the following sections.

We present a linear central and nearly optimal complexity algorithm for this problem. Let w_j be the Hermite interpolatory polynomial of degree at most $n - 1$ such that

$$(2.5) \qquad \mathfrak{N}_n(w_j) = [0, \ldots, 0, 1, 0, \ldots, 0]^t, \qquad j = 1, 2, \ldots, n.$$
$$\qquad\qquad\qquad\qquad\qquad j$$

Note that $D_p^r w_j = 0$ since $n - 1 < r$. Define the algorithm

$$(2.6) \qquad \varphi(\mathfrak{N}_n(f)) = \sum_{i=1}^{s} \sum_{k=0}^{k_i} f^{(k)}(x_i) S w_{k_0 + \cdots + k_{i-1} + i + k},$$

where $f^{(0)}(x_i) = f(x_i)$ and $k_0 = 0$. Due to Lemma 2.1 of Chapter 3, φ is a linear central interpolatory algorithm and

$$(2.7) \qquad e(\varphi) = r(\mathfrak{N}_n,S,D_p^r) = \tfrac{1}{2}d(\mathfrak{N}_n,S,D_p^r) = \sup\{Sh : h \in \ker \mathfrak{N}_n \cap W_p^r\}.$$

We specify the algorithm φ for several linear functionals S.

(i)   Interpolation: $Sf = f(x_0)$, $x_0 \in [-1,1]$

If x_0 coincides with one of the points $x_1, x_2, \ldots, x_s$, then $r(\mathfrak{N}_n,S,D_p^r) = 0$ and $\varphi(\mathfrak{N}_n(f)) = f(x_0)$. We exclude this case by the assumption that $x_0 \neq x_i$, $i = 1, 2, \ldots, s$.

Hence, we approximate $f(x_0)$ knowing the values of f and possibly its derivatives at s distinct points different from x_0.

If $n < r$, set $h(x) = \prod_{i=1}^s (x - x_i)^{k_i+1}$. Then $ch \in \ker \mathfrak{N}_n \cap W_p^r$ for any arbitrary real c. From this and (2.7), we conclude

$$(2.8) \qquad r(\mathfrak{N}_n,S,D_p^r) = +\infty \qquad \forall n < r.$$

If $n = r$, then any function h from $\ker \mathfrak{N}_n \cap W_p^r$ is of the form $h(x) = g(x) \prod_{i=1}^s (x - x_i)^{k_i+1}$, where $g(x) = h(x,x_1,\ldots,x_1,\ldots,x_s,\ldots,x_s)$ is the nth divided difference of h. For $p = +\infty$, we have $\|g\|_\infty \leq 1/r!$ which yields

$$(2.9) \qquad r(\mathfrak{N}_r,S,D_\infty^r) = \prod_{i=1}^s |x_0 - x_i|^{k_i+1}/r!.$$

For $s = 1$ and arbitrary p, we have

$$h(x) = \frac{1}{(r-1)!} \int_{x_1}^x (x - t)^{r-1} h^{(r)}(t)\, dt,$$

and from the Hölder inequality,

$$(2.10) \qquad |h(x_0)| \leq \frac{1}{(r-1)!} \|h^{(r)}\|_p \left\{ \int_{x_1}^{x_0} |x_0 - t|^{(r-1)q}\, dt \right\}^{1/q},$$

where $1/p + 1/q = 1$. Since (2.10) is sharp and $\|h^{(r)}\|_p \leq 1$, we get

$$(2.11) \qquad r(\mathfrak{N}_r,S,D_p^r) = |x_1 - x_0|^{r-1/p}/((r-1)!\sqrt[q]{(r-1)q+1}).$$

Note that (2.9) and (2.11) coincide for $s = 1$ and $p = +\infty$.

For $s = 1$, the algorithm φ defined by (2.6) has a simple form. Namely, $w_j(x) = (x - x_1)^{j-1}/(j-1)!$ and

$$(2.12) \qquad \varphi(\mathfrak{N}_n(f)) = \sum_{i=1}^{r-1} \frac{f^{(i)}(x_1)}{i!}(x_0 - x_1)^i$$

is the Taylor interpolatory formula. See Bojanov [75], where, in particular, optimality of (2.12) is also established.

We analyze the complexity of this problem for $s = 1$. To find an ε-approximation to $f(x_0)$ for every $f \in W_p^r$, we have to guarantee

$$r(\mathfrak{N}_r, S, D_p^r) < \varepsilon.$$

Due to (2.11), this holds for fixed r and x_1 close enough to x_0 or for fixed x_1 and large r. More precisely, we have to guarantee that

$$(2.13) \qquad |x_1 - x_0| < g_1(\varepsilon) \overset{\mathrm{df}}{=} (\varepsilon(r-1)! \sqrt[q]{(r-1)q+1})^{1/(r-1/p)}$$
$$= \Theta(\varepsilon^{1/(r-1/p)})$$

$$(2.14) \quad g_2(r) \overset{\mathrm{df}}{=} \log(r-1)! + \frac{1}{q}\log((r-1)q+1) + \left(r - \frac{1}{p}\right)\log\frac{1}{|x_1 - x_0|}$$

$$> \log\frac{1}{\varepsilon}.$$

Note that $g_2(r) = r\log r(1 + o(1))$ for large r and therefore (2.14) has a solution for any x_0, x_1, and ε.

If (2.13) or (2.14) hold, then due to the linearity of the optimal error algorithm (2.12), from (2.2) of Chapter 5 we conclude that the ε-complexity of the interpolation problem (S, D_p^r) with the information $\mathfrak{N}_r$ (briefly the ε-complexity of (S, D_p^r) with $\mathfrak{N}_r$) is given by

$$\mathrm{comp}(\mathfrak{N}_r, S, D_p^r, \varepsilon) = (c_1 + a)r - 1,$$

where c_1 is the complexity of evaluating a linear functional of the form $f^{(j)}(x_1)$, $j = 0, 1, \ldots, r-1$, and $a \in [1,2]$. From Lemma 2.1 of Chapter 5, we conclude that algorithm (2.12) is also a nearly optimal complexity algorithm for (S, D_p^r) with $\mathfrak{N}_r$.

We summarize these results in

Theorem 2.1 Consider the interpolation problem for $\mathfrak{I}_0 = W_p^r$ and $\mathfrak{N}_n$ defined by (2.3). Then:

(i) $\qquad r(\mathfrak{N}_n, S, D_p^r) = +\infty \qquad$ for $\quad n < r,$

$$r(\mathfrak{N}_r, S, D_p^r) = |x_1 - x_0|^{r-1/p}/((r-1)! \sqrt[q]{(r-1)q+1})$$
$$\text{for} \quad s = 1, \quad 1/q + 1/p = 1,$$

$$r(\mathfrak{N}_r, S, D_\infty^r) = \prod_{i=1}^{s} |x_0 - x_i|^{k_i+1}/r!.$$

(ii) For $n = r$ and $s = 1$, the Taylor interpolation formula φ defined by (2.12) is a linear central interpolatory and nearly optimal complexity algorithm for (S, D_p^r) with $\mathfrak{N}_r$, $e(\varphi) = r(\mathfrak{N}_r, S, D_p^r)$.

(iii) To find an ε-approximation for $n = r$ and $s = 1$, we have to guarantee

$$|x_1 - x_0| < g_1(\varepsilon) \qquad \text{or} \qquad g_2(r) > \log 1/\varepsilon,$$

where the g_i are defined by (2.13) and (2.14).

(iv)   The ε-complexity of (S,D_p^r) with $\mathfrak{N}_n$ for $n = r$ and $s = 1$ is

$$\text{comp}(\mathfrak{N}_r, S, D_p^r) = (c_1 + a)r - 1,$$

whenever (iii) holds.   ∎

We showed that the Taylor interpolation formula is an optimal error algorithm for W_p^r for any r and p. It is interesting to notice that it is not an optimal error algorithm for the set $\mathfrak{I}_0$ of analytic functions defined and bounded in modulus by unity on the unit disk. For such $\mathfrak{I}_0$, $x_1 = 0$, and a real x_0, Osipenko [72, 76] showed that a unique linear optimal error algorithm is of the form

$$\varphi(\mathfrak{N}_n(f)) = \sum_{i=0}^{n-1} \frac{f^{(i)}(0)}{i!}(1 - x_0^{2(n-i+1)})x_0^i$$

and $e(\varphi) = r(\mathfrak{N}_n, D, I) = |x_0|^n$. The proof is based on a generalization of Smolyak's theorem to the complex case.

The interpolatory problem with the information operator

$$\mathfrak{N}_n(f) = [f(x_1), \ldots, f^{(r-1)}(x_1), \ldots, f(x_s), \ldots, f^{(r-1)}(x_s)]^t$$

for any r and s was considered by Bojanov [75] for $\mathfrak{I}_0 = W_p^r$ and by Forst [77] for $\mathfrak{I}_0 = \tilde{W}_\infty^r$.

(ii)   Differentiation $Sf = f'(0)$

We assume that $k_i = 0$ and n is an odd number, $n = 2k + 1$. Define $x_{2i} = ih$, $i = 0, 1, \ldots, k$, and $x_{2i-1} = -ih$, $i = 1, \ldots, k$, where the parameter $h \in (0, 1/k)$. This means we consider

(2.15) $\mathfrak{N}_n(f) = [f(0), f(-h), f(h), \ldots, f(-kh), f(kh)]^t$.

Thus, for a given h we want to find an approximation of $f'(0)$ knowing function evaluations at the points jh for $j = 0, \pm 1, \ldots, \pm k$. Note that round-off error analysis indicates that h should not be too small.

Define

$$w(x) = x \prod_{i=1}^{k} (x^2 - i^2 h^2).$$

Note that $w \in \ker \mathfrak{N}_n$. If $n < r$, then w also belongs to W_p^r since $w^{(r)} \equiv 0$. Then $cw \in \ker \mathfrak{N}_n \cap W_p^r$ for an arbitrary real c. Since $cw'(0) = c(-1)^k(k!)^2 h^{2k}$ tends to infinity with $c(-1)^k$, this implies

(2.16) $r(\mathfrak{N}_n, S, D_p^r) = +\infty$ $\forall n < r$.

For $n = r$, define

(2.17) $g_j(x) = \dfrac{w(x)}{(x - jh)w'(jh)}$

for $j = 0, \pm 1, \ldots, \pm k$. Note that

(2.18) $$g'_0(0) = 0, \quad g'_j(0) = \frac{(-1)^{j+1}(k!)^2}{jh(k+j)!(k-j)!}.$$

Since $g_j(ih) = \delta_{ij}$, it is easy to show that the w_j defined by (2.5) are now given by

$$w_{2j+1} = g_j, \quad j = 0, 1, \ldots, k, \quad w_{2j} = g_{-j}, \quad j = 1, 2, \ldots, k.$$

Algorithm (2.6) now has the form

(2.19) $$\varphi(\mathfrak{N}(f)) = \sum_{j=-k}^{k} f(jh)g'_j(0)$$

$$= \frac{(k!)^2}{h} \sum_{j=1}^{k} \frac{(-1)^{j+1}}{j(k+j)!(k-j)!}(f(jh) - f(-jh)).$$

This linear central interpolatory algorithm is known as the nth central difference formula. Note that φ does not use the value $f(0)$.

We find its error for $p = +\infty$. Since $f(x) - \sum_{j=-k}^{k} f(jh)g_j(x) = g(x)w(x)$, where $g(x)$ is the rth normalized divided difference of f and $|g(x)| \leq 1/r!$, then

$$f'(0) - \varphi(\mathfrak{N}(f)) = g(0)w'(0) = g(0)(-1)^k(k!)^2h^{2k}.$$

This yields

(2.20) $$e(\varphi) = r(\mathfrak{N}_r, S, D^r_\infty) = h^{r-1}(\lfloor r/2 \rfloor!)^2/r!.$$

Werschulz [77b] considers the dependence of $e(\varphi)$ on h and says φ has order of accuracy p if $e(\varphi) = \Theta(h^p)$. Equation (2.20) agrees with his result that every algorithm which uses the information $\mathfrak{N}_r$ defined by (2.15) has order of accuracy no greater than $r - 1$.

We analyze the complexity of this problem. To find an ε-approximation to $f'(0)$ for every $f \in W^r_\infty$, we require

(2.21) $$h^{r-1}(\lfloor r/2 \rfloor!)^2/r! < \varepsilon.$$

With h, ε fixed, this determines r. Note that h might be chosen as small as possible consistent with good round-off. If (2.21) holds, then the complexity of algorithm (2.19) is

$$\text{comp}(\varphi) = c_1 r + (3r - 5)/2,$$

where c_1 is the complexity of one function evaluation. From this, we conclude that the ε-complexity of (S, D^r_∞) with $\mathfrak{N}_r$ is

$$\text{comp}(\mathfrak{N}_r, S, D^r_\infty, \varepsilon) = (c_1 + a)r - 1,$$

where $a \in [1, 1.5]$. This also shows that algorithm (2.19) is a nearly optimal complexity algorithm for (S, D^r_∞) with $\mathfrak{N}_r$.

We summarize these results in

Theorem 2.2   Consider the differentiation problem for $\mathfrak{I}_0 = W_p^r$ and $\mathfrak{N}_n$ defined by (2.15). Then:

(i) $\qquad\qquad r(\mathfrak{N}_n, S, D_p^r) = +\infty \qquad\qquad\quad$ for $\quad n < r,$

$\qquad\qquad\quad r(\mathfrak{N}_r, S, D_\infty^r) = h^{r-1}(\lfloor r/2 \rfloor!)^2/r! \quad$ for $\quad$ odd $r.$

(ii)   For odd r and $n = r$, the rth central difference formula φ defined by (2.19) is a linear central interpolatory and nearly optimal complexity algorithm for (S, D_∞^r) with $\mathfrak{N}_r$,

$$e(\varphi) = r(\mathfrak{N}_r, S, D_p^r).$$

(iii)   To find an ε-approximation for $n = r$, r odd, $p = +\infty$, we have to guarantee

$$h^{r-1}(\lfloor r/2 \rfloor!)^2/r! < \varepsilon.$$

(iv)   The ε-complexity of (S, D_p^r) with $\mathfrak{N}_n$ for $n = r$, r odd, $p = +\infty$, is

$$\text{comp}(\mathfrak{N}_r, S, D_\infty^r, \varepsilon) = (c_1 + a)r - 1,$$

wherever (iii) holds.   ∎

The differentiation problem has been studied by several people. Some results related to our approach may be found in Arestov [67, 69], Micchelli [76], Pallashke [76], Secrest [65a], Stechkin [67], and Taikov [68].

(iii)   Integration $Sf = \int_{-1}^{+1} \zeta(x) f(x)\, dx$

We consider the *integration problem*

(2.22) $$Sf = \int_{-1}^{+1} \zeta(x) f(x)\, dx,$$

where ζ is a weight function and $f \in W_p^r$. Consider the information operator

(2.23) $$\mathfrak{N}_n(f) = [f(x_1), f(x_2), \ldots, f(x_n)]^t,$$

where the x_i, assumed distinct, belong to $[-1, 1]$. Note that algorithm (2.6) is now of the form

(2.24) $$\varphi(\mathfrak{N}_n(f)) = \sum_{j=1}^{n} f(x_j) \int_{-1}^{+1} \zeta(x) \prod_{i \neq j}(x - x_i) \Big/ \prod_{i \neq j}(x_j - x_i)\, dx.$$

From Lemma 2.2 of Chapter 3, we conclude that if $e(\varphi) < +\infty$, then φ has order of exactness equal to $r - 1$, i.e., $\varphi(\mathfrak{N}_n(f)) = \int_{-1}^{+1} \zeta(x) f(x)\, dx$, whenever f is a polynomial of degree at most $r - 1$.

From (2.7), we have

(2.25) $$e(\varphi) = r(\mathfrak{N}_n, S, D_p^r) = \sup\left\{ \int_{-1}^{+1} \zeta(x) h(x)\, dx : h \in \ker \mathfrak{N}_n \cap W_p^r \right\}.$$

Suppose $n < r/2$. Then $h(x) = c \prod_{i=1}^{n} (x - x_i)^2$ belongs to ker $\mathfrak{N}_n \cap$ ker D_p^r for every c, which yields

$$r(\mathfrak{N}_n, S, D_p^r) = +\infty \qquad \forall n < r/2.$$

Assume then $n \geq r/2$. We want to establish when $r(\mathfrak{N}_n, S, D_p^r)$ is finite. This holds for $n = r$. For $n \leq r - 1$, we proceed as follows. Define

$$(2.26) \qquad q_i(x) = \prod_{j=1}^{i} (x - x_j), \qquad i = 0, 1, \ldots, n,$$

and the inner product

$$(2.27) \qquad (f, g) = \int_{-1}^{+1} \zeta(x) f(x) g(x) \, dx, \qquad f, g \in W_p^r[-1, 1].$$

We prove

$$(2.28) \quad r(\mathfrak{N}_n, S, D_p^r) < +\infty \qquad \text{iff} \quad (q_n, q_i) = 0, \qquad i = 0, 1, \ldots, r - 1 - n.$$

Indeed, suppose that $r(\mathfrak{N}_n, S, D_p^r) < +\infty$. Due to Theorem 3.1 of Chapter 2, we get ker $\mathfrak{N}_n \cap$ ker $D_p^r \subset$ ker S. Note $q_n q_i \in$ ker $\mathfrak{N}_n \cap$ ker D_p^r for $i = 0, 1, \ldots,$ $r - 1 - n$. Thus, $\int_{-1}^{+1} \zeta(x) q_n(x) q_i(x) \, dx = (q_n, q_i) = 0$ which proves (2.28).

Suppose now that $(q_n, q_i) = 0$, $i = 0, 1, \ldots, r - 1 - n$. Then ker $\mathfrak{N}_n \cap$ ker $D_p^r = \mathrm{lin}(q_n q_0, q_n q_1, \ldots, q_n q_{r-1-n}) \subset$ ker S. From Lemma 4.2 of Chapter 2 with $\mathfrak{N} = \mathfrak{N}^*$ and from (2.7), we get

$$r(\mathfrak{N}_n, S, D_p^r) = \left\| S D_p^{-r} \right\|_{D_p^r(\text{ker } \mathfrak{N}_n)}.$$

Since $S D_p^{-r}$ is continuous, $r(\mathfrak{N}_n, S, D_p^r)$ is finite, which completes the proof of (2.28).

If $p = +\infty$, then

$$(2.29) \qquad r(\mathfrak{N}_n, S, D_\infty^r) = \int_{-1}^{+1} \zeta(x) |q_n(x) q_{r-n}(x)| \, dx/r!.$$

Indeed, if $h \in$ ker $\mathfrak{N}_n \cap W_p^r$, then

$$h(x) = c_0 q_n(x) q_0(x) + \cdots + c_{r-1-n} q_n(x) q_{r-1-n}(x) + c_{r-n} q_n(x) q_{r-n}(x),$$

where $c_i = h(x_1, \ldots, x_n, x_1, \ldots, x_{i+1})$ is the $(n + i)$th divided difference of h, $i = 0, 1, \ldots, r - n - 1$, and

$$c_{r-n} = c_{r-n}(x) = h(x_1, \ldots, x_n, x_1, \ldots, x_{r-n}, x).$$

Since $(q_n, q_i) = 0$ for $i = 0, 1, \ldots, r - 1 - n$ and $\|c_{r-n}\|_\infty \leq 1/r!$, we get

$$\left| \int_{-1}^{+1} \zeta(x) h(x) \, dx \right| = \left| \int_{-1}^{+1} \zeta(x) c_{r-n}(x) q_n(x) q_{r-n}(x) \, dx \right|$$

$$\leq \int_{-1}^{+1} \zeta(x) |q_n(x) q_{r-n}(x)| \, dx/r!.$$

This proves (2.29).

We can now ask how to determine where f should be evaluated to minimize the error of algorithm (2.24). This is equivalent to minimizing (2.29), i.e., to find points $x_1, x_2, \ldots, x_n$ for which the functional

$$(2.30) \qquad \int_{-1}^{+1} \zeta(x) \prod_{i=1}^{r-n} (x - x_i)^2 \left| \prod_{i=r-n+1}^{n} (x - x_i) \right| dx$$

is minimized. Note that for $n = r/2$, the solution of (2.30) is given by the zeros of the nth degree orthogonal polynomial with respect to weight ζ.

Algorithm (2.24) is then the Gauss quadrature formula and its order of exactness is equal to $2n - 1$.

For instance, if $\zeta(x) = 1/\sqrt{1 - x^2}$, then

$$(2.31) \qquad x_i = \cos\left(\frac{\pi}{2n} + \frac{i-1}{n}\pi\right), \qquad i = 1, 2, \ldots, n.$$

are the zeros of the Chebyshev polynomial of the first kind,

$$\varphi(\mathfrak{N}_n(f)) = \frac{2}{n} \sum_{i=1}^{n} f(x_i)$$

is known as the Gauss–Chebyshev quadrature formula, and
$$(2.32) \qquad e(\varphi) = r(\mathfrak{N}_n, S, D_\infty^{2n}) = 2\pi/(4^n(2n)!).$$

We analyze the complexity of this problem. To find an ε-approximation, we have to guarantee $r(\mathfrak{N}_n, S, D_p^r) < \varepsilon$. This will hold only for $n \geq r/2$.

Due to the linearity of the optimal error algorithm (2.24), the ε-complexity of (S, D_p^r) with $\mathfrak{N}_n$ is

$$\text{comp}(\mathfrak{N}_n, S, D_p^r, \varepsilon) = (c_1 + a)n - 1,$$

where n is the smallest integer such that $r(\mathfrak{N}_n, S, D_p^r) < \varepsilon$, c_1 is the complexity of one function evaluation, and $a \in [1, 2]$. For instance, if $r = 2n$, $p = +\infty$, $\zeta(x) = 1/\sqrt{1 - x^2}$, and x_i are given by (2.31), we have to guarantee that $2\pi/(4^n(2n)!) < \varepsilon$. Algorithm (2.24) is also a nearly optimal complexity algorithm for (S, D_p^r) with $\mathfrak{N}_n$.

We summarize these results in

Theorem 2.3   Consider the integration problem for $\mathfrak{J}_0 = W_p^r$ and $\mathfrak{N}_n$ defined by (2.23). Then:

(i) $r(\mathfrak{N}_n, S, D_p^r) = +\infty \qquad \forall n < r/2.$
(ii)   For $n \geq r/2$, $r(\mathfrak{N}_n, S, D_p^r) < +\infty$ iff $n = r$ or $(q_n, q_i) = 0$, $i = 0, 1, \ldots, r - 1 - n$, where q_i is defined by (2.26).
(iii)   If $n = r$ or $(q_n, q_i) = 0$ for $i = 0, 1, \ldots, r - 1 - n$, then

$$r(\mathfrak{N}_n, S, D_\infty^r) = \int_{-1}^{+1} \zeta(x) |q_n(x)| dx/r!.$$

(iv)   For $n = r/2$ and $p = +\infty$, the radius of information is minimized if f is evaluated at the zeros of the nth degree orthogonal polynomial related to weight ζ.

(v) φ defined by (2.24) is a linear central interpolatory and nearly optimal complexity algorithm for (S, D_p^r) with $\mathfrak{N}_n$. For $n = r/2$, $p = +\infty$, and $x_1, \ldots, x_n$ equal to the zeros of the nth degree orthogonal polynomial, φ is the Gauss quadrature formula.

(vi)   To find an ε-approximation, we have to guarantee

(2.33) $$r(\mathfrak{N}_n, S, D_p^r) < \varepsilon,$$

which for $n = r/2$, $p = +\infty$, $\zeta(x) = 1/\sqrt{1 - x^2}$, and x_i given by (2.31) implies

$$2\pi/(4^n(2n)!) < \varepsilon.$$

(vii)   The ε-complexity of (S, D_p^r) with $\mathfrak{N}_n$ is

$$\text{comp}(\mathfrak{N}_n, S, D_p^r, \varepsilon) = (c_1 + a)n - 1,$$

where n is the smallest integer such that (2.33) holds.   ∎

3.   INTERPOLATION

In this section, we deal with the *interpolation problem*, i.e.,

(3.1) $$Sf = f(x_0) \qquad \forall f \in \mathfrak{I}_0,$$

where $f: X \to \mathbb{R}$ (or $\mathbb{C}$) is a scalar function.

For the interpolation problem, we shall restrict ourselves to optimal error algorithms which use the information operator

(3.2) $$\mathfrak{N}_n(f) = [f(x_1), f(x_2), \ldots, f(x_n)]^t$$

for $x_1 < x_2 < \cdots < x_n$ from X. Note that $\text{card}(\mathfrak{N}_n) = n$. If one of the points x_i coincides with x_0, then $r(\mathfrak{N}_n, S) = 0$. We exclude this case by assuming that $x_i \neq x_0$, $i = 1, 2, \ldots, n$.

We present optimal error algorithms for the class of smooth functions W_∞^r and for two classes of analytic functions. With one exception, these algorithms are linear and nearly optimal complexity algorithms.

The interpolation problem for different classes of problem elements was considered in many papers. See, among others, Bakhvalov [72], Barnhill and Wixom [68], Bojanov [75], Bojanov and Chernogorov [77], Forst [77], Gaffney [76, 77a,b], Gaffney and Powell [76], Golomb [77], Korobov [63], Melkman [77], Meyers and Sard [50b], Osipenko [72, 76], Richter-Dyn [71b], Secrest [65a], Smolyak [60], and Velikin [77].

(i)   Interpolation for W_∞^r

Let $\mathfrak{I}_1 = W_\infty^r[-1, 1]$ and $x_0 \in [-1, 1]$. Let $T = D_\infty^r$ with $\mathfrak{I}_4 = L_\infty$. Thus,

(3.3) $$\mathfrak{I}_0 = W_\infty^r.$$

Due to (2.8), $r(\mathfrak{N}_n, S, D_\infty^r) = +\infty$ for $n < r$. Therefore, we assume that

(3.4) $$n \geq r.$$

This problem was studied by Gaffney and Powell [76] and Gaffney [76, 77a,b]. They presented an algorithm, which in our terminology is central, defined as follows.

Let u and v be perfect splines of degree r with $n - r$ knots η_i and ξ_i, respectively, $i = 1, 2, \ldots, n - r$, such that

(3.5) $$u(x_i) = v(x_i) = f(x_i), \qquad i = 1, 2, \ldots, n,$$

and the rth derivatives of u and v satisfy the corresponding equations

(3.6) $\quad u^{(r)}(x) = (-1)^i \quad$ for $\quad \eta_i \le x < \eta_{i+1}, \qquad i = 0, 1, \ldots, n - r,$

where $\eta_0 = x_1$ and $\eta_{n-r+1} = x_n$,

(3.7) $\quad v^{(r)}(x) = -(-1)^i \quad$ for $\quad \xi_i \le x < \xi_{i+1}, \qquad i = 0, 1, \ldots, n - r,$

where $\xi_0 = x_1$ and $\xi_{n-r+1} = x_n$. For $r = 1$, the knots η_i and ξ_i are given by

$$\eta_i = \frac{(-1)^{i+1}}{2}(f(x_{i+1}) - f(x_i)) + \frac{x_i + x_{i+1}}{2}, \qquad i = 1, 2, \ldots, n - 1,$$

$$\xi_i = \frac{(-1)^i}{2}(f(x_{i+1}) - f(x_i)) + \frac{x_i + x_{i+1}}{2}, \qquad i = 1, 2, \ldots, n - 1.$$

For $r \ge 2$, the knots η_i and ξ_i can be obtained by solving the systems of non-linear equations

$$\sum_{j=0}^{n-r} (-1)^j \int_{\eta_j}^{\eta_{j+1}} M_{r,i}(x)\,dx = (r - 1)!\, f(x_i, \ldots, x_{i+r})$$

and

$$\sum_{j=0}^{n-r} (-1)^{j+1} \int_{\xi_j}^{\xi_{j+1}} M_{r,i}(x)\,dx = (r - 1)!\, f(x_i, \ldots, x_{i+r}),$$

where $f(x_i, \ldots, x_{i+r})$ is the rth divided difference of f and $M_{r,i}$ is a B-spline equal to the rth divided difference of $g_r(x;y) = (y - x)_+^{r-1}$, i.e., $M_{r,i}(x) = g_r(x; x_i, \ldots, x_{i+r})$. Define

(3.8) $$\underline{f}(x_0) = \min(u(x_0), v(x_0)), \qquad \bar{f}(x_0) = \max(u(x_0), v(x_0)).$$

Gaffney and Powell [76] proved that

(3.9) $$\underline{f}(x_0) \le f(x_0) \le \bar{f}(x_0)$$

and that both inequalities are sharp. In our notation, this means that the set $U(f)$ of solutions $\tilde{f}(x_0)$, where $\tilde{f}$ shares the same information as f, is

$$U(f) = [\underline{f}(x_0), \bar{f}(x_0)].$$

Then the algorithm

(3.10) $$\varphi(\mathfrak{N}(f)) = (\underline{f}(x_0) + \bar{f}(x_0))/2 = (u(x_0) + v(x_0))/2$$

is central and interpolatory. Furthermore, the local error of φ (see Remark 2.2 of Chapter 1) is

$$e(\varphi, f) = \text{rad } U(f) = \tfrac{1}{2}|u(x_0) - v(x_0)|.$$

In general, the algorithm φ is nonlinear and the computation of φ requires the solution of two nonlinear systems of size $n - r$. To solve these systems, Gaffney [76] proposed using Newton iteration and taking advantage of the band structure of the Jacobian matrix.

Due to Smolyak's theorem, there exists a linear optimal algorithm (which is not central). Gaffney and Powell [76] also derived this algorithm. Namely, let $\eta_1^* < \eta_2^* < \cdots < \eta_{n-r}^*$, $x_i < \eta_i^* < x_{i+r}$, $i = 1, 2, \ldots, n - r$, be defined by the equations

$$(3.11) \qquad \sum_{j=0}^{n-r} (-1)^j \int_{\eta_j^*}^{\eta_{j+1}^*} M_{r,i}(x)\, dx = 0, \qquad i = 1, 2, \ldots, n - r,$$

where $\eta_0^* = x_1$ and $\eta_{n-r+1}^* = x_n$. For instance, for $r = 1$, $\eta_i^* = (x_i + x_{i+1})/2$. Note that the knots η_i^* do not depend on f and therefore can be precomputed.

When the knots η_i^* have been found, define a spline function σ_i of degree $r - 1$ with $(n - k)$ knots $\eta_1^*, \ldots, \eta_{n-r}^*$ such that

$$\sigma_i(x_j) = \delta_{ij}, \qquad i, j = 1, 2, \ldots, n.$$

Then the algorithm

$$(3.12) \qquad \varphi(\mathfrak{N}(f)) = \sum_{i=1}^{n} f(x_i)\sigma_i(x_0)$$

is a linear optimal error algorithm and

$$(3.13) \qquad e(\varphi) = r(\mathfrak{N}_n, S, D_\infty^r) = |q_n(x_0)|,$$

where q_n is a perfect spline of degree r with $(n - r)$ knots $\eta_1^*, \ldots, \eta_{n-r}^*$ and satisfies the conditions

$$(3.14) \qquad \begin{aligned} q_n(x_i) &= 0, & & i = 1, 2, \ldots, n, \\ q_n^{(r)}(x) &= (-1)^i, & \eta_i^* \leq x < \eta_{i+1}^*, & i = 0, 1, \ldots, n - r. \end{aligned}$$

The radius of information depends on the points $x_0, x_1, \ldots, x_n$. We have

$$(3.15) \qquad r(\mathfrak{N}_n, S, D_\infty^r) = |q_n(x_0)| \leq \sup_{-1 \leq x \leq 1} |q_n(x)| = \|q_n\|_\infty.$$

It was shown by Micchelli, Rivlin, and Winograd [76] that $\|q_n\|_\infty$ is equal to the radius of information $\mathfrak{N}_n$ for the approximation problem with $\mathfrak{I}_0 = W_\infty^r$ in the space $\mathfrak{I}_2 = C[-1,1]$. In particular, they showed that for equidistant points $x_i = -1 + 2(i - 1)/n$, $i = 1, 2, \ldots, n$, $\|q_n\|_\infty = \Theta(n^{-r})$. We show in Section 5 that for optimally chosen points x_i,

$$\|q_n\|_\infty = (2/\pi n)^r K_r (1 + o(1)),$$

where K_r is the Favard constant defined as

$$(3.16) \qquad K_r = \frac{4}{\pi} \sum_{i=0}^{\infty} \frac{(-1)^{i(r+1)}}{(2i+1)^{r+1}}$$

Observe that $K_0 = 1$, $K_1 = \pi/2$, $K_2 = \pi^2/8$, $K_3 = \pi^3/24$, and $1 = K_0 < K_2 < \cdots < 4\pi < \cdots < K_3 < K_1 = \pi/2$.

We analyze the complexity of this problem. From (3.15), it follows that the ε-complexity of the interpolation problem is no greater than the ε-complexity of the corresponding approximation problem. For instance, consider equidistant points $x_i = -1 + 2(i-1)/n$. To find an ε-approximation, we have to guarantee that

$$|q_n(x_0)| < \varepsilon.$$

Since $\|q_n\| = \Theta(n^{-r})$, this holds for $n = n(x_0)$ such that

$$n = n(x_0) = O(\varepsilon^{-1/r}).$$

Furthermore, there exists a point x_0, $|q(x_0)| = \|q\|$, such that

$$n = n(x_0) = \Theta(\varepsilon^{-1/r}).$$

Due to the linearity of the optimal error algorithm (3.12), the ε-complexity of (S,D_∞^r) with $\mathfrak{N}_n$ is

$$\mathrm{comp}(\mathfrak{N}_n,S,D_\infty^r,\varepsilon) = (c_1 + a)n - 1,$$

where n is the smallest integer such that $|q_n(x_0)| < \varepsilon$, c_1 is the complexity of one function evaluation, and $a \in [1,2]$. Algorithm (3.12) is a nearly optimal complexity algorithm for (S,D_∞^r) with $\mathfrak{N}_n$.

We summarize these results in

Theorem 3.1   Consider the interpolation problem for $\mathfrak{J}_0 = W_\infty^r$ and $\mathfrak{N}_n$ defined by (3.2). Then:

(i) $r(\mathfrak{N}_n,S,D_\infty^r) = +\infty$ for $n < r$, and $r(\mathfrak{N}_n,S,D_\infty^r) = |q_n(x_0)|$, where q is a perfect spline defined by (3.14).

(ii) φ defined by (3.10) is a central interpolatory algorithm, $e(\varphi) = r(\mathfrak{N}_n,S,D_\infty^r)$.

(iii) φ defined by (3.12) is a linear optimal error and nearly optimal complexity algorithm for (S,D_∞^r) with $\mathfrak{N}_n$, $e(\varphi) = r(\mathfrak{N}_n,S,D_\infty^r)$.

(iv)   To find an ε-approximation, we have to guarantee $|q_n(x_0)| < \varepsilon$ which for equidistant points $x_i = -1 + 2(i-1)/n$, $i = 1, 2, \ldots, n$, implies $n = n(x_0) = O(\varepsilon^{-1/r})$.

(v)   The ε-complexity of (S,D_∞^r) with $\mathfrak{N}_n$ is

$$\mathrm{comp}(\mathfrak{N}_n,S,D_\infty^r,\varepsilon) = (c_1 + a)n - 1,$$

where n is the smallest integer such that $|q_n(x_0)| < \varepsilon$.   ∎

(ii) Interpolation for Analytic Functions

Let $\mathfrak{I}_1$ be the class of analytic functions f, $f:G \to \mathbb{C}$, where G is a simply connected region. Let $T = I$ be the identity operator and $\mathfrak{I}_4 = \mathfrak{I}_1$ equipped with the sup norm $\|f\| = \sup\{|f(z)|:z \in G\}$. Then

(3.17) $$\mathfrak{I}_0 = \{f : f \text{ is analytic in } G \text{ and } \|f\| \leq 1\}.$$

Consider the information operator

(3.18) $$\mathfrak{N}_n(f) = [f(x_1), f(x_2), \ldots, f(x_n)]^t$$

for distinct points x_j from G.

This problem was studied by Osipenko [76], and we report his results in our terminology. The radius of information is

(3.19) $$r(\mathfrak{N}_n, S, I) = \sup\{|h(x_0)| : h(x_j) = 0, j = 1, 2, \ldots, n, \|h\| \leq 1\}.$$

From the maximum modulus principle, $h(z) = g(z)\prod_{j=1}^{n} W_j(z)$, where $g \in \mathfrak{I}_0$ and W_j is a conformal mapping of G onto the unit disk such that $W_j(x_j) = 0$. Thus,

(3.20) $$r(\mathfrak{N}_n, S, I) = \prod_{j=1}^{n} |W_j(x_0)|.$$

Osipenko [76] generalizes the theorem of Smolyak to the complex case, and using this generalization he concludes that the algorithm

(3.21) $$\varphi(\mathfrak{N}_n(f)) = \sum_{j=1}^{n} f(x_j)g_j(z_0)$$

is a linear optimal error algorithm, where

(3.22) $$g_j(z) = \frac{w_j(z)}{w_j(x_j)}(1 - |W_j(z)|^2), \qquad w_j(z) = \prod_{k \neq j} W_k(z).$$

Note that $g_j(x_s) = \delta_{sj}$.

As an example, assume that $G = \{z : |z| < 1\}$ is the unit disk. Then $W_j(z) = (z - x_j)/(1 - x_j z)$ and

(3.23) $$r(\mathfrak{N}_n, S, I) = \prod_{j=1}^{n} \left| \frac{x_0 - x_j}{1 - x_0 x_j} \right|.$$

If

(3.24) $$x_0 = 0 \quad \text{and} \quad x_j = \zeta e^{2\Pi i j/n}$$

for $\zeta \in (0,1)$, $j = 1, 2, \ldots, n$, $i = \sqrt{-1}$, then

$$r(\mathfrak{N}_n, S, I) = \zeta^n.$$

We analyze the complexity of this problem. To find an ε-approximation, we have to guarantee

(3.25)
$$\prod_{j=1}^{n} |W_j(x_0)| < \varepsilon.$$

If G is the unit disk and (3.24) holds, then (3.25) implies

$$\zeta^n < \varepsilon.$$

For fixed ζ, this determines n,

$$n = \lfloor \log(1/\varepsilon)/\log(1/\zeta) \rfloor + 1.$$

For fixed n, this determines ζ,

$$\zeta < \varepsilon^{1/n}.$$

Due to the linearity of the optimal error algorithm (3.21), the ε-complexity of (S,I) with $\mathfrak{N}_n$ is

$$\text{comp}(\mathfrak{N}_n,S,I,\varepsilon) = (c_1 + a)n - 1,$$

where n is the smallest integer such that (3.25) holds, c_1 is the complexity of one function evaluation, and $a \in [1,2]$. Algorithm (3.21) is a nearly optimal complexity algorithm for (S,I) with $\mathfrak{N}_n$.

We summarize these results in

Theorem 3.2   Consider the interpolation problem for $\mathfrak{I}_0$ and $\mathfrak{N}_n$ defined by (3.17) and (3.18). Then:

   (i) $r(\mathfrak{N}_n,S,I) = \prod_{j=1}^{n} |W_j(x_0)|$.
   (ii) φ defined by (3.21) is a linear optimal error and nearly optimal complexity algorithm for (S,I) with $\mathfrak{N}_n$, $e(\varphi) = r(\mathfrak{N}_n,S,I)$.
   (iii)   To find an ε-approximation, we have to guarantee $\prod_{j=1}^{n} |W_j(x_0)| < \varepsilon$ which for the unit disk and points (3.24) means $\zeta^n < \varepsilon$.
   (iv)   The ε-complexity of (S,I) with $\mathfrak{N}_n$ is

$$\text{comp}(\mathfrak{N}_n,S,I,\varepsilon) = (c_1 + a)n - 1,$$

where n is the smallest integer such that (iii) holds. In particular, if G is the unit disk and $x_j, j = 1, 2, \ldots, n$, are defined by (3.24) with fixed ζ, then

$$\text{comp}(\mathfrak{N}_n,S,I,\varepsilon) = (c_1 + a)(\lfloor \log(1/\varepsilon)/\log(1/\zeta) \rfloor + 1) - 1. \quad \blacksquare$$

(iii)   Interpolation for a Hilbert Space with a Reproducing Kernel

Following Golomb [77], let $\mathfrak{I}_1$ be a Hilbert space of complex-valued functions f, $f: X \to \mathbb{C}$, which has a reproducing kernel $k: X \times X \to \mathbb{C}$, i.e.,

(3.26) $f(x) = (f,k(\cdot,x))$ $\forall f \in \mathfrak{I}_1, \quad \forall x \in X,$

where $(\cdot,\cdot)$ is the inner product in $\mathfrak{I}_1$. Let $T = I$ with $\mathfrak{I}_4 = \mathfrak{I}_1$. Then

(3.27)
$$\mathfrak{I}_0 = \{f : \|f\| = \sqrt{(f,f)} \le 1\}.$$

Consider the information operator

(3.28)
$$\mathfrak{N}_n(f) = [f(x_1), f(x_2), \ldots, f(x_n)]^t$$

for distinct points x_i from X. The radius of information is

$$r(\mathfrak{N}_n, S, I) = \sup_{\substack{h(x_j)=0 \\ \|h\| \le 1}} |h(x_0)| = \sup_{\substack{h(x_j)=0 \\ h \in \mathfrak{I}_1}} |h(x_0)|/\|h\| = 1 \bigg/ \inf_{\substack{g(x_0)=1 \\ g(x_j)=0,\, g \in \mathfrak{I}_1}} \|g\|.$$

Let c_{x_0} be a spline (see Definition 3.1 of Chapter 4) such that

(3.29)
$$c_{x_0}(x_0) = 1 \quad \text{and} \quad c_{x_0}(x_j) = 0, \quad j = 1, 2, \ldots, n.$$

This means that $\|c_{x_0}\| \le \|g\|$ for any g satisfying (3.29). Thus, we conclude

(3.30)
$$r(\mathfrak{N}_n, S, I) = \|c_{x_0}\|^{-1}.$$

Consider the function

(3.31)
$$\sigma(x) = \sum_{j=1}^{n} a_j k(x, x_j),$$

where a_j are chosen such that

(3.32)
$$\sigma(x_j) = f(x_j), \quad j = 1, 2, \ldots, n.$$

There always exists such a unique σ since $k(\cdot, x_1), \ldots, k(\cdot, x_n)$ are linearly independent and the Gram matrix $(k(x_i, x_j))$ is nonsingular. We show that σ is a spline, i.e.,

$$\|\sigma\| \le \|g\|$$

for any function g from $\mathfrak{I}_1$ satisfying (3.32). Indeed, let $h = g - \sigma$. Then $h(x_j) = 0$, $j = 1, 2, \ldots, n$, and $(h, \sigma) = \sum_{j=1}^{n} \bar{a}_j(h, k(\cdot, x_j))$. Due to (3.26), $(h, k(\cdot, x_j)) = h(x_j) = 0$ and $(h, \sigma) = 0$. Thus, σ is orthogonal to $\ker \mathfrak{N}$ and

$$\|g\|^2 = \|g - \sigma + \sigma\|^2 = \|g - \sigma\|^2 + \|\sigma\|^2 \ge \|\sigma\|^2.$$

This proves that σ is a spline. Define the spline algorithm (see Section 4 of Chapter 4)

(3.33)
$$\varphi(\mathfrak{N}(f)) = \sigma(x_0) = \sum_{j=1}^{n} a_j k(x_0, x_j).$$

Theorem 5.1 of Chapter 4 yields that φ is central and

$$e(\varphi, f) = (1 - \|\sigma(x_0)\|^2)^{1/2} \|c_{x_0}\|^{-1}.$$

As in Golomb [77], consider the following example. Let $X = X_R = \{z : |z| < R\}$ be the open disk. Define

$$(f,g) = \frac{1}{2\pi R} \int_{\partial X_R} f(z)\overline{g(z)}|dz|,$$

(3.34) $\mathfrak{I}_1 = \{f : f \text{ analytic in } X_R \text{ and } \|f\| = \sqrt{(f,f)} < +\infty\}.$

Then $\mathfrak{I}_1$ is a Hilbert space and $\xi_j(z) = z^j/R^j$ form a complete orthonormal system in $\mathfrak{I}_1$. Define

$$k(z,x) = \sum_{j=0}^{\infty} \left(\frac{z}{R}\right)^j \left(\frac{\overline{x}}{R}\right)^j = \frac{1}{1 - z\overline{x}/R^2}.$$

Then $(f,k(\cdot,x)) = (\sum_{j=0}^{\infty} a_j \xi_j, \sum_{j=0}^{\infty} (\overline{x}/R)^j \xi_j) = \sum_{j=0}^{\infty} a_j (x/R)^j = f(x)$ which shows that k is a reproducing kernel of $\mathfrak{I}_1$.

Golomb [77] derives the explicit formulas for $\|c_{x_0}\|$ and the spline algorithm. Namely,

(3.35) $r(\mathfrak{N}_n,S,I) = \|c_{x_0}\|^{-1} = \dfrac{R^{-n}}{(1 - |x_0|^2/R^2)^{3/4}} \prod_{j=1}^{n} \dfrac{|x_0 - x_j|}{|1 - x_0\overline{x}_j/R^2|},$

and

(3.36) $\varphi(\mathfrak{N}_n(f)) = \sum_{j=1}^{n} f(z_j) \dfrac{\prod_{i \neq j}(x_0 - x_i)\prod_i(x_i - R^2/\overline{x}_i)}{\prod_i(x_0 - R^2/\overline{x}_i)\prod_{i \neq j}(x_j - x_i)}.$

Note that if $R \to +\infty$, then

$$\lim_{R \to \infty} R^n r(\mathfrak{N}_n,S,I) = \prod_{j=1}^{n} |x_0 - x_j|,$$

$$\lim_{R \to \infty} \varphi(\mathfrak{N}_n(f)) = \sum_{j=1}^{n} f(z_j) \frac{\prod_{i \neq j}(x_0 - x_i)}{\prod_{i \neq j}(x_j - x_i)}.$$

Thus, the algorithm φ converges to the Lagrange interpolation formula.

For the points

(3.37) $x_0 = 0,$ $x_j = \zeta e^{2\pi i j/n},$ $j = 1, 2, \ldots, n,$ $i = \sqrt{-1},$

where $\zeta \in (0,R)$, we get from (3.35),

(3.38) $r(\mathfrak{N}_n,S,I) = (\zeta/R)^n.$

We analyze the complexity of this problem. To find an ε-approximation, we have to guarantee that

(3.39) $\|c_{x_0}\|^{-1} < \varepsilon.$

If $\mathfrak{I}_1$ is given by (3.34) and x_j by (3.37), then (3.39) implies

$$(\zeta/R)^n < \varepsilon.$$

For fixed ζ, this determines n,

$$n = \lfloor \log(1/\varepsilon)/\log(R/\zeta) \rfloor + 1.$$

For fixed n, this determines ζ,

$$\zeta < R\varepsilon^{1/n}.$$

Due to the linearity of the optimal error algorithm (3.33), the ε-complexity of (S,I) with $\mathfrak{N}_n$ is

$$\text{comp}(\mathfrak{N}_n,S,I,\varepsilon) = (c_1 + a)n - 1,$$

where n is the smallest integer such that (3.39) holds, c_1 is the complexity of one function evaluation, and $a \in [1,2]$. Algorithm (3.33) is a nearly optimal complexity algorithm for (S,I) with $\mathfrak{N}_n$.

We summarize these results in

Theorem 3.3 Consider the interpolation problem for $\mathfrak{I}_0$ and $\mathfrak{N}_n$ defined by (3.27) and (3.28). Then:

(i) $r(\mathfrak{N}_n,S,I) = \|c_{x_0}\|^{-1}$, where c_{x_0} is a spline defined by (3.29).
(ii) φ defined by (3.33) is a linear central spline and nearly optimal complexity algorithm for (S,I) with $\mathfrak{N}_n$, $e(\varphi) = r(\mathfrak{N}_n,S,I)$.
(iii)   To find an ε-approximation, we have to guarantee $\|c_{x_0}\|^{-1} < \varepsilon$ which for $\mathfrak{I}_1$ given by (3.34) and x_j given by (3.37) means $(\zeta/R)^n < \varepsilon$.
(iv)   The ε-complexity of (S,I) with $\mathfrak{N}_n$ is

$$\text{comp}(\mathfrak{N}_n,S,I,\varepsilon) = (c_1 + a)n - 1,$$

where n is the smallest integer such that (iii) holds. In particular, for (3.34) and (3.37) with fixed ζ,

$$\text{comp}(\mathfrak{N}_n,S,I,\varepsilon) = (c_1 + a)(\lfloor \log(1/\varepsilon)/\log(R/\zeta) \rfloor + 1) - 1. \quad \blacksquare$$

4.   INTEGRATION

In this section, we consider the *integration problem*, i.e.,

$$(4.1) \qquad Sf = \int_a^b f(x)\, dx \qquad \forall f \in \mathfrak{I}_0,$$

where $f:[a,b] \to \mathbb{R}$ is a scalar function. We shall set $[a,b] = [0,1]$ in subsections (i) and (iii) and $[a,b] = [0,2\pi]$ in subsection (ii).

Let Ψ_n^ζ be the class of all information operators $\mathfrak{N}_n$ of the form

$$(4.2) \qquad \mathfrak{N}_n(f) = [f(x_1), \ldots, f^{(\zeta)}(x_1), \ldots, f(x_m), \ldots, f^{(\zeta)}(x_m)]^t,$$
$$n = (\zeta + 1)m,$$

for distinct points x_i from $[a,b]$, $x_1 < x_2 < \cdots < x_m$.

We seek points x_i for which the radius of information is minimized. More precisely, let

(4.3) $$r(\Psi_n^\zeta, S, \mathfrak{I}_0) = \inf_{\mathfrak{R}_n \in \Psi_n^\zeta} r(\mathfrak{R}_n, S)$$

be the nth *minimal radius of information for the integration problem in the class* Ψ_n^ζ (also briefly the nth *minimal radius*). An information operator $\mathfrak{R}_n$ is called an nth *optimal information in* Ψ_n^ζ iff

$$r(\mathfrak{R}_n, S) = r(\Psi_n^\zeta, S, \mathfrak{I}_0).$$

If $\mathfrak{I}_0 = \{f : \|Tf\| \le 1\}$, where T is a linear restriction operator, then we shall sometimes write $r(\Psi_n^\zeta, S, T)$ instead of $r(\Psi_n^\zeta, S, \mathfrak{I}_0)$. Observe that

$$r(\Psi_n^0, S, \mathfrak{I}_0) \le r(\Psi_n^\zeta, S, \mathfrak{I}_0).$$

As we proved in Section 3 of Chapter 3, the problems of determining the nth minimal radius and an nth optimal information operator is equivalent to the Sard problem for $\mathfrak{I}_0 = W_2^r$ and to the Nikolskij problem for a balanced convex $\mathfrak{I}_0$.

We shall give the nth minimal radius and an nth optimal information operator for the class $\mathfrak{I}_0 = W_p^r$ assuming that $\zeta = r - 1$ or $\zeta = r - 2$ with even r, and also for the class $\mathfrak{I}_0 = \tilde{W}_\infty^r$ assuming $\zeta = 0$. We find the asymptotic behavior of the nth minimal radius for $\mathfrak{I}_0 = W_p^r$ and for any ζ. We also show that the nth minimal radius is related to the Gelfand n-width of the corresponding approximation problem.

The integration problem has been considered by very many people. We give only a partial list of papers in which the integration problem for scalar and multivariate cases was studied from a point of view related to ours: Alhimova [72], Aksen and Tureckij [66], Babenko [76], Babuška [68a,b], Bakhvalov [59, 61, 62b, 63, 64, 67, 70, 72], Barnhill [67, 68], Barnhill and Wixom [67], Barrar, Loeb, and Werner [74], Bojanov [73, 74, 76], Busarova [73], Chawla and Kaul [73], Chentsov [61], Coman [72], Coman and Micula [71], Eckardt [68], Elhay [69], Forst [75], Gaisarian [69], Haber [71], Ibragimov and Aliev [65], Ivanov [72b], Jetter [76], Johnson and Riess [71], Karlin [69, 71], Kautsky [70], Keast [73], Kiefer [57], Kornejčuk [68, 74], Kornejčuk and Lušpaj [69], Korobov [63], Krylov [62], Larkin [70], Lee [77], Levin and Giršovič [77], Levin, Giršovič, and Arro [76], Ligun [76], Lipow [73], Loeb and Werner [74], Lušpaj [66, 69, 74], Mansfield [71, 72], Maung Čžo Njun and Sharygin [75], Meyers and Sard [50a], Motornyj [73, 74, 76], Nikolskij [50, 58], Pallashke [76], Paulik [77], Pinkus [75], Richter [70], Richter-Dyn [71a], Šajdaeva [59], Sard [49, 63], Schmeisser [72], Schoenberg [64b, 65, 66, 69, 70], Secrest [64, 65b], Sharygin [63, 77], Smolyak [60], Sobol [69], Sobolev [65, 74], Stenger [78], Stern [67], Stetter [69], Sukharev [78b], Tikhonov and Gaisarian [69], Wilf [64], Žensykbaev [76, 77, 78], and Zhileikin and Kukarkin [78].

(i) Integration for W_p^r

Here we consider the integration problem of the form $Sf = \int_0^1 f(x)\,dx$. Let $\mathfrak{I}_1 = W_p^r[0,1]$. Define $Tf = D_p^r f = f^{(r)}$ with $\mathfrak{I}_4 = L_p$. Thus,

$$\mathfrak{I}_0 = \{f : \|f^{(r)}\|_p \le 1\} = W_p^r.$$

Let

$$\mathfrak{N}_n(f) = [f(x_1), \dots, f^{(\zeta)}(x_1), \dots, f(x_m), \dots, f^{(\zeta)}(x_m)]^t, \qquad n = m(\zeta + 1),$$

for $f \in \mathfrak{I}_0$. We assume that $\zeta \le r - 1$ to guarantee the existence of $f^{(\zeta)}(x_i)$. From Lemma 3.1 of Chapter 3, we have

$$r(\mathfrak{N}_n, S, D_p^r) = \sup\left\{\int_0^1 h(x)\,dx : h \in \ker \mathfrak{N}_n \cap W_p^r\right\}.$$

From Theorem 3.1 and Lemma 4.2 of Chapter 2, it follows that

(4.4) $\qquad r(\mathfrak{N}_n, S, D_p^r) < +\infty \qquad$ iff $\quad \ker \mathfrak{N}_n \cap \ker D_p^r \subset \ker S$.

Let

$$q(x) = \prod_{i=1}^m (x - x_i)^{\zeta + 1} \qquad \text{and} \qquad q_j(x) = \prod_{i=1}^j (x - x_i), \qquad j = 0, 1, \dots, m.$$

We prove:

if ζ is odd, then

(4.5) $\qquad\qquad\qquad r(\mathfrak{N}_n, S, D_p^r) < +\infty \qquad$ iff $\quad n \ge r$;

if ζ is even, then

(4.6) $\quad r(\mathfrak{N}_n, S, D_p^r) < +\infty \qquad$ iff $\quad n + m \ge r \quad$ and $\quad \int_0^1 q(x)q_i(x)\,dx = 0$,

$$i = 0, 1, \dots, r - 1 - n.$$

Indeed, let ζ be odd. Then $q \in \ker \mathfrak{N}_n$ and $q \notin \ker S$. Since $q \in \ker D_p^r$ iff $n < r$, (4.5) holds. Assume now that ζ is even and $r(\mathfrak{N}_n, S, D_p^r) < +\infty$. Note that $qq_i \in \ker \mathfrak{N}_n$ for $i = 0, 1, \dots, m$. If $n + i < r$, then $qq_i \in \ker D_p^r$ and (4.4) implies $\int_0^1 q(x)q_i(x)\,dx = 0$. Since $qq_m \notin \ker S$, we have $n + m \ge r$. This proves the right-hand side of (4.6).

Suppose finally that $n + m \ge r$ and $qq_i \in \ker S$, $i = 0, 1, \dots, r - 1 - n$. Then every $h \in \ker \mathfrak{N}_n \cap \ker D_p^r$ can be expressed as

$$h(x) = c_0 q_n(x)q_0(x) + \cdots + c_{r-1-n} q_n(x)q_{r-1-n}(x)$$

and $\int_0^1 h(x)\,dx = \sum_{i=0}^{r-1-n} c_i \int_0^1 q_n(x)q_i(x)\,dx = 0$ which shows that $h \in \ker S$. Due to (4.4), this proves that $r(\mathfrak{N}_n, S, D_p^r)$ is finite. Hence, (4.6) is proven.

From now on, we assume that $r(\mathfrak{N}_n, S, D_p^r) < +\infty$. Due to the theorem of Smolyak (Section 3 of Chapter 3), there exists a linear optimal error algorithm

φ which uses $\mathfrak{N}_n$,

(4.7) $$\varphi(\mathfrak{N}_n(f)) = \sum_{i=1}^{m} \sum_{k=0}^{\zeta} f^{(k)}(x_i) q_{i,k}, \qquad e(\varphi) = r(\mathfrak{N}_n, S, D_p^r).$$

We remark that if $p = 2$, then φ is a spline algorithm which is also central and interpolatory (see Theorem 5.1 of Chapter 4). In this case, $\varphi(\mathfrak{N}_n(f)) = \int_0^1 \sigma(\mathfrak{N}_n(f), x) \, dx$, where $\sigma = \sigma(\mathfrak{N}_n(f), \cdot)$ is the natural spline of degree $2r - 1$ with knots $x_1, \ldots, x_m$ of multiplicity $(\zeta + 1)$. (See also Schoenberg [64b].) As we mentioned in Section 3 of Chapter 3, φ is also an optimal algorithm in the sense of Sard.

We come back to the general case, i.e., $p \in [1, +\infty]$. From Lemma 2.2 of Chapter 3, we conclude that

(4.8) $$\varphi(\mathfrak{N}_n(f)) = \int_0^1 f(x) \, dx \qquad \forall f \in \ker D_p^r,$$

i.e., φ is exact for polynomials of degree at most $r - 1$. It is known (see Kornejčuk [74, pp. 139–165]) that

$$\int_0^1 f(t) \, dt - \varphi(\mathfrak{N}_n(f)) = \frac{(-1)^r}{r!} \int_0^1 G_r(t) f^{(r)}(t) \, dt,$$

where

(4.9) $$G_r(t) = (t - 1)^r - (-1)^r \sum_{i=1}^{m} \sum_{k=0}^{\zeta} \frac{r!}{(r - k - 1)!} q_{i,k} (x_i - t)_+^{r-k-1}.$$

Furthermore,

(4.10) $$e(\varphi) = r(\mathfrak{N}_n, S, D_p^r) = (1/r!) \|G_r\|_q,$$

where $1/q + 1/p = 1$.

Note that G_r is a spline function which belongs to the class $A_n^\zeta(\mathbf{x})$, $\mathbf{x} = (x_1, x_2, \ldots, x_m)$, defined as follows. A function $\varphi : [0,1] \to \mathbb{R}$ belongs to $A_n^\zeta(\mathbf{x})$ iff

$$\varphi(t) = \begin{cases} t^r, & 0 \le t \le x_1, \\ t^r - \sum_{i=0}^{r-1} a_{k,i} t^i, & x_k < t \le x_{k+1}, \quad k = 1, 2, \ldots, m - 1, \\ (t - 1)^r, & x_m < t \le 1, \end{cases}$$

and if $\zeta = r - 2$, then φ is continuous on $[0,1]$, while if $\zeta < r - 2$, then φ has continuous derivatives up to order $r - \zeta - 2$.

The coefficients $q_{i,k}$ of the optimal error algorithm φ defined by (4.7) enjoy the property

(4.11) $$\|G_r\|_q = \inf_{\varphi \in A_n^\zeta(\mathbf{x})} \|\varphi\|_q,$$

i.e., G_r is a spline with minimal L_q norm among all splines from the class $A_n^\zeta(\mathbf{x})$.

We seek points $x_1, x_2, \ldots, x_m$ which minimize the radius of information, i.e., we wish to find

(4.12) $$r(\Psi_n^\zeta, S, D_p^r) = \inf_{\mathfrak{N}_n \in \Psi_n^\zeta} r(\mathfrak{N}_n, S, D_p^r),$$

and an optimal information operator $\mathfrak{N}_n^*$ in Ψ_n^ζ,

(4.13) $$r(\mathfrak{N}_n^*, S, D_p^r) = r(\Psi_n^\zeta, S, D_p^r).$$

Due to the linearity of the optimal error algorithm and due to (4.10) and (4.11), we get

(4.14) $$r(\Psi_n^\zeta, S, D_p^r) = \inf_{\mathbf{x}} \inf_{\varphi \in A_n^\zeta(\mathbf{x})} \|\varphi\|_q.$$

Essentially, the points of an nth optimal information are the knots of a spline with minimal L_q norm and the nth minimal radius of information is the norm of this spline.

The problem of optimal information operators in Ψ_n^ζ is not solved for arbitrary values of ζ, $\zeta \le r - 1$, and n. We report the solution of this problem for $\zeta = r - 1$ and for $\zeta = r - 2$ with even r following Kornejčuk [74].

Let

(4.15) $$R_{r,q}(x) = x^r - \sum_{i=0}^{r-1} \beta_i x^i$$

be the polynomial with minimal norm in $L_q[-1,1]$, i.e.,

$$\|R_{r,q}\|_q = \left(\int_{-1}^{+1} |R_{r,q}(x)|^q \, dx \right)^{1/q} = \inf_{h \in \ker D_p^r} \|x^r - h\|_q.$$

Define

(4.16) $$h = (2(m-1) + 2[R_{r,q}(1)]^{1/r})^{-1},$$

(4.17) $$x_i^* = h(2(i-1) + [R_{r,q}(1)]^{1/r}), \qquad i = 1, 2, \ldots, m,$$

(4.18) $$q_{1,k}^* = (-1)^k q_{m,k}^* = h^{k+1} \left\{ \frac{(-1)^k}{(k+1)!} [R_{r,q}(1)]^{(k+1)/r} + \frac{1}{r!} R_{r,q}^{(r-k-1)}(1) \right\},$$

$$k = 0, 1, \ldots, \zeta,$$

(4.19) $$q_{i,2k}^* = \frac{2h^{2k+1}}{r!} R_{r,q}^{(r-2k-1)}(1),$$

$$i = 2, 3, \ldots, m-1, \quad k = 0, 1, \ldots, \lfloor (r-1)/2 \rfloor,$$

(4.20) $$q_{i,2k+1}^* = 0, \qquad i = 2, 3, \ldots, m-1, \quad k = 0, 1, \ldots, \lfloor (r-2)/2 \rfloor.$$

Then for $\zeta = r - 1$ or $\zeta = r - 2$ with even r, the information operator

(4.21) $$\mathfrak{N}_n^\zeta(f) = [f(x_1^*), \ldots, f^{(\zeta)}(x_1^*), \ldots, f(x_m^*), \ldots, f^{(\zeta)}(x_m^*)]^t$$

is a unique nth optimal information in Ψ_n^ζ, the algorithm

$$(4.22) \qquad \varphi(\mathfrak{N}_n^\zeta(f)) = \sum_{i=1}^{m} \sum_{k=0}^{\zeta} f^{(k)}(x_i^*)q_{i,k}^*$$

is a linear optimal error algorithm and

$$(4.23) \quad e(\varphi) = r(\mathfrak{N}_n^\zeta,S,D_p^r) = r(\Psi_n^\zeta,S,D_p^r) = \frac{R_{r,q}(1)h^r}{r!\sqrt[q]{rq+1}}$$

$$= \frac{R_{r,q}(1)(\zeta+1)^r}{2^r r!\sqrt[q]{rq+1}} \frac{1}{(n+(\zeta+1)([R_{r,q}(1)]^{1/r}-1))^r} = \Theta(n^{-r}).$$

The results (4.21) and (4.22) were essentially established by Ibragimov and Aliev [65] for $p = 1$, 2 or $+\infty$, by Aksen and Tureckij [66] for any p and $\zeta = r - 2$ with even r, by Lušpaj [66] for $\zeta = r - 1$, and also rediscovered by Kautsky [70].

For special values of p, (4.15)–(4.23) can be simplified as follows.

For $p = 1$ ($q = +\infty$), $R_{r,\infty}(x) = \cos(r \arccos x)/2^{r-1}$ is the normalized Chebyshev polynomial of the first kind and

$$(4.24) \qquad R_{r,\infty}(1) = 2^{1-r}, \qquad h = (2(m-1) + 2^{1/r})^{-1},$$

$$(4.25) \quad r(\Psi_n^\zeta,S,D_1^r) = \frac{2(\zeta+1)^r}{r!(4n + (\zeta+1)(2^{1+1/r}-4))^r} = \frac{2}{r!}\left(\frac{\zeta+1}{4n}\right)^r(1 + o(1)).$$

For $p = q = 2$, $R_{r,2}$ is the Legendre polynomial

$$(4.26) \quad \begin{aligned} R_{r,2}(1) &= 2^r(r!)^2/(2r)! = (1/\sqrt{2})(e/2r)^r(1 + o(1)), \\ h &= [2(m-1) + 4\sqrt[r]{(r!)^2/(2r)!}]^{-1} = [2(m-1) + e/r(1 + o(1))]^{-1}, \end{aligned}$$

$$(4.27) \qquad r(\Psi_n^\zeta,S,D_2^r) = \frac{r!2(\zeta+1)^r}{(2r)!\sqrt{2r+1}} \frac{1}{(n+(\zeta+1)(2\sqrt[r]{(r!)^2/(2r)!}-1))^r}$$

$$= \sqrt{\frac{2}{2r+1}}\left(\frac{e(\zeta+1)}{4rn}\right)^r(1 + o(1)).$$

For $p = +\infty$ ($q = 1$), $R_{r,1}(x) = 2^{-r}\sin((r+1)\arccos x)/\sqrt{1-x^2}$ is the normalized Chebyshev polynomial of the second kind and

$$(4.28) \quad \begin{aligned} R_{r,1}(1) &= 2^{-r}(r+1), \\ h &= [2(m-1) + \sqrt[r]{r+1}]^{-1} = (2m-1)^{-1}(1 + o(1)), \end{aligned}$$

$$(4.29) \quad r(\Psi_n^\zeta,S,D_\infty^r) = \frac{(\zeta+1)^r}{4^r r!} \frac{1}{(n+(\zeta+1)(\sqrt[r]{r+1}/2-1))^r}$$

$$= \frac{1}{r!}\left(\frac{\zeta+1}{4n}\right)^r(1 + o(1)).$$

Although we do not know the exact value of $r(\Psi_n^\zeta, S, D_p^r)$ for the general case, we can show that

(4.30) $$r(\Psi_n^\zeta, S, D_p^r) = \Theta(n^{-r})$$

for any fixed ζ, r ($\zeta \le r - 1$), and p with n going to infinity. To prove (4.30), let

$$\mathfrak{N}_n(f) = [f(x_1), \ldots, f^{(\zeta)}(x_1), \ldots, f(x_m), \ldots, f^{(\zeta)}(x_m)]^t, \qquad n = m(\zeta + 1),$$

belong to Ψ_n^ζ. Define

$$\mathfrak{N}_{n_1}(f) = [f(x_1), \ldots, f^{(r-1)}(x_1), \ldots, f(x_m), \ldots, f^{(r-1)}(x_m)]^t,$$

where $n_1 = rm = rn/(\zeta + 1)$. Of course, $\ker \mathfrak{N}_{n_1} \subset \ker \mathfrak{N}_n$ which due to Remark 3.1 of Chapter 2 yields $r(\mathfrak{N}_{n_1}, S, D_p^r) \le r(\mathfrak{N}_n, S, D_p^r)$ and

(4.31) $$r(\Psi_n^\zeta, S, D_p^r) \ge r(\Psi_{n_1}^{r-1}, S, D_p^r) = \Theta(n_1^{-r}) = \Theta(n^{-r}).$$

To prove the reverse inequality, define $k = \lceil (r - 1 - \zeta)/(\zeta + 1) \rceil$ and $m_1 = \lfloor m/(1 + k) \rfloor$. Let the points x_i of the information $\mathfrak{N}_n$ satisfy

(4.32) $$x_{m_1 + jk + i} \to x_{j+1}$$

for $j = 0, 1, \ldots, m_1 - 1$ and $i = 1, 2, \ldots, k$. If $h \in \ker \mathfrak{N}_n$, then $h^{(j)}(x_i) = 0$ for $i = 1, 2, \ldots, m$, $j = 0, 1, \ldots, \zeta$. For points x_i satisfying (4.32), we get in the limit

$$h^{(j)}(x_i) = 0 \qquad \text{for} \quad i = 1, 2, \ldots, m_1, \quad j = 0, 1, \ldots, r - 1,$$

since $\zeta + 1 + k(\zeta + 1) \ge r$. Thus,

$$r(\Psi_n^\zeta, S, D_p^r) = \inf_{x_1, \ldots, x_m} \sup \left\{ \int_0^1 h(x)\, dx : h \in \ker \mathfrak{N}_n \cap W_p^r \right\}$$

$$\le \inf_{x_1, \ldots, x_{m_1}} \sup \left\{ \int_0^1 h(x)\, dx : h^{(j)}(x_i) = 0, \right.$$

$$\left. i = 1, 2, \ldots, m_1, j = 0, 1, \ldots, r - 1, h \in W_p^r \right\}$$

$$= r(\Psi_{n_2}^{r-1}, S, D_p^r) = \Theta(n_2^{-r}),$$

where $n_2 = m_1 r$. Since $m_1 = \lfloor n/((\zeta + 1)(k + 1)) \rfloor$, $n_2 = \Theta(n)$ and $n_2^{-r} = \Theta(n^{-r})$. This completes the proof of (4.30).

We analyze the complexity of this problem. Let Ψ^ζ be the class of all linear information operators $\mathfrak{N}_n$ of the form (4.2) with $n = \zeta + 1, 2(\zeta + 1), \ldots$, i.e.,

$$\Psi^\zeta = \bigcup_{m=1}^{\infty} \Psi_{m(\zeta+1)}^\zeta.$$

Due to (4.30), we have

$$\inf_{\mathfrak{N} \in \Psi^\zeta} r(\mathfrak{N}, S, D_p^r) = 0.$$

We wish to find the ε-cardinality number $m(\Psi^\zeta, S, D_p^r, \varepsilon)$ for the integration problem (S, D_p^r) in the class Ψ^ζ (see (2.5) of Chapter 5). The ε-cardinality number is equal to the smallest $n = m(\zeta + 1)$ such that $r(\Psi_n^\zeta, S, D_p^r) < \varepsilon$. Due to (4.30), we have

(4.33) $$m(\Psi^\zeta, S, D_p^r, \varepsilon) = \Theta(\varepsilon^{-1/r}).$$

For $\zeta = r - 1$ and $\zeta = r - 2$ with even r, we can find the exact value of the ε-cardinality number. Namely, from (4.23), we get

(4.34) $$m(\Psi^\zeta, S, D_p^r, \varepsilon) = \left\lfloor \left(\frac{R_{r,q}(1)(\zeta + 1)^r}{2^r r! \sqrt[q]{rq + 1}} \frac{1}{\varepsilon} \right)^{1/r} + (\zeta + 1)(1 - [R_{r,q}(1)]^{1/r}) \right\rfloor + 1.$$

For special values of p, we can simplify (4.34) as follows. For $p = 1$,

(4.35) $$m(\Psi^\zeta, S, D_1^r, \varepsilon) = \left\lfloor \frac{\zeta + 1}{4} \left(\left(\frac{2}{r! \varepsilon} \right)^{1/r} + 4 - 2^{1 + 1/r} \right) \right\rfloor + 1$$

$$= \frac{e(\zeta + 1)}{4r} \varepsilon^{-1/r}(1 + o(1));$$

for $p = 2$,

(4.36) $$m(\Psi^\zeta, S, D_2^r, \varepsilon) = \left\lfloor (\zeta + 1) \left(\left(\frac{2r!}{(2r)! \sqrt{2r + 1} \varepsilon} \right)^{1/r} + 1 - 2 \left(\frac{(r!)^2}{(2r)!} \right)^{1/r} \right) \right\rfloor + 1$$

$$= \frac{e(\zeta + 1)}{4r} \varepsilon^{-1/r}(1 + o(1));$$

for $p = +\infty$,

(4.37) $$m(\Psi^\zeta, S, D_\infty^r, \varepsilon) = \left\lfloor \frac{(\zeta + 1)}{4} \left(\left(\frac{1}{r! \varepsilon} \right)^{1/r} + 4 - 2(r + 1) \right)^{1/r} \right\rfloor + 1$$

$$= \frac{e(\zeta + 1)}{4r} \varepsilon^{-1/r}(1 + o(1)).$$

As we see, for small ε and large r, the ε-cardinality number for $p = 1, 2$, and ∞ is asymptotically the same.

Due to the linearity of the optimal error algorithm (4.22) (see Lemma 2.2 of Chapter 5), the ε-complexity of the integration problem (S, D_p^r) in the class Ψ^ζ (briefly the ε-complexity of (S, D_p^r) in Ψ^ζ) is

$$\text{comp}(\Psi^\zeta, S, D_p^r, \varepsilon) = (c_1 + a)m(\Psi^\zeta, S, D_p^r, \varepsilon) - 1,$$

where c_1 is the complexity of evaluating one of the linear functionals $f(x)$, $f'(x), \ldots, f^{(\zeta)}(x)$ and $a \in [1, 2]$. Due to (4.33), we get

$$\text{comp}(\Psi^\zeta, S, D_p^r, \varepsilon) = \Theta(\varepsilon^{-1/r}).$$

From Lemma 2.2 of Chapter 5, it also follows that algorithm (4.22) is a nearly optimal complexity algorithm for (S,D_p^r) in Ψ^ζ.

We summarize these results in

Theorem 4.1   Consider the integration problem S for W_p^r. Then:

(i) $r(\mathfrak{N}_n,S,D_p^r)$ is finite iff (4.5) or (4.6) hold.

(ii)   For $\zeta = r - 1$ or $\zeta = r - 2$ with even r, $\mathfrak{N}_n^\zeta$ defined by (4.21) is the unique nth optimal information in Ψ_n^ζ and φ defined by (4.22) is a linear optimal error and nearly optimal complexity algorithm for (S,D_p^r) in Ψ^ζ,

$$e(\varphi) = r(\mathfrak{N}_n^\zeta,S,D_p^r) = r(\Psi_n^\zeta,S,D_p^r).$$

(iii) $r(\Psi_n^\zeta,S,D_p^r) = \Theta(n^{-r})$. The exact value of $r(\Psi_n^\zeta,S,D_p^r)$ for $\zeta = r - 1$ or $\zeta = r - 2$ with even r is given by (4.23) and by (4.25) for $p = 1$, (4.27) for $p = 2$, and (4.29) for $p = +\infty$.

(iv)   The ε-cardinality number of (S,D_p^r) in Ψ^ζ is

$$m(\Psi^\zeta,S,D_p^r,\varepsilon) = \Theta(\varepsilon^{-1/r}).$$

The exact value of $m(\Psi^\zeta,S,D_p^r,\varepsilon)$ for $\zeta = r - 1$ with even r is given by (4.34) and by (4.35) for $p = 1$, (4.36) for $p = 2$, and (4.37) for $p = +\infty$.

(v)   The ε-complexity of (S,D_p^r) in Ψ^ζ is

$$\mathrm{comp}(\Psi^\zeta,S,D_p^r,\varepsilon) = (c_1 + a)m(\Psi^\zeta,S,D_p^r,\varepsilon) - 1 = \Theta(\varepsilon^{-1/r}). \quad \blacksquare$$

The integration problem for $\mathfrak{I}_0 = W_p^r$ was considered by many people for particular values of r or p and for particular classes of information operators of the form (4.2). See, among others, Alhimova [72], Aksen and Tureckij [66], Bojanov [76], Coman [72], Ibragimov and Aliev [65], Karlin [71], Kautsky [70], Kornejčuk [74], Kornejčuk and Lušpaj [69], Krylov [62], Lee [77], Lipow [73], Lušpaj [66], Meyers and Sard [50a], Nikolskij [50, 58], Šajdaeva [59], Sard [49, 63], Schoenberg [64b, 65, 66, 69, 70], Secrest [64, 65b], and Stern [67].

(ii)   Integration for $\widetilde{W}_\infty^r$

In this subsection, we consider the integration problem of the form $Sf = \int_0^{2\pi} f(x)\,dx$. Let $\mathfrak{I}_1 = \widetilde{W}_\infty^r[0,2\pi]$. Define $Tf = \tilde{D}_\infty^r f = f^{(r)}$ with $\mathfrak{I}_4 \in L_\infty$. Thus,

$$\mathfrak{I}_0 = \{f : \|f^{(r)}\|_\infty \le r\} = \widetilde{W}_\infty^r.$$

Let

(4.38) $\mathfrak{N}_n(f) = [f(x_1),f(x_2),\ldots,f(x_n)]^{\mathrm{t}}$

for distinct points x_i from $[0,2\pi]$. Note that $\ker \tilde{D}_\infty^r$ is the set of periodic polynomials of degree at most $r - 1$. Thus, $\ker \tilde{D}_\infty^r = \{f : f(x) \equiv \mathrm{const}\}$.

Therefore,

$$\ker \mathfrak{N}_n \cap \ker \tilde{D}^r_\infty = \{0\}$$

which, as in (4.4), implies

$$r(\mathfrak{N}_n, S, \tilde{D}^r_\infty) < +\infty \qquad \forall n \geq 1.$$

In this subsection, we report an interesting result of Motornyj [73] on the optimality of the rectangle quadrature formula.

Recall that Ψ^0_n is the class of all linear information operators of the form (4.38). We are interested in finding an nth optimal information $\mathfrak{N}_n$ from the class Ψ^0_n, i.e., $\mathfrak{N}_n$ which satisfies the condition

(4.39) $$r(\mathfrak{N}_n, S, \tilde{D}^r_\infty) = r(\Psi^0_n, S, \tilde{D}^r_\infty).$$

Since for every information operator of the form (4.38) there exists a linear optimal error algorithm, (4.39) is equivalent to finding points x_i, $i = 1, 2, \ldots, n$, and weights q_i, $i = 1, 2, \ldots, n$, such that the algorithm

(4.40) $$\varphi(\mathfrak{N}_n(f)) = \sum_{i=1}^n f(x_i) q_i$$

has minimal error in $\tilde{W}^r_\infty$. This problem was solved by Motornyj [73] who showed that the algorithm φ of the form (4.40) with

(4.41) $$x_i = (2\pi/n)(i-1), \qquad i = 1, 2, \ldots, n,$$

(4.42) $$q_i = 2\pi/n$$

has minimal error $e(\varphi)$ given by

(4.43) $$e(\varphi) = 2\pi K_r/n^r,$$

where K_r is the Favard constant defined by (3.16) and $K_r \in [1, \pi/2]$. From this, we conclude that

(4.44) $$\mathfrak{N}_n(f) = [f(0), f(2\pi/n), \ldots, f(2\pi/n)(n-1))]^t$$

is an nth optimal information operator in the class Ψ^0_n and

(4.45) $$\varphi(\mathfrak{N}_n(f)) = \frac{2\pi}{n} \sum_{i=1}^n f\left(\frac{2\pi}{n}(i-1)\right)$$

is a linear optimal error algorithm, whose error is

(4.46) $$e(\varphi) = r(\mathfrak{N}_n, S, \tilde{D}^r_\infty) = r(\Psi^0_n, S, \tilde{D}^r_\infty) = 2\pi K_r/n^r.$$

Note that φ is the well-known rectangle quadrature formula. Indeed, Žensykbaev [76] proved that the rectangle quadrature formula is an optimal error algorithm for the class $\tilde{W}^r_p$ for arbitrary p from $[1, +\infty]$.

REMARK 4.1   We mentioned earlier that algorithms are traditionally derived by ad hoc criteria. For instance, for the integration problem, a commonly used criterion is to find a quadrature formula whose order of exactness is as high as possible, i.e., which is exact for all polynomials of as high a degree as possible. This leads to Gauss quadrature formulas regardless of the class $\mathfrak{I}_0$ of integrands.

In this section, we showed that for the class $\tilde{W}^r_\infty$ of periodic smooth functions, the rectangle formula which has zeroth order of exactness is an optimal error algorithm.

It is also known that for the class of analytic functions whose norm is bounded by unity on the disk $\{x : |x| \le r\}$, Gauss quadrature formulas, i.e., quadrature formulas of the highest order of exactness, are nearly optimal error algorithms for large r. See, for instance, Barnhill [68], Larkin [70], and Pinkus [75].

This indicates that optimal error quadrature formulas and their order of exactness are very much dependent on the class $\mathfrak{I}_0$ of integrands. This also shows that the order of exactness is not related to the optimality of quadrature formulas. ∎

We analyze the complexity of this problem. Let $\Psi^0 = \bigcup_{n=1}^{\infty} \Psi^0_n$ be the class of all linear information operators of the form (4.38) with $n = 1, 2, \ldots$. The ε-cardinality number $m(\Psi^0, S, \tilde{D}^r_\infty, \varepsilon)$ is equal to the smallest n such that $r(\Psi^0_n, S, \tilde{D}^r_\infty) < \varepsilon$. From (4.46), we get

$$m(\Psi^0, S, \tilde{D}^r_\infty, \varepsilon) = \lfloor (2\pi K_r / \varepsilon)^{1/r} \rfloor + 1.$$

The combinatory complexity of the linear optimal error algorithm (4.45) is m. Thus it differs from the minimal combinatory complexity by at most one. Therefore the ε-complexity of $(S, \tilde{D}^r_\infty)$ in Ψ^0 is

$$\text{comp}(\Psi^0, S, \tilde{D}^r_\infty, \varepsilon) = (c_1 + 1)(\lfloor (2\pi K_r / \varepsilon)^{1/r} \rfloor + 1) + a$$

where c_1 is the complexity of one function evaluation and $a \in [-1, 0]$. Algorithm (4.45) is a nearly optimal complexity algorithm for $(S, \tilde{D}^r_\infty)$ in Ψ^0.

We summarize these results in

Theorem 4.2   Consider the integration problem S for $\tilde{W}^r_\infty$. Then:

(i) $r(\Psi^0_n, S, \tilde{D}^r_\infty) = 2\pi K_r / n^r$.

(ii) $\mathfrak{N}_n$ defined by (4.44) is an nth optimal information operator in Ψ^0_n,

$$r(\mathfrak{N}_n, S, \tilde{D}^r_\infty) = r(\Psi^0_n, S, \tilde{D}^r_\infty).$$

(iii)   The rectangle quadrature formula defined by (4.45) is a linear optimal error and nearly optimal complexity algorithm for $(S, \tilde{D}^r_\infty)$ in Ψ^0,

$$e(\varphi) = r(\mathfrak{N}_n, S, \tilde{D}^r_\infty).$$

(iv)   The ε-cardinality number of $(S, \tilde{D}^r_\infty)$ in Ψ^0 is

$$m(\Psi^0, S, \tilde{D}^r_\infty, \varepsilon) = \lfloor 2\pi K_r / \varepsilon)^{1/r} \rfloor + 1.$$

(v)   The ε-complexity of $(S, \tilde{D}^r_\infty)$ in Ψ^0 is

$$\text{comp}(\Psi^0, S, \tilde{D}^r_\infty, \varepsilon) = (c_1 + 1)(\lfloor (2\pi K_r / \varepsilon)^{1/r} \rfloor + 1) + a. \quad \blacksquare$$

The integration problem for periodic functions of one or many variables for different subsets $\mathfrak{I}_0$ and for information operators of the form (4.2) was considered in many papers. See, among others, Babuška [68a,b], Busarova [73], Forst [75], Keast [73], Kornejčuk [74], Kornejčuk and Lušpaj [69], Korobov [63], Ligun [76], Lušpaj [69, 74], Motornyj [73, 74, 76], Smolyak [60], and Žensykbaev [76].

(iii)   Relation to the Gelfand n-Width

In this subsection, we show that the nth minimal radius of information for the *integration* problem is related to the Gelfand n-width for the corresponding *approximation* problem in the space L_1. More precisely, the Gelfand n-width is a lower bound on the nth minimal radius of information. Under some conditions, the two quantities are equal.

Consider the integration problem (4.1) with a balanced and convex $\mathfrak{I}_0$. Recall that

(4.47) $r(\Psi_n^0, S, \mathfrak{I}_0) = \displaystyle\inf_{0 \le x_1 < x_2 < \cdots < x_n \le 1} \sup\left\{\int_0^1 h(x)\,dx : h(x_i) = 0,\right.$

$$\left. i = 1, 2, \ldots, n, h \in \mathfrak{I}_0\right\}.$$

For any such $x_1, x_2, \ldots, x_n$, define

(4.48) $\overline{h}(x) = \sup\{h(x) : h(x_i) = 0, i = 1, 2, \ldots, n, h \in \mathfrak{I}_0\}.$

Since $\mathfrak{I}_0$ is balanced, $\overline{h}(x) \ge 0$, $\forall x \in [0,1]$. Assume that

(4.49) $\overline{h} \in \mathfrak{I}_0.$

Observe that $\overline{h}(x_i) = 0$, $i = 1, 2, \ldots, n$, and

(4.50) $\sup\left\{\int_0^1 h(x)\,dx : h(x_i) = 0, i = 1, 2, \ldots, n, h \in \mathfrak{I}_0\right\}$

$$= \sup\left\{\int_0^1 |h(x)|\,dx : h(x_i) = 0, i = 1, 2, \ldots, n, h \in \mathfrak{I}_0\right\}$$

$$= \int_0^1 \overline{h}(x)\,dx.$$

Define the approximation problem

$$S_1 f = f, \qquad \mathfrak{I}_2 = L_1[0,1].$$

Let $\mathfrak{N}(f) = [f(x_1), f(x_2), \ldots, f(x_n)]^t$. Then (4.50) states

$$r(\mathfrak{N}, S) = \tfrac{1}{2} d(\mathfrak{N}, S_1).$$

Thus, the radius of information for the integration problem is equal to one half of the diameter of information for the approximation problem in L_1. Recall that $d(\mathfrak{N},S_1) \geq d(n,S_1,\mathfrak{I}_0)$, where $d(n,S_1,\mathfrak{I}_0)$ is the nth minimal diameter of information (see (6.2) of Chapter 2). Due to Corollary 6.1 of Chapter 2,

$$d(n,S_1,\mathfrak{I}_0) = 2d^n,$$

where $d^n = d^n(S_1(\mathfrak{I}_0),L_1(0,1))$ is the Gelfand n-width of the range of the approximation operator in L_1. From this, we conclude

$$r(\mathfrak{N},S) \geq d^n,$$

and if (4.49) holds for any points $x_1 < x_2 < \cdots < x_n$, we have

(4.51) $r(\Psi_n^0,S,\mathfrak{I}_0) \geq d^n.$

Thus, the Gelfand n-width for the approximation problem is a lower bound on the nth minimal radius of information.

We now show when equality can be achieved in (4.51). Assume that there exist points $x_1^*, x_2^*, \ldots, x_n^*$ such that the linear subspace

(4.52) $A^n = \{f \in L_1[0,1] : f(x_i^*) = 0, i = 1, 2, \ldots, n\}$

is an nth extremal subspace of $S_1(\mathfrak{I}_0)$ in the sense of Gelfand. Then Corollary 6.2 of Chapter 2 states that the information operator

(4.53) $\mathfrak{N}_n^*(f) = [f(x_1^*),f(x_2^*), \ldots, f(x_n^*)]^t$

is an nth optimal information for the approximation problem and

$$d(\mathfrak{N}_n^*,S_1) = 2d^n.$$

Note that $\mathfrak{N}_n^* \in \Psi_n^0$. Applying the information operator $\mathfrak{N}^*$ for the integration problem, we get

$$r(\mathfrak{N}^*,S) = \tfrac{1}{2}d(\mathfrak{N}^*,S_1) = d^n.$$

From (4.51), we conclude

$$r(\mathfrak{N}^*,S) = r(\Psi_n^0,S,\mathfrak{I}_0) = d^n$$

which means that $\mathfrak{N}_n^*$ is an nth optimal information in Ψ_n^0 and the nth minimal radius of information is equal to d^n.

We summarize these results in

Theorem 4.3 (i) If for any points $x_1 < x_2 < \cdots < x_n$ the function $\bar{h}$ defined by (4.48) belongs to $\mathfrak{I}_0$, then the nth minimal radius of information for the integration problem in the class Ψ_n^0 is no less than the Gelfand n-width for the approximation problem in L_1,

$$r(\Psi_n^0,S,\mathfrak{I}_0) \geq d^n(S_1(\mathfrak{I}_0),L_1[0,1]).$$

(ii)   If, additionally, the linear subpsace A^n defined by (4.52) is an nth extremal subspace of $S_1(\mathfrak{I}_0)$ in the sense of Gelfand, then the information operator $\mathfrak{N}_n^*$ defined by (4.53) is an nth optimal information for the integration problem and

$$r(\mathfrak{N}^*,S) = r(\Psi_n^0,S,\mathfrak{I}_0) = d^n(S_1(\mathfrak{I}_0),L_1[0,1]). \quad \blacksquare$$

We illustrate Theorem 4.3 for $\mathfrak{I}_1 = W_\infty^r[0,1]$ and $Tf = D_\infty^r f = f^{(r)}$. Thus, $\mathfrak{I}_0 = W_\infty^r$. As for the interpolation problem for points $x_1 < x_2 < \cdots < x_n$ (see Section 3), define q_n as a perfect spline of degree r with $(n - r)$ knots satisfying (3.11) and (3.14). From (3.9), we conclude that

$$|h(x)| \le |q_n(x)|$$

for any $h \in W_\infty^r$ and $h(x_i) = 0$, $i = 1, 2, \ldots, n$. Assume that q_n is of constant sign in $[0,1]$. Then

$$\bar{h}(x) = |q_n(x)|$$

satisfies (4.48) and (4.49). Thus, Theorem 4.3 states that

(4.54) $$r(\Psi_n^0,S,W_\infty^r) \ge d^n(W_\infty^r,L_1).$$

The exact value of $d^n(W_\infty^r,L_1)$ and the extremal subspaces are not known. However, the asymptotic behavior of $d^n(W_\infty^r,L_1)$ is known (see Tikhomirov [76, p. 249]) to be

(4.55) $$d^n(W_\infty^r,L_1) = \Theta(n^{-r}).$$

Furthermore, it is easy to find information operators which are nearly optimal. Namely, let

$$\mathfrak{N}_n(f) = [f(x_1),f(x_2),\ldots,f(x_n)]^t, \qquad x_i = (i - 1)/(n - 1),$$
$$i = 1, 2, \ldots, n.$$

Then

(4.56) $$r(\mathfrak{N}_n,S,D_\infty^r) = \int_0^1 |q_n(x)|\,dx \le \|q_n\|_\infty.$$

Micchelli, Rivlin, and Winograd [76] showed that for equidistant points x_i,

(4.57) $$\|q_n\|_\infty = \Theta(n^{-r}).$$

Since $r(\mathfrak{N}_n,S,D_\infty^r) \ge r(\Psi_n^0,S,W_\infty^r)$, we conclude from (4.54)–(4.57) that

$$r(\Psi_n^0,S,W_\infty^r) = \Theta(n^{-r})$$

which we have already known due to (4.30). Observe that

$$r(\mathfrak{N}_n,S,D_\infty^r) = \Theta(n^{-r})$$

which shows that the information $\mathfrak{N}_n$ in the class Ψ_n^0 is nearly optimal.

5. APPROXIMATION

In this section, we deal with the *approximation problem*, i.e.,

(5.1) $$Sf = If = f,$$

for several spaces $\mathfrak{I}_1$, $\mathfrak{I}_2$ and several classes $\mathfrak{I}_0$. We shall consider this problem for a class $\mathfrak{I}_0$ in a general Hilbert space, for the class of periodic smooth functions $\tilde{W}_2^r$ in L_2, and for the class of nonperiodic smooth functions W_2^r in L_2. We shall also consider uniform approximation (i.e., $\mathfrak{I}_2 = \tilde{C}$ or C) for the class $\tilde{W}_\infty^r$ and W_∞^r.

In the previous sections, we considered solution operators which were linear functionals. For such solution operators, the problem of nth optimal information in the sense of Definition 4.1 is trivial. Indeed, since $\mathfrak{N}(f) = S(f)$ has cardinality one and $d(\mathfrak{N},S) = 0$, then $\mathfrak{N}$ is an nth optimal information operator.

In this section, we consider an infinite-dimensional linear operator $S = I$ for which the problem of nth optimal information is interesting and deep. To construct nth optimal information operators and linear optimal error algorithms, we shall extensively use the relations between the theory of approximation and the theory of analytic complexity established in Section 6 of Chapter 2 and Section 5 of Chapter 3.

The approximation problem has been studied by many people. Some relevant examples include Bojanov and Chernogorov [77], de Boor [77], Golomb [77], Ivanov [77], Kornejčuk [76], Melkman [77], Micchelli and Pinkus [77], Micchelli, Rivlin, and Winograd [76], Motornyj [76], Osipenko [76], Rice [73, 76], Schultz [74], Sukharev [78a], and Tikhomirov [76].

(i) Approximation for a General Hilbert Space

Let $H = \operatorname{lin}(\xi_1, \xi_2, \ldots)$ be an infinite-dimensional Hilbert space over the real field $\mathbb{R}$, where $(\xi_i, \xi_j) = \delta_{ij}$. Thus, $f \in H$ iff $f = \sum_{i=1}^\infty (f, \xi_i)\xi_i$ and $\sum_{i=1}^\infty (f, \xi_i)^2 < +\infty$. Let $\{\beta_i\}$ be a nonzero sequence of real numbers such that $|\beta_i| \le |\beta_{i+1}|$ for all i. Let

(5.2) $$\mathfrak{I}_1 = \left\{ f : f \in H \text{ and } \sum_{i=1}^\infty \beta_i^2 (f, \xi_i)^2 < +\infty \right\}$$

and $\mathfrak{I}_2 = H$. Let

(5.3) $$\mathfrak{I}_0 = \{ f : f \in \mathfrak{I}_1 \text{ and } \|Tf\| \le 1 \}$$

for

(5.4) $$Tf = \sum_{i=1}^\infty \beta_i (f, \xi_i)\xi_i$$

with $\mathfrak{I}_4 = T(\mathfrak{I}_1)$ equipped with the H norm. Of course, $\mathfrak{I}_4$ is a Hilbert space.

We first find the index $n^* = \text{index}(I,T)$. Since ker $I = \{0\}$, (3.3) of Chapter 2 yields $A(T,I) = \ker T$ and $n^* = \dim(\ker T)$. Let i_0 be the largest index such that $\beta_i = 0$ for $i = 1, 2, \ldots, i_0$ ($\beta_{i_0+1} \neq 0$). Then $n^* = i_0 < +\infty$. Note that

$$\mathfrak{N}^*(f) = [(f,\xi_1),(f,\xi_2), \ldots ,(f,\xi_{n^*})]^t$$

satisfies Lemma 4.1 of Chapter 2, i.e., $\ker \mathfrak{N}^* \cap \ker T = \{0\}$. From (4.3) of Chapter 2, we easily find the inverse of T,

$$T^{-1}f = \sum_{i=n^*+1}^{\infty} \frac{1}{\beta_i}(f,\xi_i)\xi_i$$

and $K = ST^{-1} = T^{-1}$. Note that K is self-adjoint, $K_1 = K^2$, and $K\xi_i = (1/\beta_i)\xi_i$ for $i > i_0$. Thus, K is compact iff $\lim_i |\beta_i| = +\infty$. From Corollary 5.1 of Chapter 2, it follows that the problem (I,T) is convergent iff $\lim_i |\beta_i| = +\infty$.

We want to find an nth optimal information $\mathfrak{N}_n$ for $n \geq n^*$. Since $K(\mathfrak{I}_4) \subset \mathfrak{I}_4$ and $(Tf,\xi_i) = \beta_i(f,\xi_i)$, the nth optimal information defined by (5.7) of Chapter 2 is equivalent to

(5.5) $$\mathfrak{N}_n(f) = [(f,\xi_1),(f,\xi_2), \ldots ,(f,\xi_n)]^t.$$

Since $T(\ker \mathfrak{N}_n)$ is closed in the Hilbert space $\mathfrak{I}_4$, we conclude from Theorem 5.2 of Chapter 2 and (4.16) of Chapter 3 that

(5.6) $$r(\mathfrak{N}_n,I,T) = \tfrac{1}{2}d(\mathfrak{N}_n,I,T) = \tfrac{1}{2}d(n,I,T) = 1/|\beta_{n+1}|.$$

The linear optimal error algorithm φ defined by (4.10) in Theorem 4.1 of Chapter 3 is given by

(5.7) $$\varphi(\mathfrak{N}_n(f)) = \sum_{i=1}^{n} (f,\xi_i)\xi_i, \qquad e(\varphi) = 1/|\beta_{n+1}|.$$

Due to Theorem 5.1 and Remark 5.1 of Chapter 4, this is a central spline algorithm. Note that φ is the initial section of $f = \sum_{i=1}^{\infty} (f,\xi_i)\xi_i$.

We analyze the complexity of this problem. Recall that Ψ_U is the class of *all* linear information operators with finite cardinality. It is easy to see that the ε-cardinality number $m = m(\Psi_U,I,T,\varepsilon)$ (see (2.5) of Chapter 5) is equal to the smallest number n such that $1/|\beta_{n+1}| < \varepsilon$. Since there exists a linear optimal error algorithm for any ε, we get the ε-complexity of the problem (I,T) in the class Ψ_U,

(5.8) $$\text{comp}(\Psi_U,I,T,\varepsilon) = (c_1 + a)m(\Psi_U,I,T,\varepsilon) - 1,$$

where c_1 is the complexity of evaluating a linear functional of the form (f,ξ_i) and $a \in [1,2]$. Note that m depends only on how fast $|\beta_i|$ goes to infinity and due to Theorem 2.1 of Chapter 5 can be essentially any decreasing function of ε. Algorithm (5.7) is a nearly optimal complexity algorithm for (I,T) in Ψ_U.

As a by-product, we get the Gelfand and the linear Kolmogorov n-widths of the solution set. Since $q(I,\mathfrak{I}_0)$ defined by (6.6) of Chapter 2 is now equal to unity, Remark 6.1 of Chapter 2 and Remark 5.3 of Chapter 3 yield

$$d^n(\mathfrak{I}_0,\mathfrak{I}_2) = \lambda_n(\mathfrak{I}_0,\mathfrak{I}_2) = 1/|\beta_{n+1}|.$$

We summarize these results in

Theorem 5.1 Consider the approximation problem for $\mathfrak{I}_0$ in the space H defined by (5.3). Then:

(i) $\mathfrak{N}_n$ defined by (5.5) is an nth optimal information,

$$r(\mathfrak{N}_n,I,T) = \tfrac{1}{2}d(n,I,T) = 1/|\beta_{n+1}|.$$

(ii) φ defined by (5.7) is a linear central spline and nearly optimal complexity algorithm for (I,T) in Ψ_U, $e(\varphi) = 1/|\beta_{n+1}|$.

(iii) The ε-cardinality number of (I,T) in Ψ_U is

$$m(\Psi_U,I,T,\varepsilon) = \min\{n: 1/|\beta_{n+1}| < \varepsilon\}.$$

(iv) The ε-complexity of (I,T) in Ψ_U is

$$\mathrm{comp}(\Psi_U,I,T,\varepsilon) = (c_1 + a)m(\Psi_U,I,T,\varepsilon) - 1. \quad \blacksquare$$

(ii) Approximation for $\tilde{W}_2^r$ in L_2

We specialize the results of the previous subsection to a particular Hilbert space of periodic smooth functions. Let

(5.9) $\mathfrak{I}_1 = \tilde{W}_2^r[0,2\pi]$ and $\mathfrak{I}_2 = L_2.$

Define

(5.10) $Tf = \tilde{D}_2^r f = f^{(r)}$

with $\mathfrak{I}_4 = T(\mathfrak{I}_1)$ equipped with the L_2 norm. Thus,

(5.11) $\mathfrak{I}_0 = \{f: \|f^{(r)}\|_2 \leq 1\} = \tilde{W}_2^r.$

Note that for $r = 0$, we have $d(\mathfrak{N},I,I) = 2 \sup\{\|h\|_2 : h \in \ker \mathfrak{N}$ and $\|h\|_2 \leq 1\} = 2$ for any information operator with finite cardinality. This shows that the approximation problem for $r = 0$ is ε-noncomputable for $\varepsilon \leq 1$. From now on, we assume that

(5.12) $r \geq 1.$

To find the index $n^* = \dim(\ker \tilde{D}_2^r)$, observe that if $f^{(r)} = 0$ and f is periodic, then $f \equiv \mathrm{const}$. Thus, $\ker \tilde{D}_2^r = \{f: f \equiv \mathrm{const}\}$ and $n^* = 1$. The information

$$\mathfrak{N}^*(f) = (f,1)$$

satisfies the necessary condition $\ker \mathfrak{N}^* \cap \ker \tilde{D}_2^r = \{0\}$. It is easy to verify that the operator K_1 defined by (5.4) of Chapter 2 is now given by $K_1 = (-1)^r [\tilde{D}_2^r]^{-2} : \mathfrak{I}_4 \to \mathfrak{I}_4$ and

$$(5.13) \qquad K_1 z = \sum_{j=1}^{\infty} j^{-2r}(a_j \sin jx + b_j \cos jx),$$

where $z = \sum_{j=1}^{\infty} (a_j \sin jx + b_j \cos jx) \in \mathfrak{I}_4$.

To find an nth optimal information, we need the eigenvalues and eigenfunctions of K_1. Observe that

$$(5.14) \qquad \begin{aligned} K_1 \sin jx &= j^{-2r} \sin jx \\ K_1 \cos jx &= j^{-2r} \cos jx \end{aligned} \qquad \forall j \geq 1,$$

which proves that j^{-2r} is a double eigenvalue of K_1 and $\sin jx$, $\cos jx$ are the corresponding eigenfunctions.

For $n = 1, 2, \ldots$, define the information operator

$$(5.15) \quad \mathfrak{N}_n(f) = [(f,1),(f, \sin x), \ldots, (f, \sin n'x),(f \cos x), \ldots,(f, \cos n'x)]^t,$$

where $n' = \lfloor (n-1)/2 \rfloor$. Note that for odd n, $\text{card}(\mathfrak{N}_n) = n$, and for even n, $\text{card}(\mathfrak{N}_n) = n - 1$. Using the equivalence relation notation of Section 2 of Chapter 2, we observe that

$$[(\tilde{D}_2^r f, \sin jx), (\tilde{D}_2^r f, \cos jx)]^t \asymp [(f, \sin jx),(f, \cos jx)]^t$$

for every j. From this, we conclude that $\mathfrak{N}_n$ defined by (5.15) is equivalent to the nth optimal information defined by (5.7) of Chapter 2.

The linear optimal error algorithm φ defined by (4.10) in Theorem 4.1 of Chapter 3 is now given by

$$(5.16) \quad \varphi(\mathfrak{N}_n(f)) = \frac{1}{\pi}(f,1) + \frac{2}{\pi} \sum_{j=1}^{n'} ((f, \sin jx)\sin jx + (f,\cos jx)\cos jx).$$

It is also a central spline algorithm. Its error is equal to $\sqrt{\lambda_n}$, where $\{\lambda_n\}$ is a nonincreasing sequence of eigenvalues of K_1, i.e.,

$$(5.17) \qquad e(\varphi) = r(\mathfrak{N}_n, I, \tilde{D}_2^r) = \tfrac{1}{2}d(n, I, \tilde{D}_2^r) = \lceil n/2 \rceil^{-r}.$$

We analyze the complexity of this problem. The ε-cardinality number $m = m(\Psi_U, I, \tilde{D}_2^r, \varepsilon)$ is the smallest n such that $\lceil n/2 \rceil^{-r} < \varepsilon$. Thus, $m = 2\lfloor \varepsilon^{-1/r} \rfloor + 1$. Due to the linearity of the optimal error algorithm (5.16), the ε-complexity of $(I, \tilde{D}_2^r)$ in Ψ_U is given by

$$(5.18) \qquad \text{comp}(\Psi_U, I, \tilde{D}_2^r, \varepsilon) = (c_1 + a)(2\lfloor \varepsilon^{-1/r} \rfloor + 1) - 1,$$

where c_1 is the complexity of evaluating a linear functional of the form $(f, \sin jx)$ or $(f, \cos jx)$ and $a \in [1,2]$. Algorithm (5.16) is a nearly optimal complexity algorithm for $(I, \tilde{D}_2^r)$ in Ψ_U.

As a by-product, we get

$$d^n(\tilde{W}^r_2, L_2) = \lambda_n(\tilde{W}^r_2, L_2) = \lceil n/2 \rceil^{-r}$$

which is already known in the theory of approximation.

We summarize these results in

Theorem 5.2 Consider the approximation problem for $\mathfrak{I}_0 = \tilde{W}^r_2$ in the space L_2. Then:

(i) $\mathfrak{N}_n$ defined by (5.15) is an nth optimal information,

$$r(\mathfrak{N}_n, I, \tilde{D}^r_2) = \tfrac{1}{2} d(n, I, \tilde{D}^r_2) = \lceil n/2 \rceil^{-r}.$$

(ii) φ defined by (5.16) is a linear central spline and nearly optimal complexity algorithm for $(I, \tilde{D}^r_2)$ in Ψ_U, $e(\varphi) = \lceil n/2 \rceil^{-r}$.

(iii) The ε-cardinality number of $(I, \tilde{D}^r_2)$ in Ψ_U is

$$m(\Psi_U, I, \tilde{D}^r_2, \varepsilon) = 2\lfloor \varepsilon^{-1/r} \rfloor + 1.$$

(iv) The ε-complexity of $(I, \tilde{D}^r_2)$ in Ψ_U is

$$\mathrm{comp}(\Psi_U, I, \tilde{D}^r_2, \varepsilon) = (c_1 + a)(2\lfloor \varepsilon^{-1/r} \rfloor + 1) - 1. \quad \blacksquare$$

(iii) Approximation for W^r_2 in L_2

We now deal with a Hilbert space of smooth nonperiodic functions. Let

(5.19) $$\mathfrak{I}_1 = W^r_2[-1,1] \quad \text{and} \quad \mathfrak{I}_2 = L_2.$$

Define

(5.20) $$Tf = D^r_2 f = f^{(r)}$$

with $\mathfrak{I}_4 = T(\mathfrak{I}_1) = L_2$. Thus,

(5.21) $$\mathfrak{I}_0 = \{f : \|f^{(r)}\|_2 \le 1\} = W^r_2.$$

For $r = 0$, $d(\mathfrak{N}, I, I) = 2$ for any information operator with finite cardinality, and the approximation problem is ε-noncomputable for $\varepsilon \le 1$. From now on, we assume $r \ge 1$.

Observe that $\ker D^r_2 = \Pi_{r-1}$ is the class of polynomials of degree at most $r - 1$. Thus,

$$n^* = \mathrm{index}(I, D^r_2) = r.$$

Let $\xi^*_1, \xi^*_2, \ldots, \xi^*_r$ be the orthonormal polynomials from Π_{r-1}, $(\xi^*_i, \xi^*_j) = \delta_{ij}$. Define the information operator

$$\mathfrak{N}^*(f) = [(f, \xi^*_1), (f, \xi^*_2), \ldots, (f, \xi^*_r)]^t.$$

Then $\ker \mathfrak{N}^* \cap \ker D^r_2 = \{0\}$.

We find an nth optimal information operator using an nth extremal subspace of $\mathfrak{I}_0$ in the sense of Gelfand as was shown in Corollary 6.1 of Chapter 2. Consider the eigenvalue problem

(5.22) $$f^{(2r)}(x) + (-1)^{r+1}\lambda^2 f(x) = 0$$

with boundary conditions

(5.23) $$f^{(r)}(\pm 1) = \cdots = f^{(2r-1)}(\pm 1) = 0.$$

Let $\{\lambda_i\}$, $\lambda_i \le \lambda_{i+1}$, be the eigenvalues of (5.22) and (5.23) and $\{\xi_i\}$ be the corresponding normalized eigenfunctions. It is known (see, for instance, Tikhomirov [76, p. 128]) that

(5.24) $$\lambda_1 = \cdots = \lambda_r = 0, \qquad \lambda_{r+1} > 0, \qquad \lambda_n = (\pi n/2)^r(1 + o(1)),$$

and the eigenfunctions ξ_i form an orthonormal basis of L_2. Without loss of generality, we can set $\xi_i = \xi_i^*$ for $i = 1, 2, \ldots, r$. Furthermore,

$$d^n(W_2^r, L_2) = \lambda_n(W_2^r, L_2) = 1/\lambda_n$$

and $A^n = \{f : (f, \xi_i) = 0, i = 1, 2, \ldots, n\}$ is an nth extremal subspace of $\mathfrak{I}_0$ in the sense of Gelfand. From (6.13) of Chapter 2, we conclude that

(5.25) $$\mathfrak{N}_n(f) = [(f, \xi_1), (f, \xi_2), \ldots, (f, \xi_n)]^t$$

is an nth optimal information operator, $r(\mathfrak{N}_n, I, D^r) = 1/\lambda_n$, and

(5.26) $$\varphi(\mathfrak{N}_n(f)) = \sum_{i=1}^{n} (f, \xi_i)\xi_i$$

is an optimal error algorithm, $e(\varphi) = 1/\lambda_n$, which is also a linear central spline algorithm. For instance, for $r = 1$, $\lambda_n = \pi n/2$, $\xi_{2n}(t) = \pi^{-1}\cos n\pi t$, and $\xi_{2n+1}(t) = \pi^{-1}\sin(n + \frac{1}{2})\pi t$. For $r = 2$, the positive eigenvalues λ_n satisfy the nonlinear equation $|\tan\sqrt{\lambda}| = \tanh\sqrt{\lambda}$.

The ε-cardinality number $m = m(\Psi_U, I, D_2^r, \varepsilon)$ is the smallest n such that $1/\lambda_n < \varepsilon$. Due to (5.24), for small ε we get

$$m = (2/\pi)\varepsilon^{-1/r}(1 + o(1)).$$

The ε-complexity is given by

$$\text{comp}\ (\Psi_U, I, D_2^r, \varepsilon) = (c_1 + a)m - 1,$$

where c_1 is the complexity of evaluating a linear functional of the form (f, ξ_i) and $a \in [1,2]$. Algorithm (5.26) is a nearly optimal complexity algorithm for (I, D_2^r) in Ψ_U.

We summarize these results in

Theorem 5.3 Consider the approximation problem for $\mathfrak{I}_0 = W_2^r$ in the space L_2. Then:

(i) $\mathfrak{N}_n$ defined by (5.25) is an nth optimal information operator,

$$r(\mathfrak{N}_n, I, D_2^r) = \tfrac{1}{2}d(n, I, D_2^r) = 1/\lambda_n = (2/\pi n)^r(1 + o(1)).$$

(ii) φ defined by (5.26) is a linear central spline and nearly optimal complexity algorithm for (I,D_2^r) in Ψ_U, $e(\varphi) = 1/\lambda_n$.

(iii)   The ε-cardinality number of (I,D_2^r) in Ψ_U is

$$m(\Psi_U,I,D_2^r,\varepsilon) = (2/\pi)\varepsilon^{-1/r}(1 + o(1)).$$

(iv)   The ε-complexity of (I,D_2^r) in Ψ_U is

$$\text{comp}(\Psi_U,I,D_2^r,\varepsilon) = (c_1 + a)(2/\pi)\varepsilon^{-1/r}(1 + o(1)). \quad \blacksquare$$

(iv)   Approximation for $\tilde{W}_\infty^r$ in C

We now consider the uniform approximation problem for periodic functions. Let

$$\mathfrak{I}_1 = \tilde{W}_\infty^r[0,2\pi] \quad \text{and} \quad \mathfrak{I}_2 = \tilde{C} = \tilde{C}[0,2\pi].$$

Let $Tf = \tilde{D}_\infty^r f$ with $\mathfrak{I}_4 = T(\mathfrak{I}_1) \subset L_\infty$. Thus, $\mathfrak{I}_0 = \{f : \|f^{(r)}\|_\infty \le 1\} = \tilde{W}_\infty^r$. It is known (see Tikhomirov [76]) that

$$(5.27) \qquad d^n(\tilde{W}_\infty^r,\tilde{C}) = \lambda_n(\tilde{W}_\infty^r,\tilde{C}) = K_r\lceil n/2 \rceil^{-r},$$

where K_r is the Favard constant defined by (3.16) and $K_r \in [1,\pi/2]$. There are two known nth extremal subspaces for $\tilde{W}_\infty^r$, $A_1^n = \{f : (f, \sin jx) = (f, \cos jx) = 0, j = 0, 1, \ldots, n' = \lfloor (n-1)/2 \rfloor\}$, $A_2^n = \{f : f(t_i) = 0, i = 1, 2, \ldots, n'' = 2\lfloor n/2 \rfloor\}$, where

$$(5.28) \qquad t_i = \begin{cases} (2i - 1)(\pi/2n) & \text{for} \quad r = 2, 4, \ldots, \\ (i - 1)(\pi/n) & \text{for} \quad r = 1, 3, \ldots. \end{cases}$$

Following the general prescription given in the construction for the optimal information operators from the nth extremal subspace in Section 6 of Chapter 2, we find that

$$(5.29) \qquad \mathfrak{N}_{n,1} = [(f,1), (f, \sin x), \ldots,$$
$$(f, \sin n'x),(f, \cos x), \ldots,(f, \cos n'x)]^t,$$

$$\text{card}(\mathfrak{N}_{n,1}) = 2n' + 1 \le n,$$

and

$$(5.30) \qquad \mathfrak{N}_{n,2} = [f(t_1),f(t_2), \ldots,f(t_{n''})]^t, \qquad \text{card}(\mathfrak{N}_{n,2}) = n'' \le n,$$

are both nth optimal information operators,

$$r(\mathfrak{N}_{n,1},I,\tilde{D}_\infty^r) = r(\mathfrak{N}_{n,2},I,\tilde{D}_\infty^r) = \tfrac{1}{2}d(n,I,\tilde{D}_\infty^r) = K_r/\lceil n/2 \rceil^{-r}.$$

From Kornejčuk [76, pp. 109–113], we find an optimal error algorithm for $\mathfrak{N}_{n,1}$. Namely, let

$$(5.31) \qquad \varphi_1(\mathfrak{N}_{n,1}(f)) = \frac{1}{2\pi}(f,1)$$

$$+ \frac{1}{\pi} \sum_{j=1}^{n'} \lambda_j((f, \sin jx) \sin jx + (f, \cos jx) \cos jx),$$

where

$$\lambda_j = \begin{cases} 1 - j^r \sum\limits_{i=1}^{\infty} (-1)^{i+1} \left[\dfrac{1}{(2in-j)^r} - \dfrac{1}{(2in+j)^r} \right], & r = 2, 4, \ldots, \\[4mm] 1 - j^r \sum\limits_{i=1}^{\infty} \left[\dfrac{1}{(2in-j)^r} - \dfrac{1}{(2in+j)^r} \right], & r = 1, 3, \ldots, \end{cases}$$

for $j = 1, 2, \ldots, n'$. In particular, for $r = 1$, $\lambda_j = (j\pi/2n)\cot(j\pi/2n)$. Thus, the algorithm φ_1 is a truncated Fourier series of f with coefficients $(f, \sin jx)$ and $(f, \cos jx)$ multiplied by λ_j. Since $e(\varphi_1) = K_r/\lceil n/2 \rceil^r$, φ_1 is a linear optimal error algorithm. Note that φ is not interpolatory because $\lambda_j \neq 1$.

To find an optimal error algorithm for $\mathfrak{N}_{n,2}$ define (as in Kornejčuk [76, pp. 287–290]) a unique 2π periodic spline g_i such that g_i is a 2π periodic function on R, $g_i^{(r-2)}$ is continuous, g_i is a polynomial of degree $\leq r - 1$ on every interval $[2(i-1)\pi/n'', 2i\pi/n'']$, $i = 1, 2, \ldots, n''$, and

$$g_i(t_j) = \delta_{ij}.$$

Define the algorithm

(5.32) $$\varphi_2(\mathfrak{N}_{n,2}(f)) = \sum_{i=1}^{n''} f(t_i)g_i.$$

Since $e(\varphi_2) = K_r/\lceil n/2 \rceil^r$, φ_2 is a linear optimal error algorithm.

We analyze the complexity of this problem. The ε-cardinality number $m = m(\Psi_U, I, \tilde{D}_\infty^r, \varepsilon)$ is the smallest n such that $K_r/\lceil n/2 \rceil^r < \varepsilon$. Thus,

$$m = 2\lfloor \sqrt[r]{K_r} \varepsilon^{-1/r} \rfloor + 1.$$

Note that $K_r \in [1, \pi/2]$ and $\sqrt[r]{K_r} \sim 1$. Due to the linearity of the optimal error algorithms (5.31) and (5.32), the ε-complexity of $(I, \tilde{D}_\infty^r)$ in Ψ_U is given by

$$\mathrm{comp}(\Psi_U, I, \tilde{D}_\infty^r, \varepsilon) = (c_1 + a)m - 1,$$

where c_1 is the complexity of evaluating a linear functional of the form $(f, \sin jx)$, $(f, \cos jx)$ or $f(t_j)$ and $a \in [1, 2]$. Algorithms (5.31) and (5.32) are nearly optimal complexity algorithms for $(I, \tilde{D}_\infty^r)$ in Ψ_U.

We summarize these results in

Theorem 5.4 Consider the approximation problem for $\mathfrak{I}_0 = \tilde{W}_\infty^r$ in the space $\tilde{C}$. Then:

(i) $\mathfrak{N}_{n,1}$ and $\mathfrak{N}_{n,2}$ defined by (5.29) and (5.30) are nth optimal information,

$$r(\mathfrak{N}_{n,i}, I, \tilde{D}_\infty^r) = \tfrac{1}{2}d(n, I, \tilde{D}_\infty^r) = K_r/\lceil n/2 \rceil^r, \qquad i = 1, 2.$$

(ii) φ_1 and φ_2 are linear optimal error and nearly optimal complexity algorithms for $(I, \tilde{D}_\infty^r)$ in Ψ_U,

$$e(\varphi_1) = e(\varphi_2) = K_r/\lceil n/2 \rceil^r.$$

(iii)   The ε-cardinality number of $(I, \tilde{D}_\infty^r)$ in Ψ_U is

$$m(\Psi_U, I, \tilde{D}_\infty^r, \varepsilon) = 2\lfloor \sqrt[r]{K_r} \varepsilon^{-1/r} \rfloor + 1.$$

(iv)   The ε-complexity of $(I, \tilde{D}_\infty^r)$ in Ψ_U is

$$\text{comp}(\Psi_U, I, \tilde{D}_\infty^r, \varepsilon) = (c_1 + a)(2\lfloor \sqrt[r]{K_r} \varepsilon^{-1/r} \rfloor + 1) - 1. \quad \blacksquare$$

We remark that the information $\mathfrak{N}_{n,2}$ and the algorithm φ_2 enjoy certain optimality properties for $\tilde{W}_\infty^r$ in the space L_1. More precisely, let $\mathfrak{I}_0 = \tilde{W}_\infty^r$ and $\mathfrak{I}_2 = L_1$. It is known (Kornejčuk [76, p. 290]) that

$$\|f - \varphi_2(\mathfrak{N}_{n,2}(f))\|_1 \le \frac{4K_{r+1}}{\lceil n/2 \rceil^r} \qquad \forall f \in \mathfrak{I}_0.$$

It is also known that $\lambda_n(\tilde{W}_\infty^r, L_1) = 4K_{r+1}/\lceil n/2 \rceil^r$ which means that φ_2 has minimal error among all linear algorithms which use a linear information operator with cardinality at most n. Unfortunately, we do not know the exact value of the Gelfand n-width and therefore we cannot conclude optimality of $\mathfrak{N}_{n,2}$. However, we know (see Tikhomirov [76, p. 249]) that

$$d^n(\tilde{W}_\infty^r, L_1) = \Theta(n^{-r}).$$

From this, it follows that $\mathfrak{N}_{n,2}$ is nearly an nth optimal information operator.

(v)   Approximation for W_∞^r in C

We deal with the uniform approximation problem for nonperiodic smooth functions. Let

(5.33) $\mathfrak{I}_1 = W_\infty^r[-1,1]$ and $\mathfrak{I}_2 = C = C[-1,1]$.

Then $Tf = D_\infty^r f = f^{(r)}$, $\mathfrak{I}_4 = L_\infty$. Thus,

(5.34) $\mathfrak{I}_0 = W_\infty^r$.

Without loss of generality, we assume $r \ge 1$. Note that $n^* = \dim(\ker D_\infty^r) = r$. We find an nth optimal information, $n \ge r$, using the Gelfand n-width of W_∞^r in C. Following Tikhomirov [76, pp. 130–135, 261–263], define $X_{k,r}$ as the class of perfect splines $f: [-1,1] \to \mathbb{R}$ of degree r which have k knots. That is, for every $f \in X_{k,r}$, there exist $t_i = t_i(f)$, $-1 \le t_1 \le t_2 \le \cdots \le t_k \le 1$, and $a_i = a_i(f) \in \mathbb{R}$, $i = 1, 2, \ldots, r$, such that

(5.35) $f(t) = \dfrac{(t+1)^r}{r!} + \sum\limits_{i=1}^{r} a_i t^{i-1} + \dfrac{2}{r!} \sum\limits_{i=1}^{k} (-1)^i (t - t_i)_+^r.$

Thus, $|f^{(r)}(t)| \equiv 1$, $f^{(r)}$ changes sign for at most k points and $f^{(r)}(-1) = 1$. Define $f_{k,r}$ as a perfect spline from $X_{k,r}$ with the minimal sup norm, i.e.,

(5.36) $\|f_{k,r}\| = \inf\limits_{f \in X_{k,r}} \|f\|,$

where $\|f\| = \sup\{|f(t)| : t \in [-1,1]\}$. It is known that such a $f_{k,r}$ exists. For instance,

$$f_{0,r}(t) = \frac{1}{r!2^{r-1}}\cos(r \arccos t), \qquad\qquad \|f_{0,r}\| = \frac{1}{r!2^{r-1}},$$

$$f_{k,1}(t) = t + 1 + 2\sum_{i=1}^{k}(-1)^i\left(t + 1 - \frac{2i}{k+1}\right)_+, \qquad \|f_{k,1}\| = \frac{1}{k}.$$

For fixed r and large k, we have

$$\|f_{k,r}\| = (2/\pi k)^r K_r(1 + o(1)),$$

where K_r is the Favard constant defined by (3.16). Furthermore, $f_{k,r}$ has $(k + r)$ distinct zeros in $[-1,1]$. Denote by

(5.37) $$x_i^* = x_i(n - r, r), \qquad i = 1, 2, \ldots, n,$$

the zeros of the function $f_{n-r,r}$ and let

$$A^n = \{f : f(x_i^*) = 0, i = 1, 2, \ldots, n\}.$$

Then

(5.38) $$d^n(W_\infty^r, C) = \|f_{n-r,r}\| = (2/\pi n)^r K_r(1 + o(1))$$

and A^n is an nth extremal subspace of W_∞^r in the sense of Gelfand. From this, we conclude that

(5.39) $$\mathfrak{N}_n(f) = [f(x_1^*), f(x_2^*), \ldots, f(x_n^*)]^t$$

is an nth optimal information operator and

(5.40) $$r(\mathfrak{N}_n, I, D_\infty^r) = \tfrac{1}{2}d(n, I, D_\infty^r) = \|f_{n-r,r}\|.$$

We turn to the question how to construct an optimal error algorithm for the nth optimal information $\mathfrak{N}_n$. In fact, using Micchelli, Rivlin, and Winograd [76], we give a linear optimal error algorithm for any information operator of the form

(5.41) $$\mathfrak{N}(f) = [f(x_1), f(x_2), \ldots, f(x_n)]^t$$

for points x_i not necessarily equal to x_i^* of (5.39). We assume that $-1 \le x_1 \le x_2 \le \cdots \le x_n \le 1$, where $x_i < x_{i+r}, i = 1, 2, \ldots, n - r$, with the convention that if $x_{i-1} < x_i = x_{i+1} = \cdots = x_{i+s} < x_{i+s+1}$, then $f(x_{i+j})$ is replaced by $f^{(j)}(x_i)$, $j = 1, 2, \ldots, s$. There exists a unique (up to multiplication by -1) perfect spline q,

(5.42) $$q(t) = \frac{(t+1)^r}{r!} + \sum_{i=1}^{r} a_i t^{i-1} + \frac{2}{r!}\sum_{i=1}^{n-r}(-1)^i(t - t_i)_+^r$$

with knots t_i satisfying $0 < t_1 < t_2 < \cdots < t_{n-r} < 1$ such that $q(x_i) = 0$, $i = 1$, $2, \ldots, n$. Note that $\|q^{(r)}\| = 1$.

To define a linear optimal error algorithm, we must compute the knots t_i of the perfect spline (5.42). Since the $t_i = t_i(n,r)$ are independent of f, they may be precomputed.

There exists a unique spline $\sigma = \sigma(\mathfrak{N}(f))$ of the form

$$\sigma(t) = \sum_{i=1}^{r} a_i t^{i-1} + \sum_{i=1}^{n-r} b_i(t - t_i)_+^{r-1}$$

(where t_i are the knots of (5.42)) such that

$$\sigma(x_i) = f(x_i), \qquad i = 1, 2, \ldots, n.$$

The coefficients a_i and b_i can be obtained by solving a system of linear equations which shows that σ depends linearly on $\mathfrak{N}(f)$. Let f_i be an element of W_∞^r such that $f_i(x_j) = \delta_{ij}$ and let $\sigma_i = \sigma(\mathfrak{N}(f_i))$. The splines σ_i can also be precomputed. Define the algorithm

(5.43)
$$\varphi(\mathfrak{N}(f)) = \sum_{i=1}^{n} f(x_i)\sigma_i.$$

Note that φ is linear but not interpolatory since φ produces an element of $W_\infty^{r-1}[-1,1]$, not necessarily an element in W_∞^r. A Fortran subroutine for the construction of algorithm (5.43) can be found in de Boor [77].

From Theorem 2 of Micchelli, Rivlin, and Winograd [76], it follows that

(5.44)
$$\|f - \varphi(\mathfrak{N}(f))\| \leq \sup_{h \in \ker \mathfrak{N} \cap \mathfrak{I}_0} \|h\| = r(\mathfrak{N}_n, I, D_\infty^r) = \|q\|.$$

Thus, $e(\varphi) = r(\mathfrak{N}_n, I, D_\infty^r)$ which proves that φ is an optimal error algorithm. The radius of information depends on the points at which f is evaluated. For $x_i = x_i^*$, (see (5.39) and (5.40)), we get a linear optimal error algorithm for the nth optimal information $\mathfrak{N}_n$. For equidistant points

(5.45)
$$x_i = -1 + 2(i - 1)/n, \qquad i = 1, 2, \ldots, n,$$

we get from Theorem 5 of Micchelli, Rivlin, and Winograd [76]

$$r(\mathfrak{N}_n, I, D_\infty^r) \leq \frac{(r-1)^{r-1}}{r!} \left(\frac{2}{n-1}\right)^r = \Theta(n^{-r}).$$

Comparing this with (5.38) and (5.40), we conclude that the information operator $\mathfrak{N}_n$ with equidistant points (5.45) is nearly an nth optimal information operator.

We analyze the complexity of this problem. The ε-cardinality number $m = m(\Psi_U, I, D_\infty^r, \varepsilon)$ is the smallest n such that $\|f_{n-r,r}\| < \varepsilon$. From (5.38), we conclude

$$m = (2/\pi)(K_r/\varepsilon)^{1/r}(1 + o(1)).$$

Due to the linearity of the optimal error algorithm (5.43), the ε-complexity of (I,D_∞^r) in Ψ_U is given by

$$\text{comp}(\Psi_U, I, D_\infty^r, \varepsilon) = (c_1 + a)m - 1,$$

where c_1 is the complexity of evaluating the linear functional $f(x)$ and $a \in [1,2]$. Algorithm (5.43) is a nearly optimal complexity algorithm for (I,D_∞^r) in Ψ_U.

We summarize these results in

Theorem 5.5   Consider the approximation problem for $\mathfrak{I}_0 = W_\infty^r$ in the space C. Then:

(i) $\mathfrak{N}_n$ defined by (5.39) is an nth optimal information,

$$r(\mathfrak{N}_n, I, D_\infty^r) = \tfrac{1}{2}d(n, I, D_\infty^r) = \|f_{n-r,r}\| = (2/\pi n)^r K_r(1 + o(1)).$$

(ii) φ defined by (5.43) is a linear optimal error and nearly optimal complexity algorithm for (I,D_∞^r) in Ψ_U, $e(\varphi) = \|f_{n-r,r}\|$.

(iii)   The ε-cardinality number of (I,D_∞^r) in Ψ_U is

$$m(\Psi_U, I, D_\infty^r, \varepsilon) = (2/\pi)(K_r/\varepsilon)^{1/r}(1 + o(1)).$$

(iv)   The ε-complexity of (I,D_∞^r) in Ψ_U is

$$\text{comp}(\Psi_U, I, D_\infty^r, \varepsilon) = (c_1 + a)(2/\pi)(K_r/\varepsilon)^{1/r}(1 + o(1)). \quad \blacksquare$$

6.   LINEAR PARTIAL DIFFERENTIAL EQUATIONS

In this section, we deal with partial differential equations of parabolic, hyperbolic, and elliptic type. For the sake of simplicity, we shall only consider rather easy examples of these differential equations.

We find new algorithms which use nth optimal information operators and which are linear central spline and nearly optimal complexity algorithms. We show that these new algorithms are "infinitely" better than the optimal error algorithms based on commonly used (nonoptimal) information operators.

Our analysis will be based on subsection (i), where nth optimal information operators and linear central spline and nearly optimal complexity algorithms are found for the approximation of a linear operator in a Hilbert space.

There have not been as many papers on the optimal solution of differential or integral equations as on the problems discussed in the previous sections. The list of relevant papers includes Bakhvalov [62a,b, 63, 70, 71b], Chzhan Guan-Tszynan [62], Emelyanov and Ilin [67], Gaisarian [70], and Weinberger [72].

(i)   A Linear Operator in a Hilbert Space

We must generalize the results obtained in subsection 5(i). As in subsection 5(i), let $H = \text{lin}(\xi_i, \xi_2, \ldots)$ be an infinite-dimensional Hilbert space over the real field $\mathbb{R}$, where $(\xi_i, \xi_j) = \delta_{ij}$. Thus, $f \in H$ iff $f = \sum_{i=1}^{\infty}(f, \xi_i)\xi_i$ and $\sum_{i=1}^{\infty}$ $(f, \xi_i)^2 < +\infty$. Let $\{\beta_i\}$ be a sequence of real numbers such that $|\beta_i| \leq |\beta_{i+1}|$ for all i. For the sake of simplicity, we assume that $\beta_i \neq 0$, $\forall i$. Let

$$(6.1) \qquad \mathfrak{I}_1 = \left\{ f : f \in H \text{ and } \sum_{i=1}^{\infty} \beta_i^2(f, \xi_i)^2 < +\infty \right\}$$

and $\mathfrak{I}_2 = H$. Define

$$(6.2) \qquad \mathfrak{I}_0 = \{ f \in \mathfrak{I}_1 : \|Tf\| \leq 1 \},$$

where

$$(6.3) \qquad Tf = \sum_{i=1}^{\infty} \beta_i(f, \xi_i)\xi_i$$

with $\mathfrak{I}_4 = T(\mathfrak{I}_1) = H$. Since $\beta_i \neq 0$, the operator T is invertible,

$$(6.4) \qquad T^{-1}f = \sum_{i=1}^{\infty} \frac{1}{\beta_i}(f, \xi_i)\xi_i.$$

Let u be an integer and denote

$$(6.5) \qquad v = v(i) = (i - 1)u, \qquad i = 1, 2, \ldots.$$

In this section, we consider the solution operator $S : \mathfrak{I}_1 \to \mathfrak{I}_2$ such that

$$(6.6) \qquad SA_v \subset A_v,$$

where

$$(6.7) \qquad A_v = \text{lin}(\xi_{v+1}, \xi_{v+2}, \ldots, \xi_{v+u})$$

for any $v = (i - 1)u$, $i = 1, 2, \ldots.$

Thus, the linear subspace A_v of dimension u is an invariant subspace of S. From (6.6), we get

$$(6.8) \qquad S\xi_{v+j} = \sum_{k=1}^{u} c_v(j,k)\xi_{v+k}$$

for some real constant $c_v(j,k)$.

As we know from Chapter 2, the nth minimal diameter and an nth optimal information depend on the eigenvalues and eigenelements of the operator $K_1 = K^*K : H \to H$, where $K = ST^{-1}$. To find the operator K_1 in terms of β_i and $c_v(j,k)$, define the $u \times u$ matrix $M_v = (m_v(i,j))$ by

$$(6.9) \qquad m_v(i,j) = c_v(j,i)/\beta_{v+j}, \qquad i, j = 1, 2, \ldots, u,$$

and the $u \times u$ matrix

$$(6.10) \qquad F_v = M_v^t M_v.$$

Finally, let

(6.11) $$\mathbf{f}_v = [(f,\xi_{v+1}), \ldots ,(f,\xi_{v+u})]^t.$$

By $(F_v\mathbf{f}_v)_j$, we mean the jth component of the vector $F_v\mathbf{f}_v$. Then the operator K_1 can be expressed as

(6.12) $$K_1 f = \sum_{i=1}^{\infty} \sum_{j=1}^{u} (F_{v(i)}\mathbf{f}_{v(i)})_j \xi_{v(i)+j}.$$

Note that the matrix F_v is symmetric and nonnegative definite, $F_v = F_v^t \geq 0$. Let $\mathbf{x}_{v+j}$ and λ_{v+j} be the eigenvectors and the eigenvalues of F_v,

(6.13) $$F_v\mathbf{x}_{v+j} = \lambda_{v+j}\mathbf{x}_{v+j}, \qquad j = 1, 2, \ldots, u,$$

where $(\mathbf{x}_{v+j},\mathbf{x}_{v+k}) = \delta_{jk}$, and define $x_v(j,i)$ by

(6.14) $$\mathbf{x}_{v+j} = [x_v(j,1),x_v(j,2), \ldots ,x_v(j,u)]^t.$$

Define

(6.15) $$\eta_{v+j} = \sum_{k=1}^{u} x_v(j,k)\xi_{v+k}$$

for $j = 1, 2, \ldots, u$, $v = (i-1)u$, and $i = 1, 2, \ldots$. From (6.12), (6.13), and (6.15), we get

(6.16) $$K_1\eta_{v+j} = \lambda_{v+j}\eta_{v+j}$$

which means that η_{v+j} is an eigenelement of K_1 and λ_{v+j} is the corresponding eigenvalue. Furthermore,

(6.17) $$(\eta_i,\eta_j) = \delta_{ij}, \qquad H = \text{lin}(\eta_1,\eta_2, \ldots).$$

Let λ_{i_j} be the jth largest eigenvalue of K_1, i.e.,

(6.18) $$\lambda_{i_1} \geq \lambda_{i_2} \geq \cdots.$$

Then K_1 is compact iff $\lim_j \lambda_{i_j} = 0$. From Section 5 of Chapter 2, it follows that the problem (S,T) is convergent iff $\lim_j \lambda_{i_j} = 0$.

To find an nth optimal information, observe that the index $n^* = \text{index}(S,T)$ is equal to zero. Then from Theorem 5.2 of Chapter 2, we conclude that the information

(6.19) $$\mathfrak{N}_n(f) = [(Tf,\eta_{i_1}),(Tf,\eta_{i_2}), \ldots ,(Tf,\eta_{i_n})]^t$$

is an nth optimal information and

(6.20) $$d(\mathfrak{N}_n,S,T) = d(n,S,T) = 2\sqrt{\lambda_{i_{n+1}}}.$$

We now find an optimal error algorithm which uses (6.19). Let

(6.21) $$\varphi(\mathfrak{N}_n(f)) = \sum_{j=1}^{n} (Tf,\eta_{i_j})ST^{-1}\eta_{i_j}.$$

We show that φ is a spline algorithm. Let $\sigma = \sum_{j=1}^{n}(Tf,\eta_{i_j})T^{-1}\eta_{i_j}$. Then $\varphi(\mathfrak{R}_n(f)) = S\sigma$ and it is enough to show that σ is a spline. Note that $(T\sigma,\eta_{i_j}) = (Tf,\eta_{i_j})$, due to (6.17). Furthermore, $\|T\sigma\|^2 = \sum_{j=1}^{n}(Tf,\eta_{i_j})^2$. If $g \in \mathfrak{I}_1$ and $(Tg,\eta_{i_j}) = (Tf,\eta_{i_j})$, then

$$\|Tg\|^2 = \sum_{i=1}^{\infty}(Tg,\eta_i)^2 \geq \sum_{j=1}^{n}(Tg,\eta_{i_j})^2 = \sum_{j=1}^{n}(Tf,\eta_{i_j})^2 = \|T\sigma\|^2.$$

This proves that σ is a spline interpolating $\mathfrak{R}_n(f)$ and φ is a spline algorithm. Theorem 5.1 of Chapter 4 yields that φ is a linear central spline algorithm and

(6.22) $$e(\varphi) = r(\mathfrak{R}_n,S,T) = \tfrac{1}{2}d(\mathfrak{R}_n,S,T) = \sqrt{\lambda_{i_{n+1}}}.$$

The algorithm φ can be expressed in terms of $\mathbf{x}_i$, β_i, and $c_\nu(j,k)$ as follows. Let

$$v = u\lfloor (i_j - 1)/u \rfloor, \qquad m = i_j - v.$$

Then from (6.15), we have

$$\eta_{i_j} = \eta_{v+m} = \sum_{k=1}^{u} x_\nu(m,k)\xi_{v+k}.$$

Since $T\xi_i = \beta_i\xi_i$, (6.8) yields

(6.23) $$ST^{-1}\eta_{i_j} = \sum_{p=1}^{u}\left(\sum_{k=1}^{u}\frac{c_\nu(k,p)}{\beta_{v+k}}x_\nu(m,k)\right)\xi_{v+p}.$$

REMARK 6.1   For $u = 1$, we can simplify the preceding formulas. Indeed, (6.8) can be now rewritten as

$$S\xi_i = c_{i-1}(1,1)\xi_i$$

which yields

$$K_1 f = \sum_{i=1}^{\infty}(c_{i-1}(1,1)/\beta_i)^2\xi_i$$

and

$$\lambda_i = (c_{i-1}(1,1)/\beta_i)^2, \qquad \eta_i = \xi_i.$$

The nth optimal information $\mathfrak{R}_n$ is now given by

$$\mathfrak{R}_n(f) = [\beta_{i_1}(f,\xi_{i_1}), \ldots, \beta_{i_n}(f,\xi_{i_n})]^t \asymp [(f,\xi_{i_1}), \ldots, (f,\xi_{i_n})]^t,$$

and the linear central spline algorithm φ is of the form

$$\varphi(\mathfrak{R}_n(f)) = \sum_{j=1}^{n}(f,\xi_{i_j})c_{i_j-1}(1,1)\xi_{i_j}. \quad \blacksquare$$

We analyze the complexity of this problem. The ε-cardinality number $m = m(\Psi_U,S,T,\varepsilon)$ (see (2.5) of Chapter 5) is equal to the smallest number n such that $\lambda_{i_{n+1}} < \varepsilon^2$. Due to the linearity of the optimal error algorithm (6.21), the ε-complexity of the problem (S,T) in the class Ψ_U is

$$\text{comp}(\Psi_U,S,T,\varepsilon) = (c_1 + a)m(\Psi_U,S,T,\varepsilon) - 1,$$

where c_1 is the complexity of evaluating a linear functional of the form (Tf,η_i) and $a \in [1,2]$. Algorithm (6.21) is a nearly optimal complexity algorithm for (S,T) in Ψ_U.

We can also find the Gelfand and the linear Kolmogorov n-widths of the solution set. For simplicity, assume that $q(S,\mathfrak{I}_0) = 1$ (see (6.6) of Chapter 2). This holds if, for instance, S is one-to-one. Then Remark 6.1 of Chapter 2 and Remark 5.3 of Chapter 3 yield

$$d^n(S(\mathfrak{I}_0),H) = \lambda_n(S(\mathfrak{I}_0),H) = \sqrt{\lambda_{i_{n+1}}}.$$

We summarize these results in

Theorem 6.1   Consider the approximation of the linear operator S for $\mathfrak{I}_0$ in the space H defined by (6.1)–(6.8). Then:

(i) $\mathfrak{N}_n$ defined by (6.19) is an nth optimal information,

$$r(\mathfrak{N}_n,S,T) = \tfrac{1}{2}d(n,S,T) = \sqrt{\lambda_{i_{n+1}}}.$$

(ii) φ defined by (6.21) is a linear central spline and nearly optimal complexity algorithm for (S,T) in Ψ_U,

$$e(\varphi) = \sqrt{\lambda_{i_{n+1}}}.$$

(iii)   The ε-cardinality number of (S,T) in Ψ_U is

$$m(\Psi_U,S,T,\varepsilon) = \min\{n : \lambda_{i_{n+1}} < \varepsilon^2\}.$$

(iv)   The ε-complexity of (S,T) in Ψ_U is

$$\text{comp}(\Psi_U,S,T,\varepsilon) = (c_1 + a)m(\Psi_U,S,T,\varepsilon) - 1. \quad \blacksquare$$

(ii)   Parabolic Differential Equation

We specialize the results of the previous subsection to the operator S defined as the solution operator for a parabolic differential equation. Let

$$\xi_k(x) = \sqrt{2/\pi}\sin kx, \qquad x \in [0,\pi], \quad k = 1, 2, \ldots,$$

and $\mathfrak{I}_2 = H = \text{lin}(\xi_1,\xi_2,\ldots)$ with the inner product

$$(f,g) = \int_0^\pi f(x)g(x)\,dx, \qquad \|f\|_2 = \sqrt{(f,f)}.$$

Let $\beta_k = (-1)^{r/2}k^r$ for an even nonnegative integer r. Then $f \in H$ means that f is an odd function and

$$f = \frac{2}{\pi}\sum_{k=1}^{\infty} (f, \sin kx) \sin kx$$

and $\mathfrak{I}_1 = \{f : \sum_{k=1}^{\infty} k^{2r}(f, \sin kx)^2 < +\infty\}$. Observe that

(6.24) $$Tf = \frac{2(-1)^{r/2}}{\pi} \sum_{k=1}^{\infty} k^r (f, \sin kx) \sin kx = f^{(r)}$$

and

(6.25) $$\mathfrak{I}_0 = \left\{ f \in \mathfrak{I}_1 : \|f^{(r)}\|_2 = \left(\frac{2}{\pi} \sum_{k=1}^{\infty} k^{2r}(f, \sin kx)^2 \right)^{1/2} \leq 1 \right\}.$$

Consider the parabolic differential equation

$$\frac{\partial u(x,t)}{\partial t} = \frac{\partial^2 u(x,t)}{\partial x^2} \qquad \text{for} \quad x \in (0,\pi), \quad t > 0,$$

(6.26)
$$u(0,t) = u(\pi,t) = 0,$$
$$u(x,0) = f(x),$$

where $f \in \mathfrak{I}_0$. We define the solution operator as

(6.27) $$(Sf)(x) = u(x,t_0), \qquad t_0 > 0,$$

i.e., S is the solution of (6.26) for a fixed $t = t_0$. As in Weinberger [72], we get by separation of variables that

(6.28) $$(Sf)(x) = \frac{2}{\pi} \sum_{k=1}^{\infty} (f, \sin kx)e^{-k^2 t_0} \sin kx.$$

Since

$$S\xi_k = e^{-k^2 t_0}\xi_k,$$

(6.6) and (6.8) hold with $u = 1$ and

$$c_{k-1}(1,1) = e^{-k^2 t_0}.$$

From Remark 6.1, we get

$$\lambda_k = e^{-2k^2 t_0}/k^{2r}.$$

Since $\{\lambda_k\}$ is a decreasing sequence, we get $\lambda_{i_k} = \lambda_k$, and again using Remark 6.1, we conclude that the information operator

(6.29) $$\mathfrak{N}_n(f) = [(f, \sin x), (f, \sin 2x), \ldots, (f, \sin nx)]^t$$

is an nth optimal information, the algorithm

(6.30) $$\varphi(\mathfrak{N}_n(f)) = \frac{2}{\pi} \sum_{k=1}^{n} (f, \sin kx)e^{-k^2 t_0} \sin kx$$

is a linear central spline algorithm, and

(6.31) $$e(\varphi) = r(\mathfrak{N}_n, S, T) = \tfrac{1}{2}d(n, S, T) = e^{-(n+1)^2 t_0}/(n+1)^r.$$

Note that φ is the initial section of Sf.

We analyze the complexity of this problem. Observe that the radius consists of two favorable factors $\exp(-(n + 1)^2 t_0)$ and $(n + 1)^{-r}$. This is an exponentially decreasing function of n and r. The ε-cardinality number $m = m(\Psi_U, S, T, \varepsilon)$ is the smallest integer n such that

$$e^{-(n+1)^2 t_0}/(n + 1)^r < \varepsilon.$$

From this, we get the asymptotic expression

$$m = \sqrt{\frac{\ln 1/\varepsilon}{t_0}}\left(1 - \frac{r \ln \ln 1/\varepsilon}{4 \ln 1/\varepsilon}(1 + o(1))\right).$$

Due to the linearity of the optimal error algorithm (6.30), the ε-complexity of (S, T) in Ψ_U is

$$\text{comp}(\Psi_U, S, T, \varepsilon) = (c_1 + a)m(\Psi_U, S, T, \varepsilon) - 1,$$

where c_1 is the complexity of evaluating a linear functional of the form $(f, \sin kx)$ and $a \in [1, 2]$. Algorithm (6.30) is a nearly optimal complexity algorithm for (S, T) in Ψ_U.

We summarize these results in

Theorem 6.2 Consider the solution of the parabolic differential equation defined by (6.24)–(6.27). Then:

(i) $\mathfrak{N}_n$ defined by (6.29) is an nth optimal information,

$$r(\mathfrak{N}_n, S, T) = \tfrac{1}{2}d(n, S, T) = e^{-(n+1)^2 t_0}/(n + 1)^r.$$

(ii) φ defined by (6.30) is a linear central spline and nearly optimal complexity algorithm for (S, T) in Ψ_U, $e(\varphi) = r(\mathfrak{N}_n, S, T)$.

(iii)   The ε-cardinality number of (S, T) in Ψ_U is

$$m(\Psi_U, S, T, \varepsilon) = \min\{n : e^{-(n+1)^2 t_0}/(n + 1)^r < \varepsilon\}$$

$$= \sqrt{\frac{\ln 1/\varepsilon}{t_0}}\left(1 - \frac{r \ln \ln 1/\varepsilon}{4 \ln 1/\varepsilon}(1 + o(1))\right).$$

(iv)   The ε-complexity of (S, T) in Ψ_U is

$$\text{comp}(\Psi_U, S, T, \varepsilon) = (c_1 + a)m(\Psi_U, S, T, \varepsilon) - 1. \quad \blacksquare$$

We wish to stress that the nth diameter is surprisingly small and it tends to zero extremely fast as n tends to infinity. Furthermore, even for small r, i.e., the class $\mathfrak{I}_0$ consists of functions of low regularity, the ε-cardinality number is asymptotically equal to $\sqrt{(\ln 1/\varepsilon)/t_0}$. For instance, if $\varepsilon = e^{-100} \cong 10^{-44}$ and $t_0 = 1$, it is enough to compute only *ten* functionals $(f, \sin x), \ldots, (f, \sin 10x)$ to solve this problem to within ε accuracy no matter how small the value of r. This may be contrasted with problems from the previous sections, where the ε-cardinality, roughly equal to $\varepsilon^{-1/r}$, depends strongly on r.

We proved that the information $\mathfrak{N}_n$ defined by (6.29) is optimal. This information is not commonly used in numerical practice for problem (6.26). Instead, the information

$$\bar{\mathfrak{N}}_n(f) = \left[f\left(\frac{\pi}{n+1}\right), f\left(\frac{2\pi}{n+1}\right), \ldots, f\left(\frac{n\pi}{n+1}\right) \right]^t$$

is used. It is interesting to ask how strong this information is and how much one loses using the information $\bar{\mathfrak{N}}_n$ instead of the nth optimal information $\mathfrak{N}_n$. For $r = 2$, i.e., $Tf = f''$, it easily follows from Weinberger [72] that

$$r(\bar{\mathfrak{N}}_n, S, T) = c_n(e^{-t_0}/(n+1)^2), \qquad c_n \in [\tfrac{1}{3}, 1].$$

Thus $r(\bar{\mathfrak{N}}_n, S, T) = \Theta(e^{-t_0}/(n+1)^2)$ which is significantly larger than $r(\mathfrak{N}_n, S, T)$. To find an ε-approximation using the information $\bar{\mathfrak{N}}_n$, we have to perform

$$n = n(\varepsilon) = \Theta((\varepsilon e^{t_0})^{-1/2})$$

function evaluations. The ε-penalty measured as the ratio of the cardinality of $\bar{\mathfrak{N}}_{n(\varepsilon)}$ and the ε-cardinality number is

$$\frac{n(\varepsilon)}{m(\Psi_U, S, T, \varepsilon)} = \Theta\left(\sqrt{\frac{t_0}{e^{t_0}\varepsilon \ln 1/\varepsilon}}\right) \to +\infty \qquad \text{as} \quad \varepsilon \to 0.$$

Thus, asymptotically in ε, the penalty is unbounded. For instance, if $\varepsilon = e^{-28}$ and $t_0 = 1$, this ratio is larger than 10^4!. This shows that $\bar{\mathfrak{N}}_n$ is very inefficient.

We want to stress that in our model of computation we assume that every linear functional costs the same. Thus, we assume that the evaluation of $\bar{\mathfrak{N}}_n(f) = [f(\pi/(n+1)), \ldots, f(n\pi/(n+1))]^t$ costs the same as the evluation of $\mathfrak{N}_n(f) = [(f, \sin x), \ldots, (f, \sin nx)]^t$. We want to discuss whether this assumption is realistic.

For many f arising in practice, such as polynomials, trigonometric, and special functions, the integration of $(f, \sin kx)$ can be done symbolically and the costs of evaluating $(f, \sin kx)$ and $f(x)$ can be comparable. In this case, the assumption looks reasonable. On the other hand, suppose that the assumption is violated and suppose that the evaluation of $(f, \sin kx)$ is j times more expensive than the evaluation of $f(x)$. Then the computation of $\bar{\mathfrak{N}}_{jn}(f)$ costs as much as the computation of $\mathfrak{N}_n(f)$. Since $r(\bar{\mathfrak{N}}_{jn}, S, T) = \Theta(e^{-t_0}/(jn)^2)$ is still proportional to n^{-2}, the asymptotic behavior of the ε-penalty is not changed. Thus, $\bar{\mathfrak{N}}_n$ is much more efficient than $\bar{\mathfrak{N}}_{jn}$ for small ε. This indicates that the information $\bar{\mathfrak{N}}_n$ should be used even if the evaluation of $(f, \sin kx)$ is much harder than the evaluation of $f(x)$.

We wish to stress that the preceding conclusion holds only if the integration $(f, \sin kx)$ is done symbolically. If one wants to approximate $(f, \sin kx)$ by evaluating f at a number of points, then instead of the information $\bar{\mathfrak{N}}_n$, one uses the information $\bar{\mathfrak{N}}_m(f) = [f(x_1), f(x_2), \ldots, f(x_m)]^t$ for some $x_1, x_2, \ldots, x_m$ and m. Then, the error of any algorithm has to be at least $r(\bar{\mathfrak{N}}_m, S, T)$ which, as can be shown, is at least $\Theta(m^{-2})$. Thus, the exponential decrease of the error has disappeared.

We end this subsection by a remark on a problem related to problem (6.27). So far, we have been interested in finding an optimal approximation to $u(x,t_0)$ for all $x \in [0,\pi]$ and for a *fixed* t_0. Suppose now that we want to approximate $u(x,t)$ for all $x \in [0,\pi]$ and *all* $t \in [0,\pi]$. That is,

(6.32) $Sf = u.$

We can solve (6.32) by the algorithm $\varphi(\mathfrak{N}_n(f),x,t)$ defined by (6.30) with $t_0 = t$. We measure the error of φ as

$$e(\varphi) = \sup_{f \in \mathfrak{I}_0} \sup_{0 \le t \le \pi} \left(\int_0^\pi (u(x,t) - \varphi(\mathfrak{N}_n(f),x,t))^2 \, dx \right)^{1/2}.$$

It is easy to show that $e(\varphi) = \Theta(n^{-r})$ and that the nth diameter of problem (6.32) is also $\Theta(n^{-r})$. Thus, the algorithm φ is nearly an optimal error algorithm for problem (6.32). However, the exponential decrease of its error as a function of n has disappeared.

For $r = 2$, the information operators $\overline{\mathfrak{N}}_n$ and $\mathfrak{N}_n$ are both nearly optimal since

$$r(\overline{\mathfrak{N}}_n, S, T) = \Theta(r(\mathfrak{N}_n, S, T)) = \Theta(n^{-2}).$$

(iii)   Elliptic Differential Equation

We now specialize the results of subsection (i) to the operator S defined as the solution operator for an elliptic differential equation.

Let $\mathfrak{I}_0, \mathfrak{I}_1, \mathfrak{I}_2$, and the operator T be defined as in subsection (ii). Consider the elliptic differential equation

$$\frac{\partial^2 u(x,y)}{\partial x^2} + \frac{\partial^2 u(x,y)}{\partial y^2} = 0 \qquad \text{for} \quad x, y \in (0,\pi),$$

(6.33) $u(0, y) = u(\pi, y) = u(x,0) = 0$ for $x, y \in [0,\pi],$

$\qquad\qquad u(x,\pi) = f(x)$ for $x \in [0,\pi].$

where $f \in \mathfrak{I}_0$. We define the solution operator as

(6.34) $(Sf)(x) = u(x,y_0), \qquad y_0 \in (0,\pi),$

i.e., S is the solution of (6.33) for a fixed $y = y_0$. As in Weinberger [72], we obtain through separation of variables that

(6.35) $(Sf)(x) = \dfrac{2}{\pi} \displaystyle\sum_{k=1}^\infty (f, \sin kx) \dfrac{\sinh ky_0}{\sinh k\pi} \sin kx.$

Thus,

$$S\xi_k = \frac{\sinh ky_0}{\sinh k\pi} \xi_k, \qquad \xi_k = \sqrt{\frac{2}{\pi}} \sin kx \qquad \forall k,$$

and (6.6), (6.8) hold with $u = 1$ and

$$c_{k-1}(1,1) = \sinh(ky_0)/\sinh k\pi.$$

From Remark 6.1, we get

$$\lambda_k = (\sinh(ky_0)/\sinh k\pi)^2/k^{2r}.$$

Since $\sinh x$ is an increasing function, $\{\lambda_k\}$ is a decreasing sequence, and $\lambda_{i_k} = \lambda_k$. Note that

$$\sqrt{\lambda_k} = \frac{1}{k^r} \frac{e^{ky_0} - e^{-ky_0}}{e^{k\pi} - e^{-k\pi}} = e^{-k(\pi - y_0)}(1 + o(1))/k^r.$$

We conclude from Remark 6.1 that the information operator

$$(6.36) \qquad \mathfrak{N}_n(f) = [(f, \sin x), (f, \sin 2x), \ldots, (f, \sin nx)]^t$$

is an nth optimal information, the algorithm

$$(6.37) \qquad \varphi(\mathfrak{N}_n(f)) = \frac{2}{\pi} \sum_{k=1}^{n} (f, \sin kx) \frac{\sinh ky_0}{\sinh k\pi} \sin kx$$

is a linear central spline algorithm, and

$$e(\varphi) = r(\mathfrak{N}_n, S, T) = \tfrac{1}{2} d(n, S, T)$$

$$= \frac{1}{(n+1)^r} \frac{\sinh(n+1)y_0}{\sinh(n+1)\pi} = e^{-(n+1)(\pi - y_0)} \frac{1 + o(1)}{(n+1)^r}.$$

Note that φ is the initial section of Sf. As in the parabolic case, the radius consists of two favorable factors and is an exponentially decreasing function of n and r.

We analyze the complexity of this problem. The ε-cardinality number $m = m(\Psi_U, S, T, \varepsilon)$ is the smallest integer n such that

$$\frac{1}{(n+1)^r} \frac{\sinh(n+1)y_0}{\sinh(n+1)\pi} < \varepsilon.$$

From this, we get the asymptotic expression

$$m = \frac{\ln 1/\varepsilon}{\pi - y_0} \left(1 - \frac{r \ln \ln 1/\varepsilon}{\ln 1/\varepsilon}(1 + o(1))\right).$$

Due to the linearity of the optimal error algorithm (6.37), the ε-complexity of (S, T) in Ψ_U is

$$\mathrm{comp}(\Psi_U, S, T, \varepsilon) = (c_1 + a)m(\Psi_U, S, T, \varepsilon) - 1,$$

where c_1 is the complexity of evaluating a linear functional of the form $(f, \sin kx)$ and $a \in [1,2]$. Algorithm (6.37) is a nearly optimal complexity algorithm for (S, T) in Ψ_U.

We summarize these results in

Theorem 6.3 Consider the solution of the elliptic differential equation defined by (6.34). Then:

(i) $\mathfrak{N}_n$ defined by (6.36) is an nth optimal information,

$$r(\mathfrak{N}_n,S,T) = \tfrac{1}{2}d(n,S,T) = \frac{1}{(n+1)^r}\frac{\sinh(n+1)y_0}{\sinh(n+1)\pi}$$

(ii) φ defined by (6.37) is a linear central spline and nearly optimal complexity algorithm for (S,T) in Ψ_U.

(iii) The ε-cardinality number of (S,T) in Ψ_U is

$$m(\Psi_U,S,T,\varepsilon) = \min\{n:\sinh((n+1)y_0)/(\sinh((n+1)\pi)(n+1)^r) < \varepsilon\}$$

$$= \frac{\ln 1/\varepsilon}{\pi - y_0}\left(1 - \frac{r\ln\ln 1/\varepsilon}{\ln 1/\varepsilon}(1 + o(1))\right).$$

(iv) The ε-complexity of (S,T) in Ψ_U is

$$\text{comp}(\Psi_U,S,T,\varepsilon) = (c_1 + a)m(\Psi_U,S,T,\varepsilon) - 1. \quad \blacksquare$$

We have shown that the nth diameter tends exponentially to zero and that for every r the ε-cardinality number is asymptotically equal to $\ln(1/\varepsilon)(\pi - y_0)$. For instance, if $\varepsilon = e^{-100}$ and $y_0 = 1$, it is enough to compute 50 functionals $(f, \sin x), \ldots, (f, \sin 50x)$ to solve this problem to within ε accuracy.

As we remarked in subsection (ii), the nth optimal information $\mathfrak{N}_n$ defined by (6.36) is not commonly used in numerical practice for the problem (6.33). Rather, one considers the information

$$\bar{\mathfrak{N}}_n(f) = \left[f\left(\frac{\pi}{n+1}\right), f\left(\frac{2\pi}{n+1}\right), \ldots, f\left(\frac{n\pi}{n+1}\right)\right]^{\mathrm{t}}.$$

From Weinberger [72], it follows that for $r = 2$,

$$r(\bar{\mathfrak{N}}_n,S,T) = \Theta(n^{-2})$$

which is significantly larger than $r(\mathfrak{N}_n,S,T)$. Let $n = n(\varepsilon)$ be the smallest integer such that $r(\bar{\mathfrak{N}}_n,S,T) < \varepsilon$. Then the ε-penalty which is the ratio of the cardinality of $\bar{\mathfrak{N}}_{n(\varepsilon)}$ and the ε-cardinality number is

$$\frac{n(\varepsilon)}{m(\Psi_U,S,T,\varepsilon)} = \Theta\left(\frac{1}{\sqrt{\varepsilon}\ln 1/\varepsilon}\right) \to +\infty \qquad \text{as} \quad \varepsilon \to 0.$$

This shows that $\bar{\mathfrak{N}}_n$ is very inefficient. From the discussion at the end of subsection (ii), it follows that one should use $\mathfrak{N}_n$ even if the evaluation of $(f, \sin kx)$ is much harder than the evaluation of $f(x_k)$.

As in subsection (ii), we remark on a problem related to problem (6.32). This is defined as the approximation of the solution $u(x,y)$ of (6.33) for *all* x, $y \in [0,\pi]$, i.e.,

$$Sf = u.$$

Define the algorithm $\varphi(\mathfrak{N}_n(f)) = \varphi(\mathfrak{N}_n(f),x,y)$ as in (6.37) with $y_0 = y$. We measure the error of φ as

$$e(\varphi) = \sup_{f \in \mathfrak{I}_0} \sup_{0 \leq y \leq \pi} \left(\int_0^\pi (u(x,y) - \varphi(\mathfrak{N}_n(f),x,y))^2 \, dx \right)^{1/2}.$$

It is easy to show that $e(\varphi) = \Theta(n^{-r})$ and the nth diameter of this generalized problem is also $\Theta(n^{-r})$. Thus, the algorithm φ is nearly an optimal error algorithm for this generalized problem. However, as in subsection (ii), the exponential decrease of its error as a function of n has disappeared. For $r = 2$, both information operators $\tilde{\mathfrak{N}}_n$ and $\mathfrak{N}_n$ are nearly optimal since

$$r(\tilde{\mathfrak{N}}_n,S,T) = \Theta(r(\mathfrak{N}_n,S,T)) = \Theta(n^{-2}).$$

(iv) Hyperbolic Differential Equation

We apply the results of subsection (i) for a hyperbolic differential equation defined as follows. Let

$$\xi_{2k-1}(x) = \sqrt{2/\pi} \sin kx, \qquad \xi_{2k}(x) = \sqrt{2/\pi} \cos kx, \qquad k = 1, 2, \ldots,$$

and $H = \text{lin}(\xi_1,\xi_2,\ldots)$ with the inner product

$$(f,g) = \int_0^\pi f(x)g(x)\,dx, \qquad \|f\|_2 = \sqrt{(f,f)}.$$

Thus $f \in H$ means that

$$f = \frac{2}{\pi} \sum_{k=1}^\infty ((f, \sin kx) \sin kx + (f, \cos kx) \cos kx)$$

and

$$\sum_{k=1}^\infty ((f, \sin kx)^2 + (f, \cos kx)^2) < +\infty.$$

Let $\beta_{2k-1} = \beta_{2k} = (-1)^{r/2}k^r$, $\forall k$, for an even r. Then

$$(6.38) \quad \mathfrak{I}_1 = \left\{ f : \sum_{k=1}^\infty k^{2r}((f, \sin kx)^2 + (f, \cos kx)^2) < +\infty \right\},$$

$$(6.39) \quad Tf = \frac{2(-1)^{r/2}}{\pi} \sum_{k=1}^\infty k^r((f, \sin kx) \sin kx + (f, \cos kx) \cos kx)$$

$$= f^{(r)},$$

and

$$(6.40) \quad \mathfrak{I}_0 = \left\{ f \in \mathfrak{I}_1 : \|f^{(r)}\|_2 = \left(\frac{2}{\pi} \sum_{k=1}^{\infty} k^{2r}((f, \sin kx)^2 + (f, \cos kx)^2) \right)^{1/2} \le 1 \right\}.$$

Consider the hyperbolic differential equation

$$(6.41) \qquad \frac{\partial u(x,t)}{\partial t} = \frac{\partial u(x,t)}{\partial x} \qquad \text{for} \quad x \in \mathbb{R}, \quad t > 0,$$

$$u(x,0) = f(x), \qquad x \in \mathbb{R},$$

where $f \in \mathfrak{I}_0$. We define the solution operator as

$$(6.42) \qquad\qquad (Sf)(x) = u(x,t_0), \qquad t_0 > 0, \quad x \in [0,\pi],$$

i.e., S is the solution of (6.41) for a fixed $t = t_0$. Since $(Sf)(x) = f(x + t_0)$, we get

$$(6.43) \quad (Sf)(x) = \frac{2}{\pi} \sum_{k=1}^{\infty} \{ [(f, \sin kx) \cos kt_0 - (f, \cos kx) \sin kt_0] \sin kx$$

$$+ [(f, \sin kx) \sin kt_0 + (f, \cos kx) \cos kt_0] \cos kx \}.$$

Observe that

$$S \sin kx = \cos kt_0 \sin kx + \sin kt_0 \cos kx,$$
$$S \cos kx = -\sin kt_0 \sin kx + \cos kt_0 \cos kx.$$

This implies that (6.6) holds with $u = 2$ and the 2×2 matrix M_v, $v = 2(k - 1)$, is given by

$$M_v = \frac{1}{k^r} \begin{bmatrix} \cos kt_0, & \sin kt_0 \\ -\sin kt_0, & \cos kt_0 \end{bmatrix}.$$

Since M_v has orthogonal rows, we find

$$F_v = (1/k^{2r})I,$$

where I is the 2×2 identity matrix. Thus,

$$\lambda_{2k-1} = \lambda_{2k} = k^{-2r},$$

and we may choose

$$\mathbf{x}_{2k-1} = [1,0]^t, \qquad \mathbf{x}_{2k} = [0,1]^t$$

as the eigenvectors of F_v. Then

$$\eta_{2k-1} = \xi_{2k-1} = \sqrt{2/\pi} \sin kx, \qquad \eta_{2k} = \xi_{2k} = \sqrt{2/\pi} \cos kx$$

are the eigenfunctions of K_1.

Since $\{\lambda_k\}$ is a nonincreasing sequence, we can set $\lambda_{i_k} = \lambda_k$. Let $n_1 = \lfloor n/2 \rfloor$, and $n_2 = 0$ if n is even and $n_2 = (n + 1)/2$ if n is odd. From (6.19), we get that the information operator

$$(6.44) \quad \mathfrak{N}_n(f) = [(f, \sin x), (f, \cos x), \ldots, (f, \sin n_1 x), (f, \cos n_1 x), (f, \sin n_2 x)]^t$$
$$\asymp [(Tf, \eta_1), (Tf, \eta_2), \ldots, (Tf, \eta_n)]^t$$

is an nth optimal information operator and

$$r(\mathfrak{N}_n, S, T) = \tfrac{1}{2}d(n, S, T) = \sqrt{\lambda_{n+1}} = \lceil (n+1)/2 \rceil^{-r}.$$

Since $r(\mathfrak{N}_{2n}, S, T) = r(\mathfrak{N}_{2n+1}, S, T) = (n+1)^{-r}$, the $2n$th optimal information $\mathfrak{N}_{2n}$ is also an $(2n+1)$th optimal information. From (6.21), we obtain that the algorithm

$$(6.45) \quad \varphi(\mathfrak{N}_{2n}(f)) = \frac{2}{\pi} \sum_{k=1}^{n} \{[(f, \sin kx) \cos kt_0 - (f, \cos kx) \sin kt_0] \sin kx$$
$$+ [(f, \sin kx) \sin kt_0 + (f, \cos kx) \cos kt_0] \cos kx\}$$

is a linear central spline algorithm,

$$e(\varphi) = r(\mathfrak{N}_{2n}, S, T) = r(\mathfrak{N}_{2n+1}, S, T) = (n+1)^{-r}.$$

Note that φ is the initial section of Sf. The radius of information is independent of t_0. The exponential factors $\exp(-(n+1)^2 t_0)$ and $\exp(-(n+1)(\pi - y_0))$ which appeared for the parabolic and elliptic equations are no longer present for the hyperbolic case. However, notice that the class $\mathfrak{I}_0$ for the hyperbolic case is larger than the corresponding class for the parabolic and elliptic cases.

We analyze the complexity of this problem. The ε-cardinality number $m = m(\Psi_U, S, T, \varepsilon)$ is equal to the smallest number n such that $\lceil (n+1)/2 \rceil^{-r} < \varepsilon$. Thus

$$m = 2\lfloor \varepsilon^{-1/r} \rfloor.$$

In contrast to the parabolic and elliptic cases, the ε-cardinality now depends critically on r. Due to the linearity of the optimal error algorithm (6.45), the ε-complexity of (S, T) in Ψ_U is

$$\mathrm{comp}(\Psi_U, S, T, \varepsilon) = (c_1 + a)m(\Psi_U, S, T, \varepsilon) - 1,$$

where c_1 is the complexity of evaluating a linear functional of the form $(f, \sin kx)$, $(f, \cos kx)$, and $a \in [1, 2]$. Algorithm (6.45) is a nearly optimal complexity algorithm for (S, T) in Ψ_U.

We summarize these results in

Theorem 6.4 Consider the solution of the hyperbolic differential equation defined by (6.41). Then:

(i) $\mathfrak{N}_{2n}(f) = [(f, \sin x), (f, \cos x), \ldots, (f, \sin nx), (f, \cos nx)]^t$ is both a $(2n)$th and a $(2n+1)$th optimal information,

$$r(\mathfrak{N}_{2n}, S, T) = \tfrac{1}{2}d(2n, S, T) = \tfrac{1}{2}d(2n+1, S, T) = (n+1)^{-r}.$$

(ii) φ defined by (6.45) is a linear central spline and nearly optimal complexity algorithm for (S, T) in Ψ_U, $e(\varphi) = r(\mathfrak{N}_{2n}, S, T)$.

(iii)   The ε-cardinality number of (S,T) in Ψ_U is
$$m(\Psi_U,S,T,\varepsilon) = 2\lfloor\varepsilon^{-1/r}\rfloor.$$

(iv)   The ε-complexity of (S,T) in Ψ_U is
$$\mathrm{comp}(\Psi_U,S,T,\varepsilon) = (c_1 + a)m(\Psi_U,S,T,\varepsilon) - 1. \quad \blacksquare$$

We want to stress that the nth diameters do not depend on the point t_0 in (6.42). This means, for every point t_0, problem (6.42) has the same complexity. This may be contrasted with the parabolic and elliptic differential equations (6.26) and (6.32), where the complexity decreases with increasing t_0 for the parabolic equation and with decreasing y_0 for the elliptic equation.

7.   SUMMARY AND OPEN PROBLEMS

We have applied the theoretical results of Chapters 1–5 to establish optimal error and nearly optimal complexity algorithms and optimal information operators for a number of important practical problems. These problems include examples from interpolation, differentiation, integration, approximation, and partial differential equations. They have been studied for certain classes $\mathfrak{I}_0$. In most cases, $\mathfrak{I}_0$ has been defined by a restriction operator T of the form $T = D_p^r$ and often p has been equal to two or infinity.

For the reader's convenience, we summarize and compare some of the optimality and complexity results obtained in this chapter. For many solution operators, we have considered information operators

$$(7.1) \qquad \mathfrak{N}_n(f) = [f(x_1), f(x_2), \ldots, f(x_n)]^t.$$

For some problems, such as interpolation and differentiation, the points x_i have been assumed given in advance. For other problems, such as integration and approximation, we have found the optimal points x_i, i.e., points for which the radius of information is minimized.

Information operators (7.1) are commonly used in practice. For some solution operators, such as integration and approximation, they enjoy certain optimality properties. In Tables 7.1–7.4 we summarize optimality and complexity results of four problems for the information operators (7.1) and for $\mathfrak{I}_0 = \{f : \|Tf\| \leq 1\}$ for different restriction operators.

TABLE 7.1

INTERPOLATION: $Sf = f(x_0)$

Restriction operator	Radius	Complexity	Comments	Reference
$T = D_\infty^r$	$O(n^{-r})$	$O(\varepsilon^{-1/r})$	x_i-equidistant	Th. 3.1
Analytic functions in a Hilbert space $T = I$	$(\zeta/R)^n$	$\Theta(\log 1/\varepsilon)$	$x_0 = 0$ $x_j = \zeta e^{2\pi i j/n}$	Th. 3.3

TABLE 7.2

DIFFERENTIATION: $Sf = f'(0)$

Restriction operator	Radius	Complexity	Comments	Reference
$T = D_\infty^n$	$\Theta(h^{n-1})$	$\Theta(n)$	n-odd, h-small $x_{2i} = ih, x_{2i-1} = -ih$	Th. 2.2

TABLE 7.3

INTEGRATION: $Sf = \int_a^b f(x)\,dx$

Restriction operator	Radius	Complexity	Comments	Reference
$T = D_p^r$	$\Theta(n^{-r})$	$\Theta(\varepsilon^{-1/r})$	$[a,b] = [0,1]$ x_i-optimal	Th. 4.1
$T = \tilde{D}_\infty^r$	$2\pi K_r/n^r$	$\Theta(\varepsilon^{-1/r})$	$[a,b] = [0,2\pi]$ $x_i = \dfrac{2\pi(i-1)}{n}$-optimal	Th. 4.2

TABLE 7.4

APPROXIMATION: $Sf = f$

Restriction operator	Radius	Complexity	Comments	Reference
$T = \tilde{D}_\infty^r$	$K_r\lceil n/2\rceil^{-r}$	$\Theta(\varepsilon^{-1/r})$	x_i-optimal	Th. 5.4
$T = D_\infty^r$	$(2/\pi n)^r K_r(1 + o(1))$	$\Theta(\varepsilon^{-1/r})$	x_i-optimal	Th. 5.5

For these problems, whenever there is a regularity parameter r, then we see that the complexity is asymptotically $\varepsilon^{-1/r}$, which is a decreasing function of r. This agrees with our intuition that it is easier to solve problems which are more regular. Open questions are:

(i) Is it true in general that more regular problems have lower complexity?

(ii) What characterizes the class of solution operators for which complexity is asymptotically $\varepsilon^{-1/r}$ for information operators (7.1) and for $\mathfrak{I}_0 = W_p^r$?

For analytic functions, the complexity of interpolation is asymptotically $\log 1/\varepsilon$. It is also known (Bojanov [74]) that for analytic functions the complexity of integration is asymptotically $(\log 1/\varepsilon)^2$. An open question is:

(iii) For analytic functions, what is the general form of solution operators for which complexity is asymptotically a power of $\log 1/\varepsilon$?

For the approximation problem, the information operator (7.1) with optimally chosen points is an nth optimal information operator for the class $\tilde{W}_\infty^r$

or W_∞^r. Open questions are:

(iv)   Is this true for the approximation problem with $\mathfrak{I}_0 = \tilde{W}_p^r$ or W_p^r for an arbitrary value of p?

(v)   What is the general form of solution operators for which an nth optimal information is of the form (7.1)?

Most of the results reported above depend critically on the particular norm and on the class $\mathfrak{I}_0$. For different choices of norms than those listed above, we sometimes know only the asymptotic behavior of the radius of information; its exact value is unknown. At best, we know only *nearly* optimal information operators. It is important to extend our knowledge of the exact value of the radius of information and optimal information operators to different choices of norm and/or to different choices of the class $\mathfrak{I}_0$.

TABLE 7.5

PARABOLIC EQUATION (6.24): $Sf = u(\cdot, t_0)$
AN nth OPTIMAL INFORMATION OPERATOR IS GIVEN BY

$$\mathfrak{N}_n(f) = [(f, \sin x), \ldots, (f, \sin nx)]^t$$

Restriction operator	nth Radius	Complexity	Comments	Reference
$T = D_2^r$	$\dfrac{\exp(-(n+1)^2 t_0)}{(n+1)^r}$	$\Theta(\sqrt{\log 1/\varepsilon})$	f-odd r-even	Th. 6.2

TABLE 7.6

ELLIPTIC EQUATION (6.32): $Sf = u(\cdot, y_0)$
AN nth OPTIMAL INFORMATION OPERATOR IS GIVEN BY

$$\mathfrak{N}_n(f) = [f, \sin x), \ldots, (f, \sin nx)]^t$$

Restriction operator	nth Radius	Complexity	Comments	Reference
$T = D_2^r$	$\dfrac{\exp(-(n+1)(\pi - y_0))}{(n+1)^r}(1 + o(1))$	$\Theta(\log 1/\varepsilon)$	f-odd r-even	Th. 6.3

TABLE 7.7

HYPERBOLIC EQUATION (6.41): $Sf = u(\cdot, t_0)$
AN nth OPTIMAL INFORMATION OPERATOR IS GIVEN BY

$$\mathfrak{N}_n(f) = [(f, \sin x), (f, \cos x), \ldots, (f, \sin n_1 x), (f, \cos n_1 x), (f, \sin n_2 x)]^t$$

Restriction operator	nth Radius	Complexity	Comments	Reference
$T = D_2^r$	$[(n+1)/2]^{-r}$	$\Theta(\varepsilon^{-1/r})$	r-even	Th. 6.4

In Section 6, we have dealt with one example each of a parabolic, elliptic, and hyperbolic differential equation. For the parabolic and elliptic problems, we have obtained the surprising results that the information operators (7.1) significantly differ from nth optimal information operators, the nth radius of information tends extremely fast to zero and complexity is asymptotically independent of regularity of functions from $\mathfrak{J}_0$. The results of Section 6 are summarized in Tables 7.5–7.7, where the nth optimal information operators, their radii, and complexities are listed.

There is a surprising difference in the complexity of these problems. We do not know if this is a property of only these specially chosen examples or whether it holds in general. A very significant research problem is:

(vi) For how general a parabolic equation, an elliptic equation or a hyperbolic equation is complexity asymptotically equal to $\sqrt{\ln 1/\varepsilon}$, $\ln 1/\varepsilon$, $\varepsilon^{-1/r}$, respectively?

In many cases, we have been able to establish the optimality of information operators by using deep results from approximation theory. The relations between n-widths and nth diameters of information have been especially fruitful. Although n-widths in the sense of Gelfand or Kolmogorov are currently known in only a few cases, there has recently been intensive research in this area. We anticipate that progress in the theory of n-widths will lead to further progress in analytic complexity. Indeed, we believe that research in both analytic complexity theory and approximation theory will lead to many new optimality results in both fields in the near future.

Finally, we again stress that the examples presented in this chapter serve primarily as illustrations of analytic complexity theory. We anticipate that for many practically important solution operators the problem of optimal information operators and optimal algorithms will be extensively studied in the future. This research should supply powerful and reliable algorithms used in practical computations.

Chapter 7

Theory of Nonlinear Information

1. INTRODUCTION

In this chapter, we study *nonlinear* information operators. We generalize the concepts of cardinality and optimality of nonlinear information operators with fixed cardinality. We show that for *any* information operator with cardinality n, there exists a nonlinear information operator with cardinality one and with the same diameter. We also prove that a problem S (not necessarily linear) can be solved exactly knowing only one suitably chosen nonlinear functional iff the cardinality of the set $S(\Im_0)$ is at most that of the continuum.

These results indicate that the class of nonlinear information operators of cardinality one supplies enough information to solve many problems exactly. However, the optimal nonlinear information operators are usually not regular and in many cases are not permissible. This shows that although it is possible to extend many results from Chapters 2–5 to nonlinear information operators, this is of only theoretical interest.

We summarize the results of this chapter. The concept of cardinality of *nonlinear* information is introduced in Section 2. In the following section, we show (Theorem 3.2) that there always exists a nonlinear information operator with cardinality one whose diameter of information is essentially as small as the nth minimal diameter. The class of problems which can be solved exactly knowing one suitably chosen functional is completely characterized in Theorem 3.3. In Section 4, relations are established between Kolmogorov n-widths and the errors of n-dimensional algorithms as well as between ε-entropy and the cardinality of information operators with radius of information less than ε.

2. CARDINALITY OF NONLINEAR INFORMATION

We generalize the theory of the cardinality of linear information given in Section 2 of Chapter 2 to a theory of the cardinality of nonlinear information. Let $\mathfrak{N}$ be an information operator. In Section 2 of Chapter 1, we show that the dependence of the radius and diameter of information $\mathfrak{N}$ for the problem S is only through the sets

$$(2.1) \qquad V(f) = V(f,\mathfrak{N}) = \{\tilde{f} : \mathfrak{N}(\tilde{f}) = \mathfrak{N}(f) \text{ and } \tilde{f} \in \mathfrak{I}_0\}.$$

This suggests we should not distinguish between two information operators with the same sets $V(f)$, $\forall f \in \mathfrak{I}_0$.

Let $\mathfrak{N}_1 : D_{\mathfrak{N}_1} \to \mathfrak{I}_3$ and $\mathfrak{N}_2 : D_{\mathfrak{N}_2} \to \mathfrak{I}_3'$ be two information operators, where $\mathfrak{I}_0 \subset D_{\mathfrak{N}_1} \cap D_{\mathfrak{N}_2}$ and $\mathfrak{I}_3'$ is a linear space not necessarily equal to $\mathfrak{I}_3$.

Definition 2.1   We shall say $\mathfrak{N}_1$ *is contained in* $\mathfrak{N}_2$ (briefly $\mathfrak{N}_1 \subset \mathfrak{N}_2$) iff

$$V(f,\mathfrak{N}_2) \subset V(f,\mathfrak{N}_1) \qquad \forall f \in \mathfrak{I}_0.$$

We shall say $\mathfrak{N}_1$ is equivalent to $\mathfrak{N}_2$ (briefly $\mathfrak{N}^1 \asymp \mathfrak{N}_2$) iff

$$V(f,\mathfrak{N}_2) = V(f,\mathfrak{N}_1) \qquad \forall f \in \mathfrak{I}_0. \qquad \blacksquare$$

Note that $\mathfrak{N}_1 \subset \mathfrak{N}_2$ means that $\mathfrak{N}_2(\tilde{f}) = \mathfrak{N}_2(f)$ implies $\mathfrak{N}_1(\tilde{f}) = \mathfrak{N}_1(f)$ for all $\tilde{f}, f \in \mathfrak{I}_0$. The relation "$\asymp$" is an equivalence relation. We verify that Definition 2.1 coincides with Definition 2.1 of Chapter 2 for the linear case.

Lemma 2.1   Let $\mathfrak{N}_i$ be a linear information operator $D_{\mathfrak{N}_i} = \mathfrak{I}_1$ for $i = 1$ and 2. Let $\mathfrak{I}_0$ be an absorbing set of $\mathfrak{I}_1$. Then

$$(2.2) \qquad \begin{array}{lll} \mathfrak{N}_1 \subset \mathfrak{N}_2 & \text{iff} & \ker \mathfrak{N}_2 \subset \ker \mathfrak{N}_1, \\ \mathfrak{N}_1 \asymp \mathfrak{N}_2 & \text{iff} & \ker \mathfrak{N}_2 = \ker \mathfrak{N}_1. \end{array}$$

PROOF   Suppose that $\mathfrak{N}_1 \subset \mathfrak{N}_2$ and let $h \in \ker \mathfrak{N}_2$. Since $\mathfrak{I}_0$ is absorbing, there exists a positive number $c = c(h)$ such that $ch \in \mathfrak{I}_0$ and $ch \in V(0,\mathfrak{N}_2)$, since $0 \in \mathfrak{I}_0$. Thus, $ch \in V(0,\mathfrak{N}_1)$ which means $\mathfrak{N}_1(ch) = c\mathfrak{N}_1(h) = 0$. Hence, $h \in \ker \mathfrak{N}_1$ which proves that $\ker \mathfrak{N}_2 \subset \ker \mathfrak{N}_1$.

Suppose now that $\ker \mathfrak{N}_2 \subset \ker \mathfrak{N}_1$ and let $\tilde{f} \in V(f,\mathfrak{N}_2)$. Then $h = \tilde{f} - f$ satisfies $\mathfrak{N}_2(h) = 0$, i.e., $h \in \ker \mathfrak{N}_2$ and $h \in \ker \mathfrak{N}_1$. Thus, $0 = \mathfrak{N}_1(h) = \mathfrak{N}_1(\tilde{f}) - \mathfrak{N}_1(f)$ which shows that $\tilde{f} \in V(f,\mathfrak{N}_1)$. Hence, $V(f,\mathfrak{N}_2) \subset V(f,\mathfrak{N}_1)$ which means $\mathfrak{N}_1 \subset \mathfrak{N}_2$.

To prove the second part of (2.2), observe that $\mathfrak{N}_1 \asymp \mathfrak{N}_2$ means $\mathfrak{N}_1 \subset \mathfrak{N}_2$ and $\mathfrak{N}_2 \subset \mathfrak{N}_1$ which due to the first part of (2.2) is equivalent to $\ker \mathfrak{N}_2 = \ker \mathfrak{N}_1$. $\quad \blacksquare$

We now show the relationship between the radii and diameters of contained or equivalent information operators.

Lemma 2.2   If

$$(2.3) \quad \mathfrak{N}_1 \subset \mathfrak{N}_2, \qquad \text{then} \quad r(\mathfrak{N}_2,S) \leq r(\mathfrak{N}_1,S) \quad \text{and} \quad d(\mathfrak{N}_2,S) \leq d(\mathfrak{N}_1,S),$$

$$(2.4) \quad \mathfrak{N}_1 \asymp \mathfrak{N}_2, \qquad \text{then} \quad r(\mathfrak{N}_2,S) = r(\mathfrak{N}_1,S) \quad \text{and} \quad d(\mathfrak{N}_2,S) = d(\mathfrak{N}_1,S). \quad \blacksquare$$

PROOF   Let $\mathfrak{N}_1 \subset \mathfrak{N}_2$. Then $V(f,\mathfrak{N}_2) \subset V(f,\mathfrak{N}_1)$ for any $f \in \mathfrak{J}_0$. From this, $U(f,\mathfrak{N}_2) = S(V(f,\mathfrak{N}_2)) \subset U(f,\mathfrak{N}_1) = S(V(f,\mathfrak{N}_1))$ and rad $U(f,\mathfrak{N}_2) \leq$ rad $U(f,\mathfrak{N}_1)$, diam $U(f,\mathfrak{N}_2) \leq$ diam $U(f,\mathfrak{N}_1)$. Taking the supremum with respect to $f, f \in \mathfrak{J}_0$, we see that (2.9) and (2.10) of Chapter 1 yield (2.3). Equation (2.4) follows easily from (2.3).   ∎

Let

$$(2.5) \qquad\qquad n = \dim \mathrm{lin}\ \mathfrak{N}(\mathfrak{J}_0), \qquad 0 \leq n \leq +\infty,$$

i.e., n is the smallest nonnegative integer, or infinity, such that there exists a linear subspace A of dimension n and $\mathfrak{N}(\mathfrak{J}_0) \subset A$. If $A = \mathrm{lin}(\xi_i)_{i \in I}$, where ξ_i is a basis of A, and the cardinality of the set I is n, $\mathrm{card}(I) = n$, then

$$(2.6) \qquad\qquad \mathfrak{N}(f) = \sum_{i \in I} L_i(f)\xi_i \qquad \forall f \in \mathfrak{J}_0,$$

where L_i is, in general, a nonlinear functional. (If $n = +\infty$ and there is no topology in $\mathfrak{J}_3$, it is assumed that only a finite number of $L_i(f)$ are nonzero for every $f \in \mathfrak{J}_0$.)

We show that any information operator may be represented by $n = \dim \mathrm{lin}\ \mathfrak{N}(\mathfrak{J}_0)$ functionals. Define the information operator $\mathfrak{N}_1 : \mathfrak{J}_0 \to \mathbb{C}^n$ as

$$(2.7) \qquad\qquad \mathfrak{N}_1(f) = [L_i(f)]^t_{i \in I}, \qquad n = \mathrm{card}(I),$$

where n and L_i are given by (2.5) and (2.6).

Lemma 2.3   Let $\mathfrak{N}$ and $\mathfrak{N}_1$ be defined as above. Then

$$(2.8) \qquad\qquad \mathfrak{N} \asymp \mathfrak{N}_1.   ∎$$

PROOF   Let $\mathfrak{N}(\tilde{f}) = \mathfrak{N}(f)$. Due to (2.6), this is equivalent to $L_i(\tilde{f}) = L_i(f)$ for $i \in I$. Thus, $\mathfrak{N}(\tilde{f}) = \mathfrak{N}(f)$ iff $\mathfrak{N}_1(\tilde{f}) = \mathfrak{N}_1(f)$ which proves (2.8).   ∎

Lemma 2.3 states that an information operator is equivalent to an information operator generated by n functionals and n is the smallest integer, or infinity, with this property. This leads us to the following definition of the cardinality of $\mathfrak{N}$ in $\mathfrak{J}_0$.

Definition 2.2   We shall say that $\mathrm{card}(\mathfrak{N},\mathfrak{J}_0)$ is the *cardinality of the information* $\mathfrak{N}$ in $\mathfrak{J}_0$ iff

$$(2.9) \qquad\qquad \mathrm{card}(\mathfrak{N},\mathfrak{J}_0) = \dim \mathrm{lin}\ \mathfrak{N}(\mathfrak{J}_0).   ∎$$

We verify that Definition 2.2 coincides with Definition 2.2 of Chapter 2 for a linear case.

Lemma 2.4   Let $\mathfrak{N}$ be a linear information operator $D_{\mathfrak{N}} = \mathfrak{J}_1$ and let $\mathfrak{J}_0$ be an absorbing set of $\mathfrak{J}_1$. Then

$$(2.10) \qquad\qquad \mathrm{card}(\mathfrak{N},\mathfrak{J}_0) = \dim \mathfrak{N}(\mathfrak{J}_1) = \mathrm{codim} \ker \mathfrak{N}.   ∎$$

PROOF   Since dim $\mathfrak{N}(\mathfrak{I}_1) = $ codim ker $\mathfrak{N}$, it is enough to prove the first part of (2.10). Let $g \in \mathfrak{N}(\mathfrak{I}_1)$, i.e., $g = \mathfrak{N}(f)$ for $f \in \mathfrak{I}_1$. There exists a positive number $c = c(f)$ such that $cf \in \mathfrak{I}_0$ and $\mathfrak{N}(cf) \in \mathfrak{N}(\mathfrak{I}_0) \subset$ lin $\mathfrak{N}(\mathfrak{I}_0)$. Thus, $g \in$ lin $\mathfrak{N}(\mathfrak{I}_0)$ which yields $\mathfrak{N}(\mathfrak{I}_1) \subset$ lin $\mathfrak{N}(\mathfrak{I}_0)$. Since the opposite inclusion is trivial, we get (2.10).   ∎

3.   OPTIMAL NONLINEAR INFORMATION

In Sections 4 and 6 of Chapter 2, we define the nth minimal diameter and a nth optimal information for the class of *linear* information operators with cardinality at most n and for *linear* solution operators. Here we extend these concepts to the class of *nonlinear* information operators $\mathfrak{N}$ with card$(\mathfrak{N},\mathfrak{I}_0) \le n$ for *nonlinear* solution operators. In Section 6 of Chapter 2, we define the class Ψ_n as the class of all linear information operators with card$(\mathfrak{N}) \le n$. We now extend the definition of Ψ_n by assuming that Ψ_n is the class of all *linear or nonlinear* information operators with card$(\mathfrak{N},\mathfrak{I}_0) \le n$.

Definition 3.1   We shall say $d(n,S)$ is the nth minimal diameter of information for the problem S iff

$$(3.1) \qquad\qquad d(n,S) = \inf_{\mathfrak{N} \in \Psi_n} d(\mathfrak{N},S).$$

We shall say $\mathfrak{N}^{oi}$ is an nth optimal information iff

$$(3.2) \qquad\qquad d(\mathfrak{N}_n^{oi},S) = d(n,S), \qquad \mathfrak{N}_n^{oi} \in \Psi_n. \quad ∎$$

For the linear case, we have shown that the sequence of the nth minimal diameters can be any nonincreasing sequence. For the nonlinear case, we shall prove that the sequence of nth minimal diameters is *constant*, $d(n,S) \equiv d(1,S)$. To show this rather surprising result, we first prove the following theorem.

Theorem 3.1   For any information operator $\mathfrak{N}$ with card$(\mathfrak{N},\mathfrak{I}_0) = n$, $n < +\infty$, there exists an information operator $\mathfrak{N}_1$ such that

$$(3.3) \qquad\qquad \mathfrak{N}_1 \asymp \mathfrak{N} \qquad \text{and} \qquad \text{card}(\mathfrak{N}_1,\mathfrak{I}_0) = 1. \quad ∎$$

PROOF   From Lemma 2.2, it follows that $\mathfrak{N} \asymp [L_1', L_2', \ldots, L_n']^t$ for certain functionals L_j'. Define

$$L_j(f) = \begin{cases} \text{Re } L_j'(f), & j = 1, 2, \ldots, n, \\ \text{Im } L_{j-n}'(f), & j = n + 1, \ldots, 2n. \end{cases}$$

Let $\mathbf{c} = [c_0, c_1, \ldots]$, where c_j is either zero or unity and only a finite number of c_j for odd j is equal to unity. Set

$$(3.4) \qquad\qquad g(\mathbf{c}) = \sum_{k=0}^{\infty} c_{2k} 2^{-k} + \sum_{k=1}^{\infty} c_{2k-1} 2^k.$$

The functional $L_j(f) = \sum_{k=-\infty}^{\infty} c_{k,j} 2^k$ with $c_{k,j} = c_{k,j}(f) \in \{0,1\}$ may be expressed as

$$L_j(f) = g([c_{0,j}, c_{1,j}, c_{-1,j}, \dots]).$$

Knowing $L_1(f), L_2(f), \dots, L_{2n}(f)$, we define the functional L as

$$L(f) = g([\mathbf{c}_0, \mathbf{c}_1, \mathbf{c}_{-1}, \dots]),$$

where $\mathbf{c}_k = [c_{k,1}, c_{k,2}, \dots, c_{k,2n}]$, $\forall k$.
   Let

(3.5) $\mathfrak{N}_1(f) = L(f)$ $\forall f \in \mathfrak{I}_0.$

Of course, $\mathrm{card}(\mathfrak{N}_1, \mathfrak{I}_0) = 1$. We prove that $\mathfrak{N}_1 \asymp \mathfrak{N}$. Indeed, $L(\tilde{f}) = L(f)$ means that $\mathbf{c}_k(\tilde{f}) = \mathbf{c}_k(f)$, $\forall k$, which is equivalent to $L'_j(\tilde{f}) = L'_j(f)$ for $j = 1, 2, \dots, n$. Thus, $\mathfrak{N}_1 \asymp [L'_1, L'_2, \dots, L'_n]^t \asymp \mathfrak{N}$. Hence, (3.3) is proven. ∎

   Note that Lemma 2.1 yields that $r(\mathfrak{N}_1, S) = r(\mathfrak{N}, S)$ and $d(\mathfrak{N}_1, S) = d(\mathfrak{N}, S)$. Thus, Theorem 3.1 states that even for an information operator $\mathfrak{N}$ of arbitrary cardinality, there exists one suitably chosen functional which supplies exactly the same information as $\mathfrak{N}$.
   Using Lemma 2.1, we can easily show that the sequence of nth minimal diameters is constant.

Theorem 3.2   For any solution operator S,

(3.6) $d(n,S) = d(1,S)$ $\forall n < +\infty.$ ∎

   PROOF   Let $\delta > 0$. Choose an information operator $\mathfrak{N}$ such that $\mathrm{card}(\mathfrak{N}, \mathfrak{I}_0) \le n$ and $d(\mathfrak{N}, S) \le d(n,S) + \delta$. From Theorem 3.1, there exists $\mathfrak{N}_1$ such that $\mathrm{card}(\mathfrak{N}_1, \mathfrak{I}_0) = 1$ and $\mathfrak{N}_1 \asymp \mathfrak{N}$. Thus, $d(1,S) \le d(\mathfrak{N}_1, S) = d(\mathfrak{N}, S) \le d(n,S) + \delta$. Since δ is arbitrary, we get $d(1,S) \le d(n,S)$. The opposite inequality is obvious which proves (3.6). ∎

   We now want to characterize solution operators S for which $d(1,S) = 0$, i.e., which problems can be solved exactly knowing one suitably chosen functional.

Theorem 3.3

(3.7) $d(1,S) = 0$

iff the cardinality of the set $S(\mathfrak{I}_0)$ is at most that of the continuum, i.e., $\mathrm{card}(S(\mathfrak{I}_0)) \le \mathfrak{C}$. ∎

   PROOF   Suppose first that $d(1,S) = 0$. This means that for every positive integer i, there exists an information operator $\mathfrak{N}_i$ such that $\mathrm{card}(\mathfrak{N}_i, \mathfrak{I}_0) = 1$ and $d(\mathfrak{N}_i, S) \le 2^{-i}$. Without loss of generality, we can assume that $\mathfrak{N}_i(f) = L_i(f)$ for a certain functional L_i. Let $f \in \mathfrak{I}_0$ and $y = [L_1(f), L_2(f), \dots]$. Choose

$f_i \in V(f,\mathfrak{N}_i) = \{\tilde{f} \in \mathfrak{I}_0 : L_i(\tilde{f}) = L_i(f)\}$ and define the function

$$(3.8) \qquad \varphi(y) = \sum_{i=0}^{\infty} (S(f_{i+1}) - S(f_i)) \qquad \text{with} \quad S(f_0) = 0.$$

The function $\varphi : D_\varphi \subset \mathbb{C}^\infty \to \mathfrak{I}_2$ is well defined since

$$\|S(f_{i+1}) - S(f_i)\| \le \|S(f_{i+1}) - S(f)\| + \|S(f_i) - S(f)\|$$
$$\le d(\mathfrak{N}_{i+1}, S) + d(\mathfrak{N}_i, S) \le 3 \cdot 2^{-(i+1)} \qquad \forall i > 1,$$

and the series in (3.8) is convergent. Furthermore,

$$\|\varphi(y) - S(f)\| = \left\| S(f_k) - S(f) + \sum_{i=k}^{\infty} (S(f_{i+1}) - S(f_i)) \right\|$$

$$\le \|S(f_k) - S(f)\| + 3 \sum_{i=k}^{\infty} 2^{-(i+1)} \le 4 \cdot 2^{-k}$$

which tends to zero with $k \to +\infty$. Thus, $\varphi(y) = S(f)$. This yields

$$\operatorname{card}(S(\mathfrak{I}_0)) = \operatorname{card}(\varphi(D_\varphi)) \le \operatorname{card}(D_\varphi) \le \operatorname{card}(\mathbb{C}^\infty) = \mathfrak{S}.$$

This completes the first part of the proof.

Suppose now that $\operatorname{card}(S(\mathfrak{I}_0)) \le \mathfrak{S}$. This means there exists a one-to-one function $g : S(\mathfrak{I}_0) \to \mathbb{R}$. Define the information operator

$$(3.9) \qquad \mathfrak{N}(f) = g(S(f)).$$

Obviously, $\operatorname{card}(\mathfrak{N}, \mathfrak{I}_0) \le 1$. To prove that $d(\mathfrak{N}, S) = 0$, observe that the algorithm

$$(3.10) \qquad \varphi(\mathfrak{N}(f)) = g^{-1}(\mathfrak{N}(f)) = S(f)$$

has the error $e(\varphi) = 0$. Hence, $d(1, S) \le d(\mathfrak{N}, S) \le 2e(\varphi) = 0$ which completes the proof. ∎

The assumption that $\operatorname{card}(S(\mathfrak{I}_0)) \le \mathfrak{S}$ holds for many practical problems. For instance, it holds if $S(\mathfrak{I}_0)$ is included in the algebraic sum $A + B$, where A is a finite-dimensional linear subspace of $\mathfrak{I}_2$ and the closure of B is compact in $\mathfrak{I}_2$.

We now show that if $\operatorname{card}(S(\mathfrak{I}_0)) > \mathfrak{S}$, then $d(1, S)$ can even be equal to infinity.

Example 3.1   Let $\mathfrak{I}_0 = \mathfrak{I}_1 = \mathfrak{I}_2$ be the set of complex functions defined on the complex plane $\mathbb{C}$ such that $\|f\| = \sup_{t \in \mathbb{C}} |f(t)| < +\infty$. Define $Sf = f$. We show

$$(3.11) \qquad d(1, S) = +\infty.$$

Indeed, suppose on the contrary there exists an information operator $\mathfrak{N}$, $\operatorname{card}(\mathfrak{N}, \mathfrak{I}_0) = 1$, such that $d(\mathfrak{N}, S) = d < +\infty$. Without loss of generality, we can assume $\mathfrak{N}(f) = L(f)$, where L is a functional. Let $X = L(\mathfrak{I}_0)$. For any $t \in X$, choose a function f_t such that $L(f_t) = t$. Then for any $f \in \mathfrak{I}_0$ with $L(f) = t$, we have

(3.12) $$\|f - f_t\| \leq \text{diam } V(f_t) \leq d(\mathfrak{N},S) = d,$$

where $V(f_t)$ is defined by (2.1). Define

(3.13) $$f(t) = \begin{cases} f_t(t) + d + 1 & \text{if} \quad t \in X \quad \text{and} \quad |f_t(t)| \leq d + 1, \\ 0 & \text{otherwise.} \end{cases}$$

Clearly $f \in \mathfrak{I}_0$, since $\|f\| \leq 2d + 1 < +\infty$. Let $y = L(f)$. Then

$$\|f - f_y\| \geq |f(y) - f_y(y)| \geq d + 1$$

which contradicts (3.12). Hence, $d(\mathfrak{N},S) = +\infty$ for any $\mathfrak{N}$ with $\text{card}(\mathfrak{N},\mathfrak{I}_0) \leq 1$, which implies $d(1,S) = 0$. Hence, (3.11) is proven.

Due to Theorem 3.2, $d(n,S) = +\infty$, $\forall n < +\infty$. This means that there does not exist a nonlinear information operator, even with arbitrarily large cardinality, which supplies enough information to solve this problem with finite error. ∎

We now turn to a model of computation. As always, we need to specify the set P of primitive operations. In Chapter 5, we used an idealized model which assumed that every linear functional is a primitive operation. Here we consider information operators generated by nonlinear functionals. If we assume that *every* nonlinear functional is primitive, then every nonlinear information operator with finite cardinality is permissible. However, it seems to us that this assumption is unrealistic. To illustrate this, we consider the following example.

Example 3.2 Let $\mathfrak{I}_0 = \mathfrak{I}_1 = \mathfrak{I}_2$ be the space of real continuous functions defined on $[0,1]$ with norm $\|f\| = \max_{0 \leq t \leq 1} |f(t)|$. Set $Sf = f$. It is well known that $\text{card}(S(\mathfrak{I}_0)) = \mathfrak{S}$ and Theorem 3.3 yields $d(1,S) = 0$. Furthermore, from the proof of Theorem 3.3, we know that the algorithm

(3.14) $$\varphi(\mathfrak{N}(f)) = g^{-1}(\mathfrak{N}(f)), \qquad \mathfrak{N}(f) = g(S(f)),$$

where $g: S(\mathfrak{I}_0) \to \mathbb{R}$ is a one-to-one function, has error $e(\varphi) = 0$. (See (3.9) and (3.10).) We specify algorithm (3.14) for our problem. Let $\{w_i\}_{i=1}^{\infty}$ be a sequence of all rational numbers from $[0,1]$. Let $f_i = f(w_i) = \sum_{i=-\infty}^{+\infty} c_{k,i} 2^{-k}$, where $c_{k,i} = c_{k,i}(f) \in \{0,1\}$. Since the set

$$A = \{(\mathbf{c}_1(f),\mathbf{c}_2(f), \ldots) : \mathbf{c}_i(f) = (\ldots,c_{-1,i},c_{0,i},c_{1,i}, \ldots) \text{ for } f \in \mathfrak{I}_0\}$$

has cardinality $\mathfrak{S}$, there exists a one-to-one function $h: A \to \mathbb{R}$. Define

(3.15) $$g(f) = h((\mathbf{c}_1(f),\mathbf{c}_2(f), \ldots)).$$

The function g is one-to-one. Indeed, $g(\tilde{f}) = g(f)$ implies $\mathbf{c}_i(\tilde{f}) = \mathbf{c}_i(f)$, $\forall i$, since h is one-to-one. Thus, $\tilde{f}(w_i) = f(w_i)$. Since $\{w_i\}$ is dense in $[0,1]$ and $\tilde{f}$, f are continuous and equal on $\{w_i\}$, then $\tilde{f} = f$. Note that to compute $\mathfrak{N}(f) = g(f)$ (i.e., to compute the nonlinear functional g), we have to know $\mathbf{c}_i(f)$, i.e., the values of f at *all* rational points of $[0,1]$. This cannot be done except for trivial cases of f.

Thus, it is unrealistic to include the nonlinear functional $\mathfrak{N}(f) = g(f)$ in the set of primitive operations. ∎

We hope that this example makes it clear that the class of nonlinear information operators with finite cardinality is rarely of practical interest. This indicates that even for nonlinear problems the study of the class of linear information operators and/or the class of suitably restricted nonlinear information operators is interesting. See Chapter 8, where we discuss linear information operators and some nonlinear information operators for nonlinear equations, and Part B, where we study linear "iterative" information operators for nonlinear problems.

4. RELATIONS TO KOLMOGOROV n-WIDTHS AND ENTROPY

This is the third section in which we show the relations between approximation theory and analytic complexity theory.

(i) We begin with a relation between Kolmogorov n-widths and the errors of n-dimensional algorithms. We remind the reader of the definition of the Kolmogorov n-width $d_n(X, \mathfrak{I}_2)$ of a balanced subset X of a linear normed space $\mathfrak{I}_2$,

$$(4.1) \qquad d_n(X, \mathfrak{I}_2) = \inf_{A_n} \sup_{x \in X} \inf_{a \in A_n} \|x - a\|,$$

where A_n is a subspace of $\mathfrak{I}_2$ of dimension $\leq n$. Thus, $A_n = \mathrm{lin}(\xi_1, \xi_2, \ldots, \xi_n)$ and $\inf_{a \in A_n} \|x - a\| = \inf_{c_1, \ldots, c_n} \|x - \sum_{i=1}^{n} c_i \xi_i\|$ denotes the best approximation of x by elements from A_n. The Kolmogorov n-width is the minimal error of the approximation of X by n-dimensional linear spaces. See Kolmogorov [36], where this concept was defined. See also Lorentz [66].

The Kolmogorov n-widths are related to the errors of n-dimensional algorithms. To show this, let $\mathfrak{N}$ be an information operator (not necessarily linear) and let

$$(4.2) \qquad \Phi_n(\mathfrak{N}) = \{\varphi : \dim \mathrm{lin}\, \varphi(\mathfrak{N}(\mathfrak{I}_0)) \leq n\}$$

be the class of algorithms which use the information operator $\mathfrak{N}$ and whose range is at most n dimensional. Let

$$(4.3) \qquad \Phi_n = \bigcup_{\mathfrak{N}} \Phi_n(\mathfrak{N})$$

be the union of such algorithms for all possible information operators. Finally, let

(4.4) $$d_n = d_n(S(\mathfrak{I}_0),\mathfrak{I}_2)$$

be the Kolmogorov n-width for the solution set $S(\mathfrak{I}_0)$ in $\mathfrak{I}_2$.

Theorem 4.1

(4.5) $$\inf_{\varphi \in \Phi_n} e(\varphi) = d_n. \quad \blacksquare$$

PROOF Let $\varphi \in \Phi_n$. Then there exists an information operator $\mathfrak{N}$ such that $\dim \operatorname{lin} \varphi(\mathfrak{N}(\mathfrak{I}_0)) \le n$. This means

$$\varphi(\mathfrak{N}(f)) = \sum_{i=1}^{n} c_i(\mathfrak{N}(f))\xi_i \qquad \forall f \in \mathfrak{I}_0,$$

for some elements $\xi_1, \xi_2, \ldots, \xi_n$ from $\mathfrak{I}_0$. Furthermore,

$$e(\varphi) = \sup_{f \in \mathfrak{I}_0} \left\| S(f) - \varphi(\mathfrak{N}(f)) \right\| \ge \sup_{f \in \mathfrak{I}_0} \inf_{c_1,\ldots,c_n} \left\| S(f) - \sum_{i=1}^{n} c_i\xi_i \right\|$$

$$\ge \inf_{\xi_1,\ldots,\xi_n} \sup_{f \in \mathfrak{I}_0} \inf_{c_1,\ldots,c_n} \left\| S(f) - \sum_{i=1}^{n} c_i\xi_i \right\| = d_n.$$

Since φ is arbitrary, $\inf\{e(\varphi):\varphi \in \Phi_n\} \ge d_n$. To prove the equality in (4.5), let A_n be a linear subspace of dimension $\le n$ such that

$$\sup_{x \in S(\mathfrak{I}_0)} \inf_{a \in A_n} \|x - a\| \le d_n + \delta,$$

where δ is a positive number. This means that for every Sf, $f \in \mathfrak{I}_0$, there exists an element $a = \sum_{i=1}^{n} c_i(Sf)\xi_i \in A_n$ such that

(4.6) $$\|S(f) - a\| \le d_n + \delta.$$

Define the information operator (in general nonlinear)

$$\mathfrak{N}(f) = [c_1(S(f)),c_2(S(f)), \ldots,c_n(S(f))]^{\mathrm{t}}$$

and the algorithm

$$\varphi(\mathfrak{N}(f)) = \sum_{i=1}^{n} c_i(S(f))\xi_i.$$

Then $\operatorname{card}(\mathfrak{N},\mathfrak{I}_0) \le n$ and $\dim \operatorname{lin} \varphi(\mathfrak{N}(\mathfrak{I}_0)) \le n$. Thus, $\varphi \in \Phi_n$. From (4.6), $e(\varphi) \le d_n + \delta$. Since δ is arbitrary, we get $\inf\{e(\varphi):\varphi \in \Phi_n\} \le d_n$. This completes the proof. $\blacksquare$

Theorem 4.1 states that any n-dimensional algorithm has error at least equal to the Kolmogorov n-width of the solution set. Furthermore, this lower bound is sharp. As we mentioned in Section 5 of Chapter 3, this result was obtained by many people for particular linear problems and usually for the class of linear algorithms.

(ii)   We now discuss a relation between the ε-entropy of the solution set $S(\mathfrak{I}_0)$ and the cardinality of an information operator $\mathfrak{N}$ with radius $r(\mathfrak{N},S) < \varepsilon$.

We remind the reader of the definition of the ε-entropy of a subset X of a linear normed space $\mathfrak{I}_2$. Let k denote the minimum number of subsets $U_1, U_2,$ $\ldots, U_k$ of $\mathfrak{I}_2$ which form an ε-covering of X, i.e., the diameter of each U_i does not exceed 2ε and the sets U_i cover X, $X \subset \bigcup_1^k U_i$. The ε-entropy of X is defined as

(4.7) $$H(\varepsilon, X) = \log k,$$

where log is the logarithm to base 2. See Kolmogorov [55], who defined this concept. See also Lorentz [66].

We now show that knowing the ε-entropy of $S(X)$, where $X \subset \mathfrak{I}_0$, we can sometimes estimate the cardinality of an information operator with radius $r(\mathfrak{N},S) < \varepsilon$ from below. Namely, let S be a linear or nonlinear solution operator and let $\mathfrak{N}$ be a linear or nonlinear information operator with $r(\mathfrak{N},S) < \varepsilon$. Of course, $\mathfrak{N} = \mathfrak{N}(\varepsilon)$. Without loss of generality, we can assume that

$$\mathfrak{N} = [L_1, L_2, \ldots, L_n]^t$$

for certain (in general nonlinear) real functionals $L_1, L_2, \ldots, L_n$, where $\text{card}(\mathfrak{N}, \mathfrak{I}_0) = n$. Assume that for a subset X of $\mathfrak{I}_0$, there exists a real number $M = M(\varepsilon) > 0$ such that

(4.8) $$\mathfrak{N}(X) \subset [-M, M]^n.$$

Then it is known that $H(\varepsilon, \mathfrak{N}(X)) \le n \log(M/\varepsilon)$. Since, by assumption, $r(\mathfrak{N}, S) < \varepsilon$, there exists an algorithm φ which uses $\mathfrak{N}$ and $e(\varphi) < \varepsilon$. Define

(4.9) $$K = K(\varepsilon) = \sup_{(y_1, y_2) \in Y} \|\varphi(y_1) - \varphi(y_2)\|,$$

where Y is the set of all (y_1, y_2), $y_i \in \mathfrak{N}(X)$ such that $\|y_1 - y_2\|_\infty \le 2\varepsilon$. Thus, $K(\varepsilon)$ measures a type of continuity of the algorithm φ. If φ is a Lipschitz operator, then $K(\varepsilon) \le C(\varphi)\varepsilon$, $\forall \varepsilon > 0$. We assume that $K(\varepsilon)$ is finite. We are ready to prove

Theorem 4.2   Let $\mathfrak{N}$ be an information operator such that $n = \text{card}(\mathfrak{N}, \mathfrak{I}_0)$ and $r(\mathfrak{N}, S) < \varepsilon$. Let $M(\varepsilon)$ and $K(\varepsilon)$ be defined by (4.8) and (4.9). Then

(4.10) $$n \ge \frac{H(\varepsilon + K(\varepsilon), S(X))}{\log M(\varepsilon)/\varepsilon}. \quad \blacksquare$$

PROOF   Let $U_1, U_2, \ldots, U_k$ be an ε-covering of $\mathfrak{N}(X)$. Due to (4.8), $k \le (M/\varepsilon)^n$. Note that $U_i \cap \mathfrak{N}(X)$ is nonempty. Let $y_i \in U_i \cap \mathfrak{N}(X)$ and $V_i = \{x : \|x - \varphi(y_i)\| \le \zeta\}$, where $\zeta = \varepsilon + K(\varepsilon)$. We show that $V_1, V_2, \ldots, V_k$ form an

ζ-covering of $S(X)$. It is enough to prove that $S(X) \subset \bigcup_1^k V_i$. For every $f \in X$, there exists y_i such that $\|\mathfrak{N}(f) - y_i\| \leq 2\varepsilon$. Then

$$\|S(f) - \varphi(y_i)\| \leq \|S(f) - \varphi(\mathfrak{N}(f))\| + \|\varphi(\mathfrak{N}(f)) - \varphi(y_i)\| \leq \varepsilon + K(\varepsilon) = \zeta$$

due to $e(\varphi) \leq \varepsilon$ and (4.9). Thus, $S(f) \in V_i$ which implies that $S(X) \subset \bigcup_1^k V_i$. Hence, $H(\zeta, S(X)) \leq \log k \leq n \log M/\varepsilon$. This proves (4.10). ∎

For nonlinear information operators, it may happen that $K(\varepsilon)/\varepsilon$ tends to infinity as ε tends to zero. Indeed, suppose that $\mathrm{card}(S(\mathfrak{I}_0)) \leq \mathfrak{S}$ and $H(\varepsilon, S(\mathfrak{I}_0))/\log 1/\varepsilon$ goes to infinity as ε tends to zero. In Section 3, we showed that there exists a (nonlinear) functional L_1 which supplies enough information to solve the problem S exactly. Furthermore, $L_1(\mathfrak{I}_0)$ may lie in a finite interval which means that (4.8) holds with a positive constant M independent of ε. We also showed an algorithm φ which uses L_1 and $e(\varphi) = 0$. Since $n = \mathrm{card}(\mathfrak{N}, \mathfrak{I}_0) = 1$, (4.10) with $X = \mathfrak{I}_0$ yields $K(\varepsilon)/\varepsilon \to +\infty$ with $\varepsilon \to 0$. Thus, algorithm φ is *not a Lipschitz operator*.

This may be contrasted with linear information operators, where (4.8) and (4.9) often hold when $M = 1$ and $K(\varepsilon) \leq C\varepsilon$. Indeed, let $\mathfrak{N} = [L_1, L_2, \ldots, L_n]^t$ be a linear information operator. Assume that the L_i are continuous on a balanced convex set X. Then, without loss of generality, we can assume that $\|L_i\|_X = 1$ which implies $\mathfrak{N}(X) \subset [-1.1]^n$. Thus, (4.8) holds with $M = 1$. Let φ be any algorithm which uses $\mathfrak{N}$ with $e(\varphi) \leq \varepsilon$. Then

$$\|\varphi(y_1) - \varphi(y_2)\| \leq \|\varphi(y_1) - S(f_1)\| + \|\varphi(y_2) - S(f_2)\| + \|S(f_1) - S(f_2)\|$$
$$\leq 2\varepsilon + \|S(f_1) - S(f_2)\|,$$

where $\mathfrak{N}(f_1) = y_1$, $\mathfrak{N}(f_2) = y_2$, and $f_1, f_2 \in X$. Define

$$G(\varepsilon) = \sup_{y_1, y_2 \in \mathfrak{N}(X)} \inf\{\|S(f_1) - S(f_2)\| : f_1, f_2 \in X, \mathfrak{N}(f_1) = y_1, \mathfrak{N}(f_2) = y_2\}.$$

Note that G depends only on S and $\mathfrak{N}$. If $G(\varepsilon) \leq c_1\varepsilon$, then (4.9) holds with

$$K(\varepsilon) = (2 + c_1)\varepsilon = c_2\varepsilon.$$

For many sets $S(X)$, $H((1 + c_2)\varepsilon, S(X)) = H(\varepsilon, S(X))(1 + o(1))$. In this case, Theorem 4.2 states that the cardinality of any information operator with radius less than ε has to satisfy

(4.11) $$n \geq \frac{H(\varepsilon, S(X))}{\log(1/\varepsilon)}(1 + o(1)) \qquad \text{for small} \quad \varepsilon.$$

We illustrate Theorem 4.2 and (4.11) by the following example.

Example 4.1   Let $\mathfrak{I}_1 = C^\infty$ be the class of real functions which are infinitely many times differentiable on $[-1,1]$. Let $\mathfrak{I}_0 = \mathfrak{I}_0(\zeta)$ be the subset of functions f from $\mathfrak{I}_1$ which can be extended to an analytic function bounded in modulus by unity in E_ζ. The region E_ζ of the complex plane is the interior of the ellipse with foci ± 1 on the real axis and with the sum of semimajor and semiminor axes ζ, $\zeta > 1$.

Let $Sf = f$ be the approximation operator and

$$\mathfrak{N}(f) = [f(x_1), f(x_2), \ldots, f(x_n)]^t$$

with $x_j = \cos((2j - 1)\pi/(2n))$, $j = 1, 2, \ldots, n$. From Osipenko [72] follows the existence of a linear optimal error algorithm φ which satisfies (4.9) with $K(\varepsilon) = O(\varepsilon)$ and whose error is

$$e(\varphi) = r(\mathfrak{N}, S) = 2\zeta^{-n} + O(\zeta^{-5n}).$$

Note that $\mathfrak{N}(\mathfrak{I}_0) = [-1, 1]^n$ which means that (4.8) holds for $X = \mathfrak{I}_0$ with $M = 1$. To assure $r(\mathfrak{N}, S) < \varepsilon$, we need

$$(4.12) \qquad n \geq \frac{\log(1/\varepsilon)}{\log \zeta}(1 + o(1)) \qquad \text{for small} \quad \varepsilon.$$

The ε-entropy of $S(\mathfrak{I}_0)$ was found by Vitushkin [59],

$$(4.13) \qquad H(\varepsilon, S(\mathfrak{I}_0)) = \frac{(\log(1/\varepsilon))^2}{2 \log \zeta}(1 + o(1)).$$

Thus, Theorem 4.2 yields that any information operator with radius less than ε for which $M = O(1)$ and $K(\varepsilon) = O(\varepsilon)$ has cardinality n such that

$$n \geq \frac{1}{2} \frac{\log(1/\varepsilon)}{\log \zeta}(1 + o(1)).$$

Note that (4.12) and (4.13) differ by a factor of $\frac{1}{2}$. This proves that the cardinality of the information based on function evaluation at Chebyshev points is close to optimal. ∎

In Theorem 4.2, we consider the entropy of $S(X)$ rather than $S(\mathfrak{I}_0)$. The reason for this is the observation that for many practical problems $\mathfrak{I}_0 = A + B$, where A is a linear subspace of finite dimension and B is a bounded set. It often happens that $S(\mathfrak{I}_0)$ and $\mathfrak{N}(\mathfrak{I}_0)$ are unbounded, where $\mathfrak{N}$ is a linear information operator with cardinality n and radius less than ε. Then $H(\varepsilon, \mathfrak{N}(\mathfrak{I}_0)) = H(\varepsilon, S(\mathfrak{I}_0)) = +\infty$, $\forall \varepsilon > 0$, and Theorem 4.2 is not applicable. However, setting $X = B$, we can often assure that (4.8) and (4.9) hold, and applying Theorem 4.2, we can find a lower bound on n.

As a final remark, note that Theorem 4.2 can be used as a tool to bound the entropy of $S(X)$ from above. Indeed, knowing the cardinality n of an information operator $\mathfrak{N}$ with radius less than ε and two numbers $M(\varepsilon)$ and $K(\varepsilon)$, (4.10) can be rewritten as

$$H(\varepsilon + K(\varepsilon), S(X)) \leq n \log M(\varepsilon)/\varepsilon.$$

The idea of using the ε-entropy of the solution set to establish how many evaluations are needed to find an ε-approximation may be found in several papers for different solution operators. Examples include Babuška and Sobolev [65] and Bakhvalov [62a]. See also Vitushkin [59], in which the problem of optimal coding for a given class of functions is studied in terms of the ε-entropy of the solution set.

The ε-entropy of many sets of practical interest may be found in Vitushkin [59], Kolmogorov [55], Kolmogorov and Tikhomirov [59], and Lorentz [66].

Chapter 8

Applications for Nonlinear Problems

1. INTRODUCTION

We show how the ideas of Chapters 1–7 may be applied to solve selected nonlinear problems using certain linear or nonlinear information operators. In Sections 2–5, we discuss the solution of a nonlinear equation $f = 0$, where f is a scalar function in Sections 2–4 and a multivariate function or operator in Section 5. In Sections 6 and 7, we discuss the search for the maximum of scalar unimodal functions.

In Section 7 of Chapter 2, we prove that adaptive linear information is not more powerful than nonadaptive linear information for linear problems. In sharp contrast to this, we shall show that adaptive linear information can be *exponentially* better than nonadaptive linear information for the nonlinear problems considered in this chapter.

Usually, nonlinear equations are solved by iterative algorithms using iterative information. Such information is studied in Part B. Here we show that nonlinear equations can also be solved in a Part A setting.

Kiefer [53] initiated research on optimal search for the maximum of a unimodal function. We shall show that "Kiefer information" is *not* optimal in the class of adaptive linear information using n linear functionals. We propose (Conjecture 7.1) that in this class the minimal radius of information is $(b - a)2^{-(n+1)}$, where $[a,b]$ is the initial interval of uncertainty. This may be contrasted with Kiefer's result of $(b - a)/(2F_n)$, where F_n is the nth Fibonacci number.

In Chapter 6, we present linear optimal error algorithms for a number of *linear* problems. The combinatory complexity of such algorithms is always proportional to n, where n is the cardinality of the information operator used by the algorithm. For the *nonlinear* problems considered in this chapter, we present several optimal error algorithms whose combinatory complexity is proportional to $\log n$, or is constant, i.e., the combinatory complexity is essentially less than the cardinality of the information operator.

We summarize the results of this chapter and give some references to the enormous literature.

In Section 2, we exhibit an nth optimal nonadaptive linear information for the solution of a nonlinear equation $f(x) = 0$, where f belongs to the class of scalar real continuous functions defined on $[a,b]$ such that $f(a) \leq 0$, $f(b) \geq 0$ and f has exactly one zero in $[a,b]$. This nth optimal information is $\mathfrak{N}(f) = [f(x_1), \ldots, f(x_n)]^t$, where $x_i = a + i(b - a)/(n + 1)$ and the radius $r(\mathfrak{N},S) = (b - a)/(2(n + 1))$. We also present an interpolatory central algorithm using $\mathfrak{N}$ whose combinatory complexity is proportional to $\log n$. This algorithm is an asymptotically optimal complexity algorithm.

In Section 3, we deal with adaptive linear information for the same solution operator as in Section 2. It is known that there exists adaptive information $\mathfrak{N}^a(f) = [f(y_1), \ldots, f(y_n)]^t$, where $y_i = y_i(y_1, f(y_1), \ldots, f(y_{i-1}))$ is the midpoint of the smallest interval containing the zero of f, such that $r(\mathfrak{N}^a, S) = (b - a)2^{-(n+1)}$. We show that the bisection algorithm is an interpolatory central algorithm and it is essentially an optimal complexity algorithm. We conjecture that the information $\mathfrak{N}^a$ is nth optimal in the class Ψ^a of adaptive linear information operators with cardinality at most n.

Section 4 deals with the solution of nonlinear scalar equations by means of a nonlinear information operator. This information operator consists of $(n - 2)$ function evaluations and lower and upper bounds on f' for the interval containing a zero of f. Based on Micchelli and Miranker [75], we present an interpolatory central algorithm whose combinatory complexity is proportional to $\log n$. This algorithm is an asymptotically optimal complexity algorithm.

Section 5 deals with the solution of nonlinear multivariate or operator equations in Banach spaces. We show that one step of some known and commonly used iterations is "asymptotically" an optimal error algorithm.

There are a number of papers in which the solution of nonlinear equations using noniterative algorithms is considered and in which optimal (or nearly optimal) algorithms are found. Examples include Booth [67, 69], Chernousko [68], Eichhorn [68], Gross and Johnson [59], Hyafil [77], Kiefer [57], Micchelli and Miranker [75], and Sukharev [75, 76], who considered the scalar case, and Majstrovskij [72], Sonnevend [77], and Todd [76], who considered the multivariate case.

In Sections 6 and 7, we discuss the search for the maximum for the class of scalar unimodal functions. Section 6 deals with nonadaptive linear information operators. We prove that the information $\mathfrak{N}(f) = [f(x_1), f(x_2), \ldots, f(x_n)]^t$ for specially chosen points x_i is nearly nth optimal with the radius of information roughly equal to $(b - a)/n$. We present an interpolatory central algorithm using $\mathfrak{N}$ whose combinatory complexity is proportional to $\log n$. This algorithm is an asymptotically optimal complexity algorithm.

Section 7 deals with Kiefer adaptive information which uses n function evaluations at points generated by the Fibonacci sequence. We modify Kiefer's prescription so as to obtain an interpolatory central algorithm which enjoys linear combinatory complexity. Due to the linearity of the combinatory complexity, this algorithm enjoys a type of optimal complexity. We report the classic result of Kiefer [53] about the optimality of his information for adaptive linear information based on n function evaluations. We show that the Kiefer information is *not* optimal in a wider class Ψ_n^a of adaptive linear information based on n linear functionals. We conjecture the value of the minimal radius of information for the class Ψ_n^a.

Kiefer's result initiated a stream of research concerning optimal search for the maximum of a function belonging to $\mathfrak{I}_0$ for different classes $\mathfrak{I}_0$. Examples include Adamski, Korytowski, and Mitkowski [77], Aphanasjev [74], Aphanasjev and Novikov [77], Avriel and Wilde [66], Beamer and Wilde [69–71], Chernousko [70a,b], Danilin [71], Eichhorn [68], Fine [66], Gal [72], Ganshin [76, 77], Ivanov [72a], Johnson [56], Judin and Nemirovsky [76a,b, 77], Karp and Miranker [68], Kiefer [57], Krolak [66, 68], Krolak and Cooper [63], Kuzovkin and Tikhomirov [67], Levin [65], Mockus [72], Newman [65], Piavsky [72], Sonnevend [77], Strongin [78], Sukharev [71, 72, 75], Tarassova [78], Wilde [64], Zaliznyak and Ligun [78], and Zhilinskas [75].

2. OPTIMAL NONADAPTIVE LINEAR INFORMATION FOR NONLINEAR SCALAR EQUATIONS

In this section, we deal with the solution of a nonlinear equation

$$(2.1) \qquad f(x) = 0,$$

where $f: [a,b] \to \mathbb{R}$ belongs to $\mathfrak{I}_0$ defined as

$$(2.2) \qquad \mathfrak{I}_0 = \{ f : f \text{ is continuous, } f(a) \le 0, f(b) \ge 0,$$
$$\text{and } f \text{ has exactly one zero in } [a,b] \}.$$

Thus, the solution operator is given by

$$(2.3) \qquad S(f) = f^{-1}(0).$$

Consider first a nonadaptive linear information operator

$$(2.4) \qquad \mathfrak{N}_n(f) = [f(x_1), \ldots, f(x_n)]^t, \qquad x_i = a + i(b - a)/(n + 1),$$
$$i = 1, 2, \ldots, n.$$

Let $j = j(\mathfrak{N}_n(f))$ be the index such that

$$f(x_j) \leq 0 \qquad \text{and} \qquad f(x_{j+1}) \geq 0,$$

where $x_0 = a$ and $x_{n+1} = b$. Define the algorithm

$$(2.5) \quad \varphi(\mathfrak{N}_n(f)) = \begin{cases} (x_j + x_{j+1})/2 & \text{if} \quad f(x_j)f(x_{j+1}) \neq 0, \\ x_i & \text{if} \quad f(x_i) = 0 \quad (i = j \quad \text{or} \quad i = j + 1). \end{cases}$$

If $f(x_j)f(x_{j+1}) \neq 0$, then it is clear that a zero of f lies in the interval (x_j, x_{j+1}) and any point of this interval can be a zero of f. This proves that the algorithm φ is interpolatory and central since $(x_j + x_{j+1})/2$ is a center of $U(f) = (x_j, x_{j+1})$ (see Section 2 of Chapter 1). Furthermore,

$$(2.6) \qquad\qquad e(\varphi) = r(\mathfrak{N}_n, S) = \max_j \frac{x_{j+1} - x_j}{2} = \frac{b - a}{2(n + 1)}.$$

We now show that the information (2.4) is an nth optimal information in the class of nonadaptive linear information operators with cardinality at most n. More precisely, let Ψ_n be the class of nonadaptive linear information operators of the form $\mathfrak{N}(f) = [L_1(f), L_2(f), \ldots, L_n(f)]^t$, where $L_1, L_2, \ldots, L_n$ are linear real functionals. An information operator $\mathfrak{N}_n^*$ is an nth *optimal nonadaptive linear information* iff

$$r(\mathfrak{N}_n^*, S) = r(n, S) = \inf_{\mathfrak{N} \in \Psi_n} r(\mathfrak{N}, S), \qquad \mathfrak{N}_n^* \in \Psi_n.$$

We are ready to prove

Theorem 2.1 The information $\mathfrak{N}_n$ defined by (2.4) is an nth optimal nonadaptive linear information and

$$(2.7) \qquad\qquad r(\mathfrak{N}_n, S) = r(n, S) = (b - a)/2(n + 1). \quad \blacksquare$$

PROOF   Let $\mathfrak{N} = [L_1, L_2, \ldots, L_n]^t$. Define the functions

$$h_i(x) = \begin{cases} 0 & \text{if} \quad x \notin [x_i, x_{i+1}], \\ x - x_i & \text{if} \quad x \in [x_i, z_i], \\ x_{i+1} - x & \text{if} \quad x \in [z_i, x_{i+1}], \end{cases}$$

$i = 0, 1, \ldots, n$, where $z_j = (x_j + x_{j+1})/2$ and $x_j = a + j(b - a)/(n + 1)$ for $j = 0, 1, \ldots, n + 1$. Let $h = \sum_{i=0}^n a_i h_i$. Choose a_i such that $h \in \ker \mathfrak{N}$. Since $h \in \ker \mathfrak{N}$ means $\sum_{i=0}^n a_i L_j(h_i) = 0$, $j = 1, 2, \ldots, n$, we get n linear homogeneous equations in $(n + 1)$ unknowns. This has a nonzero solution and we can assume that $\max|a_i| = 1$ and

$$a_0 = a_1 = \cdots = a_{k-1} = 0, \qquad a_k > 0,$$

for an integer k. Let $\delta \in (0, \mu/3)$, where $\mu = (b - a)/(n + 1)$ and $\alpha = x_{k+1} - \delta$. Define the function

$$f(x) = \begin{cases} (a_k\delta/(\mu - 2\delta))(x - \alpha) & \text{if } x \le \alpha, \\ x - \alpha & \text{if } x \ge \alpha. \end{cases}$$

Note that f is continuous, $f(a) < 0$, $f(b) > 0$ and f has exactly one zero α. Thus $f \in \mathfrak{I}_0$. Let $f_1(x) = f(x) + h(x)$. Since $h(a) = h(b) = 0$, we have $f_1(a) < 0$ and $f_1(b) > 0$. For $x \le x_k$, $h(x) \equiv 0$ and $f_1(x) = f(x) < 0$. For $x \in [x_k, z_k]$, $f_1(x) = a_k\delta(x - \alpha)/(\mu - 2\delta) + a_k(x - x_k)$ and f_1 has exactly one zero $x = x_k + \delta$. For $x \in [z_k, \alpha]$, we have $f_1(x) = a_k\delta(x - \alpha)/(\mu - 2\delta) + a_k(x_{k+1} - x)$. Since $\delta \le \mu/3$, $f'_1(x) \le 0$ and $f_1(x) \ge f_1(\alpha) = a_k\delta > 0$. Furthermore, for $x \in (\alpha, b]$, $f_1(x) = x - \alpha + h(x) \ge f_1(\alpha)$ which proves that f_1 does not have zeros in $[z_k, b]$. Thus, f_1 has exactly one zero in $[a, b]$ and $f_1 \in \mathfrak{I}_0$. Since $h \in \ker \mathfrak{N}$, $\mathfrak{N}(f_1) = \mathfrak{N}(f)$ and

$$S(f) - S(f_1) = x_{k+1} - \delta - x_k - \delta = (b - a)/(n + 1) - 2\delta.$$

Thus, $r(\mathfrak{N}, S) \ge \frac{1}{2}d(\mathfrak{N}, S) \ge \frac{1}{2}|S(f) - S(f_1)| \ge (b - a)/(2(n + 1)) - \delta$. Since δ is arbitrary, we get

$$r(\mathfrak{N}, S) \ge (b - a)/(2(n + 1))$$

for any nonadaptive linear information operator $\mathfrak{N}$, $\operatorname{card}(\mathfrak{N}) \le n$. From this and (2.6), we conclude (2.7) which completes the proof. ∎

We now analyze the complexity of problem (2.3).

If we want to solve the nonlinear equation (2.1) to within ε accuracy by the use of a nonadaptive linear information operator, we have to guarantee that

$$r(n, S) < \varepsilon.$$

As in Chapter 5, define the ε-cardinality number $m(\Psi_U, S, \varepsilon)$ of the problem S in the class $\Psi_U = \bigcup_{n=1}^{\infty} \Psi_n$ as the smallest integer n such that $r(n, S) < \varepsilon$. From Theorem 2.1, we get

(2.8) $$m(\Psi_U, S, \varepsilon) = \lfloor (b - a)/2\varepsilon \rfloor.$$

We discuss the combinatory complexity of the central algorithm φ defined by (2.5). To compute $\varphi(\mathfrak{N}_n(f))$, we need to know the index $j = j(\mathfrak{N}_n(f))$ such that $f(x_j) \le 0$ and $f(x_{j+1}) \ge 0$. It is obvious that this can be done in $\lceil \log n \rceil$ comparison operations and $\lceil \log n \rceil$ arithmetic operations. If comparison and the arithmetic operations are primitives, then the combinatory complexity of φ is proportional to $\log n$,

(2.9) $$\sup_{f \in \mathfrak{I}_0} \operatorname{comp}(\varphi(\mathfrak{N}_n(f))) = \Theta(\log n).$$

Note that for large n, the combinatory complexity of φ is significantly smaller than the cardinality of the information operator $\mathfrak{N}_n$. This may be contrasted with linear problems for which linear optimal error algorithms have combinatory complexity proportional to n. (See Chapters 5 and 6.)

From (2.8) and (2.9), we conclude that the ε-complexity of the problem S in the class Ψ_U (see Definition 3.2 of Chapter 1) is

(2.10) $\mathrm{comp}(\Psi_U,S,\varepsilon) = (c_1 + o(1))\lfloor (b - a)/2\varepsilon \rfloor,$

where c_1 is the complexity of one function evaluation. Note that the complexity of the algorithm φ using the information operator $\mathfrak{N}_n$ with $n = m(\Psi_U,S,\varepsilon)$ is also of the form (2.10). Furthermore,

(2.11) $\dfrac{\mathrm{comp}(\varphi)}{\mathrm{comp}(\Psi_U,S,\varepsilon)} = 1 + o(\varepsilon) \qquad \text{as} \quad \varepsilon \to 0.$

We shall say φ^{aoc} is an *asymptotically optimal complexity algorithm for S in* Ψ_U iff φ^{aoc} satisfies (2.11).

We summarize these results in

Theorem 2.2   Consider the solution of the nonlinear equation (2.1) for $\mathfrak{I}_0$ defined by (2.2). Then:

(i) $\mathfrak{N}_n$ defined by (2.4) is an nth optimal nonadaptive linear information,

$$r(\mathfrak{N},S) = r(n,S) = (b - a)/(2(n + 1)).$$

(ii) φ defined by (2.5) is an interpolatory central and asymptotically optimal complexity algorithm for S in Ψ_U, $e(\varphi) = r(\mathfrak{N}_n,S)$.

(iii)   The ε-cardinality number of S in Ψ_U is $m(\Psi_U,S,\varepsilon) = \lfloor (b - a)/2\varepsilon \rfloor$.

(iv)   The ε-complexity of S in Ψ_U is $\mathrm{comp}(\Psi_U,S,\varepsilon) = (c_1 + o(1))\lfloor (b - a)/2\varepsilon \rfloor$.

∎

3.   ADAPTIVE LINEAR INFORMATION
FOR NONLINEAR SCALAR EQUATIONS

We proved in Section 7 of Chapter 2 that adaptive linear information operators do not supply more knowledge about *linear* problems than nonadaptive ones. Here we show that for the *nonlinear* problem (2.3) an adaptive information operator $\mathfrak{N}_n^{\mathrm{a}}$, $\mathrm{card}(\mathfrak{N}_n^{\mathrm{a}}) \leq n$, supplies significantly more knowledge than the nth optimal nonadaptive linear information operators, i.e.,

$$r(\mathfrak{N}_n^{\mathrm{a}},S) \ll r(n,S).$$

To show this, let

(3.1) $\mathfrak{N}_n^{\mathrm{a}}(f) = [f(y_1),f(y_2),\ldots,f(y_n)]^{\mathrm{t}},$

where the $y_i = y_i(f(y_1),\ldots,f(y_{i-1}))$ are inductively defined as follows.

Let $a_0 = a$ and $b_0 = b$. Suppose that a_j, b_j are defined for $j = 0, 1, \ldots, i - 1$. Then

$$y_i = \tfrac{1}{2}(a_{i-1} + b_{i-1}),$$

(3.2) $\qquad a_i = \begin{cases} a_{i-1} & \text{if } f(y_i) > 0, \\ y_i & \text{if } f(y_i) \leq 0, \end{cases} \qquad b_i = \begin{cases} b_{i-1} & \text{if } f(y_i) < 0, \\ y_i & \text{if } f(y_i) \geq 0. \end{cases}$

Note that $[a_i, b_i]$ is the interval containing the zeros of the functions from $\mathfrak{I}_0$ which agree with f at $y_1, y_2, \ldots, y_i$. Furthermore,

$$b_i - a_i = \begin{cases} (b - a)2^{-i} & \text{if } f(y_j) \neq 0, \quad j = 1, 2, \ldots, i, \\ 0 & \text{otherwise.} \end{cases}$$

It is clear that the bisection algorithm defined as

(3.3) $\qquad\qquad \varphi(\mathfrak{N}_n(f)) = \tfrac{1}{2}(a_n + b_n)$

is an interpolatory central algorithm and

(3.4) $\qquad\qquad e(\varphi) = r(\mathfrak{N}_n^a, S) = (b - a)2^{-(n+1)}.$

Since the minimal nth radius of nonadaptive information operators is $(b - a)/(2(n + 1))$, we have obtained an exponential improvement. To find an ε-approximation, it is now enough to perform n function evaluations, where $n = n(\varepsilon) = \lfloor \log(b - a)/\varepsilon \rfloor$. The information complexity of $\mathfrak{N}_{n(\varepsilon)}^a$ (see (3.4) of Chapter 1) is

$$\text{comp}(\mathfrak{N}_{n(\varepsilon)}^a) = (c_1 + 3)n(\varepsilon) - 1,$$

where c_1 is the complexity of one function evaluation, and the complexity of addition, division, and comparison is taken as unity. Observe that the combinatory complexity of the bisection algorithm is two. From this, we conclude that the ε-complexity of the problem S for the information $\mathfrak{N}_{n(\varepsilon)}^a$ (see Definition 3.1 of Chapter 1) is

$$\text{comp}(\mathfrak{N}_{n(\varepsilon)}^a, S, \varepsilon) = (c_1 + 3)\lfloor \log(b - a)/\varepsilon \rfloor + a_1,$$

where $a_1 \in [-1, 1]$. This also proves that the complexity of the bisection algorithm differs from the ε-complexity of S with $\mathfrak{N}_{n(\varepsilon)}^a$ by at most two.

We summarize these results in

Theorem 3.1 Consider the problem S and the information $\mathfrak{N}_n^a$ defined by (2.2), (2.3), and (3.1). Then:

(i) The bisection algorithm defined by (3.3) is an interpolatory central algorithm. This algorithm is essentially an optimal complexity algorithm for S with $\mathfrak{N}_{n(\varepsilon)}^a$, since its complexity differs from the ε-complexity of S with $\mathfrak{N}_{n(\varepsilon)}^a$ by at most two.

(ii)   The radius of information is

$$r(\mathfrak{N}_n^a, S) = (b - a)2^{-(n+1)}.$$

(iii)   The ε-complexity of S with $\mathfrak{N}_{n(\varepsilon)}^a$ is

$$\text{comp}(\mathfrak{N}_{n(\varepsilon)}, S, \varepsilon) = (c_1 + 3)\lfloor \log(b - a)/\varepsilon \rfloor + a_1. \quad \blacksquare$$

It is natural to inquire about the optimality of the information $\mathfrak{N}_n^a$ defined by (3.1) in the class of adaptive linear information operators. It is known that $\mathfrak{N}_n^a$ is optimal in the class of information operators of the form $\mathfrak{N}^a(f) = [f(x_1), f(x_2), \ldots, f(x_n)]^t$, where $x_i = x_i(f(x_1), \ldots, f(x_{i-1}))$ is an arbitrary function. See, for instance, Kung [76]. It seems plausible to us to conjecture that $\mathfrak{N}_n^a$ remains optimal even if arbitrary adaptive information operators are permitted. More precisely, let Ψ_n^a be the class of adaptive linear information operators of the form $\mathfrak{N}^a(f) = [L_1(f), L_2(f; L_1(f)), \ldots, L_n(f; L_1(f), \ldots)]^t$, where L_i linearly depends on the first argument. We propose the following conjecture.

Conjecture 3.1 $\mathfrak{N}_n^a$ defined by (3.1) is an nth optimal adaptive linear information in Ψ_n^a, i.e.,

$$\inf_{\mathfrak{N} \in \Psi_n^a} r(\mathfrak{N}, S) = r(\mathfrak{N}_n^a, S) = (b - a)2^{-(n+1)}. \quad \blacksquare$$

4. NONLINEAR INFORMATION

FOR NONLINEAR SCALAR EQUATIONS

We present an example of a nonlinear information operator for the solution of nonlinear scalar equations. This example is based on Micchelli and Miranker [75]. Let $\mathfrak{I}_1$ be the space of real functions $f:[a,b] \to \mathbb{R}$ such that f is absolutely continuous and $f' \in L_\infty$. Let m_0, M_0 be two given numbers such that $0 < m_0 \le M_0 < +\infty$. Define

(4.1) $\mathfrak{I}_0 = \{f \in \mathfrak{I}_1 : f(a) \le 0, f(b) \ge 0, \text{ and } f'(x) \in [m_0, M_0]$
                    almost everywhere for $x \in [a,b]\}$.

Note that every f from $\mathfrak{I}_0$ has exactly one zero in $[a,b]$. Therefore, the solution operator

(4.2) $Sf = f^{-1}(0) \qquad \forall f \in \mathfrak{I}_0,$

is well defined. Let $n \ge 4$ and

$$x_{i+1} = a + i(b - a)/(n - 3), \qquad i = 0, 1, \ldots, n - 3.$$

Consider the information operator

(4.3) $\mathfrak{N}_n(f) = [f(x_1), f(x_2), \ldots, f(x_{n-2}), m(f), M(f)]^t$

for

(4.4)
$$m(f) = \inf_{x \in [x_j, x_{j+1}]} \text{ess } f'(x), \qquad M(f) = \sup_{x \in [x_j, x_{j+1}]} \text{ess } f'(x),$$

where $f(x_j) \le 0$ and $f(x_{j+1}) \ge 0$. Thus, $\mathfrak{N}_n$ is nonlinear since $m(f)$ and $M(f)$ depend nonlinearly on f. The information $\mathfrak{N}_n$ is also adaptive since the interval $[x_j, x_{j+1}]$ which affects $m(f)$ and $M(f)$ depends on the computed values $f(x_1), \ldots, f(x_{n-2})$. Due to (4.1), $m(f) \ge m_0 > 0$ and $M(f) \le M_0 < +\infty$.

We wish to find a central algorithm and the radius of information. For $i = 1, 2, \ldots, n-3$, let

$$M_i = \begin{cases} M_0 & \text{for } i \neq j, \\ M(f) & \text{for } i = j, \end{cases} \qquad m_i = \begin{cases} m_0 & \text{for } i \neq j, \\ m(f) & \text{for } i = j. \end{cases}$$

Further, let

$$\underline{z}_i = x_{i+1} - \delta_i, \qquad \bar{z}_i = x_i + \delta_i,$$

where $\delta_i = (x_{i+1} - x_i)((f_{i+1} - f_i)/(x_{i+1} - x_i) - m_i)/(M_i - m_i)$, $f_i = f(x_i)$, with the convention $0/0 = 0$. Thus $\underline{z}_i, \bar{z}_i \in [x_i, x_{i+1}]$.

For any $i \in [1, n-2]$ and any $x \in [x_{i-1}, x_{i+1}]$ (with $x_0 = a$ and $x_{n-1} = b$), we have

$$f(x) = f(x_i) + \int_{x_i}^x f'(x)\,dx.$$

This yields

(4.5)
$$\underline{f}(x) \le f(x) \le \bar{f}(x) \qquad \forall x \in [a, b],$$

where

$$\underline{f}(x) = \begin{cases} f(x_i) + m_i(x - x_i) & \text{if } x \in [x_i, \underline{z}_i], \\ f(x_{i+1}) + M_i(x - x_{i+1}) & \text{if } x \in [\underline{z}_i, x_{i+1}], \end{cases}$$

$$\bar{f}(x) = \begin{cases} f(x_i) + M_i(x - x_i) & \text{if } x \in [x_i, \bar{z}_i], \\ f(x_{i+1}) + m_i(x - x_{i+1}) & \text{if } x \in [\bar{z}_i, x_{i+1}]. \end{cases}$$

Note that $\underline{f}$ and $\bar{f}$ belong to $\mathfrak{I}_0$ and $\mathfrak{N}_n(\underline{f}) = \mathfrak{N}_n(\bar{f}) = \mathfrak{N}_n(f)$. Let $\alpha_1 = \alpha_1(\mathfrak{N}_n(f))$ be the zero of $\bar{f}$ and $\alpha_2 = \alpha_2(\mathfrak{N}_n(f))$ be the zero of $\underline{f}$. Of course, α_1 and α_2 belong to $[x_j, x_{j+1}]$. From (4.5), we get that for any $\tilde{f}$ from $\mathfrak{I}_0$ such that $\mathfrak{N}_n(\tilde{f}) = \mathfrak{N}_n(f)$,

$$\alpha_1 \le S(\tilde{f}) \le \alpha_2$$

and any point from the interval $[\alpha_1, \alpha_2]$ can be the value of $S(\tilde{f})$. Thus, the algorithm

(4.6)
$$\varphi(\mathfrak{N}_n(f)) = \tfrac{1}{2}(\alpha_1(\mathfrak{N}_n(f)) + \alpha_2(\mathfrak{N}_n(f)))$$

is an interpolatory central algorithm.

The local error of φ (see (2.2) of Chapter 4) is

(4.7) $e(\varphi,f) = \text{rad}([\alpha_1,\alpha_2]) = \frac{1}{2}(\alpha_2 - \alpha_1).$

It is possible to verify that

(4.8) $e(\varphi,f) \leq \dfrac{M(f) - m(f)}{M(f) + m(f)} \dfrac{x_{j+1} - x_j}{2}$

and there exists a function f from $\mathfrak{I}_0$ such that the equality holds in (4.8). Note that for regular f, (4.8) implies a kind of "quadratic" bound of the error of the central algorithm φ. Namely, if f' is a Lipschitz function in $[x_j,x_{j+1}]$, then

$$\frac{M(f) - m(f)}{M(f) + m(f)} = O(x_{j+1} - x_j)$$

and

$$e(\varphi,f) = O((x_{j+1} - x_j)^2) = O(n^{-2}).$$

To find the radius of information, observe that $(M(f) - m(f))/(M(f) + m(f)) \leq (M_0 - m_0)/(M_0 + m_0)$. Since there exists a function f from $\mathfrak{I}_0$ such that $m(f) = m_0$ and $M(f) = M_0$, we conclude

(4.9) $r(\mathfrak{R}_n,S) = \dfrac{M_0 - m_0}{M_0 + m_0} \dfrac{b - a}{2(n - 3)}.$

We analyze the complexity of this problem. To find an ε-approximation, we have to compute $(n - 2)$ function evaluations and two nonlinear functionals (4.4), where

(4.10) $n = n(\varepsilon) = \left\lfloor \dfrac{(M_0 - m_0)(b - a)}{(M_0 + m_0)2\varepsilon} \right\rfloor + 4.$

Assume that c_1 is the complexity of one function evaluation and c_2 is the complexity of the evaluation of the nonlinear functionals $m(f)$ and $M(f)$. Then the information complexity of $\mathfrak{R}_{n(\varepsilon)}$ is $\text{comp}(\mathfrak{R}_{n(\varepsilon)}) = c_1(n(\varepsilon) - 2) + 2c_2$. Let the arithmetic operations and comparison be primitives. As in Section 2, it is easy to verify that the combinatory complexity of the algorithm φ defined by (4.6) is proportional to $\log n(\varepsilon)$. Thus, the complexity of φ is

$$\text{comp}(\varphi) = c_1(n(\varepsilon) - 2) + 2c_2 + \Theta(\log n(\varepsilon)).$$

From this, we conclude that the ε-complexity of the problem S with the information $\mathfrak{R}_{n(\varepsilon)}$ is

$$\text{comp}(\mathfrak{R}_{n(\varepsilon)},S,\varepsilon) = (c_1 + o(1))n(\varepsilon) + 2(c_2 - c_1).$$

This also shows that the algorithm φ is an asymptotically optimal complexity algorithm for S with $\mathfrak{R}_{n(\varepsilon)}$. (See (2.11).)

We summarize these results in

Theorem 4.1 Consider the problem S and the information $\mathfrak{N}_n$ defined by (4.1)–(4.4). Then:

(i) φ defined by (4.6) is an interpolatory central and asymptotically optimal complexity algorithm for S with $\mathfrak{N}_{n(\varepsilon)}$.

(ii) The radius of information is

$$r(\mathfrak{N}_n,S) = \frac{M_0 - m_0}{M_0 + m_0} \frac{b - a}{2(n - 3)}.$$

(iii) The ε-complexity of S with $\mathfrak{N}_{n(\varepsilon)}$ is

$$\text{comp}(\mathfrak{N}_{n(\varepsilon)},S,\varepsilon) = (c_1 + o(1))n(\varepsilon) + 2(c_2 - c_1),$$

where $n(\varepsilon)$ is given by (4.10). ∎

5. MULTIVARIATE AND ABSTRACT NONLINEAR EQUATIONS

In this section, we show that one step of certain known iterations is "asymptotically" an optimal error algorithm.

Let

$$(5.1) \qquad\qquad f:D \subset B_1 \to B_2,$$

where $D = \{x:\|x\| < 2q\}$ and B_1, B_2 are Banach spaces over the real or complex fields of dimension m, $m = \dim(B_1) = \dim(B_2)$, $1 \le m \le +\infty$. Let $\mathfrak{J}_1$ be the class of all operators f which are k-times differentiable in the Fréchet sense on D, $k \ge 2$. Define

$$(5.2)\quad \mathfrak{J}_0 = \mathfrak{J}_0(A_2,A_k) = \left\{ f : f \in \mathfrak{J}_1 \text{ and there exists } \alpha = \alpha(f), \|\alpha\| < q, \right.$$

$$\text{such that } f(\alpha) = 0 \text{ and } \left\| f'(\alpha)^{-1} \frac{f''(x)}{2} \right\| \le A_2$$

$$\left. \text{and } \left\| f'(\alpha)^{-1} \frac{f^{(k)}(x)}{k!} \right\| \le A_k \text{ for all } x \in D \right\}$$

for constants A_2 and A_k which satisfy the condition

$$(5.3) \qquad\qquad 2kA_k(3q)^{k-1} + 2A_2q < 1.$$

The solution of nonlinear equations in $\mathfrak{J}_0$ is described by

$$(5.4) \qquad\qquad S(f) = f^{-1}(0), \qquad \mathfrak{J}_2 = B_2,$$

where $\|f^{-1}(0)\| < q$.

We first show that S is well defined. It suffices to prove that $f(x) = 0$ for $\|x\| < q$ has a unique solution $\alpha = f^{-1}(0)$, for $f \in \mathfrak{I}_0$. Let

$$(5.5) \qquad R_j(x,y;f) = \int_0^1 f^{(j)}(y + t(x - y))(x - y)^j \frac{(1 - t)^{j-1}}{(j - 1)!} \, dt$$

for $x, y \in D$ and $j \leq k$. Then

$$(5.6) \qquad\qquad f(x) = f(\alpha) + f'(\alpha)(x - \alpha) + R_2(x,\alpha;f)$$

and $f(x) = 0$ is equivalent to the equation

$$x - \alpha = -f'(\alpha)^{-1} R_2(x,\alpha;f).$$

Note that $\|\alpha\| < q$. From (5.2) and (5.5), we get for $\|x\| < q$,

$$\|x - \alpha\| \leq A_2 \|x - \alpha\|^2 \leq 2A_2 q \|x - \alpha\|.$$

Due to (5.3), $2A_2 q < 1$ which implies $x = \alpha$. Thus, S is well defined by (5.4) and $\alpha = S(f)$ satisfies the nonlinear equation $f(x) = 0$.

Define the information operator

$$(5.7) \qquad\qquad \mathfrak{N}(f) = [y(f), f(y(f)), \ldots, f^{(k-1)}(y(f))]^t,$$

where $y = y(f)$ is an approximation to the solution $\alpha = S(f)$, $\|y\| < q$.

We want to find $d(\mathfrak{N},S)$, the diameter of information $\mathfrak{N}$ for the problem S.

Lemma 5.1

$$(5.8) \qquad\qquad d(\mathfrak{N},S) \leq \frac{2A_k}{1 - A_2 q} \sup_{f \in \mathfrak{I}_0} \left(\frac{3}{2} \|y(f) - S(f)\| \right)^k. \quad \blacksquare$$

PROOF Note that $\mathfrak{N}(\tilde{f}) = \mathfrak{N}(f)$ implies $\tilde{f}^{(j)}(y) = f^{(j)}(y)$ for $j = 0, 1, \ldots, k-1$ and $\tilde{f}, f \in \mathfrak{I}_0$. Then

$$f(x) - \tilde{f}(x) = R_k(x, y; f - \tilde{f}).$$

Since $\tilde{f}(x) = 0$ is equivalent to $f(x) = R_k(x, y; f - \tilde{f})$ and f satisfies (5.6), we get

$$(5.9) \qquad x = H(x) \stackrel{\text{df}}{=} \alpha + f'(\alpha)^{-1} \{ R_k(x, y; f - \tilde{f}) - R_2(x,\alpha;f) \}.$$

We show that H is a contraction on $J = \{x : \|x - \alpha\| \leq \frac{1}{2} \|y - \alpha\| \}$. Indeed,

$$\|H(x) - \alpha\| \leq 2A_k \|x - y\|^k + A_2 \|x - \alpha\|^2$$
$$\leq 2A_k (\tfrac{3}{2})^k \|y - \alpha\|^k + \tfrac{1}{4} A_2 \|y - \alpha\|^2$$
$$\leq \tfrac{1}{2} \|y - \alpha\| (6A_k (3q)^{k-1} + A_2 q) \leq \tfrac{1}{2} \|y - \alpha\|$$

due to (5.2) and (5.3). Furthermore,

$$\|H'(x)\| \leq 2kA_k \|x - y\|^{k-1} + 2A_2 \|x - \alpha\| \leq 2kA_k (3q)^{k-1} + 2A_2 q < 1$$

due to (5.3). Thus, equation (5.9) has a unique solution $\tilde{\alpha}$, $\|\tilde{\alpha} - \alpha\| \leq \frac{1}{2}\|y - \alpha\| \leq q$. Set $x = \tilde{\alpha}$ in (5.9). Then $\|\tilde{\alpha} - \alpha\| \leq 2A_k\|\tilde{\alpha} - y\|^k + A_2\|\tilde{\alpha} - \alpha\|^2$ which yields

$$\|\tilde{\alpha} - \alpha\| \leq \frac{2A_k(\frac{3}{2}\|y - \alpha\|)^k}{1 - A_2q}.$$

This proves (5.8) and completes the proof. ∎

We want to prove that (5.8) is, in general, sharp with respect to $\|y(f) - S(f)\|^k$.

Lemma 5.2 If $y(f)$ approaches $\alpha = S(f)$, then

(5.10) $$d(\mathfrak{N},S) = 2A_k \sup_{f \in \mathfrak{I}_0} \|y(f) - S(f)\|^k(1 + o(1)). \quad \blacksquare$$

PROOF Equation (5.9) for $x = \tilde{\alpha}$ yields

(5.11) $$\|\tilde{\alpha} - \alpha\| \leq 2A_k\|y(f) - \tilde{\alpha}\|^k + A_2\|\tilde{\alpha} - \alpha\|^2$$
$$\leq 2A_k(\|y(f) - \alpha\| + \|\alpha - \tilde{\alpha}\|)^k + A_2\|\tilde{\alpha} - \alpha\|^2.$$

Since $\|\tilde{\alpha} - \alpha\| = O(\|y(f) - \alpha\|^k)$, (5.11) can be rewritten as

$$\|\tilde{\alpha} - \alpha\| \leq 2A_k\|y(f) - \alpha\|^k(1 + o(1)).$$

Since this bound is sharp, we have proven (5.10). ∎

Lemma 5.2 states that the diameter $d(\mathfrak{N},S)$ is roughly equal to

$$2A_k \sup_{f \in \mathfrak{I}_0} \|y(f) - S(f)\|^k,$$

where k is the first omitted derivative in the information (5.7).

We establish asymptotically optimal error algorithms for the problem S if $y(f)$ approaches α. Let

(5.12) $$\tilde{f}(x) = f(y) + f'(y)(x - y) + \cdots + \frac{1}{(k-1)!} f^{(k-1)}(y)(x - y)^{k-1}.$$

Note that $\mathfrak{N}(\tilde{f}) = \mathfrak{N}(f)$. From (3.10) in Traub and Woźniakowski [77b], we know that

$$\left\|\tilde{f}'(\tilde{\alpha})^{-1} \frac{\tilde{f}''(x)}{2}\right\| \leq \frac{A_2 + (k(k-1)/2)A_k(2\|y - \alpha\|)^{k-2}}{1 - A_2\|y - \alpha\| - kA_k(\frac{3}{2}\|y - \alpha\|)^{k-1}} \overset{\text{df}}{=} \tilde{A}_2(y).$$

It is also known that $\tilde{f}$ has a unique zero $\tilde{\alpha}$ such that $\|\tilde{\alpha}\| < q$. Thus $\tilde{f} \in \mathfrak{I}_0(\tilde{A}_2(y),0)$ and $\tilde{A}_2(y) = A_2 + O(\|y - \alpha\|)$. Define the algorithm

(5.13) $$\varphi(\mathfrak{N}(f)) = S(\tilde{f}),$$

i.e., $\varphi(\mathfrak{N}(f))$ is a unique solution of the nonlinear equation $\tilde{f}(x) = 0$ in the ball $\{x: \|x\| < q\}$. Algorithm (5.13) is known as the interpolatory iteration I_k and was considered by Traub and Woźniakowski [76b, 77a,b]. Note that for $k = 2$, we get one step of Newton iteration since $\tilde{f}(x) = f(y) + f'(y)(x - y)$ and $\varphi(\mathfrak{N}(f)) = \tilde{\alpha} = f(y) - f'(y)^{-1}f(y)$.

Lemma 5.3 If $y(f)$ approaches $\alpha = S(f)$, then the algorithm φ is an asymptotically optimal error algorithm, i.e.,

$$(5.14) \quad e(\varphi) = r(\mathfrak{N},S)(1 + o(1)) = A_k \sup_{f \in \mathfrak{I}_0} \|y(f) - S(f)\|^k(1 + o(1)). \quad \blacksquare$$

PROOF We repeat a part of the proof of Lemma 5.1. Note that $R_k(x, y; f - \tilde{f})$ in (5.9) for $\tilde{f}$ defined by (5.12) has the bound $\|R_k(x, y; f - \tilde{f})\| \leq A_k\|x - y\|^k$. Similarly, we show that

$$\|\varphi(\mathfrak{N}(f)) - S(f)\| \leq A_k\|y(f) - S(f)\|^k(1 + o(1)).$$

From Lemma 5.2, we conclude

$$e(\varphi) = A_k \sup_{f \in \mathfrak{I}_0} \|y(f) - S(f)\|^k(1 + o(1))$$

$$= \tfrac{1}{2}d(\mathfrak{N},S)(1 + o(1)) = r(\mathfrak{N},S)(1 + o(1)).$$

This proves (5.14). $\blacksquare$

Algorithm (5.13) is known to have maximal order of convergence among all iterations using the information of (5.7) (see Traub and Woźniakowski [76a] and Part B). Lemma 5.3 states that this algorithm has asymptotically optimal error in the class $\mathfrak{I}_0$.

Complexity of algorithm (5.13) and its dependence on k were considered in detail by Traub and Woźniakowski [77b].

6. NEARLY OPTIMAL NONADAPTIVE LINEAR INFORMATION FOR THE SEARCH FOR THE MAXIMUM OF UNIMODAL FUNCTIONS

In this and the next section, we deal with the search for the maximum in the class of scalar unimodal functions $f:[a,b] \to \mathbb{R}$. We remind the reader that f is unimodal if there exists a point $\alpha = \alpha(f) \in [a,b]$ such that for any $x, y \in [a,b]$, we have

$$(6.1) \qquad \begin{array}{lll} x < y \leq \alpha & \text{implies} & f(x) < f(y), \\ \alpha \leq x < y & \text{implies} & f(x) > f(y). \end{array}$$

Note that f need not be continuous. Let

$$(6.2) \qquad \mathfrak{I}_0 = \{f : f \text{ is unimodal on } [a,b]\}$$

and define the solution operator as

$$(6.3) \qquad\qquad S(f) = \alpha(f),$$

where $\alpha = \alpha(f)$ is the point satisfying (6.1).

We first consider a nonadaptive linear information operator

(6.4) $$\mathfrak{N}_n(f) = [f(x_1), \ldots, f(x_n)]^t,$$

where $a = x_0 \le x_1 < x_2 < \cdots < x_n \le x_{n+1} = b$.

Observe that if $f(x_i) \le f(x_{i+1})$, then $\alpha \ge x_{i+1}$; if $f(x_i) \ge f(x_{i+1})$, then $\alpha \le x_{i+1}$.
Let $j = j(\mathfrak{N}_n(f))$ be the minimal index such that $f(x_j) = \max_{1 \le i \le n} f(x_i)$. Then α belongs to the interval $[a_1, b_1]$, where

(6.5) $$a_1 = \begin{cases} x_{j-1} & \text{if } f(x_{j+1}) < f(x_j), \\ x_j & \text{if } f(x_{j+1}) = f(x_j), \end{cases}$$

$$b_1 = x_{j+1}.$$

Define the algorithm

(6.6) $$\varphi(\mathfrak{N}_n(f)) = \tfrac{1}{2}(a_1 + b_1).$$

It is clear that any point of the interval (a_1, b_1) can be the solution of our problem. This yields that the algorithm φ is interpolatory and central. Furthermore,

(6.7) $$e(\varphi) = r(\mathfrak{N}_n, S) = \max_{0 \le i \le n} (\tfrac{1}{2}(x_{i+1} - x_i)).$$

We want to find points $x_1, x_2, \ldots, x_n$ which minimize the radius of information (6.7). Suppose first that n is odd, $n = 2k - 1$. Then $\sum_{i=1}^{k} (x_{2i} - x_{2(i-1)}) = b - a$. This yields that $\max_i (x_{2i} - x_{2(i-1)}) = (b-a)/k$ and from (6.7) we have

(6.8) $$r(\mathfrak{N}_n, S) \ge \frac{b-a}{2k} = \frac{b-a}{2\lceil (n+1)/2 \rceil}.$$

We now exhibit points $x_1, \ldots, x_n$ for which $r(\mathfrak{N}_n, S)$ is equal to $(b-a)/2k$. Let

(6.9) $$x_i = a + ((b-a)/(n+1))i, \qquad i = 1, 2, \ldots, n.$$

Then $x_{i+1} - x_{i-1} = 2(b-a)/(n+1) = (b-a)/k$ which proves that the minimal radius of information for odd n is $(b-a)/2k$.

Suppose now that n is even, $n = 2k$. Of course,

(6.10) $$\inf_{x_1, \ldots, x_n} r(\mathfrak{N}_n, S) \ge \min_{x_1, \ldots, x_{n+1}} r(\mathfrak{N}_{n+1}, S) = \frac{b-a}{2(k+1)} = \frac{b-a}{2\lceil (n+1)/2 \rceil}.$$

We prove that equality holds in (6.10). Choose an arbitrary positive δ, $\delta < (b-a)/(n+2)$, and define

(6.11) $$x_{2i} = a + \frac{b-a}{n+2} 2i, \qquad x_{2i-1} = x_{2i} - \delta, \qquad i = 1, 2, \ldots, k.$$

Then $\max_i(x_{i+1} - x_{i-1}) = x_{n+1} - x_{n-1} = b - x_n + \delta = ((b-a)/(k+1)) + \delta$.
From (6.7), we get equality in (6.10). Thus, (6.8) and (6.10) state that

$$(6.12) \qquad \inf_{x_1,\ldots,x_n} r(\mathfrak{N}_n,S) = \frac{b-a}{2\lceil (n+1)/2 \rceil}$$

It is natural to ask if the information (6.4) is nth optimal in the class Ψ_n of nonadaptive linear information operators of the form $\mathfrak{N}(f) = [L_1(f),\ldots,L_n(f)]^t$, where $L_1, L_2, \ldots, L_n$ are linear real functionals well defined for every unimodal function.

Although we do not know the answer to this problem, we show that

$$(6.13) \qquad r(n,S) = \inf_{\mathfrak{N} \in \Psi_n} r(\mathfrak{N},S)$$

satisfies the inequality

$$(6.14) \qquad \frac{b-a}{2(n+1)} \le r(n,S) \le \frac{b-a}{2\lceil (n+1)/2 \rceil}.$$

The right-hand side of (6.14) follows from (6.12). To establish the left-hand side of (6.14), we use Theorem 2.1. Let $\bar{\mathfrak{I}}_0$ be the subset of $\mathfrak{I}_0$ consisting of all functions f such that f' is continuous, $f'(a) \ge 0$, $f'(b) \le 0$ and f' has exactly one zero in $[a,b]$. Then $\alpha = S(f)$ is a unique solution of the equation $f'(x) = 0$. Note that the set $G_0 = \{g : g = -f' \text{ for } f \in \bar{\mathfrak{I}}_0\}$ coincides with (2.2) and the problem (6.3) for $f \in \bar{\mathfrak{I}}_0$ coincides with the problem (2.3) for $g \in G_0$. Let $S_1(g) = S(f)$ for $g = -f' \in G_0$. Then, of course, $r(\mathfrak{N},S) \ge r(\mathfrak{N},S_1)$ for any information operator $\mathfrak{N}$ from Ψ_n. From Theorem 2.1, we get

$$r(\mathfrak{N},S_1) \ge (b-a)/2(n+1)$$

which proves the left-hand side of (6.14).

The lower and upper bounds (6.14) are fairly tight since they differ at most by a factor of two. This establishes that the information operator of the form (6.4) with x_i given by (6.9) for odd n and by (6.11) with small δ for even n is nearly an nth optimal information.

We now analyze the complexity of the problem (6.3). To find an ε-approximation, we have to guarantee that $r(n,S) < \varepsilon$. Let $m(\Psi_U,S,\varepsilon)$ be the ε-cardinality number defined as in Section 2. From (6.14), we get

$$m(\Psi_U,S,\varepsilon) = \lfloor a_1((b-a)/2\varepsilon) \rfloor, \qquad a_1 \in [1,2].$$

We proved that the algorithm φ defined by (6.6) is central. We now discuss its combinatory complexity. To compute $\varphi(\mathfrak{N}_n(f))$, we need to know the index $j = j(\mathfrak{N}_n(f))$. This can be done in $\Theta(\log n)$ comparison and arithmetic operations. Thus, the complexity of the algorithm φ and the ε-complexity of

S in Ψ_U is

$$\text{comp}(\Psi_U, S, \varepsilon) = \text{comp}(\varphi)(1 + o(1))$$
$$= (c_1 + o(1))\lfloor a_1((b - a)/2\varepsilon) \rfloor \qquad \text{as} \quad \varepsilon \to 0,$$

where c_1 is the complexity of one function evaluation. Thus, the algorithm φ is an asymptotically optimal complexity algorithm for S in Ψ_U.

We summarize these results in

Theorem 6.1   Consider the search for the maximum in the class of unimodal functions defined by (6.2) and (6.3). Then:

(i) $\mathfrak{N}_n$ defined by (6.4) with x_i defined by (6.9) for odd n and by (6.11) with small δ for even n is nearly an nth optimal nonadaptive linear information,

$$r(\mathfrak{N}_n, S) = \frac{b - a}{2\lceil (n + 1)/2 \rceil} + \delta_n \le 2r(n, S) + \delta_n,$$

where $\delta_n = 0$ for odd n and $\delta_n = \delta$ for even n.

(ii) φ defined by (6.6) is an interpolatory central and asymptotically optimal complexity algorithm for S in Ψ_U, $e(\varphi) = r(\mathfrak{N}_n, S)$.

(iii)   The ε-cardinality number of S in Ψ_U is

$$m(\Psi_U, S, \varepsilon) = \lfloor a_1(b - a)/2\varepsilon \rfloor.$$

(iv)   The ε-complexity of S in Ψ_U is

$$\text{comp}(\Psi_U, S, \varepsilon) = (c_1 + o(1))\lfloor a_1(b - a)/2\varepsilon \rfloor. \quad \blacksquare$$

7.   ADAPTIVE LINEAR INFORMATION FOR THE SEARCH FOR THE MAXIMUM OF UNIMODAL FUNCTIONS

We deal with the problem defined by (6.2) and (6.3). We show that for this nonlinear problem there exists an adaptive information operator $\mathfrak{N}_n^a$, $\text{card}(\mathfrak{N}_n^a) \le n$, which is significantly more efficient than any nonadaptive information operator $\mathfrak{N}$ of cardinality n, i.e., $r(\mathfrak{N}_n^a, S) \ll r(n, S)$ for large n.

In a classic paper, Kiefer [53] posed and solved the problem of obtaining the points y_i at which f should be adaptively evaluated. However, Kiefer presented only an implicit prescription for the selection of the points y_i. We modify Kiefer's prescription so as to obtain an algorithm which enjoys small combinatory complexity.

Let

(7.1) $\mathfrak{N}_n^a(f) = [f(y_1), f(y_2), \ldots, f(y_n)]^t,$

where $y_i = y_i(f(y_1), \ldots, f(y_{i-1}))$ are defined as follows.

Let $\{F_n\}$ be the Fibonacci sequence, $F_0 = F_1 = 1$, $F_n = F_{n-1} + F_{n-2}$, $n = 2, 3, \ldots$. Let

(7.2) $$g(x) = f(a + x(b - a)/F_n), \qquad x \in [0, F_n].$$

Let δ be a given positive number, $\delta \leq 1$.

For $n = 2$, define $x'_2 = 1 - \delta$, $x_2 = 1$ and evaluate $g(x'_2)$ and $g(x_2)$. Thus

(7.3) $$y_1 = a + (1 - \delta)(\tfrac{1}{2}(b - a)), \qquad y_2 = a + \tfrac{1}{2}(b - a).$$

Then the solution $S(f)$ lies in the interval $[c, d]$, where

(7.4)
$$
\begin{aligned}
c = a, \quad & d = y_2 \quad && \text{if } g(x'_2) > g(x_2), \\
c = y_1, \quad & d = y_2 \quad && \text{if } g(x'_2) = g(x_2), \\
c = y_1, \quad & d = b \quad && \text{if } g(x'_2) < g(x_2).
\end{aligned}
$$

For $n \geq 3$, we define the sequences $\{a_i\}$, $\{b_i\}$, $\{x_i\}$, and $\{z_i\}$ by (7.5) and (7.6), and next we motivate their definition. Let $a_2 = 0, b_2 = F_n, x_2 = F_{n-2}, z_2 = F_{n-1}$ and evaluate $g(x_2)$ and $g(z_2)$. For $i = 2, 3, \ldots, n - 1$, define:

if $g(x_i) > g(z_i)$, then

(7.5) $a_{i+1} = a_i, \qquad b_{i+1} = z_i, \qquad x_{i+1} = a_i + F_{n-i-1}, \qquad z_{i+1} = x_i;$

if $g(x_i) \leq g(z_i)$, then

(7.6) $a_{i+1} = x_i, \qquad b_{i+1} = b_i, \qquad x_{i+1} = z_i, \qquad z_{i+1} = x_i + F_{n-1}.$

The motivation of (7.5) and (7.6) is as follows. Suppose that $[a_i, b_i]$ denotes a set in which the maximum of g lies and that $a_1 < x_i < z_i < b_i$ for $i < n$. Observe that this holds for $i = 2$. Knowing the values of $g(x_i)$ and $g(z_i)$, we determine the new interval containing the maximum of g. Suppose first that $g(x_i) > g(z_i)$ as is shown in Figure 1. Then $[a_i, z_i]$ is the new interval and the next point at which g will be evaluated is $x_{i+1} = a_i + F_{n-i-1}$. This motivates the formulas (7.5).

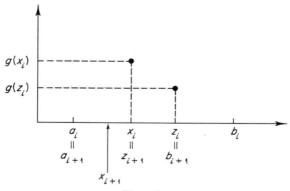

Figure 1

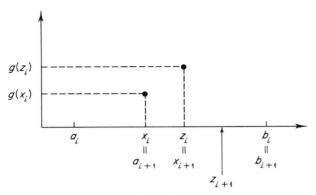

Figure 2

Suppose now that $g(x_i) \leq g(z_i)$ as is shown in Figure 2. In this case, $[x_i, b_i]$ is the new interval and the next point at which g will be evaluated is $z_{i+1} = x_i + F_{n-i}$. This motivates formulas (7.6). From this, it also follows that if $g(x_i) \neq g(z_i)$, then $[x_i, b_i]$ is the smallest closed set containing the maximum of g. If $g(x_i) = g(z_i)$, then $[x_i, z_i]$ contains the maximum of g. For the sake of simplicity, we define $b_{i+1} = b_i$ instead of $b_{i+1} = z_i$ in this case also. Observe that this simplification does not change the radius of information, since there exists a function g such that $g(x_i)$ differs from $g(z_i)$ for every i.

It can be easily verified by induction that

(7.7)
$$b_i - a_i = F_{n+2-i}, \qquad z_i - a_i = b_i - x_i = F_{n+1-i},$$
$$x_i - a_i = b_i - z_i = F_{n-i}, \qquad z_i - x_i = F_{n-1-i},$$

for $i = 2, 3, \ldots, n$.

For $i \leq n - 1$, $x_i < z_i$ and we evaluate the function g at x_i or z_i, respectively. For points x_n and z_n, we get

$$z_n = x_n = \begin{cases} x_{n-1} & \text{if } g(x_{n-1}) > g(z_{n-1}), \\ z_{n-1} & \text{if } g(x_{n-1}) \leq g(z_{n-1}). \end{cases}$$

Thus we know the value of $g(z_n) = g(x_n)$. This shows that the execution of (7.5) and (7.6) requires $n - 1$ evaluations of g. Define the nth point at which g is evaluated as $x'_n = x_n - \delta$. Note that $x_n = a_n + 1$ and $b_n - a_n = 2$.

We now show at which points y_i the function f is evaluated (see (7.1)). From (7.2), (7.5), and (7.6), we conclude that

(7.8)
$$y_i = a + ((b - a)/F_n)\xi_i,$$

where

$$\xi_1 = F_{n-2}, \qquad \xi_2 = F_{n-1}, \qquad \xi_{i+1} = \begin{cases} x_{i+1} & \text{if } g(x_i) > g(z_i), \\ z_{i+1} & \text{if } g(x_i) \leq g(z_i), \end{cases}$$

for $i = 2, 3, \ldots, n - 2$ and $\xi_n = x'_n$.

Then the solution $S(f)$ lies in the interval $[c,d]$, where

(7.9) $c = a + ((b - a)/F_n)a_{n+1}, \qquad d = a + ((b - a)/F_n)b_{n+1},$

and

$$a_{n+1} = a_n, \qquad b_{n+1} = a_n + 1 \qquad \text{if} \quad g(x'_n) > g(x_n),$$
$$a_{n+1} = a_n + 1 - \delta, \qquad b_{n+1} = a_n + 1 \qquad \text{if} \quad g(x'_n) = g(x_n),$$
$$a_{n+1} = a_n + 1 - \delta, \qquad b_{n+1} = b_n \qquad \text{if} \quad g(x'_n) < g(x_n).$$

We defined the algorithm φ as

(7.10) $\varphi(\mathfrak{N}_n^a(f)) = \tfrac{1}{2}(c + d),$

where c, d are defined by (7.4) for $n = 2$ and by (7.9) for $n \geq 3$. Since $[c,d]$ contains the solution $S(f)$ and any point of (c,d) can be the solution, we conclude that φ is an interpolatory central algorithm. Furthermore,

(7.11) $e(\varphi) = r(\mathfrak{N}_n^a, S)$

$$= \sup_{f \in \mathfrak{I}_0} \frac{d - c}{2} = \frac{b - a}{2F_n} \sup_{f \in \mathfrak{I}_0} (b_{n+1} - a_{n+1}) = \frac{b - a}{2F_n}(1 + \delta).$$

Recall that the minimal nth radius $r(n,s)$ of nonadaptive information is equal to $(b - a)/(2(n + 1))$. Since $1/F_n = \Theta((2/(\sqrt{5} + 1))^n)$, we obtained an exponential improvement.

To find an ε-approximation, it is now enough to perform $n = n(\varepsilon)$ function evaluations, where n is the smallest integer such that

(7.12) $F_n > ((b - a)(1 + \delta))/2\varepsilon.$

Since

$$F_n = \frac{\sqrt{5}}{5} \left(\frac{\sqrt{5} + 1}{2} \right)^{n+1} (1 + o(1)),$$

we get

(7.13) $n(\varepsilon) = \log \frac{\sqrt{5}(b - a)(1 + \delta)}{2\varepsilon} \bigg/ \log\left(\frac{\sqrt{5} + 1}{2} \right)(1 + o(1)).$

The complexity of the central algorithm (7.10) is

(7.14) $\text{comp}(\varphi) = (c_1 + O(1))n(\varepsilon),$

where c_1 is the complexity of one function evaluation and the complexity of addition, subtraction, multiplication, division, and comparison are primitive operations. It is obvious that the ε-complexity of the problem S with the information $\mathfrak{N}_{n(\varepsilon)}^a$ is also of the form (7.14).

We summarize these results in

Theorem 7.1   Consider the search for the maximum for the class of unimodal functions defined by (6.2) and (6.3). Then:

(i)   The algorithm defined by (7.10) is an interpolatory central algorithm. This algorithm enjoys a type of optimal complexity of S with $\mathfrak{N}^a_{n(\varepsilon)}$, since its complexity is $\Theta(\text{comp}(\mathfrak{N}^a_{n(\varepsilon)},S,\varepsilon)$.

(ii)   The radius of information is

$$r(\mathfrak{N}^a_n,S) = ((b - a)/2F_n)(1 + \delta).$$

(iii)   The ε-complexity of S with $\mathfrak{N}^a_{n(\varepsilon)}$ is

$$\text{comp}(\mathfrak{N}^a_{n(\varepsilon)},S,\varepsilon) = (c_1 + O(1))n(\varepsilon),$$

where $n(\varepsilon)$ is given by (7.12) and (7.13).   ∎

It is natural to inquire about the optimality of the information $\mathfrak{N}^a_n$ defined by (7.1) in the class of adaptive linear information operators. In our terminology, Kiefer [53] proved that

(7.15)
$$\inf_{\mathfrak{N}^a_n \in \Psi^a_f} r(\mathfrak{N}^a_n,S) = (b - a)/2F_n,$$

where Ψ^a_f is the class of adaptive linear information

$$\mathfrak{N}^a_n(f) = [f(x_1),f(x_2), \ldots ,f(x_n)]^t.$$

where $x_i = x_i(f(x_1), \ldots ,f(x_{i-1}))$ is an arbitrary function.

Note that the information (7.1) has a radius which differs from (7.15) by a factor of $1 + \delta$. Since δ can be arbitrarily small, we conclude that $\mathfrak{N}^a_n$ is nearly optimal in Ψ^a_f.

We discuss the optimality of $\mathfrak{N}^a_n$ in the class $\bar{\Psi}^a_n$ of adaptive linear information operators of the form $\mathfrak{N}(f) = [L_1(f),L_2(f;L_1(f)), \ldots ,L_n(f;L_1(f), \ldots)]^t$, where L_i depends linearly on the first argument. Note that the classes Ψ^a_n defined in Section 3 and $\bar{\Psi}^a_n$ differ by the domains of functionals $L_1, L_2, \ldots , L_n$. For Ψ^a_n, functionals are defined for continuous functions; for $\bar{\Psi}^a_n$, they are defined for unimodal functions.

In the light of the results of Section 3, it comes as no surprise that $\mathfrak{N}^a_n$ is *not* optimal in the class $\bar{\Psi}^a_n$. To show this, suppose for a moment that f has a continuous first derivative and $f'(a) \geq 0$, $f'(b) \leq 0$. Then $\alpha = S(f)$ is a unique solution of the equation $f'(x) = 0$. The information

(7.16)
$$\mathfrak{N}(f) = [f'(y_1),f'(y_2), \ldots ,f'(y_n)]^t,$$

with y_i defined by (3.2) and with $f(y_i)$ replaced by $-f'(y_i)$, has radius of information equal to $(b - a)2^{-(n+1)}$. The information (7.16) is not well defined for the class of unimodal functions, since f' does not exist in general. Therefore, instead of (7.16), we define

(7.17)
$$\mathfrak{N}^a_\delta(f) = [f(y_1,\delta),f(y_2,\delta), \ldots ,f(y_n,\delta)]^t,$$

where $f(y_i,\delta) = (f(y_i) - f(y_i - \delta))/\delta$ is the first divided difference and δ is a sufficiently small positive number. The points y_i are defined as follows. Let $a_0 = a$ and $b_0 = b$. Suppose that a_j, b_j are defined for $j = 0, 1, \ldots, i - 1$. Then

$$y_i = \tfrac{1}{2}(a_{i-1} + b_{i-1}),$$

$$(7.18) \quad a_i = \begin{cases} a_{i-1} & \text{if } f(y_i,\delta) < 0, \\ y_i - \delta & \text{if } f(y_i,\delta) \geq 0, \end{cases} \qquad b_i = \begin{cases} b_{i-1} & \text{if } f(y_i,\delta) > 0, \\ y_i & \text{if } f(y_i,\delta) \leq 0. \end{cases}$$

Compare with (3.2). It is clear that $[a_j,b_j]$ contains all solution elements of functions $\tilde{f}$ from $\mathfrak{I}_0$ which have the same information as f. Furthermore,

$$b_i - a_i \leq \tfrac{1}{2}(b_{i-1} - a_{i-1}) + \delta \leq (b - a)2^{-i} + 2\delta(1 - 2^{-i})$$

and there exists a function $f \in \mathfrak{I}_0$ such that this inequality is sharp. This yields the radius of information

$$r(\mathfrak{N}_\delta^a, S) = (b - a)2^{-(n+1)} + \delta(1 - 2^{-n}).$$

Since δ can be arbitrarily small,

$$(7.19) \qquad \inf_{\mathfrak{N} \in \Psi_n^a} r(\mathfrak{N}, S) \leq (b - a)2^{-(n+1)}.$$

This shows that the Kiefer information operator is not optimal in Ψ_n^a.

The actual value of $\inf_{\mathfrak{N} \in \Psi_n^a} r(\mathfrak{N}, S)$ is not known. It is plausible to conjecture that equality holds in (7.19).

Conjecture 7.1

$$\inf_{\mathfrak{N} \in \Psi_n^a} r(\mathfrak{N}, S) = (b - a)2^{-(n+1)}. \quad \blacksquare$$

It is obvious that Conjecture 3.1 implies Conjecture 7.1. Furthermore, $\inf_{\mathfrak{N} \in \Psi_n^a} r(\mathfrak{N}, S_1) \leq \inf_{\mathfrak{N} \in \Psi_n^a} r(\mathfrak{N}, S)$, where S_1 now stands for the solution of non-linear equations defined by (2.2) and (2.3).

As in Section 6 of Chapter 6, we wish to comment on the assumption that the evaluation of every linear functional costs unity. Under this assumption, the evaluation of $f(y_i,\delta)$ in (7.17) has the same complexity as the evaluation of $f(y_i)$. Suppose that this assumption is violated and let the evaluation of $f(y_i,\delta)$ cost c times more than the evaluation of $f(y_i)$. Disregarding two arithmetic operations, we assume that $c \in [1,2]$. Then the cost of the computation of $\mathfrak{N}_\delta^a$ is approximately the same as the cost of the computation of $\mathfrak{N}_n^a$ with $m = \lfloor cn \rfloor$. Since $r(\mathfrak{N}_m^a, S) = \Theta((2/(\sqrt{5} + 1))^{cn})$, then $r(\mathfrak{N}_\delta^a, S)$ is less than $r(\mathfrak{N}_m^a, S)$ for large n and small δ iff

$$(7.20) \qquad c < 1/\log \tfrac{1}{2}(\sqrt{5} + 1) \doteq 1.44.$$

This shows that the information operator $\mathfrak{N}_\delta^a$ is more effective than $\mathfrak{N}_n$ even if the evaluation of $f(y_i,\delta)$ is more expensive than the evaluation of $f(y_i)$ as long as c satisfies (7.20).

Chapter 9

Complexity Hierarchy

1. INTRODUCTION

Problems can be arranged into a hierarchy depending on their complexity. We stress that a problem complexity hierarchy depends essentially on the class of permissible information operators. Different classes of information usually yield different complexity hierarchies.

We can provide only partial hierarchies now. We anticipate that in the course of future research these hierarchies will be expanded.

In the iterative information model of Part B, only two types of complexity functions can occur. (See Part B, Section 11.)

We summarize the results of this chapter. In Section 2, we define our complexity relations and the problems which appear in the hierarchy. In Sections 3–7, we construct complexity hierarchies for five classes of information operators. In Section 8, we construct complexity hierarchies by fixing the problem and varying the classes of information. The concluding section presents a tableau summarizing our results.

2. BASIC DEFINITIONS

Let S, S_1, and S_2 be solution operators and let Ψ be a class of permissible information operators. Recall that comp(Ψ,S,ε), the ε-complexity of the problem S in the class Ψ, is defined by (3.9) of Chapter 1.

Definition 2.1   (i)   We shall say S_1 *has greater complexity than* S_2 *in the class* Ψ *for the set* A, denoted by $S_1 > S_2$, iff

$$(2.1) \qquad\qquad \text{comp}(\Psi, S_1, \varepsilon) > \text{comp}(\Psi, S_2, \varepsilon) \qquad \forall \varepsilon \in A.$$

(ii)   We shall say S_1 *and* S_2 *have the same* Θ-*complexity in the class* Ψ, denoted by $S_1 \asymp S_2$, iff

$$(2.2) \qquad\qquad \text{comp}(\Psi, S_1, \varepsilon) = \Theta(\text{comp}(\Psi, S_2, \varepsilon)) \qquad \varepsilon \to 0.$$

(iii)   We shall say S_1 *has essentially greater complexity than* S_2 *in the class* Ψ, denoted by $S_1 \succ S_2$, iff

$$(2.3) \qquad\qquad \text{comp}(\Psi, S_2, \varepsilon) = o(\text{comp}(\Psi, S_1, \varepsilon)) \qquad \text{as} \quad \varepsilon \to 0. \quad \blacksquare$$

Thus, $S_1 > S_2$ means that it is harder to find an ε-approximation for the problem S_1 than for the problem S_2 using information operators from the class Ψ for all ε from the set A. Usually, A is an interval $(0, \varepsilon_0)$ for some positive ε_0. $S_1 \asymp S_2$ means that $\text{comp}(\Psi, S_1, \varepsilon)$ and $\text{comp}(\Psi, S_2, \varepsilon)$ differ only by a positive constant factor for all small ε. $S_1 \succ S_2$ means that $\text{comp}(\Psi, S_1, \varepsilon)$ tends to infinity as ε tends to zero essentially faster than $\text{comp}(\Psi, S_2, \varepsilon)$. Note that "$\asymp$" is an equivalence relation and $S_1 \succ S_2$, $S_2 \succ S_3$ imply $S_1 \succ S_3$. Observe that to establish the relation $S_1 \succ S_2$, we need only know the asymptotic behavior of the ε-complexities of S_1 and S_2.

The relation "$>$" is not always of interest, since it may depend on some unimportant problem details. For instance, consider the two integration problems $S_1 f = \int_a^b f(x)\,dx$ and $S_2 f = \int_c^d f(x)\,dx$, where $f \in \mathfrak{I}_0 = W_p^r$ and Ψ is the class of information operators consisting of function evaluations. Since the ε-complexity depends on the length of the interval of integration, then $d - c > b - a$ implies $S_2 > S_1$ for any interval $A = (0, \varepsilon_0)$. Thus, the essentially identical problems S_1 and S_2 occupy different positions in the complexity hierarchy under the relation "$>$." Of course, $S_1 \asymp S_2$ since their ε-complexities differ by a constant factor. Thus, S_1 and S_2 occupy the same position in the complexity hierarchy under the relation "$\asymp$."

Therefore, we shall primarily use the relations $\asymp$ and $\succ$ rather than the relation $>$.

In Chapters 6 and 8, we have analyzed a number of solution operators for different classes of permissible information operators. We deal with the following operators:

DIF(r)   the *differentiation* operator for $\mathfrak{I}_0 = W_\infty^r$, r odd and $r \geq 3$, studied in subsection (2ii) of Chapter 6.

INP(r)   the *interpolation* operator for $\mathfrak{I}_0 = W_\infty^r$ studied in subsection (3i) of Chapter 6.

INP(A)   the *interpolation* operator for the class $\mathfrak{I}_0$ of analytic bounded functions studied in subsection (3ii) of Chapter 6.

INP(H)   the *interpolation* operator for the class $\mathfrak{I}_0$ of analytic bounded functions in a Hilbert space studied in subsection (3iii) of Chapter 6.

INT(W_p^r)   the *integration* operator for $\mathfrak{I}_0 = W_p^r$ studied in subsection (4i) of Chapter 6.

INT($\tilde{W}_p^r$)   the *integration* operator for $\mathfrak{I}_0 = \tilde{W}_p^r$ studied in subsection (4ii) of Chapter 6.

APP($\tilde{W}_2^r,L_2$)   the *approximation* operator for $\mathfrak{I}_0 = \tilde{W}_2^r$ and $\mathfrak{I}_2 = L_2$ studied in subsection (5ii) of Chapter 6.

APP(W_2^r,L_2)   the *approximation* operator for $\mathfrak{I}_0 = W_2^r$ and $\mathfrak{I}_2 = L_2$ studied in subsection (5iii) of Chapter 6.

APP($\tilde{W}_\infty^r,\tilde{C}$)   the *approximation* operator for $\mathfrak{I}_0 = \tilde{W}_\infty^r$ and $\mathfrak{I}_2 = \tilde{C}$ studied in subsection (5iv) of Chapter 6.

APP(W_∞^r,C)   the *approximation* operator for $\mathfrak{I}_0 = W_\infty^r$ and $\mathfrak{I}_2 = C$ studied in subsection (5v) of Chapter 6.

PAR(r)   the *parabolic* operator for the class $\mathfrak{I}_0$ of odd functions such that $\|f^{(r)}\|_2 \leq 1$, r even, studied in subsection (6ii) of Chapter 6.

ELL(r)   the *elliptic* operator for the class $\mathfrak{I}_0$ of odd functions such that $\|f^{(r)}\|_2 \leq 1$, r even, studied in subsection (6iii) of Chapter 6.

HYP(r)   the *hyperbolic* operator for $\mathfrak{I}_0 = W_2^r$, n even, studied in subsection (6iv) of Chapter 6.

NON   the *nonlinear scalar equation* operator for the class $\mathfrak{I}_0$ studied in Sections 2 and 3 of Chapter 8.

UNI   the *maximum search* operator for the class $\mathfrak{I}_0$ of *unimodal* functions studied in Sections 6 and 7 of Chapter 8.

We denote any of the interpolation problems INP(r), INP(A), or INP(H) by INP. Similarly, we denote INT(W_p^r) or INT($\tilde{W}_p^r$) by INT and APP($\tilde{W}_2^r,L_2$), APP(W_2^r,L_2), APP($\tilde{W}_\infty^r,\tilde{C}$), or APP($W_\infty^r,C$) by APP.

We present complexity hierarchies for several classes Ψ of information operators. We assume that the set of primitives is defined in such a way that Ψ is a permissible class and there exists a permissible optimal error algorithm for every information operator from Ψ. We also assume that the information complexity of every information operator $\mathfrak{N}$, $\mathfrak{N} \in \Psi$, and card($\mathfrak{N}$) $= n$, is equal to nc_1, where c_1 depends only on the class Ψ.

3.   COMPLEXITY HIERARCHY FOR CLASS Ψ_f^{non}

Let Ψ_f^{non} be the class of all nonadaptive information operators $\mathfrak{N}_n^{non}(f) = [f(x_1),f(x_2),\ldots,f(x_n)]^t$ for some n and distinct points $x_1, x_2, \ldots, x_n$ given in advance.

The differentiation and interpolation problems are trivial for the class Ψ_f^{non}. Indeed, from (2.20) of Chapter 6, we conclude that $r(\mathfrak{N}_r,\mathrm{DIF}(r),D_\infty^r)$ can be arbitrarily small for small h. This implies that

$$\text{(3.1)} \qquad\qquad \mathrm{comp}(\Psi_f^{non},\mathrm{DIF}(r),\varepsilon) = (c_1 + a)r - 1 \qquad \forall \varepsilon,$$

where $a \in [1,2]$. For the interpolation problem, it is enough to observe that $\mathfrak{N}(f) = f(x_0)$ belongs to Ψ_f^{non} and has radius equal to zero. Thus,

$$\text{(3.2)} \qquad\qquad \mathrm{comp}(\Psi_f^{non},\mathrm{INP},\varepsilon) = c_1 \qquad \forall \varepsilon > 0.$$

From (3.1) and (3.2), we obtain

$$\text{(3.3)} \qquad\qquad\qquad \mathrm{DIF}(r) > \mathrm{INP}$$

for the set $A = \{\varepsilon : \varepsilon > 0\}$.

We have shown that the ε-complexity of the integration problems $\mathrm{INT}(W_p^r)$ and $\mathrm{INT}(\tilde{W}_p^r)$, and the approximation problems $\mathrm{APP}(\tilde{W}_\infty^r,\tilde{C})$ and $\mathrm{APP}(W_\infty^r,C)$ in the class Ψ_f^{non}, is asymptotically equal to $\varepsilon^{-1/r}$. This proves that these four problems have the same Θ-complexity.

We do not know which of these problems has greater complexity (see Definition 2.1(i)) since we do not know the exact value of the ε-complexity for arbitrary values p and r. For $p = +\infty$, however, it is possible to show that there exists a positive ε_0 such that

$$\text{(3.4)} \qquad\qquad \mathrm{APP}(\tilde{W}_\infty^r,\tilde{C}) > \mathrm{APP}(W_\infty^r,C)$$

for the set $A = (0,\varepsilon_0)$. Furthermore,

$$\text{(3.5)} \qquad\qquad \mathrm{APP}(\tilde{W}_\infty^r,\tilde{C}) > \mathrm{INT}(\tilde{W}_\infty^r)$$

for the set $A = (0,0.5)$, whenever $c_1 \geq 19/(15 - 8\sqrt[r]{2\pi})$ for $r \geq 3$.

For the nonlinear scalar equation operator NON and the maximum search operator UNI, we have shown that their ε-complexities are asymptotically equal to ε^{-1}. Hence,

$$\text{(3.6)} \qquad\qquad\qquad \mathrm{NON} \asymp \mathrm{UNI}.$$

Note that these two nonlinear solution operators have essentially higher complexity than any of the above linear solution operators for $r \geq 2$.

These results are summarized in

Theorem 3.1   For the class of nonadaptive information operators Ψ_f^{non}, we have the following complexity hierarchy:

$$\mathrm{NON} \asymp \mathrm{UNI} \asymp \mathrm{INT}(W_p^1) \asymp \mathrm{INT}(\tilde{W}_p^1) \asymp \mathrm{APP}(\tilde{W}_\infty^1,\tilde{C}) \asymp \mathrm{APP}(W_\infty^1,C)$$
$$\succ \mathrm{INT}(W_p^r) \asymp \mathrm{INT}(\tilde{W}_p^r) \asymp \mathrm{APP}(\tilde{W}_\infty^r,\tilde{C}) \asymp \mathrm{APP}(W_\infty^r,C)$$
$$\succ \mathrm{DIF}(r) \asymp \mathrm{INP} \qquad \forall r \geq 2. \quad \blacksquare$$

4.   COMPLEXITY HIERARCHY FOR CLASS Ψ_f^a

Let Ψ_f^a be the class of all adaptive information operators

$$\mathfrak{N}_n^a(f) = [f(x_1), f(x_2), \ldots, f(x_n)]^t$$

for some n and points x_i such that $x_i = x_i(x_1, f(x_1), \ldots, f(x_{i-1}))$, $i = 2, 3, \ldots, n$. As we have proven in Section 7 of Chapter 2, the adaptive information operators cannot help for *linear* problems S. More precisely, we have shown that for every adaptive information operator $\mathfrak{N}_n^a$, there exists a nonadaptive information operator $\mathfrak{N}_n^{non}$ such that $d(\mathfrak{N}_n^a, S) \geq d(\mathfrak{N}_n^{non}, S)$. From this, we conclude that $r(\mathfrak{N}_n^a, S) \geq 0.5 \ \mathrm{rad}(\mathfrak{N}_n^{non}, S)$. From the construction of $\mathfrak{N}_n^{non}$, it follows that $\mathfrak{N}_n^a \in \Psi_f^a$ implies that $\mathfrak{N}_n^{non} \in \Psi_f^{non}$. Therefore,

$$\mathrm{comp}(\Psi_f^{non}, S, \varepsilon/2) \leq \mathrm{comp}(\Psi_f^a, S, \varepsilon) \leq \mathrm{comp}(\Psi_f^{non}, S, \varepsilon)$$

for every linear solution operator S. Since

$$\mathrm{comp}(\Psi_f^{non}, S, \varepsilon/2) = \Theta(\mathrm{comp}(\Psi_f^{non}, S, \varepsilon))$$

for all linear solution operators S considered in this chapter, we conclude that

$$\mathrm{comp}(\Psi_f^a, S, \varepsilon) = \Theta(\mathrm{comp}(\Psi_f^{non}, S, \varepsilon)).$$

This means that for *linear* problems the complexity hierarchy in the class Ψ_f^a is the same as in the class Ψ_f^{non}.

For the two nonlinear solution operators NON and UNI, we have shown a dramatic difference between their ε-complexities in the two classes Ψ_f^{non} and Ψ_f^a,

$$\mathrm{comp}(\Psi_f^{non}, \mathrm{NON}, \varepsilon) = \Theta(\mathrm{comp}(\Psi_f^{non}, \mathrm{UNI}, \varepsilon)) = \Theta(\varepsilon^{-1}),$$
$$\mathrm{comp}(\Psi_f^a, \mathrm{NON}, \varepsilon) = \Theta(\mathrm{comp}(\Psi_f^a, \mathrm{UNI}, \varepsilon)) = \Theta(\log 1/\varepsilon).$$

Thus, the problems NON and UNI, which have the greatest essential complexity (see Theorem 3.1) among the problems in Ψ_f^{non}, have one of the lowest complexities (see Theorem 4.1) among the same problems in Ψ_f^a.

From Sections 3 and 7 of Chapter 8, it easily follows that there exists a positive ε_0 such that

$$\mathrm{UNI} > \mathrm{NON}$$

for the set $A = (0, \varepsilon_0)$.

We summarize these results in

Theorem 4.1   For the class of adaptive information operators Ψ_f^a we have the following complexity hierarchy:

$$\mathrm{INT}(W_p^1) \asymp \mathrm{INT}(\tilde{W}_p^1) \asymp \mathrm{APP}(\tilde{W}_\infty^1, \tilde{C}) \asymp \mathrm{APP}(W_\infty^1, C)$$
$$\succ \mathrm{INT}(W_p^r) \asymp \mathrm{INT}(\tilde{W}_p^r) \asymp \mathrm{APP}(\tilde{W}_\infty^r, \tilde{C}) \asymp \mathrm{APP}(W_\infty^r, C)$$
$$\succ \mathrm{NON} \asymp \mathrm{UNI}$$
$$\succ \mathrm{DIF}(r) \asymp \mathrm{INP} \qquad \forall r \geq 2. \ \blacksquare$$

REMARK 4.1   The ε-complexity of the problem NON in the class Ψ_f^a is asymptotically log $1/\varepsilon$. That is, we can find an ε-approximation for the equation $f(x) = 0$ for every $f \in \Im_0$ with complexity asymptotically equal to log $1/\varepsilon$. Here the class $\Im_0$ is defined by (2.2) of Chapter 8 as the class of continuous functions on $[a,b]$ such that $f(a) \leq 0$, $f(b) \geq 0$ and f has exactly one zero in $[a,b]$. Note that the bisection algorithm which is an interpolatory central algorithm is globally convergent, i.e., we do not need to know a close initial approximation to a solution.

In Part B, we shall study locally convergent algorithms for a much more general class $\Im_0$ of nonlinear equations. Local convergence means that we know a sufficiently close initial approximation to a solution. The complexity of locally convergent algorithms with order greater than one is very low and is asymptotically equal to log log $1/\varepsilon$. The assumption of *local* convergence is very crucial. Recently, Wasilkowski [79] showed that any *globally* convergent iteration using a linear information operator with finite cardinality has infinite complexity even for the class of polynomial equations with simple zeros.   ∎

5.   COMPLEXITY HIERARCHY FOR CLASS Ψ_L^{non}

Let Ψ_L^{non} be the class of all linear nonadaptive information operators $\mathfrak{N}_n^{non}(f) = [L_1(f), L_2(f), \ldots, L_n(f)]^t$ for some n and linear functionals L_1, $L_2, \ldots, L_n$.

Since $\Psi_f^{non} \subset \Psi_L^{non}$, comp$(\Psi_f^{non}, S, \varepsilon) \leq$ comp$(\Psi_L^{non}, S, \varepsilon)$ for every solution operator S. Note that the integration problem INT in the class Ψ_L^{non} becomes trivial since $\mathfrak{N}_1^{non}(f) = \int_a^b f(t)\, dt$ belongs to Ψ_L^{non} and $r(\mathfrak{N}_1, \text{INT}) = 0$. Thus,

$$\text{comp}(\Psi_L^{non}, \text{INT}, \varepsilon) = c_1.$$

We have shown in Section 5 of Chapter 6 that the ε-complexity of approximation problems APP$(\tilde{W}_2^r, L_2)$, APP(W_2^r, L_2), APP$(\tilde{W}_\infty^r, \tilde{C})$, and APP$(W_\infty^r, C)$ in the class Ψ_L^{non} is equal to $\Theta(\varepsilon^{-1/r})$. Thus, they have the same Θ-complexity. It is also easy to show that there exists a positive ε_0 such that

$$\text{APP}(\tilde{W}_2^r, L_2) > \text{APP}(W_2^r, L_2)$$

for the set $A = (0, \varepsilon_0)$.

For the partial differential solution operators PAR(r), ELL(r), and HYP(r), we have proven in Section 6 of Chapter 6 that

$$
\begin{aligned}
\text{comp}(\Psi_L^{non}, \text{PAR}(r), \varepsilon) &= \Theta(\sqrt{\log 1/\varepsilon}), \\
\text{comp}(\Psi_L^{non}, \text{ELL}(r), \varepsilon) &= \Theta(\log 1/\varepsilon), \\
\text{comp}(\Psi_L^{non}, \text{HYP}(r), \varepsilon) &= \Theta(\varepsilon^{-1/r}),
\end{aligned}
$$

(5.1)

for every even r. It can be shown that (5.1) also holds for odd r. Hence,

$$\mathrm{HYP}(r) \succ \mathrm{ELL}(r) \succ \mathrm{PAR}(r) \qquad \forall r.$$

For the nonlinear solution operators NON and UNI we have shown in Sections 2 and 6 of Chapter 8 that

$$\mathrm{comp}(\Psi_L^{non},\mathrm{NON},\varepsilon) = \Theta(\mathrm{comp}(\Psi_L^{non},\mathrm{UNI},\varepsilon)) = \Theta(\varepsilon^{-1}).$$

Thus, $\mathrm{NON} \asymp \mathrm{UNI}$. We summarize these results in

Theorem 5.1 For the class of linear nonadaptive information operators Ψ_L^{non}, we have the following complexity hierarchy:

$$\begin{aligned}
\mathrm{NON} \asymp \mathrm{UNI} &\asymp \mathrm{APP}(\tilde{W}_2^1,L_2) \asymp \mathrm{APP}(W_2^1,L_2) \\
&\asymp \mathrm{APP}(\tilde{W}_\infty^1,\tilde{C}) \asymp \mathrm{APP}(W_\infty^1,C) \asymp \mathrm{HYP}(1) \\
&\succ \mathrm{APP}(\tilde{W}_2^r,L_2) \asymp \mathrm{APP}(W_2^r,L_2) \asymp \mathrm{APP}(\tilde{W}_\infty^r,C) \\
&\asymp \mathrm{APP}(W_\infty^r,C) \asymp \mathrm{HYP}(r) \\
&\succ \mathrm{ELL}(r) \\
&\succ \mathrm{PAR}(r) \\
&\succ \mathrm{DIF}(r) \asymp \mathrm{INP} \asymp \mathrm{INT} \qquad \forall r \geq 2. \quad \blacksquare
\end{aligned}$$

6. COMPLEXITY HIERARCHY FOR CLASS Ψ_L^a

Let $\Psi = \Psi_L^a$ be the class of all linear adaptive information operators $\mathfrak{N}_n^a(f) = [L_1(f),L_2(f;L_1(f)),\ldots,L_n(f;L_1(f)),\ldots)]^t$. As in Section 4, we have $\mathrm{comp}(\Psi_L^a,S,\varepsilon) = \Theta(\mathrm{comp}(\Psi_L^{non},S,\varepsilon)$ for all linear solution operators S considered in this chapter.

For the nonlinear solution operators NON and UNI, we know only that

(6.1)
$$\begin{aligned}
\mathrm{comp}(\Psi_L^a,\mathrm{NON},\varepsilon) &= O(\log 1/\varepsilon), \\
\mathrm{comp}(\Psi_L^a,\mathrm{UNI},\varepsilon) &= O(\log 1/\varepsilon).
\end{aligned}$$

If Conjectures 3.1 and 7.1 of Chapter 8 are true, they imply that (6.1) is sharp, i.e., O can be replaced by Θ. Then

(6.2) $$\mathrm{NON} \asymp \mathrm{UNI} \asymp \mathrm{ELL}(r).$$

In either case, we know that the approximation problems APP and the hyperbolic problem $\mathrm{HYP}(r)$ have essentially greater complexity than NON and UNI.

We summarize these results in

Theorem 6.1   For the class of linear adaptive information operators Ψ_L^a, we have the following complexity hierarchy:

$$\mathrm{APP}(\tilde{W}_2^r,L_2) \asymp \mathrm{APP}(W_2^r,L_2) \asymp \mathrm{APP}(\tilde{W}_\infty^r,\tilde{C}) \asymp \mathrm{APP}(W_\infty^r,C) \asymp \mathrm{HYP}(r)$$
$$\succ \mathrm{ELL}(r)$$
$$\succ \mathrm{PAR}(r)$$
$$\succ \mathrm{DIF}(r) \asymp \mathrm{INP} \asymp \mathrm{INT} \qquad \forall r.$$

Furthermore,

$$\mathrm{HYP}(r) \succ \mathrm{NON} \text{ and } \mathrm{HYP}(r) \succ \mathrm{UNI} \qquad \forall r. \quad \blacksquare$$

7. COMPLEXITY HIERARCHY FOR CLASS Ψ_{NON}

Let Ψ_{NON} be the class of all nonlinear information operators $\mathfrak{N}$ of cardinality one, i.e., $\mathfrak{N}(f) = L(f)$, where L is a nonlinear functional.

We have proven in Section 3 of Chapter 7 that there exists an information operator $\mathfrak{N}$ from Ψ_{NON} such that $r(\mathfrak{N},S) = 0$ iff the set $S(\mathfrak{I}_0)$ has cardinality at most $\mathfrak{S}$. Note that the set $S(\mathfrak{I}_0)$ for all solution operators S considered in this chapter shares that property, i.e., $S(\mathfrak{I}_0)$ has cardinality at most $\mathfrak{S}$. Thus, each of the problems S can be solved exactly using one suitable chosen nonlinear functional. Hence,

$$\mathrm{comp}(\Psi_{\mathrm{NON}},S,\varepsilon) = \mathrm{const},$$

and all problems S have the same Θ-complexity. This again shows that the class Ψ_{NON} is so "rich" that all problems S become trivial. We summarize this in

Theorem 7.1   For the class of nonlinear information operators Ψ_{NON}, we have

$$\mathrm{DIF}(r) \asymp \mathrm{INP} \asymp \mathrm{INT} \asymp \mathrm{APP} \asymp \mathrm{PAR}(r) \asymp \mathrm{ELL}(r)$$
$$\asymp \mathrm{HYP}(r) \asymp \mathrm{NON} \asymp \mathrm{UNI}. \quad \blacksquare$$

8. COMPLEXITY HIERARCHY FOR A FIXED PROBLEM S

In the previous sections, we have considered the complexity hierarchy for different problems for a fixed class of information operators. Here we consider the complexity hierarchy for different classes of information operators for a fixed problem S.

Let Ψ_1 and Ψ_2 be two classes of information operators.

Definition 8.1   We shall say Ψ_1 *and* Ψ_2 *are* Θ-*equivalent for the problem S,* denoted by $\Psi_1 \asymp \Psi_2$, iff

$$\text{comp}(\Psi_1, S, \varepsilon) = \Theta(\text{comp}(\Psi_2, S, \varepsilon)) \qquad as \quad \varepsilon \to 0.$$

We shall say Ψ_1 *is essentially more efficient than* Ψ_2 *for the problem S,* denoted $\Psi_1 \succ \Psi_2$, iff

$$\text{comp}(\Psi_1, S, \varepsilon) = o(\text{comp}(\Psi_2, S, \varepsilon)) \qquad as \quad \varepsilon \to 0. \quad \blacksquare$$

Thus, $\Psi_1 \asymp \Psi_2$ means that the ε-complexity of the problem S is asymptotically the same in both the classes Ψ_1 and Ψ_2. $\Psi_1 \succ \Psi_2$ means that the ε-complexity of S in the class Ψ_1 is essentially less than the ε-complexity of S in the class Ψ_2.

The results of Chapters 6 and 8 enable us to built complexity hierarchies for all problems considered in this chapter. Although we have not discussed such problems as ELL(r) and PAR(r) in the classes Ψ_f^{non} and Ψ_f^a, it is relatively easy to show that their ε-complexities are $\Theta(\varepsilon^{-1/r})$.

We summarize these results in

Theorem 8.1   (i)   Let S be DIF(r) or INP. Then

$$\Psi_{NON} \asymp \Psi_L^a \asymp \Psi_L^{non} \asymp \Psi_f^a \asymp \Psi_f^{non}.$$

(ii)   Let S be INT. Then

$$\Psi_{NON} \asymp \Psi_L^a \asymp \Psi_L^{non} \succ \Psi_f^a \asymp \Psi_f^{non}.$$

(iii)   Let S be APP or HYP(r). Then

$$\Psi_{NON} \succ \Psi_L^a \asymp \Psi_L^{non} \asymp \Psi_f^a \asymp \Psi_f^{non}.$$

(iv)   Let S be ELL(r) or PAR(r). Then

$$\Psi_{NON} \succ \Psi_L^a \asymp \Psi_L^{non} \succ \Psi_f^a \asymp \Psi_f^{non}.$$

(v)   Let S be NON or UNI. Then

$$\Psi_{NON} \succ \Psi_f^a \succ \Psi_L^{non} \asymp \Psi_f^{non}, \qquad \Psi_L^a \succ \Psi_L^{non} \asymp \Psi_f^{non}. \quad \blacksquare$$

Note that since we do not know the ε-complexity of the problems NON or UNI in the class Ψ_L^a, we do not know whether Ψ_L^a and Ψ_f^a are Θ-equivalent, and whether Ψ_{NON} is essentially more efficient than Ψ_L^a. If, however, Conjectures 3.1 and 7.1 of Chapter 8 are true, then

$$\Psi_{NON} \succ \Psi_L^a \asymp \Psi_f^a.$$

9.   SUMMARY

For the reader's convenience, the Θ-complexity results of this chapter are exhibited in the tableau of Table 9.1.

TABLE 9.1

	DIF(r) INP	INT	APP HYP(r)	ELL(r)	PAR(r)	NON UNI
Ψ_f^{non}	1	$\varepsilon^{-1/r}$	$\varepsilon^{-1/r}$	$\varepsilon^{-1/r}$	$\varepsilon^{-1/r}$	ε^{-1}
Ψ_f^{a}	1	$\varepsilon^{-1/r}$	$\varepsilon^{-1/r}$	$\varepsilon^{-1/r}$	$\varepsilon^{-1/r}$	$\log 1/\varepsilon$
Ψ_L^{non}	1	1	$\varepsilon^{-1/r}$	$\log 1/\varepsilon$	$\sqrt{\log 1/\varepsilon}$	ε^{-1}
Ψ_L^{a}	1	1	$\varepsilon^{-1/r}$	$\log 1/\varepsilon$	$\sqrt{\log 1/\varepsilon}$	?
Ψ_{NON}	1	1	1	1	1	1

Scanning Table 9.1 vertically, we get complexity hierarchies of classes of information operators for a fixed problem. Scanning Table 9.1 horizontally, we get complexity hierarchies of problems for a fixed class of information operators.

Chapter 10

Different Models of Analytic Complexity

1. INTRODUCTION

We have been studying a worst case model of analytic computational complexity. In this chapter, we briefly discuss four other models: average case, relative, perturbed, and asymptotic.

We summarize the results of this chapter. In Section 2, we pose the question of analyzing an average case model. In the next section, we show that a relative error model would not be of interest for linear problems and briefly discuss a relative η-error model. In Section 4, we argue that a certain perturbed information model is not of greater utility than our unperturbed model. In the concluding section, we define concepts for an asymptotic model and list interesting problems arising from this model.

2. AVERAGE CASE MODEL

The model we have discussed so far is a *worst case* model with respect to the problem elements in a class. An *average case* model should be defined and studied.

We conjecture that if S is linear and $\mathfrak{N}$ is linear and nonadaptive, then the average case results will be the same, or nearly the same, as the worst case results. This may be contrasted with results elsewhere in complexity theory, e.g., combinatorial complexity, where the average and worst case results are vastly different.

We have proven that the class of linear adaptive information operators does not supply more knowledge about linear problems than the class of nonadaptive linear information operators. That is, for every nonadaptive information operator, there exists a problem element for which adaption cannot help. It is important to know for what proportion of problem elements in the class adaption will help. If this proportion is large, then adaption will help significantly in an average case model for linear problems.

It would also be of interest to investigate an average case model for nonlinear solution operators and nonlinear information operators.

3. RELATIVE MODEL

We have been analyzing the model in which the absolute error of an algorithm φ, $e(\varphi) = \sup_{f \in \mathfrak{I}_0} \|S(f) - \varphi(\mathfrak{N}(f))\|$, is minimized. One immediate objection to this model might be that in many cases we want to minimize the relative error of an algorithm rather than the absolute one, i.e., we want to find

$$(3.1) \qquad \text{rel } r(\mathfrak{N},S) = \inf_{\varphi} \sup_{f \in \mathfrak{I}_0} \frac{\|S(f) - \varphi(\mathfrak{N}(f))\|}{\|S(f)\|}$$

with the convention that $0/0 = 0$.

We shall show that this does not lead to an interesting model, at least for the linear case. We prove that for linear operators S and $\mathfrak{N}$, and for a balanced and convex $\mathfrak{I}_0$,

$$(3.2) \qquad \text{rel } r(\mathfrak{N},S) = \begin{cases} 0 & \text{if } r(\mathfrak{N},S) = 0, \\ 1 & \text{if } r(\mathfrak{N},S) > 0. \end{cases}$$

Indeed, if $r(\mathfrak{N},S) = 0$, then there exists an algorithm φ such that $e(\varphi) = 0$ which implies that $\varphi(\mathfrak{N}(f)) = S(f)$, $\forall f \in \mathfrak{I}_0$, and rel $r(\mathfrak{N},S) = 0$. Assume then that $r(\mathfrak{N},S) > 0$ and choose an arbitrary algorithm φ. Note that if $\varphi(0) \neq 0$, then setting $f = 0 \in \mathfrak{I}_0$, we get $\|0 - \varphi(0)\|/\|0\| = +\infty$. Thus, let $\varphi(0) = 0$. Then $r(\mathfrak{N},S) > 0$ implies the existence of $f \in \ker \mathfrak{N} \cap \mathfrak{I}_0$ and $Sf \neq 0$. Observe that $\|Sf - \varphi(0)\|/\|Sf\| = 1$ which implies rel $r(\mathfrak{N},S) \geq 1$. The algorithm $\varphi(\mathfrak{N}(f)) \equiv 0$ establishes that rel $r(\mathfrak{N},S) = 1$. Hence, (3.2) is proven.

Thus, except for the trivial case $r(\mathfrak{N},S) = 0$, we cannot find an approximation to a solution element with relative error less than one no matter how many linear functionals are evaluated.

One may argue that the relative error defined by $\|S(f) - \varphi(\mathfrak{N}(f))\|/\|S(f)\|$ is not a reasonable measure of error since it requires us to have small relative error even for elements f for which $\|S(f)\|$ is extremely small. Therefore, we propose another measure of relative error which does lead to a useful model.

Let $\eta > 0$ be a given (small) number. We propose the relative model in which

$$(3.3) \qquad \|S(f) - \varphi(\mathfrak{N}(f))\|/(\|S(f)\| + \eta)$$

is minimized. If $\|S(f)\|$ is large compared to η, then this is close to the relative error, whereas if $\|S(f)\|$ is small compared to η, this is close to the absolute error. Define

$$(3.4) \qquad \text{rel } r(\mathfrak{N},S,\eta) = \inf_{\varphi} \sup_{f \in \mathfrak{I}_0} \frac{\|S(f) - \varphi(\mathfrak{N}(f))\|}{\|S(f)\| + \eta}.$$

Assuming that S and $\mathfrak{N}$ are linear and $\mathfrak{I}_0$ is balanced and convex, we prove

$$(3.5) \qquad \frac{d(\mathfrak{N},S)}{d(\mathfrak{N},S) + 2\eta} \le \text{rel } r(\mathfrak{N},S,\eta) \le \min\,(1,r(\mathfrak{N},S)/\eta).$$

Indeed, setting $\varphi(\mathfrak{N}(f)) = 0$, we get rel $r(\mathfrak{N},S,\eta) \le 1$. For an arbitrary positive δ, let φ be an algorithm such that $e(\varphi) \le r(\mathfrak{N},S) + \delta$. Then

$$\|S(f) - \varphi(\mathfrak{N}(f))\|/(\|S(f)\| + \eta) \le (r(\mathfrak{N},S) + \delta)/\eta$$

which proves the right-hand side of (3.5).

To prove the left-hand side of (3.5), let $f \in \ker \mathfrak{N} \cap \mathfrak{I}_0$ and

$$\|Sf\| \ge \sup_{h \in \ker \mathfrak{N} \cap \mathfrak{I}_0} \|Sh\| - \delta = \tfrac{1}{2}d(\mathfrak{N},S) - \delta.$$

Choose an algorithm φ and observe that

$$2\|Sf\| \le \|Sf - \varphi(0)\| + \|S(-f) - \varphi(0)\|$$

which yields

$$\|S(cf) - \varphi(0)\| \ge \|Sf\|,$$

where $c = -1$ or $c = +1$. From this, we get

$$\frac{\|S(cf) - \varphi(0)\|}{\|S(cf)\| + \eta} \ge \frac{\tfrac{1}{2}d(\mathfrak{N},S) - \delta}{\tfrac{1}{2}d(\mathfrak{N},S) + \eta}.$$

Since δ and φ are arbitrary, we obtain the left-hand side of (3.5) which completes the proof.

From (3.5), we have

(i)   if $\eta \ll d(\mathfrak{N},S)$, then

$$\text{rel } r(\mathfrak{N},S,\eta) \cong 1,$$

(ii)   if $2\eta \le d(\mathfrak{N},S)$, then

$$\text{rel } r(\mathfrak{N},S,\eta) \in [0.5,1],$$

(iii)   if $d(\mathfrak{N},S) \ll \eta$ and $r(\mathfrak{N},S) = \tfrac{1}{2}d(\mathfrak{N},S)$, then

$$\text{rel } r(\mathfrak{N},S,\eta) = \frac{r(\mathfrak{N},S)}{\eta}(1 - c), \qquad c \in \left[0, \frac{r(\mathfrak{N},S)}{\eta}\right].$$

This means that if 2η is less than or comparable to $d(\mathfrak{N},S)$, then the relative error is of order unity. If η is significantly larger than $d(\mathfrak{N},S)$ and $r(\mathfrak{N},S) = \frac{1}{2}d(\mathfrak{N},S)$ (which holds in most practical cases), then rel $r(\mathfrak{N},S,\eta)$ is roughly equal to the radius of information divided by η. It is clear that for linear problems, the theory developed for the absolute error model is mutatis mutandis valid for the relative η-error model with $r(\mathfrak{N},S)$ replaced by $r(\mathfrak{N},S)/\eta$.

It would also be of interest to investigate a relative model for nonlinear S or $\mathfrak{N}$.

4. PERTURBED MODEL

Some papers mention that in practical computations the value $\mathfrak{N}(f)$ of an information operator $\mathfrak{N}$ cannot be exactly known. Therefore, they assume that they know only an element y such that $\|y - \mathfrak{N}(f)\| \le \delta$, where δ is a small positive number given in advance.

We have decided to deal with unperturbed information operators, i.e., $\delta = 0$, for three reasons: (1) it is easier to analyze problems with unperturbed information operators; (2) even for this case, there exist problems with arbitrarily high complexity; (3) [this is the most important reason] we think that the assumption that the element y satisfies $\|y - \mathfrak{N}(f)\| \le \delta$ is not of greater utility for applications than the unperturbed model.

To motivate our viewpoint consider the following example. Suppose that $\mathfrak{I}_0 = W_p^r$ is the class of scalar functions defined on $[a,b]$ whose $(r-1)$th derivative is absolutely continuous and $\|f^{(r)}\|_p \le 1$ with $r \ge 1$. Note that $\mathfrak{I}_0$ is unbounded, i.e., $\sup_{f \in \mathfrak{I}_0} \|f\|_p = +\infty$, since any polynomial of degree $r-1$ belongs to $\mathfrak{I}_0$. Let $\mathfrak{N}(f) = [f(x_1), f(x_2), \ldots, f(x_n)]^t$ for distinct points x_i and let $\|\mathfrak{N}(f)\| = \max\{|f(x_i)| : i = 1, 2, \ldots, n\}$. Thus, the element $y = [y_1, y_2, \ldots, y_n]^t$ satisfies the condition

$$(4.1) \qquad |y_i - f(x_i)| \le \delta \qquad \forall f \in \mathfrak{I}_0, \quad \forall i \in [1,n].$$

This implies that no matter what the value of the norm of $\mathfrak{N}(f)$, we know its approximation with absolute error no greater than δ. However, even if we make the idealistic assumption that the only source of error is the representation of $f(x_i)$ in floating point arithmetic, then, at best, we have $y_i = (1 + \eta_i)f(x_i)$ with $|\eta_i| \le 2^{-t}$, where t is the number of mantissa bits. Thus,

$$y_i - f(x_i) = \eta_i f(x_i)$$

and $|y_i - f(x_i)|$ can be of order $2^{-t}|f(x_i)|$. Since $|f(x_i)|$ is unbounded in $\mathfrak{I}_0$, condition (4.1) is violated even in this very favorable situation.

We hope that this example illustrates that the assumption of knowing the exact information $\mathfrak{N}(f)$ or the element y satisfying $\|y - \mathfrak{N}(f)\| \le \delta$ is of the same utility for applications.

Although we do not intend to analyze here the effects of errors in the information $\mathfrak{N}(f)$, we comment on how this problem can be attacked. The errors in the information $\mathfrak{N}(f)$ should be studied together with the errors of an algorithm implementation. Following the Wilkinson-type model used in numerical linear algebra, the concepts of condition number and algorithm stability should be defined. The condition number is a measure of problem sensitivity to small change in problem elements. Algorithm stability guarantees that the error of algorithm implementation causes an error in the solution proportional to the product of the condition number and the norm of information error. We believe that algorithm stability can be established for many problems.

5. ASYMPTOTIC MODEL

Our model assumes *one* information operator $\mathfrak{N}$ is used to approximate a solution operator S. Here we consider an asymptotic model of analytic complexity which assumes that a *sequence* of information operators is used to approximate S. The following example motivates this model.

Example 5.1 Let $\mathfrak{I}_0 = \mathfrak{I}_1$ be the class of scalar functions whose second derivatives are continuous in $[0,1]$. Let

$$(5.1) \qquad Sf = \int_0^1 f(t)\,dt, \qquad \mathfrak{N}_n(f) = [f(x_1),f(x_2),\ldots,f(x_n)]^t,$$

where $x_i = (i-1)h$, $h = 1/(n-1)$, $i = 1, 2, \ldots, n$. Since the restriction operator $T = 0$, we have

$$r(\mathfrak{N}_n,S) = \sup\{|Sh| : h \in \ker \mathfrak{N}_n\} = +\infty \qquad \forall n.$$

This means there exists no algorithm using $\mathfrak{N}_n$ with finite error.

Despite the infinite radius of information, for fixed f we can proceed as follows. We approximate Sf by the trapezoidal algorithm,

$$(5.2) \qquad \varphi_n(\mathfrak{N}_n(f)) = \tfrac{1}{2}h(f(0) + f(1)) + h\sum_{i=1}^{n-2} f(ih).$$

It is known that

$$(5.3) \qquad Sf - \varphi_n(\mathfrak{N}_n(f)) = f''(\xi)/(12(n-1)^2)$$

for $\xi \in (0,1)$. Since $|f''(\xi)|$ can be arbitrarily large, this explains why

$$e(\varphi_n) = \sup_{f \in \mathfrak{I}_0} |Sf - \varphi(\mathfrak{N}_n(f))| = +\infty \qquad \forall n.$$

However, for every fixed f, the sequence $\{\varphi_n(\mathfrak{N}_n(f))\}$ tends to Sf and the speed of convergence is proportional to n^{-2}. This means that for fixed f, we can approximate Sf with arbitrary accuracy by the sequence of algorithms φ_n using the sequence of information operators $\mathfrak{N}_n$.

Note that for the restriction operator $Tf = f''$, for *every* f in the class $\{f : \|f''\|_\infty \le 1\}$, one obtains a good approximation of Sf, namely,

$$|Sf - \varphi_n(\mathfrak{N}_n(f))| \le 12/(n-1)^2 \qquad \forall n \ge n_0.$$

If $T = 0$, as in this example, we get a good approximation of Sf for a *fixed* f only as n goes to infinity. ∎

This example suggests the study of a model of analytic complexity defined as follows.

Let $\bar\varphi = \{\varphi_n\}$ denote a sequence of algorithms φ_n and let $\bar{\mathfrak{N}} = \{\mathfrak{N}_n\}$ denote a sequence of linear information operators $\mathfrak{N}_n$ with $\operatorname{card}(\mathfrak{N}_n) = n$. We write $\bar\varphi \in \Phi(\bar{\mathfrak{N}})$ iff $\bar\varphi = \{\varphi_n\}$ and every algorithm φ_n uses the information $\mathfrak{N}_n$. Let S be a solution operator. Define

(5.4) $$e(\varphi_n, f) = \|S(f) - \varphi_n(\mathfrak{N}_n(f))\|$$

as the error of φ_n for an element f. For a fixed element f, we are interested in the behavior of $e(\varphi_n, f)$ as n tends to infinity.

We say $\bar\varphi \in \Phi(\bar{\mathfrak{N}})$ is an *asymptotically convergent sequence of algorithms for the problem S* iff

(5.5) $$\lim_{n \to \infty} e(\varphi_n, f) = 0 \qquad \forall f \in \mathfrak{I}_0.$$

We say $\bar{\mathfrak{N}} = \{\mathfrak{N}_n\}$ is an *asymptotically convergent sequence of information operators for the problem S* iff there exists $\bar\varphi \in \Phi(\bar{\mathfrak{N}})$ such that $\bar\varphi$ is asymptotically convergent for S.

In Example 5.1, the sequence $\{\mathfrak{N}_n\}$ defined by (5.1) is asymptotically convergent for S, although $r(\mathfrak{N}_n, S) = +\infty$, $\forall n$.

If $\bar\varphi$, $\bar\varphi \in \Phi(\bar{\mathfrak{N}})$, is asymptotically convergent for S, then we are interested in how fast the sequence $\{e(\varphi_n, f)\}$ tends to zero.

We shall say that the *function* $\rho = \rho(\bar{\mathfrak{N}}, \bar\varphi)$ *measures the asymptotic speed of convergence* of $\bar\varphi$, $\bar\varphi \in \Phi(\bar{\mathfrak{N}})$, iff for every $f \in \mathfrak{I}_0$,

(5.6) $$e(\varphi_n, f) = O(\rho(n)) \qquad \text{as} \quad n \to +\infty,$$

and for some f, (5.6) is sharp, i.e.,

(5.7) $$e(\varphi_n, f) = \Theta(\rho(n)) \qquad \text{as} \quad n \to +\infty.$$

Note that if f is any twice differentiable function, then for the sequence of trapezoidal algorithms (see (5.2)), $\rho(n) = n^{-2}$. To stress the dependence on $\bar{\mathfrak{N}}$ and $\bar\varphi$, we shall sometimes write $\rho(n) = \rho(\bar{\mathfrak{N}}, \bar\varphi, n)$.

As always, we are interested in optimal algorithms. Here they are defined as follows. We say $\bar\varphi^* = \{\varphi_n^*\} \in \Phi(\bar{\mathfrak{N}})$ is an *asymptotically optimal sequence of algorithms for the problem S* iff for every $\bar\varphi \in \Phi(\bar{\mathfrak{N}})$,

(5.8) $$\rho(\bar{\mathfrak{N}}, \bar\varphi^*, n) = O(\rho(\bar{\mathfrak{N}}, \bar\varphi, n)) \qquad \text{as} \quad n \to +\infty.$$

Let Ψ be a class of sequences of information operators $\mathfrak{N} = \{\mathfrak{N}_n\}$. Then $\mathfrak{N}^* = \{\mathfrak{N}_n^*\} \in \Psi$ is called an *asymptotically optimal sequence of information operators in Ψ for the problem S* iff for every $\mathfrak{N} \in \Psi$ and every $\bar{\varphi} \in \Phi(\mathfrak{N})$,

(5.9) $$\rho(\mathfrak{N}^*,\bar{\varphi}^*,n) = O(\rho(\mathfrak{N},\bar{\varphi},n)) \qquad \text{as} \quad n \to +\infty.$$

Here $\bar{\varphi}^*$ is an asymptotically optimal sequence of algorithms such that $\bar{\varphi}^* \in \Phi(\mathfrak{N}^*)$.

Algorithms which are asymptotically convergent are widely used in numerical analysis. Examples include Gauss, Romberg, and composite quadrature formulas. However, if asymptotically convergent algorithms are applied to a problem for which the radius of information is infinity, then we can never be sure if we have solved the problem to any degree of approximation in a given class of problem elements. Thus, for fixed n, $S(f) - \varphi_n(\mathfrak{N}_n(f))$ will be arbitrarily large for some elements from $\mathfrak{I}_0$.

We want to stress that the asymptotic model can also be utilized for problems with small radius of information. In this case, an asymptotically optimal sequence of algorithms φ_n may produce approximations for which $S(f) - \varphi_n(\mathfrak{N}_n(f))$ is small for every n and every f from $\mathfrak{I}_0$.

We provide a partial list of interesting problems arising from this asymptotic model:

(i) For which problems S does there exist a sequence of linear information operators which is asymptotically convergent for S?

(ii) How can an asymptotically optimal sequence of algorithms and an asymptotically optimal sequence of information operators be obtained for a problem S?

(iii) For a given function ρ, what is the class of solution operators for which ρ measures the asymptotic speed of convergence of the asymptotically optimal sequence of algorithms using the asymptotically optimal sequence of information operators?

The complexity model can be formulated as follows. Assume that we want to find an ε-approximation to $S(f)$ for every $f \in \mathfrak{I}_0$. Let $\bar{\varphi} \in \Phi(\mathfrak{N})$ be asymptotically convergent for the problem S and let $\rho = \rho(\mathfrak{N},\bar{\varphi})$ measure the asymptotic speed of convergence of $\bar{\varphi}$. Suppose, additionally, that ρ is a one-to-one function and $\rho^{-1}(cn) = \Theta(\rho^{-1}(n))$ for every fixed positive c as n tends to infinity. Let $n = n(f,\varepsilon)$ be the smallest integer such that

$$e(\varphi_n,f) < \varepsilon.$$

Since $e(\varphi_n,f) = O(\rho(n))$, then from the assumed properties of ρ, we get

(5.10) $$n(f,\varepsilon) = O(\rho^{-1}(\varepsilon)) \qquad \text{as} \quad \varepsilon \to 0.$$

Note that (5.10) is sharp, i.e., there exists an element $f \in \mathfrak{I}_0$ such that O can be replaced by Θ. If the complexity of computing $\varphi_n(\mathfrak{N}_n(f))$ is proportional to the cardinality of $\mathfrak{N}_n$, then

(5.11) $\mathrm{comp}(f,\varepsilon) = O(\rho^{-1}(\varepsilon))$ as $\varepsilon \to 0$.

Observe that the constant in (5.11) depends on f and can be arbitrarily large for some elements f.

   We anticipate that further investigation of this asymptotic model will lead to a rich theory.

Bibliography

ADAMSKI, A., KORYTOWSKI, A., AND MITKOWSKI, W.
[77]   A conception of optimality for algorithms and its application to the optimal search for a minimum, *Zastos. Mat.* **14** (1977), 499–509.

AHLBERG, J. H., AND NILSON, E. N.
[66]   The approximation of linear functionals, *SIAM J. Numer. Anal.* **3** (1966), 173–182.

AHO, A. V., HOPCROFT, J. E., AND ULLMAN, J. D.
[74]   "The Design and Analysis of Computer Algorithms." Addison-Wesley, Reading, Massachusetts, 1974.

AKSEN, M. B., AND TURECKIJ, A. H.
[66]   Best quadrature formulas for certain classes of functions (in Russian), *Dokl. Akad. Nauk SSSR* **166** (1966), 1019–1021 [*English transl.: Soviet Math. Dokl.* **7** (1966), 203–205].

ALHIMOVA, V. M.
[72]   Best quadrature formulas with equidistant nodes (in Russian), *Dokl. Akad. Nauk SSSR* **204** (1972), 263–266 [*English transl.: Soviet Math. Dokl.* **13** (1972), 619–623].

ANSELONE, P. M., AND LAURENT, P. J.
[68]   A general method for the construction of interpolating or smoothing spline functions, *Numer. Math.* **12** (1968), 66–82.

APHANASJEV, A. YU
[74]   On the search of minimum function with limited second derivative (in Russian), *Zh. Vychisl. Mat. Mat. Fiz.* **14** (1974), 1018–1021 [*English transl.:* Afanas'ev, A. Yu., The search for the minimum of a function with a bounded second derivative, *U.S.S.R. Computational Math. and Math. Phys.* **14** (1974), 191–195].

APHANASJEV, A. YU, AND NOVIKOV, V. A.
[77]   On the search of minimum of a function with the limited third derivative (in Russian), *Zh. Vychisl. Mat. Mat. Fiz.* **17** (1977), 1031–1034.

ARESTOV, V. V.

[67]   On the best approximation of differentiation operators (in Russian), *Mat. Zametki* **1** (1967), 149–154 [*English transl.*: *Math. Notes* **1** (1967), 100–103].

[69]   On the best uniform approximation of differentiation operators (in Russian), *Mat. Zametki* **5** (1969), 273–284 [*English transl.*: *Math. Notes* **5** (1969), 167–173].

ATTEIA, M.

[65]   Fonctions-spline généralisées, *C.R. Acad. Sci. Paris* **261** (1965), 2149–2152.

AVRIEL, M., AND WILDE, D. J.

[66]   Optimal search for a maximum with sequences of simultaneous function evaluations, *Management Sci.* **12** (1966), 722–731.

BABENKO, V. F.

[76]   Asymptotically sharp bounds for the remainder for the best quadrature formulas for several classes of functions (in Russian), *Mat. Zametki* **19** (1976), 313–322 [*English transl.*: *Math. Notes* **19** (1976), 187–193].

BABUŠKA, I.

[68a] Problems of optimization and numerical stability in computations, *Apl. Mat.* **1** (1968), 3–26.

[68b] Über universal optimale quadraturformeln, *Apl. Mat.* **4** (1968), 305–338; **5** (1968), 388–404.

BABUŠKA, I., AND SOBOLEV, S. L.

[65]   Optimization of numerical methods (in Russian), *Apl. Mat.* **1** (1965), 96–130.

BAKHVALOV, N. S.

[59]   On the approximate calculation of multiple integrals (in Russian), *Vestn. MGU. Ser. of Math. Mech. Astron. Phys. Chem.* **4** (1959), 3–18.

[61]   An estimate of the mean remainder in quadrature formulae (in Russian), *Zh. Vychisl. Mat. Mat. Fiz.* **1** (1961), 64–77 [*English transl.*: *U.S.S.R. Computational Math. and Math. Phys.* **1** (1961), 68–82].

[62a] On optimal methods of specifying information in the solution of differential equations (in Russian), *Zh. Vychisl. Mat. Mat. Fiz.* **2** (1962), 569–592 [*English transl.*: *U.S.S.R. Computational Math. and Math. Phys.* **2** (1962), 608–640].

[62b] On the estimate of the amount of computational labor necessary in approximate solutions (in Russian), Appendix IV in the book of S. K. Godunov and W. S. Riabenki, "Theory of Difference Schemes—An Introduction" pp. 316–329. Moscow, English translation of the book published by American Elsevier, New York, 1964, pp. 268–279.

[63]   Optimal properties of Adams and Gregory formulae of numerical integration (in Russian), *in* "Problems of Computational Mathematics and Computational Technique" (L. A. Ljusternik, ed.), pp. 9–26. Mashgiz, Moscow, 1963.

[64]   On optimal bounds for the convergence of quadrature formulas and Monte Carlo type integration methods for classes of functions (in Russian), *in* "Numerical Methods for the Solution of Differential and Integral Equations and Quadrature Formulas," pp. 5–63. Moscow, 1964.

[67]   On the optimal speed of integrating analytic functions (in Russian), *Zh. Vychisl. Mat. Mat. Fiz.* **7** (1967), 1011–1020 [*English transl.*: *U.S.S.R. Computational Math. and Math. Phys.* **7** (1967), 63–75].

[68]   On optimal methods for the solution of problems (in Russian), *Apl. Mat.* **1** (1968), 27–38.

[70]   Properties of optimal methods for the solution of problems of mathematical physics (in Russian), *Zh. Vychisl. Mat. Mat. Fiz.* **10** (1970), 555–568 [*English transl.*: *U.S.S.R. Computational Math. and Math. Phys.* **10** (1970), 1–20].

[71a] On the optimality of linear methods for operator approximation in convex classes of functions (in Russian), *Zh. Vychisl. Mat. Mat. Fiz.* **11** (1971), 1014–1018 [*English transl.*: *U.S.S.R. Computational Math. and Math. Phys.* **11** (1971), 244–249].

[71b]   Optimization of methods of solving ordinary differential equations with strongly oscillating solutions (in Russian), *Zh. Vychisl. Mat. Mat. Fiz.* **11** (1971), 1318–1322 [*English transl.*: *U.S.S.R. Computational Math. and Math. Phys.* **11** (1971), 287–292].

[72]   A lower bound for the asymptotic characteristics of classes of functions with dominating mixed derivative (in Russian), *Mat. Zametki* **12** (1972), 655–664.

BARNHILL, R. E.
[67]   Optimal quadratures in $L^2(E_\zeta)$. I and II, *SIAM J. Numer. Anal.* **4** (1967), 390–397, 534–541.

[68]   Asymptotic properties of minimum norm and optimal quadratures, *Numer. Math.* **12** (1968), 384–393.

BARNHILL, R. E., AND WIXOM, J. A.
[67]   Quadratures with remainders of minimum norm. I and II, *Math. Comp.* **21** (1967), 66–75, 382–387.

[68]   An error analysis for interpolation of analytic functions, *SIAM J. Numer. Anal.* **5** (1968), 522–528.

BARRAR, R. B., LOEB, H. L., AND WERNER, M.
[74]   On the existence of optimal integration formulas for analytic functions, *Numer. Math.* **23** (1974), 105–117.

BEAMER, J. H., AND WILDE, D. J.
[69]   Time delay in minimax optimization of unimodal functions of one variable, *Management Sci.* **15** (1969), 528–538.

[70]   Minimax optimization of unimodal functions by variable block search, *Management Sci.* **16** (1970), 529–541.

[71]   Minimax optimization of a unimodal function by variable block derivative search with time delay, *J. Comb. Theory* **10** (1971), 160–173.

BOJANOV, B. D.
[73]   Optimal rate of integration and ε-entropy of a class of analytic functions (in Russian), *Mat. Zametki* **14** (1973), 3–10 [*English transl.*: *Math. Notes* **19** (1973), 551–556].

[74]   Best quadrature formula for a certain class of analytic functions, *Zastos. Mat.* **14** (1974), 441–447.

[75]   Best methods of interpolation for certain classes of differentiable functions (in Russian), *Mat. Zametki* **17** (1975), 511–524 [*English transl.*: *Math. Notes* **17** (1975), 301–309].

[76]   Optimal methods of integration in the class of differentiable functions, *Zastos. Mat.* **15** (1976), 105–115.

BOJANOV, B. D., AND CHERNOGOROV, V. G.
[77]   An optimal interpolation formula, *J. Approx. Theory* **20** (1977), 264–274.

BOOTH, R. S.
[67]   Location of zeros of derivatives, *SIAM J. Appl. Math.* **15** (1967), 1496–1501.
[69]   Location of zeros of derivatives. II, *SIAM J. Appl. Math.* **17** (1969), 409–415.

BORODIN, A.
[72]   Complexity classes of recursive functions and the existence of complexity gaps, *J. Assoc. Comput. Mach.* **19** (1972), 158–174, 576.

BRENT, R. P., AND KUNG, H. T.
[78]   Fast algorithms for manipulating formal series, *J. Assoc. Comput. Mach.* **25** (1978), 581–595.

BRENT, R. P., WINOGRAD, S., AND WOLFE, P.
[73]   Optimal iterative processes for root-finding. *Numer. Math.* **20** (1973), 327–341.

BUSAROVA, T. N.
[73]   Best quadrature formulae for a class of differentiable and periodic functions (in Russian), *Ukr. Mat. Z.* **25** (1973), 291–301.

CHAWLA, M. M., AND KAUL, V.
[73] Optimal rules for numerical integration round the unit circle, *BIT* **13** (1973), 145–152.

CHENTSOV, N. N.
[61] On quadrature formulae for functions of an infinitely large number of variables (in Russian), *Zh. Vychisl. Mat. Mat. Fiz.* **1** (1961), 418–424 [*English transl.*: *U.S.S.R. Computational Math. and Math. Phys.* **1** (1961), 455–464].

CHERNOUSKO, F. L.
[68] An optimal algorithm for finding the roots of an approximately computed function (in Russian), *Zh. Vychisl. Mat. Mat. Fiz.* **8** (1968), 705–724 [*English transl.*: *U.S.S.R. Computational Math. and Math. Phys.* **8** (1968), 1–24].
[70a] Optimal search for the minimum of convex functions (in Russian), *Zh. Vychisl. Mat. Mat. Fiz.* **10** (1970), 1355–1366 [*English transl.*: *U.S.S.R. Computational Math. and Math. Phys.* **10** (1970), 20–34].
[70b] Optimal search for extrema of unimodal functions (in Russian), *Zh. Vychisl. Mat. Mat. Fiz.* **10** (1970), 922–933 [*English transl.*: *U.S.S.R. Computational Math. and Math. Phys.* **10** (1970), 146–161].

CHZHAN GUAN-TSZYNAN
[62] On the minimum number of interpolation points in the numerical integration of the heat-conduction equation (in Russian), *Zh. Vychisl. Mat. Mat. Fiz.* **2** (1962), 80–88 [*English transl.*: *U.S.S.R. Computational Math. and Math. Phys.* **2** (1962), 78–87].

COMAN, GH.
[72] Monosplines and optimal quadrature formulae in L_p, *Rend. Mat.* **5** (1972), 567–577.

COMAN, GH., AND MICULA, GH.
[71] Optimal cubature formulae, *Rend. Mat.* **4** (1971), 303–311.

DANILIN, YU. M.
[71] On one algorithm efficiency estimation of absolute minimum finding (in Russian), *Zh. Vychisl. Mat. Mat. Fiz.* **11** (1971), 1026–1030 [*English transl.*: Estimation of the efficiency of an absolute-minimum-finding algorithm, *U.S.S.R. Computational Math. and Math. Phys.* **11** (1971), 261–267].

DE BOOR, C.
[77] Computational aspects of optimal recovery, *in* "Optimal Estimation in Approximation Theory" (C. A. Micchelli and T. J. Rivlin, eds.), pp. 69–91. Plenum Press, New York, 1977.

ECKHARDT, U.
[68] Einige eigenschaften wilfscher quadraturformeln, *Numer. Math.* **12** (1968), 1–7.

EDWARDS, R. E.
[65] "Functional Analysis." Holt, New York, 1965.

EICHHORN, B. H.
[68] On sequential search, "Selected Statistical Papers," Vol. 1, pp. 81–95. Math. Centrum, Amsterdam, 1968.

ELHAY, S.
[69] Optimal quadrature, *Bull. Austral. Math. Soc.* **1** (1969), 81–108.

EMELYANOV, K. V., AND ILIN, A. M.
[67] Number of arithmetical operations necessary for the approximate solution of Fredholm integral equations of the second kind (in Russian), *Zh. Vychisl. Mat. Mat. Fiz.* **7** (1967), 905–910 [*English transl.*: *U.S.S.R. Computational Math. and Math. Phys.* **7** (1967), 259–266].

FINE, T.
[66]   Optimum search for the location of the maximum of a unimodal function, *IEEE Trans. Informat. Theory* **IT-12** (1966), 103–111.

FORST, W.
[75]   Zur optimalität interpolatorischer quadraturformeln periodischer funktionen, *Numer. Math.* **25** (1975), 15–21.
[77]   Optimale hermite-interpolation differenzierbarer periodischer funktionen, *J. Approx. Theory* **20** (1977), 333–347.

GAFFNEY, P. W.
[76]   Optimal interpolation. D. Phil. Thesis, Oxford Univ. (1976).
[77a]   The range of possible values of $f(x)$. Computer Science and Systems Division Rep., AERE, Harwell, Oxfordshire (1977).
[77b]   To compute the optimal interpolation formula. Computer Science and Systems Division Rep., AERE, Harwell, Oxfordshire (1977).

GAFFNEY, P. W., AND POWELL, M. J. D.
[76]   Optimal interpolation, *in* "Numerical Analysis" (G. A. Watson, ed.), Lecture Notes in Math., Vol. 506, pp. 90–100. Springer Verlag, Berlin and New York, 1976.

GAISARIAN, S. S.
[69]   An optimal algorithm for the approximate computation of quadratures (in Russian), *Zh. Vychisl. Mat. Mat. Fiz.* **9** (1969), 1015–1023 [*English transl.: U.S.S.R. Computational Math. and Math. Phys.* **9** (1969), 42–53].
[70]   The choice of optimal networks for the numerical solution of the Cauchy problem for a set of ordinary differential equations, *Zh. Vychisl. Mat. Mat. Fiz.* **10** (1970), 465–474 [*English transl.: U.S.S.R. Computational Math. and Math. Phys.* **10** (1970), 253–267].

GAL, S.
[72]   Multidimensional minimax search for a maximum, *SIAM J. Appl. Math.* **23** (1972), 513–526].

GAL, S., AND MICCHELLI, A. C.
[78]   Optimal sequential and non-sequential procedures for evaluating a functional. Univ. of Wisconsin—Madison Rep. 1871. To appear in *Appl. Anal.* (1978).

GANSHIN, G. S.
[76]   Calculation of the greatest value of function (in Russian), *Zh. Vychisl. Mat. Mat. Fiz.* **16** (1976), 30–39 [*English transl.: U.S.S.R. Computational Math. and Math. Phys.* **16** (1976), 26–36].
[77]   Optimal algorithms of calculation of the function highest value (in Russian), *Zh. Vychisl. Mat. Mat. Fiz.* **17** (1977), 562–572.

GAREY, M. R., AND JOHNSON, D. S.
[76]   Approximation algorithms for combinatorial problems: An annotated bibliography, *in* "Algorithms and Complexity: New Directions and Recent Results" (J. F. Traub, ed.), pp. 41–52. Academic Press, New York, 1976.

GOLOMB, M.
[77]   Interpolation operators as optimal recovery schemes for classes of analytic functions, *in* "Optimal Estimation in Approximation Theory" (C. A. Micchelli and T. J. Rivlin, eds.), pp. 93–138. Plenum Press, New York, 1977.

GOLOMB, M., AND WEINBERGER, H. F.
[59]   Optimal approximation and error bounds, *in* "On Numerical Approximation" (R. E. Langer, ed.), pp. 117–190. The University of Wisconsin Press, Madison, Wisconsin, 1959.

GREBENNIKOV, A. I.
[78] On optimal approximation of nonlinear operators (in Russian), *Zh. Vychisl. Mat. Mat. Fiz.* **18** (1978), 762–768.

GREBENNIKOV, A. I., AND MOROZOV, V. A.
[77] On optimal approximation of operators (in Russian), *Zh. Vychisl. Mat. Mat. Fiz.* **17** (1977), 3–15.

GROSS, O., AND JOHNSON, S. M.
[59] Sequential minimax search for a zero of a convex function, *MTAC* (now *Math. Comp.*) **13** (1959), 44–51.

HABER, S.
[71] The error in numerical integration of analytic functions, *Quart. Appl. Math.* **29** (1971), 411–420.

HOLMES, R.
[72] *R*-splines in Banach spaces: I. Interpolation of linear manifolds, *J. Math. Anal. Appl.* **40** (1972), 574–593.

HYAFIL, L.
[77] Optimal search for the zero of the $(n-1)$st derivative. IRIA/LABORIA Rep. No. 247 (1977).

IBRAGIMOV, I. I., AND ALIEV, R. M.
[65] Best quadrature formulas for certain classes of functions (in Russian), *Dokl. Akad. Nauk SSSR* **162** (1965), 23–25 [*English transl.: Soviet Math. Dokl.* **6** (1965), 621–623].

ISMAGILOV, R. S.
[74] Diameters of sets in normed linear spaces and the approximation of functions by trigonometric polynomials (in Russian), *Usp. Mat. Nauk* **29** (3) (1974), 161–178 [*English transl.: Russian Math. Surveys* **29** (3) (1974), 169–186].

IVANOV, V. V.
[72a] On optimal algorithms for function minimization on certain classes of functions (in Russian), *Kibernetika* **4** (1972), 81–94.
[72b] On optimal algorithms for the calculation of singular integrals (in Russian), *Dokl. Akad. Nauk SSSR* **204** (1972), 21–24 [*English transl.: Soviet Math. Dokl.* **13** (1972), 576–580].
[75] On optimal in accuracy algorithms for approximate solution of operator equations of the first kind (in Russian), *Zh. Vychisl. Mat. Mat. Fiz.* **15** (1975), 3–12 [*English transl.:* Algorithms of optimal accuracy for the approximate solution of operator equations of the first kind, *U.S.S.R. Computational Math. and Math. Phys.* **15** (1975), 1–9].
[77] On optimal algorithms approximating functions for some classes (in Russian), *in* "Theory of Approximation of Functions," pp. 195–200. Nauka, Moscow, 1977.

JETTER, K.
[76] Optimale quadraturformeln mit semidefiniten peano-kernen, *Numer. Math.* **25** (1976), 239–249.

JOHNSON, L. W., AND RIESS, R. D.
[71] Minimal quadratures for functions of low-order continuity, *Math. Comp.* **25** (1971), 831–835.

JOHNSON, S. M.
[56] Best exploration for maximum is Fibonaccian, RAND Corp. Rep. P-856 (1956).

JUDIN, D. B., AND NEMIROVSKY, A. S.
[76a] A bound of information complexity for mathematical programming problems (in Russian), *Ekon. Mat. Metody* **12** (1976), 128–142.

[76b]   Information complexity and effective methods for the solution of convex extremal problems (in Russian), *Ekon. Mat. Metody* **12** (1976), 357–369.

[77]   Information complexity for strict convex programming (in Russian), *Ekon. Mat. Metody* **13** (1977), 550–559.

KACEWICZ, B.

[75]   Integrals with a kernel in the solution of nonlinear equations in N dimensions. Computer Science Department Rep., Carnegie-Mellon Univ. (1975). See also *J. Assoc. Comput. Mach.* **26** (1979), 239–249.

[76a]   An integral-interpolation iterative method for the solution of scalar equations, *Numer. Math.* **26** (1976), 355–365.

[76b]   The use of integrals in the solution of nonlinear equations in N dimensions, *in* "Analytic Computational Complexity" (J. F. Traub, ed.), pp. 127–141. Academic Press, New York, 1976.

KARLIN, S.

[69]   Best quadrature formulas and interpolation by splines satisfying boundary conditions, and the fundamental theorem of algebra for monosplines satisfying certain boundary conditions and applications to optimal quadrature formulae, *in* "Approximations with Special Emphasis on Spline Functions" (I. J. Schoenberg, ed.), pp. 447–466, 467–484. Academic Press, New York, 1969.

[71]   Best quadrature formulas and splines, *J. Approx. Theory* **4** (1971), 59–90.

KARP, R. M., AND MIRANKER, W. L.

[68]   Parallel minimax search for a maximum, *J. Comb. Theory* **4** (1968), 19–35.

KAUTSKY, J.

[70]   Optimal quadrature formulae and minimal monosplines in L_q, *J. Austral. Mat. Soc.* **11** (1970), 48–56.

KEAST, P.

[73]   Optimal parameters for multidimensional integration, *SIAM J. Numer. Anal.* **10** (1973), 831–838.

KIEFER, J.

[53]   Sequential minimax search for a maximum, *Proc. Amer. Math. Soc.* **4** (1953), 502–505.

[57]   Optimum sequential search and approximation methods under minimum regularity assumptions, *J. Soc. Indust. Appl. Math.* **5** (1957), 105–136.

KNAUFF, W., AND KRESS, R.

[74]   Optimale approximation linearer funktionale auf periodischen funktionen, *Numer. Math.* **22** (1974), 187–205.

[76]   Optimale approximation mit nebenbedingungen an lineare funktionale auf periodischen funktionen, *Numer. Math.* **25** (1976), 149–159.

KNUTH, D. E.

[76]   Big omicron and big omega and big theta, *SIGACT News* (April 1976), 18–24.

KOLMOGOROV, A. N.

[36]   Über die beste annäherung von funktionen einer gegebenen funktionklasse, *Ann. of Math.* **37** (2) (1936), 107–110.

[55]   Evaluation of minimal number of elements of ε-nets in different functional classes and their application to the problem of representation of functions of several variables by superposition of functions of a smaller number of variables (in Russian), *Usp. Mat. Nauk* **10** (1) (1955), 192–194.

KOLMOGOROV, A. N., AND TIKHOMIROV, V. M.

[59]   The ε-entropy and ε-capacity of sets in functional spaces (in Russian), *Usp. Mat. Nauk* **14** (2) (86) (1959), 3–80.

KORNEJČUK, N. P.

[68] Best cubature formulas for some classes of functions of many variables (in Russian), *Mat. Zametki* **3** (1968), 565–576 [*English transl.: Math. Notes* **3** (1968), 360–367].

[74] New results on extremal problems of the theory of quadratures (in Russian), appendix to the second edition of S. M. Nikolskij, "Quadrature Formulae," pp. 136–223. Moscow, 1974.

[76] "Extremal Problems of Approximation Theory" (in Russian). Nauka, Moscow, 1976.

KORNEJČUK, N. P., AND LUŠPAJ, N. E.

[69] Best quadrature formulas for classes of differentiable functions and piecewise-polynomial approximation (in Russian), *Izv. Akad. Nauk SSSR Ser. Mat.* **33** (1969), 1416–1437 [*English transl.: Math. U.S.S.R.-Izv.* **3** (1969), 1335–1355].

KOROBOV, N. M.

[63] "Number Theory Methods in Approximation Analysis" (in Russian). Fizmatgiz, Moscow, 1963.

KOROTKOV, V. B.

[77] A lower bound on cubature formulas (in Russian), *Sibirskii Mat. Zh.* **17** (1977), 1188–1191.

KROLAK, P.

[66] Property of the Krolak–Cooper extension of Fibonaccian search, *SIAM Rev.* **8** (1966). 510–517.

[68] Further extensions of Fibonaccian search of nonlinear programming problems, *SIAM J. Control* **6** (1968), 258–265.

KROLAK, P., AND COOPER, L.

[63] An extension of Fibonaccian search to several variables, *Comm. ACM* **6** (1963), 639–641.

KRYLOV, V. I.

[62] "Approximate Calculation of Integrals," Chapter 8, pp. 133–149. Macmillan, New York, 1962.

KUNG, H. T.

[76] The complexity of obtaining starting points for solving operator equations by Newton's method, *in* "Analytic Computational Complexity" (J. F. Traub, ed.), pp. 35–57. Academic Press, New York, 1976.

KUNG, H. T., AND TRAUB, J. F.

[74] Optimal order of one-point and multipoint iterations, *J. Assoc. Comput. Mach.* **21** (1974), 643–651.

[76] Optimal order and efficiency for iterations with two evaluations, *SIAM J. Numer. Anal.* **13** (1976), 84–99.

KUZOVKIN, A. I., AND TIKHOMIROV, V. M.

[67] On the number of operations for finding the minimum of convex functions (in Russian), *Ekon. Mat. Metody* **3** (1967), 95–103.

LARKIN, F. M.

[70] Optimal approximation in Hilbert space with reproducing kernel functions, *Math. Comp.* **24** (1970), 911–921.

LEE, J. W.

[77] Best quadrature formulas and splines, *J. Approx. Theory* **20** (1977), 348–384.

LEVIN, A. JU.

[65] On an algorithm for the minimization of convex functions (in Russian), *Dokl. Akad. Nauk SSSR* **160** (1965), 1244–1247 [*English transl.: Soviet Math. Dokl.* **6** (1965), 286–289].

LEVIN, M. I., AND GIRŠOVIČ, JU. M.
[77]   Extremal problems for cubature formulas (in Russian), *Dokl. Akad. Nauk SSSR* **236** (1977), 1303–1306 [*English transl.: Soviet Math. Dokl.* **18** (1977), 1355–1358].

LEVIN, M. J., GIRŠOVIČ, JU. M., AND ARRO, V. K.
[76]   Best quadrature formulas on sets of functions (in Russian), *Dokl. Akad. Nauk SSSR* **226** (1976), 51–54 [*English transl.: Soviet Math. Dokl.* **17** (1976), 46–50].

LIGUN, A. A.
[76]   Exact inequalities for splines and best quadrature formulas for certain classes of functions (in Russian), *Mat. Zametki* **19** (1976), 913–926 [*English transl.: Math. Notes* **19** (1976), 533–541].

LIPOW, P. R.
[73]   Spline functions and intermediate best quadrature formulas, *SIAM J. Numer. Anal.* **10** (1973), 127–136.

LOEB, H., AND WERNER, M.
[74]   Optimal numerical quadrature in H_p spaces, *Math. Z.* **138** (1974), 111–117.

LORENTZ, G. G.
[66]   "Approximation of Functions." Holt, New York, 1966.

LUŠPAJ, N. E.
[66]   Best quadrature formulae for some classes of functions (in Russian), *Proc. Internat. Conf. Young Res. Math., Charkov* (1966), pp. 58–62.
[69]   Best quadrature formulas on classes of differentiable periodic functions (in Russian), *Mat. Zametki* **6** (1969), 475–482 [*English transl.: Math. Notes* **6** (1969), 740–744].
[74]   Best quadrature formula on the class $W^r_* L_2$ of periodic functions (in Russian), *Mat. Zametki* **16** (1974), 193–204 [*English transl.: Math. Notes* **16** (1974), 701–708].

MAJSTROVSKIJ, G. D.
[72]   On the optimality of Newton's method (in Russian), *Dokl. Akad. Nauk SSSR* **204** (1972), 1313–1315 [*English transl.: Soviet Math. Dokl.* **13** (1972), 838–840].

MANGASARIAN, O. L., AND SCHUMAKER, L. L.
[73]   Best summation formulae and discrete splines. *SIAM J. Numer. Anal.* **10** (1973), 448–459.

MANSFIELD, L. E.
[71]   On the optimal approximation of linear functionals in spaces of bivariate functions, *SIAM J. Numer. Anal.* **8** (1971), 115–126.
[72]   Optimal approximation and error bounds in spaces of bivariate functions, *J. Approx. Theory* **5** (1972), 77–96.

MARCHUK, A. G., AND OSIPENKO, K. YU
[75]   Best approximation of functions specified with an error at a finite number of points (in Russian), *Mat. Zametki* **17** (1975), 359–368 [*English transl.: Math. Notes* **17** (1975), 207–212].

MAUNG ČŽO NJUN, AND SHARYGIN, I. F.
[75]   Optimal cubature formulae on the classes $D_2^{1,c}$ and D_s^{1,L_1} (in Russian) *Problems of Numer. and Appl. Math., Tashkent* **5** (1975), 22–27.

MEERSMAN, R.
[76a]   Optimal use of information in certain iterative processes, *in* "Analytic Computational Complexity" (J. F. Traub, ed.), pp. 109–125. Academic Press, New York, 1976.
[76b]   On maximal order of families of iterations for nonlinear equations. Doctoral Thesis, Vrije Univ. Brussel, Brussels, 1976.

MEINGUET, J.
[67]   Optimal approximation and error bounds in seminormed spaces, *Numer. Math.* **10** (1967), 370–388.

MELKMAN, A. A.
[77]   *n*-widths and optimal interpolation of time- and band-limited functions, *in* "Optimal Estimation in Approximation Theory" (C. A. Micchelli and T. J. Rivlin, eds.), pp. 55–68. Plenum Press, New York, 1977.

MELKMAN, A. A., AND MICCHELLI, C. A.
[77]   Optimal estimation of linear operators in Hilbert spaces from inaccurate data. Univ. Bonn Rep. (1977). See also *SIAM J. Num. Anal.* **16** (1979), 87–105.

MEYERS, L. F., AND SARD, A.
[50a]   Best approximate integration formulas, *J. of Math. and Phys.* **28** (1950), 118–123.
[50b]   Best interpolation formulas, *J. of Math. and Phys.* **29** (1950), 198–206.

MICCHELLI, C. A.
[75]   Optimal estimation of linear functionals. IBM Research Rep. 5729 (1975).
[76]   On an optimal method for the numerical differentiation of smooth functions, *J. Approx. Theory* **18** (1976), 189–204.
[78]   Private communication (1978).

MICCHELLI, C. A., AND MIRANKER, W. L.
[75]   High order search methods for finding roots, *J. Assoc. Comput. Mach.* **22** (1975), 51–60.

MICCHELLI, C. A., AND PINKUS, A.
[77]   On a best estimator for the class M^r using only function values, *Indiana Univ. Math. J.* **26** (1977), 751–759.

MICCHELLI, C. A., AND RIVLIN, T. J.
[77]   A survey of optimal recovery, *in* "Optimal Estimation in Approximation Theory" (C. A. Micchelli and T. J. Rivlin, eds.), pp. 1–54. Plenum Press, New York, 1977.

MICCHELLI, C. A., RIVLIN, T. J., AND WINOGRAD, S.
[76]   The optimal recovery of smooth functions, *Numer. Math.* **26** (1976), 191–200.

MOCKUS, I. B.
[72]   Bayesian methods for extremum search (in Russian) *Avtomat. Vyčisl. Techn.* **3** (1972), 53–62.

MOTORNYJ, V. P.
[73]   On the best quadrature formula of the form $\sum_{k=1}^{n} p_k f(x_k)$ for some classes of periodic differentiable functions (in Russian), *Dokl. Akad. Nauk SSSR* **211** (1973), 1060–1062 [*English transl.: Soviet Math. Dokl.* **14** (1973), 1180–1183].
[74]   On the best quadrature formulae of the form $\sum_{k=1}^{n} p_k f(x_k)$ for some classes of periodic differentiable functions (in Russian), *Dokl. Akad. Nauk SSSR Ser. Math.* **38** (1974), 583–614.
[76]   Some extremal problems of theory of quadrature and approximation of functions (in Russian), *Mat. Zametki* **19** (1976), 299–311 [*English transl.: Math. Notes* **19** (1976), 176–183].

NEWMAN, D. J.
[65]   Location of the maximum on unimodal surfaces, *J. Assoc. Comput. Mach.* **12** (1965), 395–398.

NIELSON, G. M.
[73]   Bivariate spline functions and the approximation of linear functionals, *Numer. Math.* **21** (1973), 138–160.

NIKOLSKIJ, S. M.
[50] On the problem of approximation estimate by quadrature formulae, *Usp. Mat. Nauk* **5** (1950), 165–177.
[58] "Quadrature Formulae" (in Russian). Nauka, Moscow. 1st ed., 1958, 2nd ed., 1974 English translation, Hindustan Publ., Delhi, India, 1964.

OSIPENKO, K. YU
[72] Optimal interpolation of analytic functions (in Russian), *Mat. Zametki* **12** (1972), 465–476 [*English transl.: Math. Notes* **12** (1972), 712–719].
[76] Best approximation of analytic functions from information about their values at a finite number of points (in Russian), *Mat. Zametki* **19** (1976), 29–40 [*English transl.: Math. Notes* **19** (1976), 17–23].

PALLASHKE, D.
[76] Optimale differentiations und integrationsformeln in $C_0[a, b]$, *Numer. Math.* **16** (1976), 201–210.

PAN, V.
[78] Strassen's algorithm is not optimal: Trilinear technique of aggregating, uniting and canceling for construction fast algorithms for matrix operations, *19th Ann. Symp. Foundations of Comput. Sci.*, IEEE Computer Society (1978).

PAULIK, A.
[77] Zur existenz optimaler quadraturformeln mit freien knoten bei integration analytischer funktionen, *Numer. Math.* **27** (1977), 395–405.

PIAVSKY, S. A.
[72] One algorithm for the searching of global extremum of function (in Russian), *Zh. Vychisl. Mat. Mat. Fiz.* **12** (1972), 888–896 [*English transl.: U.S.S.R. Computational Math. and Math. Phys.* **12** (1972), 57–67].

PINKUS, A.
[75] Asymptotic minimum norm quadrature formulae, *Numer. Math.* **24** (1975), 163–175.

REINSCH, CH.
[74] Two extensions of the Sard–Schoenberg theory of best approximation, *SIAM J. Numer. Anal.* **11** (1974), 45–51.

RICE, J. R.
[73] On the computational complexity of approximation operators, *in* "Approximation Theory" (G. G. Lorentz, ed.), pp. 449–456. Academic Press, New York, 1973.
[76] On the computational complexity of approximation operators II, *in* "Analytic Computational Complexity" (J. F. Traub, ed.), pp. 191–204. Academic Press, New York, 1976.

RICHTER, N.
[70] Properties of minimal integration rules, *SIAM J. Numer. Anal.* **7** (1970), 67–79.

RICHTER-DYN, N.
[71a] Properties of minimal integration rules, II, *SIAM J. Numer. Anal.* **8** (1971), 497–508.
[71b] Minimal interpolation and approximation in Hilbert spaces, *SIAM J. Numer. Anal.* **8** (1971), 583–597.

RITTER, K.
[70] Two dimensional spline functions and best approximations of linear functionals, *J. Approx. Theory* **3** (1970), 352–368.

ŠAJDAEVA, T. A.
[59] Quadrature formulae with least bound for the remainder for some classes of functions (in Russian), *Trudy Mat. Inst. Steklova Akad. Nauk SSSR* **53** (1959), 313–341.

SARD, A.
[49] Best approximate integration formulas; best approximation formulas, *Amer. J. Math.* **71** (1949), 80–91.
[63] "Linear Approximation," American Math. Soc., Providence, Rhode Island, 1963.
[67] Optimal approximation, *J. Functional Anal.* **1** (1967), 222–244.
[73] Approximation based on nonscalar observations, *J. Approx. Theory* **8** (1973), 315–334.

SCHMEISSER, G.
[72] Optimale quadraturformeln mit semidefiniten kernen, *Numer. Math.* **20** (1972), 32–53.

SCHOENBERG, I. J.
[64a] On best approximations of linear operators, *Nederl. Akad. Wetensch. Indag. Math.* **67** (1964), 155–163.
[64b] Spline interpolation and best quadrature formulae, *Bull. Amer. Math. Soc.* **70** (1964) 143–148.
[65] On monosplines of least deviation and best quadrature formulae, *SIAM J. Numer. Anal. Ser. B.* **2** (1965), 144–170.
[66] On monosplines of least square deviation and best quadrature formulae II, *SIAM J. Numer. Anal.* **3** (1966), 321–328.
[69] Monosplines and quadrature formulae, *in* "Theory and Applications of Spline Functions" (T. N. E. Greville, ed.), pp. 157–207. Academic Press, New York, 1969.
[70] A second look at approximate quadrature formulae and spline interpolation, *Advances in Math.* **4** (1970), 277–300.

SCHULTZ, M. H.
[74] The complexity of linear approximation algorithms, *in* "Complexity of Computation," (R. M. Karp, ed.), pp. 135–148. American Mathematical Soc., Providence, Rhode Island, 1974.

SECREST, D.
[64] Numerical integration of arbitrarily spaced data and estimation of errors, *SIAM J. Numer. Anal. Ser. B.* **2** (1964), 52–68.
[65a] Error bounds for interpolation and differentiation by the use of spline functions, *SIAM J. Numer. Anal. Ser. B.* **2** (1965), 440–447.
[65b] Best approximate integration formulas and best error bounds, *Math. Comp.* **19** (1965), 79–83.

SHARYGIN, I. F.
[63] A lower estimate for the error of quadrature formulae for certain classes of functions (in Russian), *Zh. Vychisl. Mat. Mat. Fiz.* **3** (1963), 370–376 [*English transl.*: *U.S.S.R. Computational Math. and Math. Phys.* **3** (1963), 489–497].
[77] A lower bound for the error of a formula for approximation summation in the class $E_{s,p}(C)$ (in Russian), *Mat. Zametki* **21** (1977), 371–375 [*English transl.*: *Math. Notes* **21** (1977), 207–210].

SMOLYAK, S. A.
[60] Interpolation and quadrature formulas for the classes W_s^α and E_s^α (in Russian), *Dokl. Akad. Nauk SSSR* **131** (1960), 1028–1031 [*English transl.*: *Soviet Math. Dokl.* **1** (1960), 384–387].
[65] On optimal restoration of functions and functionals of them (in Russian). Candidate Dissertation, Moscow State Univ. (1965),

SOBOL, I. M.
[69] "Multivariate Quadrature Formulas and Haar Functions," Nauka, Moscow, 1969 (in Russian).

SOBOLEV, S. L.
[65] On the order of convergence of cubature formulas (in Russian), *Dokl. Akad. Nauk SSSR* **162** (1965), 1005–1008 [*English transl.: Soviet Math. Dokl.* (1965), 808–812].
[74] "Introduction to the Theory of Cubature Formulas." Nauka, Moscow, 1974 (in Russian).

SONNEVEND, G.
[77] On optimization of algorithms for function minimization (in English), *Zh. Vychisl. Mat. Mat. Fiz.* **17** (1977), 591–609.

STECHKIN, S. B.
[67] Best approximation of linear operators (in Russian), *Mat. Zametki* **1** (1967), 137–148.

STENGER, F.
[78] Optimal convergence of minimum norm approximations in H_p, *Numer. Math.* **29** (1978), 345–362.

STERN, M. D.
[67] Optimal quadrature formulae, *Comput. J.* **9** (1967), 396–403.

STETTER, F.
[69] On best quadrature of analytic functions, *Quart. Appl. Math.* **27** (1969), 270–272.

STRONGIN, R. G.
[78] "Numerical Methods for Multivariate Extremal Problems." Nauka, Moscow, 1978 (in Russian).

SUKHAREV, A. G.
[71] Optimal strategies of the search for an extremum (in Russian), *Zh. Vychisl. Mat. Mat. Fiz.* **11** (1971), 910–924 [*English transl.: U.S.S.R. Computational Math. and Math. Phys.* **11** (1971), 119–137].
[72] Best sequential search strategies for finding an extremum (in Russian), *Zh. Vychisl. Mat. Mat. Fiz.* **12** (1972), 35–50 [*English transl.: U.S.S.R. Computational Math. and Math. Phys.* **12** (1972), 39–59].
[75] "Optimal Search for Extremum." Moscow State Univ., 1975 (in Russian).
[76] Optimal search for a zero of function satisfying Lipschitz's condition (in Russian), *Zh. Vychisl. Mat. Mat. Fiz.* **16** (1976), 20–30 [*English transl.:* Optimal search for the roots of a function satisfying a Lipschitz condition, *U.S.S.R. Computational Math. and Math. Phys.* **16** (1976), 17–26].
[78a] The optimal method for constructing best uniform approximations for functions of certain class (in Russian), *Zh. Vychisl. Mat. Mat. Fiz.* **18** (1978), 302–313.
[78b] Optimal quadrature formulas for some functional classes. Report (1978).

TAIKOV, L. V.
[68] Kolmogorov-type inequalities and the best formulas for numerical differentiation (in Russian), *Mat. Zametki* **4** (1968), 233–238 [*English transl.: Math. Notes* **4** (1968), 631–634].

TARASSOVA, V. P.
[78] Optimal strategies of search for domain of greatest values for some classes of functions (in Russian), *Zh. Vychisl. Mat. Mat. Fiz.* **18** (1978), 886–896.

TIKHOMIROV, V. M.
[60] Diameters of sets in function spaces and the theory of best approximations (in Russian), *Usp. Mat. Nauk* **15** (3) (1960), 81–112 [*English transl.: Russ. Math. Surveys* **15** (3) (1960), 75–111].
[65] A remark on n-dimensional diameters of sets in Banach spaces (in Russian), *Usp. Mat. Nauk* **20** (1) (1965), 227–230.

[69]   Best methods of approximating and interpolating differentiable functions in the space $C[-1, 1]$ (in Russian), *Mat. Sb.* **80** (122) (1969), 290–304 [*English transl.*: *Math. U.S.S.R. Sb.* **9** (1969), 275–289].

[76]   "Some Problems of Approximation Theory." Moscow Univ. 1976 (in Russian).

TIKHONOV, A. N., AND GAISARIAN, S. S.

[69]   The choice of optimum networks in the approximate calculation of quadratures (in Russian), *Zh. Vychisl. Mat. Mat. Fiz.* **9** (1969), 1170–1176 [*English transl.*: *U.S.S.R. Computational Math. and Math. Phys.* **9** (1969), 252–262].

TODD, M. J.

[76]   Optimal dissection of simplices. Department of Operations Research Rep., Cornell Univ. (1976).

TRAUB, J. F.

[61]   On functional iteration and the calculation of roots, *Preprints of Papers 16th Nat. ACM Conf.* Session 5A-1, pp. 1–4. Los Angeles, California, 1961.

[64]   "Iterative Methods for the Solution of Equations." Prentice-Hall, Englewood Cliffs, New Jersey, 1964.

TRAUB, J. F., AND WOŹNIAKOWSKI, H.

[76a]   Optimal linear information for the solution of non-linear operator equations, *in* "Algorithms and Complexity: New Directions and Recent Results" (J. F. Traub, ed.), pp. 103–119. Academic Press, New York, 1976.

[76b]   Optimal radius of convergence of interpolatory iterations for operator equations. Dept. of Computer Science Rep., Carnegie-Mellon Univ. (1976). To appear in *Aequationes Math.*

[77a]   Convergence and complexity of Newton iteration for operator equations. Dept. of Computer Science Rep., Carnegie-Mellon Univ. (1977). See also *J. Assoc. Comput. Mach.* **26** (1979), 250–258.

[77b]   Convergence and complexity of interpolatory-Newton iteration in a Banach space. Dept. of Computer Science Rep., Carnegie-Mellon Univ. (1977). To appear in *Comp. and Maths with Appls.*

TROJAN, J. M.

[79]   Tight bounds on the complexity index of one-point iterations (1979). To appear in *Comp. and Maths. with Appls.*

VELIKIN, V. L.

[77]   Optimal interpolation of periodic differentiable functions with bounded rth derivative (in Russian), *Mat. Zametki* **22** (1977), 663–670.

VITUSHKIN, A. G.

[59]   "Estimation of the Complexity of the Tabulation Problems" Fizmatgiz, Moscow, 1959 (in Russian) [*English transl.*: Vitushkin, A. G., "Theory of the Transmission and Processing Information." Pergamon, Oxford, 1961].

WASILKOWSKI, G. W.

[79]   Any iteration for polynomial equations using linear information has infinite complexity. Dept. of Computer Science Rep., Carnegie-Mellon Univ. (1979).

WASILKOWSKI, G. W., AND WOŹNIAKOWSKI, H.

[78]   Optimality of spline algorithms. Computer Science Dept. Rep., Carnegie-Mellon Univ. (1978).

WEINBERGER, H. F.

[61]   Optimal approximation for functions prescribed at equally spaced points, *J. Res. Nat. Bur. Std. Sect. B* **65** (1961), 99–104.

[72] On optimal numerical solution of partial differential equations, *SIAM J. Numer. Anal.* **9** (1972), 182–198.

WERSCHULZ, H. G.

[77a] Maximal order and order of information for numerical quadrature. Mathematics Research Rep. 77–2, Univ. of Maryland, Baltimore County (1977). See also *J. Assoc. Comput. Mach.* **26** (1979), 527–537.

[77b] Maximal order for approximation of derivatives. Mathematics Research Rep. 77-8, Univ. of Maryland Baltimore County (1977). See also *J. of Comput. and Syst. Sci.*, **18** (1979), 213–217.

WILANSKY, A.

[78] "Modern Methods in Topological Vector Spaces, McGraw-Hill, New York, 1978.

WILDE, D. J.

[64] "Optimum Seeking Methods." Prentice-Hall, Englewood Cliffs, New Jersey, 1964.

WILF, H. S.

[64] Exactness conditions in numerical quadrature, *Numer. Math.* **6** (1964), 315–319.

WINOGRAD, S.

[76] Some remarks on proof techniques in analytic complexity, *in* "Analytic Computational Complexity," (J. F. Traub, ed.), pp. 5–14. Academic Press, New York, 1976.

WOŹNIAKOWSKI, H.

[72] On nonlinear iterative processes in numerical methods (in Polish). Ph.D. Thesis, Univ. of Warsaw (1972).

[74] Maximal stationary iterative methods for the solution of operator equations, *SIAM J. Numer. Anal.* **11** (1974), 934–949.

[75] Generalized information and maximal order of iteration for operator equations, *SIAM J. Numer. Anal.* **12** (1975), 121–135.

[76] Maximal order of multipoint iterations using n evaluations, *in* "Analytic Computational Complexity" (J. F. Traub, ed.), pp. 75–107. Academic Press, New York, 1976.

ZALIZNYAK, N. F., AND LIGUN, A. A.

[78] On optimum strategy in search of global maximum of function (in Russian), *Zh. Vychisl. Mat. Mat. Fiz.* **18** (1978), 314–321.

ŽENSYKBAEV, A. A.

[76] On the best quadrature formula on the class $W^r L_p$ (in Russian), *Dokl. Akad. Nauk SSSR*, **227** (1976), 277–279 [*English transl.: Soviet Math. Dokl.* **17** (1976), 377–380].

[77] Best quadrature formulas for some classes of nonperiodic functions (in Russian), *Dokl. Akad. Nauk SSSR* **236** (1977), 531–534.

[78] On a property of the best quadrature formulae (in Russian), *Mat. Zametki* **23** (1978), 551–562.

ZHILEIKIN, YA. M., AND KUKARKIN, A. B.

[78] On the optimal evaluation of integrals with strongly oscillating integrand (in Russian), *Zh. Vychisl. Mat. Mat. Fiz.* **18** (1978), 294–301.

ZHILINSKAS, A. G.

[75] One-step Baysian method for searching for the extremum of functions of one variable (in Russian), *Cybernetics* **1** (1975), 139–144.

Glossary

We summarize basic concepts used throughout Part A. We list a symbol, its meaning, and the chapter and section reference where this symbol appears for the first time.

Symbol	Meaning	Chapter, Section or Equation
S	the solution operator, sometimes called the problem, $S:\Im_0 \to \Im_2$ and $\Im_0 \subset \Im_1$	1, (2.1)
$\Im_0$	the domain of S	1, (2.1)
$\Im_1$	linear space, $\Im_0 \subset \Im_1$	1, 2
$\Im_2$	the range of S	1, 2
ε	error parameter, $\varepsilon > 0$	1, 2
$x = x(f)$	ε-approximation, $\|x - \alpha\| < \varepsilon$,	1, (2.2)
f	the problem element, $f \in \Im_0$	1, 2
α	the solution element $\alpha = S(f)$	1, 2
$\mathfrak{N}$	the information operator, $\mathfrak{N}:D_{\mathfrak{N}} \to \Im_3$	1, (2.3)
$\Im_3$	the range of $\mathfrak{N}$	1, (2.3)
$d(\mathfrak{N},S)$	the diameter of information $\mathfrak{N}$ for the problem S	1, Def. 2.1, (2.9)
$r(\mathfrak{N},S)$	the radius of information $\mathfrak{N}$ for the problem S	1, Def. 2.1, (2.10)
φ	algorithm, $\varphi:\mathfrak{N}(\Im_0) \to \Im_2$	1, 2
$e(\varphi)$	the error of algorithm φ	1, Def. 2.2, (2.13)
$\Phi(\mathfrak{N},S)$	the class of all algorithms using the information $\mathfrak{N}$ for the problem S	1, 2
φ^I	interpolatory algorithm for the problem S with the information $\mathfrak{N}$ (interpolatory algorithm)	1, Def. 2.3, (2.16)

Symbol	Meaning	Chapter, Section or Equation
$e(\mathfrak{N},S)$	the optimal error	1, Def. 2.4, (2.18)
φ^{oe}	optimal error algorithm for the problem S with the information $\mathfrak{N}$ (optimal error algorithm)	1, Def. 2.4, (2.19)
φ^{c}	central algorithm for the problem S with the information $\mathfrak{N}$ (central algorithm)	1, Def. 2.5, (2.23)
P	the set of primitives	1, 3
$\mathrm{comp}(\mathfrak{N}(f))$	the information complexity of computing $\mathfrak{N}(f)$, where $\mathfrak{N}$ is a permissible information operator	1, 3
$\mathrm{comp}(\varphi(y))$	the combinatory complexity of computing $\varphi(y)$, where φ is a permissible algorithm	1, 3
$\Phi(\varepsilon)$	the class of all permissible algorithms for which $e(\varphi) < \varepsilon$	1, 3
$r(\mathfrak{N},S) \geq \varepsilon$	the problem S with information $\mathfrak{N}$ is ε-noncomputable	1, 3
$r(\mathfrak{N},S) < \varepsilon$	the problem S with permissible $\mathfrak{N}$ and $\Phi(\varepsilon) \neq \varnothing$ is ε-computable with respect to P	1, 3
$\mathrm{comp}(\varphi)$	the complexity of an algorithm φ	1, (3.1)
$\mathrm{comp}(\mathfrak{N},S,\varepsilon)$	the ε-complexity of the problem S with information $\mathfrak{N}$ (the ε-complexity of S with $\mathfrak{N}$)	1, Def. 3.1, (3.2)
φ^{oc}	optimal complexity algorithm for the problem S with the information $\mathfrak{N}$ (optimal complexity algorithm for S with $\mathfrak{N}$)	1, Def. 3.1, (3.3)
$\mathrm{comp}(\mathfrak{N})$	the information complexity	1, (3.4)
Ψ	a class of permissible information operators	1, (3.8)
$\mathrm{comp}(\Psi,S,\varepsilon)$	the ε-complexity of the problem S in the class Ψ (the ε-complexity of S in Ψ)	1, Def. 3.2, (3.9)
φ^{oc}	optimal complexity algorithm for the problem S in the class Ψ (optimal complexity algorithm for S in Ψ)	1, Def. 3.2, (3.10)
$\mathfrak{N}_1 \subset \mathfrak{N}_2$	$\ker \mathfrak{N}_2 \subset \ker \mathfrak{N}_1$; see also the extended definition of $\mathfrak{N}_1 \subset \mathfrak{N}_2$ in Section 2 of Chapter 7	2, Def. 2.1
$\mathfrak{N}_1 \asymp \mathfrak{N}_2$	$\ker \mathfrak{N}_1 = \ker \mathfrak{N}_2$; see also the extended definition of $\mathfrak{N}_1 \asymp \mathfrak{N}_2$ in Section 2 of Chapter 7	2, Def. 2.1
$A^{\perp}$	algebraic complement of A	2, (2.1), (2.2)
$\mathrm{codim}\, A$	codimension of A	2, (2.2)
$\mathrm{card}(\mathfrak{N})$	the cardinality of the information $\mathfrak{N}$; see also the extended definition of $\mathrm{card}(\mathfrak{N})$ in Section 2 of Chapter 7	2, Def. 2.2, (2.5)
T	the restriction operator, $T: \mathfrak{I}_1 \to \mathfrak{I}_4$	2, (3.1)
$\mathfrak{I}_4$	the range of T	2, (3.1)
$d(\mathfrak{N},S,T)$	the diameter of information $\mathfrak{N}$ for the problem (S,T)	2, 3
$\mathrm{index}(S,T)$	the index of the problem (S,T)	2, Def. 3.1
$A(S,T)$	algebraic complement of $\ker T \cap \ker S$ in the space $\ker T$	2, (3.3)
$\xi_1^*, \ldots, \xi_{n^*}^*$	basis of $A(T,S)$, $n^* = \mathrm{index}(S,T)$	2, (3.3)
$\mathfrak{N}^*$	information operator such that $\mathrm{card}(\mathfrak{N}^*) = \mathrm{index}(S,T)$ and $\ker \mathfrak{N}^* \cap \ker T \subset \ker S$	2, (4.1)
T^{-1}	the inverse operator of T	2, (4.2)
Ψ_n	the class of all information operators $\mathfrak{N}$ such that $\mathfrak{N}^* \subset \mathfrak{N}$ and $\mathrm{card}(\mathfrak{N}) \leq n$; see also the extended definitions of Ψ_n in Section 6 of Chapter 2 and Section 2 of Chapter 7	2, 4
$d(n,S,T)$	the nth minimal diameter of information	2, Def. 4.1, (4.6)
$\mathfrak{N}_n^{oi}$	nth optimal information; see also the extended definitions of $\mathfrak{N}_n^{oi}$ in Section 6 of Chapter 2 and Section 3 of Chapter 7	2, Def. 4.1, (4.7)

Symbol	Meaning	Chapter, Section or Equation
K	the linear operator $K = ST^{-1}$	2, (4.9)
$b(m,K)$	the mth minimal norm of the linear operator K	2, (4.9)
B_m	mth minimal subspace of the linear operator K	2, (4.10), (4.11)
$d(S,T)$	the diameter of problem error	2, Def. 4.2, (4.15)
$d(S,T) = +\infty$	the problem (S,T) is strongly noncomputable	2, Def. 4.2
$d(S,T) \geq 2\varepsilon$	the problem (S,T) is ε-noncomputable	2, Def. 4.2
$d(S,T) = 0$	the problem is convergent	2, Def. 4.2
Ψ_n	the class of all linear information operators with card$(\mathfrak{N}) \leq n$ (extends definition of Ψ_n from Section 4 of Chapter 2); see also Section 2 of Chapter 7	2, 6
$d(n) = d(n,S,\mathfrak{I}_0)$	the nth minimal diameter of information	2, Def. 6.1, (6.2)
$\mathfrak{N}_n^{\text{oi}}$	nth optimal information (extends definition of $\mathfrak{N}_n^{\text{oi}}$ from Section 4 of Chapter 2; see also Section 3 of Chapter 7)	2, Def. 6.1, (6.3)
$d^n(X,\mathfrak{I}_2)$	the Gelfand n-width of X in $\mathfrak{I}_2$	2, (6.4)
$d^n = d^n(S(\mathfrak{I}_0),\mathfrak{I}_2)$	the Gelfand n-width of the range of the solution operator in $\mathfrak{I}_2$	2, (6.5)
A^n	nth extremal subspace of $S(\mathfrak{I}_0)$ in the sense of Gelfand	2, (6.12)
$\mathfrak{N}^{\text{a}}$	an adaptive linear information operator	2, (7.2)
$\Phi_L(n)$	the class of linear algorithms which use a linear information with cardinality at most n	3, 5
$\lambda(n) = \lambda(n,S)$	the nth minimal linear error	3, (5.1)
$\lambda_n(X,\mathfrak{I}_2)$	the linear Kolmogorov n-width of X in $\mathfrak{I}_2$	3, (5.2)
$d_n(X,\mathfrak{I}_2)$	the Kolmogorov n-width of X in $\mathfrak{I}_2$	3, (5.3); 7, (4.1)
$\lambda_n = \lambda_n(S(\mathfrak{I}_0),\mathfrak{I}_2)$	the linear Kolmogorov n-width of the range of the solution operator in $\mathfrak{I}_2$	3, (5.5)
$e(\varphi,f)$	the local error	4, (2.2)
$\text{dev}(\varphi)$	the deviation of the algorithm φ	4, Def. 2.1, (2.4)
$\sigma(y)$	a spline interpolating y	4, Def. 3.1, (3.2)
φ^s	a spline algorithm for the problem S with the information $\mathfrak{N}$ (a spline algorithm)	4, Def. 4.1, (4.1)
Φ^s	the class of spline algorithms	4, Def. 4.1, (4.1)
φ^{noc}	nearly optimal complexity algorithm for (S,T) with $\mathfrak{N}$	5, (1.3) and (1.4)
$m(\Psi,S,T,\varepsilon)$	the ε-cardinality number of the problem (S,T) in the class Ψ (the ε-cardinality number of (S,T) in Ψ)	5, Def. 1.1, (1.5)
Ψ_U	the class of all linear information operators $\mathfrak{N}$ such that card$(\mathfrak{N}) < +\infty$	5, 1
$L_p = L_p(X)$	the space of real scalar functions f such that $\|f\|_p < +\infty$; X is an interval of the real axis R	6, (1.1)
$W_p^r(X)$	the space of real scalar functions f such that $f^{(r-1)}$ is absolutely continuous and $f^{(r)} \in L_p(X)$	6, (1.3)
W_p^r	the space of real scalar functions f such that $f \in W_p^r(X)$ and $\|f^{(r)}\|_p \leq 1$	6, (1.5)
$\tilde{W}_p^r(X)$	the space of real scalar periodic functions f such that $f \in W_p^r(X)$	6, (1.6)
$\tilde{W}_p^r$	the space of real scalar periodic functions such that $f \in W_p^r$	6, 1
D_p^r	the linear operator from $W_p^r(X)$ into L_p, $D_p^r f = f^{(r)}$	6, (1.4)
$\tilde{D}_p^r$	the linear operator from $\tilde{W}_p^r(X)$ into L_p, $\tilde{D}_p^r f = f^{(r)}$	6, (1.7)
$C = C(X)$	the space of continuous functions equipped with the sup norm	6, 1

Symbol	Meaning	Chapter, Section or Equation
$\tilde{C} = \tilde{C}(X)$	the space of periodic continuous functions equipped with the sup norm	6, 1
Θ	the big theta notation	6, 1
K_r	the Favard constant	6, (3.16)
Ψ_n^c	the class of information operators of the form (4.2) of Chapter 6	6, (4.2)
$\mathfrak{N}_1 \subset \mathfrak{N}_2$	$V(f,\mathfrak{N}_2) \subset V(f,\mathfrak{N}_1)$, $\forall f \in \mathfrak{I}_0$ (extends definition of $\mathfrak{N}_1 \subset \mathfrak{N}_2$ from Section 2 of Chapter 2)	7, Def. 2.1
$\mathfrak{N}_1 \asymp \mathfrak{N}_2$	$V(f,\mathfrak{N}_2) = V(f,\mathfrak{N}_1)$, $\forall f \in \mathfrak{I}_0$ (extends definition of $\mathfrak{N}_1 \asymp \mathfrak{N}_2$ from Section 2 of Chapter 2)	7, Def. 2.1
$\text{card}(\mathfrak{N},\mathfrak{I}_0)$	the cardinality of the information $\mathfrak{N}$ in $\mathfrak{I}_0$ (extends definition of $\text{card}(\mathfrak{N})$ from Section 2 of Chapter 2)	7, Def. 2.2, (2.9)
Ψ_n	the class of all linear or nonlinear information operators with $\text{card}(\mathfrak{N},\mathfrak{I}_0) \le n$ (extends definition of Ψ_n from Sections 4 and 6 of Chapter 2)	7, 2
$d(n,S)$	the nth minimal diameter of information for the problem S	7, Def. 3.1, (3.1)
$\mathfrak{N}_n^{oi}$	nth optimal information (extends definition of $\mathfrak{N}_n^{oi}$ from Sections 4 and 6 of Chapter 2)	7, Def. 3.1, (3.2)
$\mathfrak{S}$	the cardinality of the set of real numbers	7, (3.7)
$\Phi_n(\mathfrak{N})$	the class of algorithms which use $\mathfrak{N}$ and whose range is at most n dimensional	7, (4.2)
Φ_n	the union of $\Phi_n(\mathfrak{N})$ for all possible $\mathfrak{N}$	7, (4.3)
$d_n = d_n(S(\mathfrak{I}_0),\mathfrak{I}_2)$	the Kolmogorov n-width of the range of the solution operator in $\mathfrak{I}_2$	7, (4.4)
$H(\varepsilon,X)$	the ε-entropy of X	7, (4.7)
φ^{aoc}	asymptotically optimal complexity algorithm for S in Ψ_U	8, (2.11)
$S_1 > S_2$	$\text{comp}(\Psi,S_1,\varepsilon) > \text{comp}(\Psi,S_2,\varepsilon)$, $\forall \varepsilon \in A$	9, Def. 2.1, (2.1)
$S_1 \asymp S_2$	$\text{comp}(\Psi,S_1,\varepsilon) = \Theta(\text{comp}(\Psi,S_2,\varepsilon))$ as $\varepsilon \to 0$	9, Def. 2.1, (2.2)
$S_1 \succ S_2$	$\text{comp}(\Psi,S_2,\varepsilon) = o(\text{comp}(\Psi,S_1,\varepsilon))$ as $\varepsilon \to 0$	9, Def. 2.1, (2.3)
$\text{DIF}(r)$	a differential operator	9, 2
$\text{INP}(r)$ $\text{INP}(A)$ $\text{INP}(H)$	interpolation operators	9, 2
$\text{INT}(W_p^r)$ $\text{INT}(\tilde{W}_p^r)$	integration operators	9, 2
$\text{APP}(\tilde{W}_2^r,L_2)$ $\text{APP}(W_2^r,L_2)$ $\text{APP}(\tilde{W}_\infty^r,\tilde{C})$ $\text{APP}(W_\infty^r,C)$	approximation operators	9, 2
$\text{PAR}(r)$	a parabolic operator	9, 2
$\text{ELL}(r)$	an elliptic operator	9, 2
$\text{HYP}(r)$	a hyperbolic operator	9, 2
NON	a nonlinear scalar equation operator	9, 2
UNI	the maximum search operator for unimodal functions	9, 2
INP	an interpolation operator	9, 2
INT	an integration operator	9, 2
APP	an approximation operator	9, 2

Symbol	Meaning	Chapter, Section or Equation
Ψ_f^{non}	the class of all nonadaptive information operators $\mathfrak{N}_n^{non}(f) = [f(x_1),f(x_2),\ldots,f(x_n)]^t$	9, 3
Ψ_f^a	the class of all adaptive information operators $\mathfrak{N}_n^a(f) = [f(x_1),f(x_2),\ldots,f(x_n)]^t$	9, 4
Ψ_L^{non}	the class of all linear nonadaptive information operators $\mathfrak{N}_n^{non}(f) = [L_1(f),L_2(f),\ldots,L_n(f)]^t$	9, 5
Ψ_L^a	the class of all linear adaptive information operators $\mathfrak{N}_n^a(f) = [L_1(f),L_2(f;L_1(f)),\ldots,L_n(f;L_1(f),\ldots,)]^t$	9, 6
Ψ_{NON}	the class of all nonlinear information operators of cardinality one	9, 7
$\Psi_1 \asymp \Psi_2$	$comp(\Psi_1,S,\varepsilon) = \Theta(comp(\Psi_2,S,\varepsilon))$ as $\varepsilon \to 0$	9, Def. 8.1
$\Psi_1 \succ \Psi_2$	$comp(\Psi_1,S,\varepsilon) = o(comp(\Psi_2,S,\varepsilon))$ as $\varepsilon \to 0$	9, Def. 8.1

PART B

ITERATIVE INFORMATION MODEL

1. INTRODUCTION

Part B is devoted to an iterative information model of analytic complexity. In this introduction, we shall compare and contrast concepts and results in Parts A and B.

In Part A, we study a general information model of analytic complexity. In particular, we show that a given information operator may not be "strong" enough to solve a problem to within a desired ε. It may, however, turn out that the information operator can be used to compute a better approximation from a given approximation. By repeating this procedure one may finally be able to solve the problem to within ε even for *arbitrarily* small ε. This informal description of an *iterative* information operator and an *iterative* algorithm is formalized in Section 2.

Part B is built on some 20 years of research on iterative complexity initiated by the work of Traub [61, 64]. It consists of the generalization of earlier work as well as many new questions and results. It is our experience that the material of Part B is conceptually and technically more difficult than that of Part A.

We limit ourselves to iterative *linear* information throughout Part B for a number of reasons.

1. Our results for the linear theory are very strong. In some cases, the questions we pose can be completely answered.

2. In practice, we can often compute only linear information.

3.   The *class* of nonlinear information is too powerful (see Part A, Chapter 7) and most problems become trivial.

We point out in the Overview that our theory uses two mathematical models which we refer to as models α and β. Much of Parts A and B are devoted to model α for the reasons given in the Overview. We remind the reader that negative results established in model α are all the stronger since they are independent of any model of computation. We now discuss some of the model α results of Part B.

The deepest question studied in Part B is what problems can be solved by iteration using iterative linear information. Traditionally, certain problems are solved by iteration; others are not. For example, a zero of a scalar nonlinear function is approximated by iteration using function evaluations; the value of a definite integral of a scalar function is not. Must this be so? More precisely, for what problems is the class of iterative algorithms (in the very general sense of this paper) empty? For one-point stationary iterations using iterative linear information, we solve this problem, showing the class is empty unless the "index" of the problem is finite. We advance a conjecture which characterizes all problems with finite index. If this conjecture is true, then essentially only *nonlinear equations can be solved by iteration.*

As we have remarked elsewhere in this monograph, the central role of information leads to great generality combined with remarkable simplicity. We show that the maximal order of any algorithm which uses certain information is the "order of information." The maximal order is independent of smoothness, structure, or any other characteristics of the algorithm. Maximal order algorithms play a role analogous to optimal error algorithms in Part A. The order of information plays a role similar to the radius (or diameter) of information since both measure the "strength" of the information operators.

As in Part A, we are interested in determining the most "relevant" information for a given problem. We present a complete answer to this question for the class of iterative linear information operators.

We describe concepts and results from model β. The "minimal complexity index algorithms" correspond to the optimal complexity algorithms of Part A. See Section 3 as to why we are interested in algorithms which minimize the complexity index rather than the complexity.

In Part A, we prove that an optimal error algorithm is a nearly optimal complexity algorithm provided its combinatory complexity is small or comparable to the information complexity. In Part B, we establish a similar relation between a maximal order algorithm and a minimal complexity index algorithm.

In Part A, we prove that problem complexity can be essentially any function and that there exist even linear problems with arbitrarily high complexity. In contrast, we show that in the Part B setting, complexity is infinity, $\Theta(\log 1/\varepsilon)$ or $\Theta(\log \log 1/\varepsilon)$. Indeed, if a problem is solved with an iterative algorithm φ

of order greater than unity, then the complexity of φ is $z(\log \log 1/\varepsilon)(1 + o(1))$ as ε goes to zero, where z is the complexity index of φ. Thus, for small ε the minimization of the complexity of φ is equivalent to finding an algorithm with minimal complexity index.

The benefit of using minimal complexity index algorithms is limited. The ratio of the complexity of the minimal complexity index algorithm to the complexity of any iterative algorithm with order greater than one is asymptotically equal to the ratio of their indices; it is independent of ε. This ratio can, however, be very large. There are cases where it goes to infinity (see Traub and Woźniakowski [77b, Section 6]).

We study *local* convergence of iterative algorithms; that is, we assume that an initial approximation is a sufficiently close approximation to a solution. Very recent results of Wasilkowski [78, 79] show that for the case of iterative linear information, we cannot achieve better than local convergence with finite complexity. He proved that for the solution of nonlinear equations the class of iterative linear information operators is too weak to obtain a globally convergent stationary iteration, or a globally convergent nonstationary iteration with finite complexity, even for relatively simple classes of problem elements such as the class of complex scalar polynomials with all simple zeros.

We have seen the difficulty of analyzing adaptive information for nonlinear problems in the Part A setting (see, for example, Part A, Chapter 8, Conjectures 3.1 and 7.1). In Part B, multipoint iterative information corresponds to adaptive information. Conjecture 10.2 involves multipoint information. We believe that new tools are needed to obtain deep results (such as settling the above conjectures) for multipoint (i.e., adaptive) information. On the other hand, one-point iterative information corresponds to nonadaptive information and here our tools are much more powerful.

We summarize the major concepts and results of Part B.

SECTION 2   We introduce the basic concepts of solution operator S, iterative information $\mathfrak{N}$, and algorithm φ. We define $d(\mathfrak{N},S)$, the limiting diameter of information $\mathfrak{N}$ for the problem S, and $p(\mathfrak{N},S)$, the order of information $\mathfrak{N}$ for the problem S. We prove that $p(\mathfrak{N},S)$ is an upper bound on the order of any algorithm. This upper bound is achieved by any "interpolatory" algorithm.

SECTION 3   We present a model of computation which, as in Part A, consists of primitive operations, permissible information, and permissible algorithms. We develop the methodology for complexity analysis and show why it is desirable to minimize the complexity index.

SECTION 4   In Sections 4–8, we consider *linear* information operators. The *cardinality* card($\mathfrak{N}$) of linear information $\mathfrak{N}$ is defined, and we show that linear information with finite cardinality equal to n can be represented by n linearly independent linear functionals.

SECTION 5   In Sections 5 and 6, we study when the class of iterative algorithms is empty. We define an iterative algorithm and prove that the class of iterative algorithms which use information $\mathfrak{N}$ is empty whenever the limiting diameter $d(\mathfrak{N},S)$ is positive.

SECTION 6   We define index(S), the index of the problem S, and prove (Theorem 6.1) that if card($\mathfrak{N}$) is less than index(S), then the limiting diameter $d(\mathfrak{N},S)$ is positive. In this case, the class of iterative algorithms using $\mathfrak{N}$ is empty. We define the basic information $\mathfrak{N}^*$ such that its cardinality is equal to the index of S and the limiting diameter $d(\mathfrak{N}^*,S)$ is zero. Furthermore, in Theorem 6.3, we prove that $d(\mathfrak{N},S) = 0$ implies that information $\mathfrak{N}$ "contains" $\mathfrak{N}^*$.

SECTION 7   For a given m, we study when the class of iterative algorithms of order m is empty. We define the mth index of S and prove (Theorem 7.1) that if the cardinality of information is less than the mth index, then the order of information is less than m. This implies that the class of iterative algorithms of order m is empty. We define the mth basic information $\mathfrak{N}_m^*$ such that its cardinality is equal to the mth index of S and its order of information is at least m. We also show that $p(\mathfrak{N},S) \geq m$ implies that information $\mathfrak{N}$ contains $\mathfrak{N}_m^*$ (Theorem 7.3).

SECTION 8   We specify our model of computation for the linear case. We obtain lower and upper bounds on the complexity index and on the nth minimal complexity index. We also show how to find the nth maximal order for the problem S and an nth maximal order information for the problem S (Theorems 8.1 and 8.2).

SECTION 9   We define information with memory and generalize our concepts and theorems on limiting diameter and order of information, interpolatory algorithm, and complexity index.

SECTION 10   We list some extensions and open problems. We propose Conjecture 10.1 which characterizes problems which can be solved by iterative algorithms using one-point linear information with finite cardinality. If this conjecture is true, then essentially only nonlinear equations can be solved by such iterations. Conjecture 10.2 generalizes the Kung–Traub conjecture. Conjecture 10.2 states that for the solution of scalar nonlinear equations, the maximal order of multipoint iterations using information with cardinality n is at most 2^{n-1} for iterations without memory and is at most 2^n for iterations with memory.

SECTION 11   We compare the results of Parts A and B in terms of the limiting diameter and order of information as well as the asymptotic dependence of the complexity on ε.

APPENDIX   We prove (Lemma A.1) that under very weak assumptions, our definition of order of an algorithm agrees with the "classical" definition.

2.   DIAMETER AND ORDER OF INFORMATION

As in Part A, we consider a linear or nonlinear solution operator S such that

(2.1) $$S:\mathfrak{I}_0 \to \mathfrak{I}_2,$$

where $\mathfrak{I}_0$ is a subset of a linear space $\mathfrak{I}_1$ over the real or complex field and $\mathfrak{I}_2$ is a linear normed space over the real or complex field. We wish to approximate the *solution element* $\alpha = S(f)$ for all *problem elements* $f \in \mathfrak{I}_0$. Let x_0 be an initial approximation to the solution α and let ε', $\varepsilon' \in (0,1)$, be a given real number. By *solving* (or *approximating*) the problem S, we mean that we seek an ε'-approximation $y = y(f)$, $y \in \mathfrak{I}_2$, to α such that

(2.2) $$\|y(f) - \alpha\| \le \varepsilon'\|x_0 - \alpha\|.$$

To find such an approximation, we need to know something about the problem element f. Let

(2.3) $$\mathfrak{N}:D_{\mathfrak{N}} \subset \mathfrak{I}_1 \times \mathfrak{I}_2 \to \mathfrak{I}_3$$

be an *iterative information operator* (not necessarily linear), where $(f,x) \in D_{\mathfrak{N}}$ for all $f \in \mathfrak{I}_0$ and all x close enough to $\alpha = S(f)$ and where $\mathfrak{I}_3$ is a given space. We call $\mathfrak{N}$ an iterative information operator, since we compute $\mathfrak{N}(f,x)$ for different x and the next approximation x_{k+1} is based on the information $\mathfrak{N}(f,x_k)$, $k = 0, 1, \ldots$. For brevity, $\mathfrak{N}$ will also be called an *information operator*. The information $\mathfrak{N}$ is *stationary* and *without memory* in the sense of Traub [64]. For most problems, the information operator $\mathfrak{N}$ is not one-to-one and $\mathfrak{N}(f,x)$ does not uniquely define the solution $\alpha = S(f)$. This means there exist many different $f \in \mathfrak{I}_0$ with the same information $\mathfrak{N}(f,x)$. Thus, for a given f, the set of problem elements $\tilde{f}(x)$ such that $\tilde{f}(x) \in \mathfrak{I}_0$ and $\mathfrak{N}(\tilde{f}(x),x) = \mathfrak{N}(f,x)$ determines the "uncertainty" of the information operator $\mathfrak{N}$. Note that $\tilde{f} = \tilde{f}(x)$ is a function of x which has the same information as a problem element f for every x close enough to α. For technical reasons, we have to assume that the function $\tilde{f}$ is "regular" at α. To formalize this idea, we define equality with respect to $\mathfrak{N}$.

Definition 2.1   We shall say $\tilde{f} = \tilde{f}(x)$ *is equal to* f *with respect to* $\mathfrak{N}$ iff

(i) $\tilde{f}:D_{\tilde{f}} \subset \mathfrak{I}_2 \to \mathfrak{I}_0$, $\tilde{f} \in W$,

where W is a given class and there exists $\Gamma = \Gamma(f) > 0$ such that

$$J(\Gamma) = \{x:\|x - \alpha\| \le \Gamma\} \subset D_{\tilde{f}}, \quad \text{where} \quad \alpha = S(f),$$

(ii) $\mathfrak{N}(\tilde{f}(x),x) = \mathfrak{N}(f,x)$ $\forall x \in D_{\tilde{f}}$.

For brevity, we write $\tilde{f} \in V(f)$, where $V(f)$ is the set of all functions $\tilde{f}$ which are equal to f with respect to $\mathfrak{N}$.   ∎

The class W describes the regularity of $\tilde{f}$ and its definition depends on the regularity of the solution operator S. We always assume that the constant functions $\tilde{f}(x) \equiv f$ belong to W.

We define the limiting diameter of information as the maximal distance between S operating on two problem elements with the same information at x as x tends to $\alpha = S(f)$. In Section 5, we prove that the limiting diameter of a linear information operator has to be zero in order to solve the problem S iteratively for arbitrarily small ε'.

Definition 2.2 We shall say $d(\mathfrak{N},S)$ is the *limiting diameter of information* $\mathfrak{N}$ *for the problem S* iff

$$(2.4) \qquad d(\mathfrak{N},S) = \sup_{f \in \mathfrak{Z}_0} \sup_{\tilde{f}_1, \tilde{f}_2 \in V(f)} \limsup_{x \to \alpha} \left\| S(\tilde{f}_1(x)) - S(\tilde{f}_2(x)) \right\|.$$

We shall say the information $\mathfrak{N}$ is *convergent* for the problem S iff

$$(2.5) \qquad d(\mathfrak{N},S) = 0.$$

$\mathfrak{N}$ is called *divergent* iff $d(\mathfrak{N},S) > 0$. ∎

Note that the limiting diameter of $\mathfrak{N}$ coincides with the diameter of $\mathfrak{N}$ introduced in Part A for information operators independent of x, i.e., $\mathfrak{N}(f,x) \equiv \mathfrak{N}(f)$.

We give a geometrical interpretation of (2.4). Assume that $S(\tilde{f}(x))$ is continuous at α for any $\tilde{f} \in V(f)$. Then

$$(2.6) \qquad d(\mathfrak{N},S) = \sup_{f \in \mathfrak{Z}_0} \sup_{\tilde{f}_1, \tilde{f}_2 \in V(f)} \left\| S(\tilde{f}_1(\alpha)) - S(\tilde{f}_2(\alpha)) \right\|.$$

Note that $U(f) = \{S(\tilde{f}(\alpha)) : \tilde{f} \in V(f)\}$ is the set of all solutions $S(\tilde{f}(\alpha))$ which share the same information as f at α. Then (2.6) yields

$$(2.7) \qquad d(\mathfrak{N},S) = \sup_{f \in \mathfrak{Z}_0} \operatorname{diam}(U(f)),$$

where $\operatorname{diam}(U(f))$ denotes the diameter of the set $U(f)$. (See Section 2 of Chapter 1 in Part A.) This can be schematized as in Figure 1.

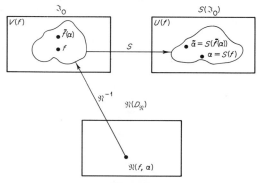

Figure 1

The information $\mathfrak{N}$ is convergent iff the set $U(f)$ contains only one element $\alpha = S(f)$. Thus Figure 1 shows a divergent information operator, i.e., $d(\mathfrak{N},S) > 0$. We illustrate the concept of limiting diameter of information by an example.

Example 2.1 Let $\mathfrak{I}_1$ be the class of analytic operators f, $f : D_f \subset B_1 \to B_2$, where B_1 and B_2 are Banach spaces and $\dim(B_1) = \dim(B_2)$. Let $\mathfrak{I}_0$ be the class of analytic operators with a unique simple zero, i.e., $f \in \mathfrak{I}_0$ iff $f \in \mathfrak{I}_1$ and there exists a unique $\alpha \in D_f$ such that $f(\alpha) = 0$ and $f'(\alpha)^{-1}$ exists and is bounded. Define

$$(2.8) \qquad S(f) = f^{-1}(0), \qquad \mathfrak{I}_2 = B_1.$$

Thus, $\alpha = S(f)$ is the solution of the nonlinear equation $f(x) = 0$. Let

$$(2.9) \qquad \mathfrak{N}(f,x) = [f^{(j)}(x), f^{(j+1)}(x)]$$

for a nonnegative integer j, and $f^{(j)}$ denotes the jth Fréchet derivative. In this example, $\tilde{f}(x) = \tilde{f}(x, \cdot)$ is an analytic operator with respect to the second argument. Let $W = C^{j+1}$ be the class of all functions $\tilde{f}$ which are $(j+1)$ times continuously differentiable at α. An example of $\tilde{f}$ which belongs to $V(f)$ is given by

$$\tilde{f}(x,t) = \begin{cases} f(t) + c & \text{for } j > 0, \\ f(t) + L(t-x)^2 & \text{for } j = 0, \end{cases}$$

where c is a suitably chosen element of B_2 and L is a bilinear operator. It is easy to verify that

$$(2.10) \qquad d(\mathfrak{N},S) = \begin{cases} +\infty & \text{for } j > 0, \\ 0 & \text{for } j = 0. \end{cases}$$

Thus, for $j = 0$, we have convergent information. ∎

REMARK 2.1 In Part A, we defined $r(\mathfrak{N},S)$, the radius of information $\mathfrak{N}$ for the problem S, where $r(\mathfrak{N},S) \in [\frac{1}{2}d(\mathfrak{N},S), d(\mathfrak{N},S)]$. Since our focus here is on convergent information operators, $d(\mathfrak{N},S) = r(\mathfrak{N},S) = 0$, we need not consider the limiting radius of information. ∎

We solve problem (2.2) by an *algorithm* φ defined as follows. Let

$$(2.11) \qquad \varphi : D_\varphi \subset \mathfrak{I}_2 \times \mathfrak{N}(D_\mathfrak{N}) \to \mathfrak{I}_2.$$

(See also the definition of "permissible algorithm" in Section 3.) Recall that x_0 is an initial approximation to the solution $\alpha = S(f)$. Then the algorithm φ generates the sequence of approximations by

$$(2.12) \qquad x_{i+1} = \varphi(x_i; \mathfrak{N}(f,x_i)), \qquad i = 0, 1, \ldots.$$

Thus, φ is a stationary algorithm and since x_{i+1} depends only on the previously computed approximation x_i, the algorithm φ is without memory in the sense of Traub [64]. In Section 5, we impose some conditions on φ and define the concept of *iterative algorithm*. Information operators with memory and stationary algorithms with memory are considered in Section 9. We shall not pursue the analysis of nonstationary algorithms here.

Let $\Phi(\mathfrak{N},S)$ be the class of *all* algorithms defined by (2.11) and (2.12). Let $\varphi \in \Phi(\mathfrak{N},S)$. We examine the convergence of the sequence $\{x_i\}$ to α. Since φ is stationary, it suffices to find how x_1 depends on x_0 and whether x_1 converges to α as x_0 tends to α. Recall that the algorithm uses the information $\mathfrak{N}(f,x)$. Suppose that $\tilde{f} \in V(f)$ which means that the information on $\alpha = S(f)$ and $\tilde{\alpha} = S(\tilde{f}(x))$ is exactly the same. Hence, any algorithm φ will produce the same approximation to the solution elements α and $\tilde{\alpha}$. Since we are unable to distinguish $f(x)$ from f, an algorithm φ should approximate not only the solution element α but also the solution element $\tilde{\alpha}$. This motivates

Definition 2.3   We shall say $e(\varphi)$ is the *limiting error of algorithm* φ iff

$$(2.13) \qquad e(\varphi) = \sup_{f \in \mathfrak{I}_0} \sup_{\tilde{f} \in V(f)} \limsup_{x \to \alpha} \left\| \varphi(x,\mathfrak{N}(f,x)) - S(\tilde{f}(x)) \right\|.$$

The algorithm φ is called *convergent* iff $e(\varphi) = 0$. ∎

We are ready to prove that $\frac{1}{2}d(\mathfrak{N},S)$ is a lower bound on $e(\varphi)$ for any algorithm φ from the class $\Phi(\mathfrak{N},S)$.

Theorem 2.1   For any algorithm φ, $\varphi \in \Phi(\mathfrak{N},S)$,

$$(2.14) \qquad\qquad\qquad e(\varphi) \geq \tfrac{1}{2}d(\mathfrak{N},S). \quad ∎$$

PROOF   Choose any $\tilde{f}_1$ and $\tilde{f}_2$ from $V(f)$, where $\alpha = S(f)$. Then

$$\overline{\lim_{x \to \alpha}} \left\| S(\tilde{f}_1(x)) - S(\tilde{f}_2(x)) \right\| \leq \overline{\lim_{x \to \alpha}} (\left\| \varphi(x,\mathfrak{N}(f,x)) - S(\tilde{f}_1(x)) \right\|$$

$$+ \left\| \varphi(x,\mathfrak{N}(f,x)) - S(\tilde{f}_2(x)) \right\|) \leq 2e(\varphi).$$

Taking the supremum with respect to f and $\tilde{f}_1$, $\tilde{f}_2$, from (2.4) we get $d(\mathfrak{N},S) \leq 2e(\varphi)$ which proves (2.14). ∎

Theorem 2.1 states that $\frac{1}{2}d(\mathfrak{N},S)$ is the inherent error of information $\mathfrak{N}$ for any algorithm φ. This is especially interesting for divergent information $\mathfrak{N}$, $d(\mathfrak{N},S) > 0$, since it is then impossible to find an algorithm whose error is less than $\frac{1}{2}d(\mathfrak{N},S)$ no matter how sophisticated an algorithm is used. We showed in Example 2.1 that it is even possible that $d(\mathfrak{N},S) = +\infty$.

We now show that $d(\mathfrak{N},S)$ is an upper bound on "interpolatory algorithms" which are defined as follows.

Definition 2.4   We shall say φ^{I}, $\varphi^{\mathrm{I}} \in \Phi(\mathfrak{N},S)$, is an *interpolatory algorithm* iff

$$(2.15) \qquad\qquad\qquad \varphi^{\mathrm{I}}(x,\mathfrak{N}(f,x)) = S(\tilde{f}(x))$$

for some $\tilde{f} \in V(f)$. ∎

This means that knowing the information $\mathfrak{N}(f,x)$, one finds a problem element $\tilde{f}(x)$ which has the same information as f at x and the next approximation is the solution of the problem $S(\tilde{f}(x))$. In practice, $\tilde{f}(x)$ is chosen to be "simpler" than f. In some cases, an assumption how to choose a unique $\tilde{f}(x)$ is added. Examples of interpolatory algorithms in the sense of this paper include Newton, secant, or any $I_{n,s}$ interpolatory algorithms for the solution of nonlinear equations. (See Traub [64] and Woźniakowski [74].)

Theorem 2.2   For any interpolatory algorithm φ^I, $\varphi^I \in \Phi(\mathfrak{N},S)$,

$$(2.16) \qquad\qquad e(\varphi^I) \leq d(\mathfrak{N},S). \quad \blacksquare$$

PROOF   Take any $f \in \mathfrak{I}_0$. Then $\varphi^I(x,\mathfrak{N}(f,x)) = S(\tilde{f}_0(x))$ for some $\tilde{f}_0$ such that $\tilde{f}_0 \in V(f)$. Hence,

$$\overline{\lim_{x \to \alpha}} \|\varphi^I(x,\mathfrak{N}(f,x)) - S(\tilde{f}(x))\| = \overline{\lim_{x \to \alpha}} \|S(\tilde{f}_0(x)) - S(\tilde{f}(x))\| \leq d(\mathfrak{N},S)$$

for any $\tilde{f} \in V(f)$. Taking the supremum with respect to f and $\tilde{f}$, we get $e(\varphi^I) \leq d(\mathfrak{N},S)$. $\blacksquare$

From Theorems 2.1 and 2.2, we get

Corollary 2.1   There exists a convergent algorithm in $\Phi(\mathfrak{N},S)$ iff the information $\mathfrak{N}$ is convergent for the problem S, i.e., $d(\mathfrak{N},S) = 0$. $\blacksquare$

For convergent information operators, $S(\tilde{f}(x))$ approaches $\alpha = S(f)$ as x tends to α for all $\tilde{f} \in V(f)$. We define the "order of information" which measures the speed of convergence of $S(\tilde{f}(x))$ to $S(f)$ for a worst case. Let A be a set of real numbers defined by

$$(2.17) \qquad A = \left\{ q : q \geq 1, \forall f \in \mathfrak{I}_0, \alpha = S(f), \text{ and } \forall \tilde{f} \in V(f) \text{ we have} \right.$$
$$\left. \lim_{x \to \alpha} \frac{\|S(\tilde{f}(x)) - S(f)\|}{\|x - \alpha\|^{q-\eta}} = 0, \forall \eta > 0 \right\}. \quad \blacksquare$$

Definition 2.5   We shall say $p(\mathfrak{N},S)$ is the *order of information* $\mathfrak{N}$ *for the problem S* iff

$$(2.18) \qquad\qquad p(\mathfrak{N},S) = \begin{cases} 0 & \text{if } A \text{ is empty,} \\ \sup A & \text{otherwise.} \quad \blacksquare \end{cases}$$

Note that for divergent information operators, $p(\mathfrak{N},S) = 0$. It is easy to verify that $p(\mathfrak{N},S)$ is an integer for a sufficiently regular function $S(\tilde{f}(x))$. Roughly speaking, the order $p = p(\mathfrak{N},S)$ measures how fast $S(\tilde{f}(x))$ tends to $S(f)$, $\|S(\tilde{f}(x)) - S(f)\| = O(\|x - \alpha\|^p)$ for all $\tilde{f} \in V(f)$. We prove that any algorithm from $\Phi(\mathfrak{N},S)$ has order no greater than the order of information. This significantly simplifies the complexity analysis since the maximal order of an algorithm is independent of the "structure" of that algorithm, depending only on the information used. (See Section 3.) We illustrate the concept of the order of information by an example.

Example 2.2 Consider the solution of nonlinear equations defined in Example 2.1. Let

$$(2.19) \qquad \mathfrak{N}(f,x) = [f(x), f'(x), \dots, f^{(n-1)}(x)], \qquad n \geq 2,$$

be *standard information*. Let $W = C^n$ be the class of all functions $\tilde{f}$ which are n times continuously differentiable at α. Then $\tilde{f} \in V(f)$ means $\tilde{f}^{(j)}(x,t)|_{t=x} = f^{(j)}(x)$ for $j = 0, 1, \dots, n-1$, where $\tilde{f}^{(j)}$ denotes the jth Fréchet derivative with respect to the second argument. Furthermore,

$$\tilde{f}(x,t) - f(t) = \int_0^1 \{\tilde{f}^{(n)}(x, \tau t + (1-\tau)x) - f^{(n)}(\tau t + (1-\tau)x)\}$$

$$\times (t-x)^n \frac{(1-\tau)^{n-1}}{(n-1)!} d\tau$$

which yields for $t = \alpha$, $\tilde{f}(x,\alpha) = O(\|x - \alpha\|^n)$. Since $0 = \tilde{f}(x,\tilde{\alpha}) = \tilde{f}(x,\alpha) + \tilde{f}'(x,\alpha)(\tilde{\alpha} - \alpha) + O(\|\tilde{\alpha} - \alpha\|^2)$ and $\tilde{f}'(x,\alpha)$ tends to $f'(\alpha)$ which is invertible, we get

$$\|\tilde{\alpha} - \alpha\| = O(\|\tilde{f}(x,\alpha)\|) = O(\|x - \alpha\|^n).$$

This bound is sharp which yields

$$p(\mathfrak{N},S) = n.$$

See also Woźniakowski [75], where the order of information for nonlinear equations was first defined and analyzed. The study of standard information can be found in Traub and Woźniakowski [76c, 77a,b]. ∎

For convergent information, Corollary 2.1 guarantees the existence of convergent algorithms. Let φ be an algorithm from $\Phi(\mathfrak{N},S)$. We want to examine how fast $\varphi(x,\mathfrak{N}(f,x))$ converges to $\alpha = S(f)$ as x tends to α. Let B be a set of real numbers defined by

$$(2.20) \qquad B = \left\{ q : q \geq 1, \forall f \in \mathfrak{I}_0, \alpha = S(f), \text{ and } \forall \tilde{f} \in V(f), \text{ we have} \right.$$

$$\left. \lim_{x \to \alpha} \frac{\|\varphi(x, \mathfrak{N}(f,x)) - S(\tilde{f}(x))\|}{\|x - \alpha\|^{q-\eta}} = 0, \forall \eta > 0 \right\}.$$

Definition 2.6 We shall say $p(\varphi)$ is the *order of algorithm* φ iff

$$(2.21) \qquad p(\varphi) = \begin{cases} 0 & \text{if } B \text{ is empty,} \\ \sup B & \text{otherwise.} \end{cases} \quad ∎$$

Note that for a divergent information operator $\mathfrak{N}$, the order of any algorithm φ from $\Phi(\mathfrak{N},S)$ is equal to zero. Definition 2.6 of the order $p(\varphi)$ differs from the "classical" definition of order, where $\varphi(x,\mathfrak{N}(f,x))$ is compared only with $\alpha = S(f)$. In the Appendix, we show that for all algorithms of practical interest the "classical" order is equal to $p(\varphi)$.

Let $g(x) = \varphi(x,\mathfrak{N}(f,x)) - S(\tilde{f}(x))$. Then (2.20) yields that $g^{(j)}(\alpha) = 0$ for $j = 0$, $1,\ldots,k$, where $k = p(\varphi) - 1$ if $p(\varphi)$ is an integer and $k = \lfloor p(\varphi) \rfloor$ otherwise. If $p(\varphi) = +\infty$ and g is analytic in a neighborhood of α, then $g(x) \equiv 0$ which means that $\varphi(x,\mathfrak{N}(f,x)) = S(\tilde{f}(x)) = \alpha$ for x close to α. Thus, the problem S can be solved in one step. Therefore, we shall say φ is *a direct algorithm if* $p(\varphi) = +\infty$.

We now prove that the order of information is an upper bound on the order of any algorithm φ and that every interpolatory algorithm achieves this bound.

Theorem 2.3   For any algorithm φ, $\varphi \in \Phi(\mathfrak{N},S)$,

$$(2.22) \qquad\qquad p(\varphi) \le p(\mathfrak{N},S). \quad \blacksquare$$

PROOF   Suppose first that $B = \varnothing$. Then $p(\varphi) = 0 \le p(\mathfrak{N},S)$. Without loss of generality, we can then assume $B \ne \varnothing$. Let q be an arbitrary element of B. Let $f \in \mathfrak{I}_0$ and $\tilde{f}_1, \tilde{f}_2 \in V(f)$. Then

$$\begin{aligned}
\|S(\tilde{f}_1(x)) - S(\tilde{f}_2(x))\| &\le \|\varphi(x,\mathfrak{N}(f,x)) - S(\tilde{f}_1(x))\| \\
&\quad + \|\varphi(x,\mathfrak{N}(f,x)) - S(\tilde{f}_2(x))\| \\
&= o(\|x - \alpha\|^{q-\eta}) \qquad \forall \eta > 0.
\end{aligned}$$

This proves that $q \in A$ (see (2.17)) and $B \subset A$. Thus, $\sup B \le \sup A$ which means $p(\varphi) \le p(\mathfrak{N},S)$. Hence, (2.22) is proven. $\blacksquare$

Theorem 2.3 states that no algorithm can approximate $S(\tilde{f}(x))$ with order higher than the order of information $p(\mathfrak{N},S)$. We show that the bound $p(\mathfrak{N},S)$ is achieved by the order of any interpolatory algorithm.

Theorem 2.4   For any interpolatory algorithm φ^I, $\varphi^I \in \Phi(\mathfrak{N},S)$,

$$(2.23) \qquad\qquad p(\varphi^I) = p(\mathfrak{N},S). \quad \blacksquare$$

PROOF   Without loss of generality, assume that $A \ne \varnothing$. Let $q \in A$. Then

$$(2.24) \qquad \|S(\tilde{f}_1(x)) - S(\tilde{f}_2(x))\| = o(\|x - \alpha\|^{q-\eta}) \qquad \forall \eta > 0,$$

for any $f \in \mathfrak{I}_0$ and $\tilde{f}_1, \tilde{f}_2 \in V(f)$. Since $\varphi^I(x,\mathfrak{N}(f,x)) = S(\tilde{f}_0(x))$ for $\tilde{f}_0 \in V(f)$, we get from (2.24)

$$\|\varphi^I(x,\mathfrak{N}(f,x)) - S(\tilde{f}(x))\| = \|S(\tilde{f}_0(x)) - S(\tilde{f}(x))\| = o(\|x - \alpha\|^{q-\eta}),$$

since $\tilde{f}, \tilde{f}_0 \in V(f)$. This proves that $q \in B$ and $A \subset B$. Thus, $\sup A \le \sup B$ and $p(\mathfrak{N},S) \le p(\varphi^I)$. From Theorem 2.3, we find $p(\varphi^I) = p(\mathfrak{N},S)$ which completes the proof. $\blacksquare$

From Theorems 2.3 and 2.4, we obtain

Corollary 2.2   An interpolatory algorithm φ^I achieves the *maximal order* $p(\mathfrak{N},S)$ in the class $\Phi(\mathfrak{N},S)$,

$$p(\varphi^I) = p(\mathfrak{N},S) = \sup_{\varphi \in \Phi(\mathfrak{N},S)} p(\varphi). \quad \blacksquare$$

The problem of maximal order algorithms was first posed by Traub [61] for nonlinear equations. It was solved for a particular class of nonstationary iterations using standard information for scalar equations by Brent, Winograd, and Wolfe [73]. Theorems 2.3 and 2.4 were established in full generality for nonlinear equations by Woźniakowski [75] and used by many people to establish the maximal order of certain iterative algorithms and/or to compare different information operators from a computational complexity point of view (see, e.g., Kacewicz [75, 76a,b], Meersman [76a,b], Traub and Woźniakowski [76b], Wasilkowski [77], and Woźniakowski [72, 74, 76]). Compare also with recent papers of Werschulz [77a,b], who uses information operators $\mathfrak{N}(f,h)$ for numerical quadrature and differentiation.

3.   COMPLEXITY OF GENERAL INFORMATION

We present our general model of computation; it is quite similar to the model in Part A. We discuss complexity in this model and derive bounds on the "complexity index."

Model of Computation

(i)   We assume that the computations are performed on a random access machine. (See Aho, Hopcroft, and Ullman. [74, Chapter 1].) Let p be a *primitive operation*. Examples of primitive operations include arithmetic operations, the evaluation of a square root, or of an integral. Let comp(p) be the complexity (the total cost) of p; comp(p) must be finite. Suppose that P *is a given collection of* primitives. The choice of P and comp(p), $p \in P$, are arbitrary and can depend on the particular problem being solved.

(ii)   Let $\mathfrak{N}$ be an information operator. We say that $\mathfrak{N}$ is a *permissible information operator with respect to P*, if there exists a program using a finite number of primitive operations from P which computes $\mathfrak{N}(f,x)$ for all (f,x) under consideration. Let comp($\mathfrak{N}(f,x)$) denote the *information complexity* of computing $\mathfrak{N}(f,x)$. We assume that if $\mathfrak{N}(f,x)$ requires the evaluation of primitives $p_1, p_2, \ldots, p_k$, then

$$\text{comp}(\mathfrak{N}(f,x)) = \sum_{i=1}^{k} \text{comp}(p_i).$$

(iii)   Let φ be an algorithm which uses the permissible information $\mathfrak{N}$. To evaluate $\varphi(\mathfrak{N}(f,x))$, we

(a)   compute $y = \mathfrak{N}(f,x)$,
(b)   compute $\varphi(x,y)$.

The complexity of computing y is given by (ii). We say that φ is a *permissible algorithm with respect to* P, if there exists a program using a finite number of primitive operations from P which computes $\varphi(x,y)$ for all $(x,\mathfrak{N}(f,x))$ under consideration. Let $\text{comp}(\varphi(x,y))$ be the *combinatory complexity* of computing $\varphi(x,y)$. We assume that if $\varphi(x,y)$ requires the evaluation of primitives q_1, $q_2, \ldots, q_j$, then

$$\text{comp}(\varphi(x,y)) = \sum_{i=1}^{j} \text{comp}(q_i). \quad \blacksquare$$

Let $\mathfrak{N}$ be a convergent permissible information operator with order of information $p(\mathfrak{N},S)$ greater than unity. Let φ be a convergent permissible algorithm from the class $\Phi(\mathfrak{N},S)$ with order $p = p(\varphi)$ greater than unity. We analyze the complexity of finding an approximation $y = y(f)$ to the solution $\alpha = S(f)$ for $f \in \mathfrak{J}_0$ using algorithm φ, where

(3.1) $$\|y(f) - \alpha\| \le \varepsilon' \|x_0 - \alpha\|$$

for a given initial approximation x_0 and a given number $\varepsilon' \in (0,1)$. The analysis is primarily based on Traub and Woźniakowski [76a, 77b], where the nonlinear equation problem is studied.

Assume that φ generates the sequence $x_i = \varphi(x_{i-1},\mathfrak{N}(f,x_{i-1})), i = 1, 2, \ldots, k$, such that

(3.2) $$e_i = G_i e_{i-1}^p, \qquad e_i = \|x_i - \alpha\|, \qquad i = 1, 2, \ldots, k,$$

where $G_i = G_i(f)$ satisfies

(3.3) $$0 < \underline{G} \le G_i \le \bar{G} < +\infty$$

and algorithm φ is terminated after k steps. From (3.2), we get

(3.4) $$e_i = \left(\frac{1}{\omega_i}\right)^{p^i - 1} e_0, \qquad \text{where} \qquad \frac{1}{\omega_i} = (G_1^{p^{i-1}} G_2^{p^{i-2}} \cdot \cdots \cdot G_i)^{1/(p^i - 1)} e_0.$$

Note that $(e_0 \omega_i)^{1-p}$ is the geometric mean of the $G_1, G_2, \ldots, G_i$. Furthermore, $e_i < e_0$ iff $\omega_i > 1$. From (3.3), we have

(3.5) $$1/\underline{\omega} = (\underline{G})^{1/(p-1)} e_0 \le 1/\omega_i \le (\bar{G})^{1/(p-1)} e_0 = 1/\bar{\omega}.$$

Assume that $\bar{\omega} > 1$. For a given ε', let k be the smallest index for which $e_k \leq \varepsilon' e_0$. Define $\varepsilon \leq \varepsilon'$ so that

(3.6) $e_k = \varepsilon e_0.$

From (3.4) and (3.6), we find

(3.7) $(1/\omega_k)^{p^k-1} = \varepsilon$ and $k = g(\omega_k)/\log p,$

where

(3.8) $g(\omega) = \log\left(1 + \dfrac{t}{\log \omega}\right), \qquad t = \log\left(\dfrac{1}{\varepsilon}\right).$

We take all logarithms for the remainder of Part B to base 2.

Let comp $= \text{comp}(\varphi, f)$ be the complexity of computing x_k starting at x_0. We do not consider the complexity of finding an initial approximation x_0. See Kung [76], where this problem is considered. The cost of the ith step is equal to $\text{comp}(\mathfrak{N}(f, x_i)) + \text{comp}(\varphi(x_i\mathfrak{N}(f, x_i)))$. For simplicity, we assume that the information complexity and the combinatory complexity do not depend on x_i. Then the cost of each step is equal to $c(\varphi, f) = \text{comp}(\mathfrak{N}(f, x)) + \text{comp}(\varphi(x, \mathfrak{N}(f, x)))$ and $\text{comp}(\varphi, f) = kc(\varphi, f)$. From (3.7), we get

(3.9) $\text{comp}(\varphi, f) = zg(\omega_k),$

where

(3.10) $z = z(\varphi, f) \overset{\text{df}}{=} \dfrac{c(\varphi, f)}{\log p(\varphi)}$

is called the *complexity index of φ for f*. By the *complexity index $z(\varphi)$ of an algorithm φ*, we mean

(3.11) $z(\varphi) = \sup_{f \in \mathfrak{I}_0} z(\varphi, f).$

REMARK 3.1   We have been considering the case $p = p(\varphi) > 1$. For completeness, we exhibit the case $p = 1$ with the additional assumption that $G_i < 1$, $i = 1, 2, \ldots, k$. Then $e_i = (1/\omega_i)^i e_0$ with $1/\omega_i = (G_1 G_2 \cdot \cdots \cdot G_i)^{1/i}$. Hence, $1/\omega_i$ is the geometric mean of $G_1, G_2, \ldots, G_i$. Assume that $e_k = \varepsilon e_0$. Then $k = (\log 1/\varepsilon)/\log \omega_k$ and the complexity $\text{comp}(\varphi, f)$ of computing x_k is given by

$$\text{comp}(\varphi, f) = z \log(1/\varepsilon),$$

where $z = z(\varphi, f, \varepsilon) = c(\varphi, f)/(\log \omega_k)$ and $z(\varphi, \varepsilon) = \sup_{f \in \mathfrak{I}_0} z(\varphi, f, \varepsilon)$ is called the *complexity index for $p = 1$*. Assume that $1/\underline{\omega} \leq 1/\omega_i \leq 1/\bar{\omega}$ with $\bar{\omega} > 1$. Then

$$c(\varphi, f)/(\log \underline{\omega}) \leq z(\varphi, f, \varepsilon) \leq c(\varphi, f)/(\log \bar{\omega}).$$

We shall not pursue the case $p = 1$ further and shall assume for the remainder of this section that $p > 1$. ∎

We analyze the complexity $\text{comp}(\varphi,f)$ defined by (3.9). Since $g(\omega)$ is a monotonically decreasing function, (3.5) yields bounds on the complexity

$$(3.12) \qquad\qquad zg(\underline{\omega}) \leq \text{comp}(\varphi,f) \leq zg(\overline{\omega}).$$

As $\varepsilon \to 0$, $g(\omega) \cong \log t$ and $\text{comp} \cong z \log t$. Furthermore, if we assume that

$$(3.13) \qquad\qquad 2 \leq \overline{\omega} \leq \underline{\omega} \leq t,$$

then (3.12) becomes

$$z(\log t - \log \log t) \leq \text{comp}(\varphi,f) \leq z \log(1 + t).$$

(See Theorem 3.1 in Traub and Woźniakowski [76a].) In this case, $z \log t$ is a good measure of complexity. However, if ω_k in (3.9) is near unity, the g factor dominates and for ε fixed, $\text{comp} \cong z \log \log \omega_k$. Thus, for $\omega_k \cong 1$, the effect of the error coefficients G_i and the initial error e_0 cannot be neglected.

Remark 3.1 and (3.12) show how the complexity $\text{comp}(\varphi,f)$ depends asymptotically on ε. Using the Θ-notation of Knuth [76] (see also Section 1 of Chapter 6 in Part A),

$$(3.14) \qquad \text{comp}(\varphi,f) = \begin{cases} \Theta(\log(1/\varepsilon)) & \text{for} \quad p(\varphi) = 1, \\ \Theta(\log\log(1/\varepsilon)) & \text{for} \quad p(\varphi) > 1. \end{cases}$$

This may be contrasted with Theorem 2.2 of Chapter 5 in Part A, where we prove that the complexity of a linear problem can be an "arbitrary" decreasing function of ε.

We want to minimize the complexity of computing x_k, i.e., we want to find a permissible algorithm φ with minimal complexity. Since we do not know the value $g(\omega_k)$ in (3.9), we are not able to minimize complexity. However, if (3.13) holds or ε is small enough then the minimal complexity is approximately achieved by an algorithm with minimal complexity index. Therefore we seek an algorithm with the smallest complexity index.

This discussion motivates the following definition.

Definition 3.1   We shall say $z(\mathfrak{N},S)$ is the *complexity index of the problem S with the information $\mathfrak{N}$* iff

$$(3.15) \qquad\qquad z(\mathfrak{N},S) = \inf\{z(\varphi) : \varphi \in \Phi_{\text{perm}}(\mathfrak{N},S)\},$$

where $\Phi_{\text{perm}}(\mathfrak{N},S)$ is the class of all permissible algorithms.

We shall say φ^{mc}, $\varphi^{\text{mc}} \in \Phi_{\text{perm}}(\mathfrak{N},S)$, is a *minimal complexity index algorithm* iff

$$(3.16) \qquad\qquad z(\varphi^{\text{mc}}) = z(\mathfrak{N},S). \quad \blacksquare$$

Let

$$(3.17) \qquad\qquad \text{comp}(\mathfrak{N}) = \sup_{(f,x) \in D_{\mathfrak{N}}} \text{comp}(\mathfrak{N}(f,x))$$

be the *information complexity of* $\mathfrak{N}$. Every algorithm φ which uses $\mathfrak{N}$ has to perform a certain number of primitive operations to produce the next approximation. More precisely, let

$$(3.18) \qquad m(\mathfrak{N},S) = \inf_{\varphi \in \Phi_{\text{perm}}(\mathfrak{N},S)} \sup_{(f,x) \in D} \text{comp}(\varphi(x,\mathfrak{N}(f,x))),$$

where D, $D \subset D_{\mathfrak{N}}$, denotes "hard problems," i.e., $(f,x) \in D$ iff $\text{comp}(\mathfrak{N}(f,x)) = \text{comp}(\mathfrak{N})$. In general, $m(\mathfrak{N},S)$ depends at least linearly on the total number of "independent pieces" of information $\mathfrak{N}$. See Sections 4 and 8 where the "cardinality" of information $\mathfrak{N}$ is introduced and its influence on the combinatory complexity of φ is discussed.

In Theorems 2.3 and 2.4, we showed that the order $p(\varphi)$ of any algorithm φ from the class $\Phi(\mathfrak{N},S)$ is no larger than the order of information $p(\mathfrak{N},S)$ and there exist algorithms such that $p(\varphi) = p(\mathfrak{N},S)$. From this and (3.14), (3.17), (3.18), we get a lower bound on the complexity index $z(\mathfrak{N},S)$,

$$(3.19) \qquad z(\mathfrak{N},S) \geq \frac{\text{comp}(\mathfrak{N}) + m(\mathfrak{N},S)}{\log p(\mathfrak{N},S)}.$$

Furthermore, if there exists a maximal order permissible algorithm φ, $p(\varphi) = p(\mathfrak{N},S)$, such that $\text{comp}(\varphi(x,\mathfrak{N}(f,x))) \ll \text{comp}(\mathfrak{N})$ for all $(f,x) \in D_{\mathfrak{N}}$, then

$$(3.20) \qquad z(\mathfrak{N},S) \cong \frac{\text{comp}(\mathfrak{N})}{\log p(\mathfrak{N},S)}.$$

Equations (3.19) and (3.20) motivate our interest in the information complexity $\text{comp}(\mathfrak{N})$ and the order of information $p(\mathfrak{N},S)$.

Suppose that the problem $\alpha = S(f)$ can be solved by the use of different information operators from a given class Ψ. We want to know which information operator is more relevant for the problem $\alpha = S(f)$. This discussion motivates the definition of an information operator with minimal complexity index.

Definition 3.2 We shall say an information operator $\mathfrak{N}_1$ is *more relevant* than an information operator $\mathfrak{N}_2$ for the problem S iff

$$(3.21) \qquad z(\mathfrak{N}_1,S) < z(\mathfrak{N}_2,S).$$

We shall say an information operator $\mathfrak{N}^{\circ}$ is *optimal in the class* Ψ, $\mathfrak{N}^{\circ} \in \Psi$, iff

$$(3.22) \qquad z(\mathfrak{N}^{\circ},S) = \inf_{\mathfrak{N} \in \Psi} z(\mathfrak{N},S). \quad \blacksquare$$

In Section 8, we will study optimal information operators for the linear case.

We compare two information operators by their complexity indices and an information operator with minimal complexity index is called optimal. Note, however, that the complexity of an algorithm using an optimal information operator also depends on error coefficients G_i and the initial approximation.

As noted earlier, the minimal complexity index can be a poor measure of complexity. See Traub and Woźniakowski [76a] for a discussion of this point.

4. CARDINALITY OF LINEAR INFORMATION

Let $\mathfrak{N}$ be an information operator such that

$$(4.1) \qquad \mathfrak{N}:\mathfrak{I}_1 \times X \to \mathfrak{I}_3,$$

where X is an open subset of $\mathfrak{I}_2$. In Sections 4–8, we deal with information operators which are linear with respect to the first argument, i.e.,

$$(4.2) \qquad \mathfrak{N}(c_1 f_1 + c_2 f_2, x) = c_1 \mathfrak{N}(f_1, x) + c_2 \mathfrak{N}(f_2, x)$$

for any elements f_1, f_2 from $\mathfrak{I}_1$, any constants c_1, c_2, and any $x \in X$. Let

$$(4.3) \qquad \ker \mathfrak{N}(\cdot, x) = \{ f : \mathfrak{N}(f, x) = 0 \}$$

be the kernel of $\mathfrak{N}(\cdot, x)$ for any $x \in X$. As we shall see in Section 5, the kernel of $\mathfrak{N}$ will play an essential role.

Let $\mathfrak{N}_1 : \mathfrak{I}_1 \times X \to \mathfrak{I}_3$ and $\mathfrak{N}_2 : \mathfrak{I}_1 \times X \to \mathfrak{I}'_3$ be two information operators, where the space $\mathfrak{I}'_3$ is not necessarily equal to $\mathfrak{I}_3$.

Definition 4.1   We shall say $\mathfrak{N}_1$ is *contained in* $\mathfrak{N}_2$ (briefly $\mathfrak{N}_1 \subset \mathfrak{N}_2$) iff $\ker \mathfrak{N}_2(\cdot, x) \subset \ker \mathfrak{N}_1(\cdot, x), \forall x \in X$.

We shall say $\mathfrak{N}_1$ is *equivalent to* $\mathfrak{N}_2$ (briefly $\mathfrak{N}_1 \asymp \mathfrak{N}_2$) iff $\ker \mathfrak{N}_1(\cdot, x) = \ker \mathfrak{N}_2(\cdot, x), \forall x \in X$.   ∎

Note that $A = \ker \mathfrak{N}(\cdot, x)$ is a linear subspace of $\mathfrak{I}_1$. Recall that there exists a linear subspace $A^\perp$ of $\mathfrak{I}_1$ such that

$$(4.4) \qquad \mathfrak{I}_1 = A \oplus A^\perp,$$

where $A^\perp$ is isomorphic to the quotient space $\mathfrak{I}_1/A$ and

$$(4.5) \qquad \operatorname{codim} A \overset{\mathrm{df}}{=} \dim A^\perp = \dim \mathfrak{I}_1/A.$$

$A^\perp$ is called an algebraic complement of A.

To simplify further considerations, we assume that the set X is chosen in such a way that $\operatorname{codim} \ker \mathfrak{N}(\cdot, x) \equiv \operatorname{const}$ for every $x \in X$.

Definition 4.2   We shall say that $\operatorname{card}(\mathfrak{N})$ is the *cardinality of the information* $\mathfrak{N}$ iff

$$(4.6) \qquad \operatorname{card}(\mathfrak{N}) = \operatorname{codim} \ker \mathfrak{N}(\cdot, x), \qquad x \in X.   ∎$$

As an example, consider the information operator $\mathfrak{N}$ defined by

$$(4.7) \qquad \mathfrak{N}(f, x) = [L_1(f, x), \ldots, L_n(f, x)],$$

where $L_j: \mathfrak{I}_1 \times X \to \mathbb{C}$ is a linear functional with respect to the first argument, $j = 1, 2, \ldots, n$. We assume that $L_1(\cdot, x), \ldots, L_n(\cdot, x)$ are linearly independent for every $x \in X$.

Lemma 4.1   Let $\mathfrak{N}$ be defined by (4.7). Then

(4.8) $$\operatorname{card}(\mathfrak{N}) = n. \quad \blacksquare$$

PROOF   Note that $A(x) = \ker \mathfrak{N}(\cdot, x) = \{f: L_j(f, x) = 0 \text{ for } j = 1, 2, \ldots, n\}$. Let $\mathfrak{I}_1 = A(x) \oplus A(x)^\perp$ and $\lim(\xi_1, \xi_2, \ldots, \xi_m) \subset A(x)^\perp$ for linearly independent $\xi_1, \xi_2, \ldots, \xi_m$. Let $f = \sum_{j=1}^{m} c_j \xi_j, f \in A(x)^\perp$. We want to find $\mathbf{c} = [c_1, c_2, \ldots, c_m]^t$ so that $f \in A(x)$. ("t" denotes the transposition of a vector.) Observe that $L_i(f, x) = \sum_{j=1}^{m} c_j L_i(\xi_j, x) = 0$ for $i = 1, 2, \ldots, n$ is equivalent to the system of homogenous linear equations,

(4.9) $$M\mathbf{c} = \mathbf{0},$$

where $M = (L_i(\xi_j, x))$.

To prove (4.8), assume that $m > n$. Then (4.9) has a nonzero solution $[c_1, c_2, \ldots, c_m]^t$ and $f = \sum_{j=1}^{m} c_j \xi_j \in A(x)^\perp \cap A(x)$. Thus $f = 0$ which contradicts the linear independence of $\xi_1, \ldots, \xi_m$. Hence, $\operatorname{card}(\mathfrak{N}) \leq n$. To prove that $\operatorname{card}(\mathfrak{N}) = n$, it suffices to observe that $L_1(\cdot, x), \ldots, L_n(\cdot, x)$ are linearly independent iff $M^t \mathbf{d} = \mathbf{0}$ has the unique solution $\mathbf{d} = \mathbf{0}$ which holds iff $m = n$. $\blacksquare$

We now show that any linear information operator $\mathfrak{N}$ may be represented by linear functionals.

Lemma 4.2   Let $\mathfrak{N}$ be a linear information operator and $n = \operatorname{card}(\mathfrak{N}) \leq +\infty$. Then there exist $L_1, L_2, \ldots, L_n$ such that

(i) $L_j: \mathfrak{I}_1 \times X \to \mathbb{C}, \qquad j = 1, 2, \ldots, n,$

(ii) $L_j(\cdot, x), \ldots, L_n(\cdot, x)$ are linearly independent linear functionals, and $\mathfrak{N} \asymp \mathfrak{N}_1$, where $\mathfrak{N}_1 = [L_1, L_2, \ldots, L_n]$. $\blacksquare$

PROOF   Let $\mathfrak{I}_1 = A(x) \oplus A(x)^\perp$, where $A(x) = \ker \mathfrak{N}(\cdot, x)$. Then $A(x)^\perp = \lim(\xi_1(x), \ldots, \xi_n(x))$, where $\xi_i(x)$ are linearly independent. Every element f, $f \in \mathfrak{I}_1$, has a unique representation $f = f_0(x) + \sum_{i=1}^{n} L_i(f, x)\xi_i(x)$ for some linearly independent functionals L_i such that $L_i(\xi_j(x), x) = \delta_{ij}$ and $f_0(x) \in A(x)$. Then the information operator $\mathfrak{N}_1 = [L_1, L_2, \ldots, L_n]$ satisfies

$$\ker \mathfrak{N}_1(\cdot, x) = \{f: L_i(f, x) = 0, i = 1, 2, \ldots, n\} = A(x) = \ker \mathfrak{N}(\cdot, x).$$

This proves that $\mathfrak{N} \asymp \mathfrak{N}_1$. $\blacksquare$

Let $A(\mathfrak{I}_1)$ be the class of all linear subspaces of $\mathfrak{I}_1$. Consider

(4.10) $$A: X \subset \mathfrak{I}_2 \to A(\mathfrak{I}_1),$$

i.e., $A(x)$ is a linear subspace of $A(\mathfrak{I}_1)$ for any $x \in X$. We show the relationship between transformations A and linear information operators.

Lemma 4.3   Let $A: X \subset \mathfrak{I}_2 \to A(\mathfrak{I}_1)$ and codim $A(x) = n$, $\forall x \in X$. Then there exists a linear information operator

(4.11) $$\mathfrak{N}(f,x) = [L_1(f,x), \ldots, L_n(f,x)],$$

where $L_1(\cdot,x), \ldots, L_n(\cdot,x)$ are linearly independent linear functionals such that

(4.12) $$\ker \mathfrak{N}(\cdot,x) = A(x). \quad \blacksquare$$

PROOF   Let $\mathfrak{I}_1 = A(x) \oplus A(x)^\perp$, where $A(x)^\perp = \mathrm{lin}(\xi_1(x), \ldots, \xi_n(x))$ for linearly independent $\xi_1(x), \ldots, \xi_n(x)$. Then every element $f \in \mathfrak{I}_1$ has a unique representation

(4.13) $$f = f_0(x) + \sum_{j=1}^{n} c_j(f,x)\xi_j(x),$$

where $f_0(x) \in A(x)$ and $c_j(f,x)$ is a linear functional with respect to f, $c_j(\xi_i(x),x) = \delta_{ij}$. Define

(4.14) $$L_j(f,x) = c_j(f,x) \qquad \text{for} \quad j = 1, 2, \ldots, n.$$

It is obvious that $L_1(\cdot,x), \ldots, L_n(\cdot,x)$ are linearly independent and

$$\ker \mathfrak{N}(\cdot,x) = \{f: L_j(f,x) = 0, j = 1, 2, \ldots, n\} = A(x),$$

which completes the proof. $\blacksquare$

In the following sections, we shall consider "regular" linear information operators defined as follows.

Let $\tilde{f}: D_f \subset \mathfrak{I}_2 \to \mathfrak{I}_0$, where $\mathfrak{I}_0 \subset \mathfrak{I}_1$. Assume that $\mathfrak{I}_1$ is a linear normed space. We shall say that $\tilde{f}$ belongs to the *class Lip(k)*, $k \geq 0$, iff the kth Fréchet derivative of $\tilde{f}$ exists at every solution element $\alpha \in S(\mathfrak{I}_0)$ and satisfies a Lipschitz condition. That is, $\tilde{f} \in \mathrm{Lip}(k)$ iff for any $\alpha \in S(\mathfrak{I}_0)$ there exist $\Gamma = \Gamma(\alpha,\tilde{f}) > 0$ and $q = q(\alpha,\tilde{f})$ such that

(4.15) $$\left\| \tilde{f}^{(k)}(x_1) - \tilde{f}^{(k)}(x_2) \right\| \leq q \|x_1 - x_2\|$$

for $\|x_1 - x_2\| \leq \Gamma$.

Let $L: \mathfrak{I}_1 \times X \to \mathbb{C}$ be a linear functional with respect to the first argument. We shall say that $L \in \mathrm{Lip}(k)$ iff $L(h,\cdot) \in \mathrm{Lip}(k)$ for any $h \in \mathfrak{I}_1$. We are ready to define what we mean by $\mathfrak{N} \in \mathrm{Lip}(k)$.

Definition 4.3   We shall say a *linear information operator* $\mathfrak{N}$ *belongs to the class* Lip(k) (briefly $\mathfrak{N} \in \mathrm{Lip}(k)$) iff there exist linearly independent linear functionals $L_j: \mathfrak{I}_1 \times X \to \mathbb{C}$, $j = 1, 2, \ldots, n = \mathrm{card}(\mathfrak{N})$ such that

(i) $\ker \mathfrak{N}(\cdot,x) = \{h: L_j(h,x) = 0, j = 1, 2, \ldots, n\}$, $\forall x \in X$,
(ii) $L_j \in \mathrm{Lip}(k)$, $j = 1, 2, \ldots, n$. $\blacksquare$

Thus, $\mathfrak{N} \in \mathrm{Lip}(k)$ means that the linear functionals which form the kernel of $\mathfrak{N}$ are k-times differentiable and the kth derivative satisfies a Lipschitz condition.

Lemma 4.4   Assume that $\mathfrak{N}:\mathfrak{I}_1 \times X \to \mathfrak{I}_3$ belongs to Lip(k) and $n = \text{card}(\mathfrak{N})$. Let $g_0 \in \ker \mathfrak{N}(\cdot,\alpha)$. Then there exists a function $h:X \to \mathfrak{I}_1$, such that

   (i) $\mathfrak{N}(h(x),x) = 0 \qquad \forall x \in X$,

   (ii) $h \in \text{Lip}(k)$,

   (iii) $h(\alpha) = g_0$. ∎

PROOF   From Definition 4.3, we have

(4.16) $\ker \mathfrak{N}(\cdot,x) = \{g:L_j(g,x) = 0, j = 1, 2, \ldots, n\}, \qquad x \in X$,

where $L_j \in \text{Lip}(k)$ and $L_1(\cdot,x), \ldots, L_n(\cdot,x)$ are linearly independent linear functionals. Let $\xi_1(x), \ldots, \xi_n(x)$ be elements of $\mathfrak{I}_1$ such that $L_j(\xi_i(x),x) = \delta_{ij}$ for $i, j = 1, 2, \ldots, n$ and $\xi_j \in \text{Lip}(k)$. Define

(4.17) $h(x) = g_0 - \sum_{i=1}^{n} L_i(g_0,x)\xi_i(x)$.

Note that $L_j(h(x),x) = L_j(g_0,x) - \sum_{i=1}^{n} L_i(g_0,x)\delta_{ij} = 0$ which means that $h(x) \in \ker \mathfrak{N}(\cdot,x)$, $\forall x \in X$. Clearly, $h \in \text{Lip}(k)$ and $h(\alpha) = g_0$, since $g_0 \in \ker \mathfrak{N}(\cdot,\alpha)$ means $L_i(g_0,\alpha) = 0$ for $i = 1, 2, \ldots, n$. This completes the proof. ∎

5.   WHEN IS THE CLASS OF ITERATIVE ALGORITHMS EMPTY?

Recall that the solution operator S transforms $\mathfrak{I}_0$, $\mathfrak{I}_0 \subset \mathfrak{I}_1$, into $\mathfrak{I}_2$. Throughout the rest of Part B, we assume that

(5.1) $\mathfrak{I}_1$ is a linear *normed* space,

(5.2) $\mathfrak{I}_0$ is *open*, i.e., for any $f \in \mathfrak{I}_0$ there exists a positive number $\delta = \delta(f)$ such that $f + h \in \mathfrak{I}_0$ for any $h \in \mathfrak{I}_1$ with $\|h\| \leq \delta$,

(5.3) $\mathfrak{I}_2$ is a *Banach* space,

(5.4) S is a Lipschitz operator at every $f \in \mathfrak{I}_0$, i.e., $\|S(f_1) - S(f_2)\| \leq q(S,f)\|f_1 - f_2\|$ for all f_1 and f_2 sufficiently close to f.

Let $\mathfrak{N}$ be an information operator. Recall that $\tilde{f}:D_{\tilde{f}} \to \mathfrak{I}_0$ belongs to $V(f)$ and $V(f) \subset W$. (See Definition 2.1.) We assume that W is the class of functions $\tilde{f}$ such that

(5.5) $\|\tilde{f}(x) - f\| < \delta(f)$,

(5.6) $\tilde{f}$ is a Lipschitz function at $\alpha = S(f)$, i.e.,
$\|\tilde{f}(x_1) - \tilde{f}(x_2)\| \leq q(\tilde{f})\|x_1 - x_2\|$,

where x, x_1, and x_2 belong to the ball $J = \{x:\|x - \alpha\| \leq \Gamma\}$ for a sufficiently small positive $\Gamma = \Gamma(\tilde{f})$.

REMARK 5.1   Since $\mathfrak{N}$ is linear, $\mathfrak{N}(\tilde{f}(x) - f, x) = 0$. This means that $h(x) = \tilde{f}(x) - f$ belongs to ker $\mathfrak{N}(\cdot, x)$. Furthermore, $\|\tilde{f}(x_1) - \tilde{f}(x_2)\| = \|h(x_1) - h(x_2)\|$ and (5.5), (5.6) state that $\|h(x)\| < \delta$ and h is a Lipschitz function at α. From Lemma 4.2, we know that $h(x) \in \ker \mathfrak{N}(\cdot, x)$ is equivalent to $L_i(h(x), x) = 0$ for $i = 1, 2, \ldots, n = \mathrm{card}(\mathfrak{N})$ for some linear functionals $L_1(\cdot, x), \ldots, L_n(\cdot, x)$. Thus, $h = h(x)$ has to satisfy n homogeneous equations.   ∎

We deal with "iterative" algorithms which are defined as follows. Recall that φ is an algorithm (see (2.3)), if $\varphi : D_\varphi \subset \mathfrak{I}_2 \times \mathfrak{N}(D_\mathfrak{N}) \to \mathfrak{I}_2$.

Definition 5.1   We shall say φ is an *iterative algorithm* (or briefly an iteration) iff for any $f \in \mathfrak{I}_0$, $\alpha = S(f)$, there exists a positive number $\Gamma = \Gamma(f)$ such that for every $x_0 \in J = \{x : \|x - \alpha\| \leq \Gamma\}$ the sequence $x_{i+1} = \varphi(x_i, \mathfrak{N}(f, x_i))$ is well defined and

$$(5.7) \qquad \lim_{i \to \infty} x_i = \alpha,$$

$$(5.8) \qquad \alpha = \varphi(\alpha, \mathfrak{N}(f, \alpha)). \quad ∎$$

Conditions (5.7) and (5.8) mean that the algorithm φ produces convergent sequences whenever an initial approximation belongs to the ball J and the solution α is a fixed point of φ. Let $IT(\mathfrak{N}, S)$ be the class of *all* iterative algorithms which use the information operator $\mathfrak{N}$. For which information operators $\mathfrak{N}$ is the class $IT(\mathfrak{N}, S)$ nonempty? We show that this problem is related to the limiting diameter of information $d(\mathfrak{N}, S)$ defined by (2.4). We need the following lemma.

Lemma 5.1   Assume there exists $f \in \mathfrak{I}_0$, $\alpha = S(f)$, such that for any $\Gamma > 0$ one can find $f_0 \in \mathfrak{I}_0$ satisfying

(i) $0 < \|\alpha_0 - \alpha\| \leq \Gamma$, where $\alpha_0 = S(f_0)$,
(ii) $\mathfrak{N}(f_0, \alpha_0) = \mathfrak{N}(f, \alpha_0)$.

Then $IT(\mathfrak{N}, S) = \varnothing$.   ∎

PROOF   Suppose to the contrary that $IT(\mathfrak{N}, S) \neq \varnothing$ and let $\varphi \in IT(\mathfrak{N}, S)$. Then for f, $\alpha = S(f)$, there exists $\Gamma = \Gamma(f) > 0$ such that $x_{i+1} = \varphi(x_i, \mathfrak{N}(f, x_i))$ is convergent to α for all $\|x_0 - \alpha\| \leq \Gamma$. Let f_0 satisfy the assumptions of Lemma 5.1. Apply the iterative algorithm φ to f_0 with the initial approximation $x_0 = \alpha_0$. Since $\alpha_0 = S(f_0)$ is a fixed point of φ, we get $\varphi(\alpha_0, \mathfrak{N}(f_0, \alpha_0)) = \alpha_0$. But $\|\alpha_0 - \alpha\| \leq \Gamma$ which means that α_0 can be used as an initial approximation of α. However, $x_1 = \varphi(\alpha_0, \mathfrak{N}(f, \alpha_0)) = \varphi(\alpha_0, \mathfrak{N}(f_0, \alpha_0)) = \alpha_0$ which yields $x_{i+1} \equiv \alpha_0 \neq \alpha$. This contradicts $\{x_i\}$ tending to α.   ∎

Lemma 5.1 states that if one can find a problem element f_0 which shares the same information as f and α_0 is sufficiently close to but different from α, then the class of iterations $IT(\mathfrak{N}, S)$ is empty. Compare with Theorem 4.1 in Kung and Traub [76b] and Lemma 3.2 in Woźniakowski [76] in which a similar proof technique is used.

We are ready to prove

Theorem 5.1 Suppose that (5.1)–(5.6) hold. Let $\mathfrak{N}$ be a linear information operator. If $d(\mathfrak{N},S) > 0$, then $IT(\mathfrak{N},S) = \varnothing$. ∎

PROOF   Since $d(\mathfrak{N},S) > 0$, there exist $f \in \mathfrak{J}_0$, $\alpha = S(f)$, and $\tilde{f} \in V(f)$ such that

$$(5.9) \qquad\qquad S(\tilde{f}(\alpha)) \neq S(f) = \alpha.$$

Let $h(x) = \tilde{f}(x) - f$ for x close to α. Then $h(x) \in \ker \mathfrak{N}(\cdot,x)$ and due to (5.6) $\tilde{f}$ is a Lipschitz function at α,

$$\|h(x_1) - h(x_2)\| = \|\tilde{f}(x_1) - \tilde{f}(x_2)\| \leq q(\tilde{f})\|x_1 - x_2\|.$$

Due to (5.2) and (5.5), $f + ch(x) \in \mathfrak{J}_0$ for $c \in [0,1]$ since $\|ch(x)\| < \delta(f)$. Consider $g(c) \stackrel{\text{df}}{=} S(f + ch(\alpha)) - S(f)$ for $c \in [0,1]$. From (5.4), it follows that g is continuous. Note that $g(0) = 0$ and $g(1) \neq 0$ due to (5.9). Choose $c_0 \in [0,1)$ such that $g(c_0) = 0$ and $g(c_0 + \delta) \neq 0$, for $\delta \in (0, 1 - c_0]$. Let $f_1 = f + c_0 h(\alpha)$. Note that $f_1 \in \mathfrak{J}_0$ and $f_1 + ch(x) \in \mathfrak{J}_0$ for $c \in [0,\delta(f_1)/\delta(f)]$. For such c, define $F(x) = S(f_1 + ch(x))$. Then $S(f_1) = \alpha$ and $F(\alpha) = S(f_1 + ch(\alpha)) \neq \alpha$ for small positive c. We consider the equation $x = F(x)$ for $x \in J = \{x : \|x - \alpha\| \leq \Gamma\}$ for small positive Γ. Note that $\|F(x) - \alpha\| = \|S(f_1 + ch(x)) - S(f_1)\| \leq cq(S,f_1) \sup_{x \in J} \|h(x)\| \leq \Gamma$ for small c and Γ. Furthermore, $\|F(x_1) - F(x_2)\| \leq cq(S,f_1)\|h(x_1) - h(x_2)\| \leq cq(S,f_1)q(\tilde{f})\|x_1 - x_2\|$. Thus, for small c, F is a contraction mapping in J. Since J is a closed ball in the Banach space $\mathfrak{J}_2$, there exist $\alpha_0 \in J$ such that $\alpha_0 = F(\alpha_0)$, i.e., $\alpha_0 = S(f_1 + ch(\alpha_0))$ and $\alpha_0 \neq \alpha$. Let $f_0 = f_1 + ch(\alpha_0)$. Since $h(\alpha_0) \in \ker \mathfrak{N}(\cdot,\alpha_0)$, $\mathfrak{N}(f_0,\alpha_0) = \mathfrak{N}(f_1,\alpha_0)$. Applying Lemma 5.1 for f_1 and f_0, we get $IT(\mathfrak{N},S) = \varnothing$. ∎

Theorem 5.1 states that $d(\mathfrak{N},S) = 0$ is a necessary condition for the class of iterations $IT(\mathfrak{N},S)$ not to be empty. We prove in the next section that unless the cardinality of the information operator $\mathfrak{N}$ is sufficiently large, $d(\mathfrak{N},S) > 0$.

Assume then that $\mathfrak{N}$ is a convergent linear information operator and let $p(\mathfrak{N},S)$ be the order of information $\mathfrak{N}$. Since S is a Lipshitz operator and W is the class of Lipschitz functions, $p(\mathfrak{N},S) \geq 1$.

Lemma 5.2 If $p(\mathfrak{N},S) > 1$, then $IT(\mathfrak{N},S)$ is nonempty. ∎

PROOF   Let φ^I, $\varphi^I \in \Phi(\mathfrak{N},S)$, be an interpolatory algorithm. From Theorem 2.4, we get $p(\varphi^I) = p(\mathfrak{N},S)$. This yields

$$(5.10) \qquad \|\varphi^I(x,\mathfrak{N}(f,x)) - \alpha\| = o(\|x - \alpha\|^{p-\eta}) \qquad \forall \eta > 0,$$

for all x sufficiently close to $\alpha = S(f)$ with $p = p(\mathfrak{N},S)$. Set $\eta = (p-1)/2$ and $q = (p+1)/2$. Then there exist positive constants C and Γ such that (5.10) can be rewritten

$$\|\varphi^I(x,\mathfrak{N}(f,x)) - \alpha\| \leq C\|x - \alpha\|^q \qquad \text{for} \quad \|x - \alpha\| \leq \Gamma.$$

Let $e_i = \|x_i - \alpha\|$, where $x_{i+1} = \varphi^1(x_i, \mathfrak{N}(f, x_i))$. Then $e_{i+1} \leq Ce_i^q$. Since $q > 1$, then for $e_0 < C^{1/(1-q)}$ the sequence $\{x_i\}$ is convergent to α. Of course, $\alpha = \varphi(\alpha, \mathfrak{N}(f, \alpha))$. This means that φ^1 is an iterative algorithm in the sense of Definition 5.1 and $\mathrm{IT}(\mathfrak{N}, S)$ is nonempty. ∎

From the proof of Lemma 5.2, it easily follows

Corollary 5.1   If $\varphi \in \Phi(\mathfrak{N}, S)$ and $p(\varphi) > 1$, then φ is an iterative algorithm. ∎

For $p(\mathfrak{N}, S) = 1$, the class of iterations $\mathrm{IT}(\mathfrak{N}, S)$ is nonempty if S is a contraction mapping, more precisely if $\|S(\tilde{f}(x_1)) - S(\tilde{f}(x_2))\| \leq q(f)\|x_1 - x_2\|$ for all x_1 and x_2 close enough to $\alpha = S(f)$ and for all $\tilde{f}_1, \tilde{f}_2 \in V(f)$, where $q(f) < 1$. For $p(\mathfrak{N}, S) = 1$ and for noncontraction mappings S, it seems plausible to conjecture that $\mathrm{IT}(\mathfrak{N}, S) = \varnothing$. We do not pursue this problem here.

6. INDEX OF THE PROBLEM S

Recall that $S: \mathfrak{I}_0 \to \mathfrak{I}_2$, where $\mathfrak{I}_0$ is a subset of a linear space $\mathfrak{I}_1$. Let $\alpha \in S(\mathfrak{I}_0)$. Define

$$(6.1) \qquad G(\alpha) = \{h \in \mathfrak{I}_1 : \forall f \in \mathfrak{I}_0, S(f) = \alpha, \text{ we have}$$
$$S(f + ch) = S(f), \forall |c| < \delta(f)/\|h\|\}.$$

That is, $G(\alpha)$ is a set of elements h which multiplied by a small c and added to f do not change the solution element $\alpha = S(f)$. Due to assumption (5.2), $f + ch \in \mathfrak{I}_0$ for $|c| < \delta(f)/\|h\|$ and therefore $S(f + ch)$ is well defined. Note that $G(\alpha)$ is a homogeneous set, i.e., $h \in G(\alpha)$ implies $ah \in G(\alpha)$ for any constant a. If S is analytic at any $f \in \mathfrak{I}_0$, then

$$(6.2) \qquad G(\alpha) = \{h \in \mathfrak{I}_1 : \forall f \in \mathfrak{I}_0, S(f) = \alpha, S^{(j)}(f)h^j = 0, \text{ for } j \geq 1\}.$$

Let $A(\mathfrak{I}_1)$ be the class of all linear subspaces of $\mathfrak{I}_1$.

Definition 6.1   We shall say that index(S) is the *index of the problem S* iff

$$(6.3) \qquad \mathrm{index}(S) = \max_{\alpha \in S(\mathfrak{I}_0)} \quad \min_{A \subset G(\alpha) \text{ and } A \in A(\mathfrak{I}_1)} \quad \mathrm{codim}\, A. \quad ∎$$

Let $A(\alpha)$ be a subset of $G(\alpha)$ such that $A(\alpha)$ is a linear subspace of $\mathfrak{I}_1$ and has minimal codimension, i.e., $A(\alpha) \subset G(\alpha)$, $A(\alpha) \in A(\mathfrak{I}_1)$, and

$$(6.4) \qquad \mathrm{codim}\, A(\alpha) = \min_{A \subset G(\alpha) \text{ and } A \in A(\mathfrak{I}_1)} \mathrm{codim}\, A.$$

If $G(\alpha)$ is linear, then clearly $A(\alpha) = G(\alpha)$. The index of the problem S is the maximal codimension of $A(\alpha)$. If $G(\alpha)$ is linear for all $a \in S(\mathfrak{I}_0)$, then (6.3) becomes

$$(6.5) \qquad \mathrm{index}(S) = \max_{\alpha \in S(\mathfrak{I}_0)} \mathrm{codim}\, G(\alpha).$$

In Section 5, we showed that if the limiting diameter of information $d(\mathfrak{N}, S)$ is nonzero, then the class of iterations $\mathrm{IT}(\mathfrak{N}, S)$ is empty. We now prove that

$d(\mathfrak{N},S) = 0$ implies that the cardinality of $\mathfrak{N}$ is at least equal to the index S. Recall that throughout the rest of Part B we assume that (5.1)–(5.6) hold. We also assume that $S(\mathfrak{I}_0)$ is an open subset of $\mathfrak{I}_2$.

Theorem 6.1 Let $\mathfrak{N}: \mathfrak{I}_1 \times S(\mathfrak{I}_0) \to \mathfrak{I}_3$ be an arbitrary linear information operator such that $\mathfrak{N} \in \text{Lip}(0)$ and $\text{card}(\mathfrak{N}) < \text{index}(S)$. Then

(6.6) $d(\mathfrak{N},S) > 0.$ ∎

PROOF   Assume that there exists an information operator $\mathfrak{N}$ such that $\mathfrak{N} \in \text{Lip}(0)$, $n = \text{card}(\mathfrak{N}) < \text{index}(S)$, and $d(\mathfrak{N},S) = 0$. This means that for any f and $\tilde{f}$ such that $\tilde{f} \in V(f)$, $\|S(\tilde{f}(x)) - \alpha\| \to 0$ as x tends to $\alpha = S(f)$. Since $\tilde{f}$ and S are continuous, $S(\tilde{f}(\alpha)) = \alpha$. The information $\mathfrak{N}$ is linear and belongs to $\text{Lip}(0)$. This yields that $h(x) = \tilde{f}(x) - f$ is an element of $\ker \mathfrak{N}(\cdot,x) = \{h: L_j(h,x) = 0, j = 1, 2, \ldots, n\}$, where the linear functionals $L_j \in \text{Lip}(0)$. Take $\alpha \in S(\mathfrak{I}_0)$ such that $\text{codim} \ker A(\alpha) = \text{index}(S)$. Let g_0 be an arbitrary element of $\ker \mathfrak{N}(\cdot,\alpha)$. From Lemma 4.4, it follows that there exists a function $h = h(x)$ such that $\mathfrak{N}(h(x),x) = 0$, $h \in \text{Lip}(0)$, and $h(\alpha) = g_0$. Define $\tilde{f}(x) = f + ch(x)$ for $|c| < \delta(f)/\|g_0\|$, where $S(f) = \alpha$. Then $\tilde{f} \in V(f)$ and $\tilde{f}(\alpha) = f + cg_0$. Thus, $S(f + cg_0) = S(f)$ which yields that $g_0 \in G(\alpha)$ and $\ker \mathfrak{N}(\cdot,\alpha) \subset G(\alpha)$. From (6.4), we get $n = \text{codim} \ker \mathfrak{N}(\cdot,\alpha) \geq \text{codim} A(\alpha) = \text{index}(S)$. This is a contradiction. Hence, $d(\mathfrak{N},S) > 0$ which completes the proof.   ∎

Theorem 6.1 states that the limiting diameter of information $d(\mathfrak{N},S)$ is nonzero for any linear information operator with cardinality less than $\text{index}(S)$. From Theorems 5.1 and 6.1, we get

Corollary 6.1   The class of iterative algorithms $\text{IT}(\mathfrak{N},S)$ is empty for any linear information operator $\mathfrak{N}$ such that $\mathfrak{N} \in \text{Lip}(0)$ and $\text{card}(\mathfrak{N}) < \text{index}(S)$.   ∎

We now show that there exists a linear information operator $\mathfrak{N}$ such that $\mathfrak{N} \in \text{Lip}(0)$, $\text{card}(\mathfrak{N}) = \text{index}(S)$, and $d(\mathfrak{N},S) = 0$. Recall that $A(\alpha)$ is defined by (6.4). We assume that the domain $\mathfrak{I}_0$ of S is chosen in such a way that $\text{codim} A(\alpha)$ does not change for $\alpha \in S(\mathfrak{I}_0)$, i.e., $\text{index}(S) = \text{codim} A(\alpha)$, $\forall \alpha \in S(\mathfrak{I}_0)$. We apply Lemma 4.3 for $A(x)$, where $x \in X = S(\mathfrak{I}_0)$. This yields a linear information operator

(6.7) $\mathfrak{N}^*(f,x) = [L_1^*(f,x), \ldots, L_{n^*}^*(f,x)]$

such that $L_1^*(\cdot,x), \ldots, L_{n^*}^*(\cdot,x)$ are linearly independent linear functionals for every x, $n^* = \text{card}(\mathfrak{N}) = \text{index}(S)$, and $\ker \mathfrak{N}^*(\cdot,x) = A(x)$, $\forall x \in S(\mathfrak{I}_0)$. We shall call $\mathfrak{N}^*$ a *basic linear information operator*.

Definition 6.2   We shall say a *solution operator S belongs to the class* $\text{Lip}(0)$ (briefly $S \in \text{Lip}(0)$) iff there exists $\mathfrak{N}^*$ of the form (6.7) such that $\mathfrak{N}^* \in \text{Lip}(0)$.   ∎

Compare with Definition 4.3. Thus, $S \in \text{Lip}(0)$ means that linear functionals whose kernels form $A(x)$ satisfy a Lipschitz condition.

Theorem 6.2 Assume that $S \in \mathrm{Lip}(0)$. Then $\mathfrak{N}^*$ defined by (6.7) is a convergent linear information operator, i.e., $d(\mathfrak{N}^*,S) = 0$. ∎

PROOF Take any $f \in \mathfrak{I}_0$, $\alpha = S(f)$. Let $\tilde{f}$ be any function, $\tilde{f} \in \mathrm{Lip}(0)$, such that $\tilde{f} \in V(f)$. Let $h(x) = \tilde{f}(x) - f$. Then $\|h(x)\| < \delta(f)$ and $h(x)$ belongs to $\ker \mathfrak{N}^*(\cdot,x) = A(x)$. Thus $\overline{\lim}_{x \to \alpha}\|S(\tilde{f}(x)) - S(f)\| = \|S(f + h(\alpha)) - S(f)\|$. Since $h(\alpha) \in A(\alpha)$ and $A(\alpha) \subset G(\alpha)$, we get $S(f + ch(\alpha)) = S(f)$ for $|c| < \delta(f)/\|h(\alpha)\|$. But $\|h(\alpha)\| < \delta(f)$ and setting $c = 1$, we get $S(\tilde{f}(\alpha)) = S(f)$. This proves that $d(\mathfrak{N}^*,S) = 0$ which means that $\mathfrak{N}^*$ is convergent. ∎

Theorems 6.1 and 6.2 state that it is necessary and sufficient to use linear information operators with cardinality at least equal to the index of the problem S if we wish to compute α by a convergent iterative algorithm.

We proved that the basic linear information operator $\mathfrak{N}^*$ is convergent. We now show that provided a certain technical assumption holds any convergent linear information operator contains $\mathfrak{N}^*$. This means that the information $\mathfrak{N}^*$ must be computed. (See Definition 4.1.)

Theorem 6.3 Let $G(\alpha)$ be a linear set for all $\alpha \in S(\mathfrak{I}_0)$ and let $\mathfrak{N}$ be a linear information operator such that $\mathfrak{N} \in \mathrm{Lip}(0)$. Then $d(\mathfrak{N},S) = 0$ implies $\mathfrak{N}^* \subset \mathfrak{N}$. ∎

PROOF Since $G(\alpha)$ is linear, $\ker \mathfrak{N}^*(\cdot,\alpha) = G(\alpha)$, $\forall \alpha \in S(\mathfrak{I}_0)$. Take any α and f such that $\alpha = S(f)$. Since $d(\mathfrak{N},S) = 0$, then $S(\tilde{f}(\alpha)) = S(f)$ for any $\tilde{f} \in V(f)$. Let g_0 be an arbitrary element of $\ker \mathfrak{N}(\cdot,\alpha)$. From Lemma 4.4, we know that there exists a function $h \in \mathrm{Lip}(0)$ such that $\mathfrak{N}(h(x),x) = 0$ and $h(\alpha) = g_0$. Define $\tilde{f}(x) = f + ch(x)$ for $|c| < \delta(f)/\|h(\alpha)\|$. Then $\tilde{f} \in V(f)$ and hence $S(f + ch(\alpha)) = S(f)$. This proves that $h(\alpha) = g_0$ belongs to $G(\alpha)$ which yields $\ker \mathfrak{N}(\cdot,\alpha) \subset G(\alpha) = \ker \mathfrak{N}^*(\cdot,\alpha)$. Hence $\mathfrak{N}^* \subset \mathfrak{N}$ which completes the proof.

We discuss the implications of Theorems 6.1–6.3 from a computational point of view. What is the most general form of linear information operators which is permissible? Consider an idealized model in which *every* linear functional is a primitive. Then every linear information operator $\mathfrak{N}$ defined by a *finite* number of linear functionals, i.e., $\mathrm{card}(\mathfrak{N}) < +\infty$, is permissible. However, even this idealized model of permissible information does not help for problems with infinite index. (See also Section 8.)

As we now show in several examples, the index may be infinite. Then linear information operators with finite cardinality do not supply enough information to solve the problem with an iterative algorithm.

Example 6.1 Assume that S is a one-to-one operator. Then $G(\alpha) = \{0\}$ and codim $G(\alpha) = \dim \mathfrak{I}_1$. Hence

$$(6.8) \qquad \qquad \mathrm{index}(S) = \dim \mathfrak{I}_1.$$

Thus, if $\dim \mathfrak{I}_1 = +\infty$, then the problem S cannot be solved by an iterative algorithm using a linear information operator with finite cardinality.

Many problems can be defined by a one-to-one operator on an infinite-dimensional space $\mathfrak{I}_1$. One instance is provided by the approximation problem $S(f) = f$ and $\mathfrak{I}_1 = C[a,b]$. As a second instance, consider the solution of any linear differential equation $D\alpha(x) = f_1(x)$ for $x \in \Omega$ and $\alpha(x) = f_2(x)$ for $x \in \partial\Omega$. Then $S(f) = \alpha$, where $f = (f_1, f_2)$ is a one-to-one operator. This shows that such approximation and differential equation problems cannot be solved iteratively by means of linear information operators with finite cardinality. ∎

Example 6.2 Let $\mathfrak{I}_0 = \mathfrak{I}_1 = C[a,b]$ and let k be a positive integer. Define

$$(6.9) \qquad\qquad S(f) = \int_0^1 [f(t)]^k \, dt.$$

Since S is analytic, (6.2) yields $h \in G(\alpha)$ iff $S^{(j)}(f)h^j = 0$ for $j = 1, 2, \ldots$, where $\alpha = S(f)$. Note that

$$S^{(k-1)}(f)h^{k-1} = k! \int_0^1 f(t)h^{k-1}(t)\, dt, \qquad S^{(k)}(f)h^k = k! \int_0^1 h^k(t)\, dt$$

and $S^{(j)}(f) \equiv 0$ for $j > k$.
Assume that $k = 1$. Then

$$G(\alpha) = \left\{ h : \int_0^1 h(t)\, dt = 0 \right\} \qquad \text{and} \qquad \text{codim } G(\alpha) = 1.$$

Thus index$(S) = 1$. This means there exists a linear information operator $\mathfrak{N}$ with card$(\mathfrak{N}) = 1$ such that $d(\mathfrak{N},S) = 0$. Indeed, the basic information $\mathfrak{N}^*$ defined by (6.7) is now given by $\mathfrak{N}^*(f,x) = L_i^*(f,x) = \int_0^1 f(t)\, dt$ and $d(\mathfrak{N}^*,S) = 0$. Furthermore, the order of information $p(\mathfrak{N}^*,S) = +\infty$ and we can solve this directly. In this case, $k = 1$, the solution operator S is a linear functional and can also be used as a linear information operator whose cardinality is equal to unity. (Of course, if we want to approximate S we rule out $\mathfrak{N}^* = S$ as a permissible information operator.)

Assume that $k \geq 2$. Then if k is even, $S^{(k)}(f)h^k = 0$ implies $h = 0$. If k is odd, then $S^{(k-1)}(f)h^{k-1} = 0$ for a positive f implies $h = 0$. Thus, $G(\alpha) = \{0\}$ and

$$\text{index}(S) = \text{codim } G(\alpha) = +\infty.$$

Hence it is impossible to iteratively approximate the value of $\int_0^1 f^k(t)\, dt$ by means of linear information operators with finite cardinality, for $k \geq 2$. ∎

Example 6.3 *Nonlinear Equations* Let $\mathfrak{I}_1$ be the class of analytic operators, f, $f: D \subset B_1 \to B_2$, where B_1 and B_2 are real or complex Banach spaces and $N = \dim(B_1) = \dim(B_2)$. We assume that D is an open set and $\|f\| = \sup_{x \in D} \|f(x)\| < +\infty$ for $f \in \mathfrak{I}_1$. Let $\mathfrak{I}_0$ be the class of analytic operators with a unique zero which is simple, i.e., $f \in \mathfrak{I}_0$ iff $f \in \mathfrak{I}_1$ and there exists $\alpha \in D$ such that $f(\alpha) = 0$ and $f'(\alpha)^{-1}$ exists and is bounded, and α is a unique zero of f.

Define

(6.10) $S(f) = f^{-1}(0), \qquad \mathfrak{I}_2 = B_1.$

Thus, $\alpha = S(f)$ means $f(\alpha) = 0$ and the problem S is that of finding the solution of nonlinear equations. It is easy to verify that $G(\alpha) = \{h \in \mathfrak{I}_1 : h(\alpha) = 0\}$ and

$$\text{index}(S) = \dim B_1 = N.$$

Thus, if $N < +\infty$ we can find a convergent linear information operator with cardinality equal to N. Indeed, the basic information $\mathfrak{N}^*$ is now given by

(6.11) $\mathfrak{N}^*(f,x) = f(x) = [f_1(x), f_2(x), \dots, f_N(x)],$

where $f_j : D \to \mathbb{C}$ are the components of f.

For $N = +\infty$, the index of the problem S is equal to infinity. This means that we have to use linear information operators with infinite cardinality to assure the existence of iterative algorithms. For instance, $\mathfrak{N}(f,x) = f(x)$ is a convergent linear information operator. ∎

7.   THE mth-ORDER ITERATIONS

In the previous section, we proved that if we use linear information operators with cardinality less than the index of the problem S, the class $IT(\mathfrak{N}, S)$ is empty.

Let m be a real number, $m \geq 1$. We now pose and answer the following question: What is the minimal cardinality which assures the existence of an mth order iteration? (See Definition 2.6.) From Theorem 2.4 follows that we seek a linear information operator $\mathfrak{N}$ with minimal cardinality such that the order of information $p(\mathfrak{N}, S)$ is at least equal to m. Let

(7.1) $k = k(m) = \lceil m \rceil - 1.$

In this section, we assume that $S(\mathfrak{I}_0)$ is an open subset of $\mathfrak{I}_2$ and W is the class of functions $\tilde{f}$ such that (5.5) holds and the condition (5.6) is strengthened by the assumption that $\tilde{f} \in \text{Lip}(k)$.

Let H be the class of functions h such that $h : S(\mathfrak{I}_0) \to \mathfrak{I}_1$ and $h \in \text{Lip}(k)$. Define

$$(7.2) \quad G(m) = \left\{ h : h \in H \text{ and } \forall f \in \mathfrak{I}_0, \lim_{x \to S(f)} \frac{\|S(f + ch(x)) - S(f)\|}{\|x - S(f)\|^{m-\eta}} = 0, \right.$$

$$\left. \forall \eta > 0, \forall |c| < \delta(f) / \|h(S(f))\| \right\},$$

where $\delta(f)$ is defined by (5.2). Note that $S(f + ch(x))$ is well defined for x close to $\alpha = S(f)$ since $\|ch(x)\| < \delta(f)\|h(x)\| / \|h(\alpha)\| = \delta(f)(1 + o(1))$. Let $g(x) = S(f + ch(x)) - S(f)$. Then (7.2) implies that $g^{(j)}(\alpha) = 0$ for $j = 0, 1, \dots, k(m)$. If S is sufficiently regular, then $g^{(j)}(\alpha) = 0$ involves conditions on $h(\alpha), h'(\alpha), \dots,$

$h^{(k)}(\alpha)$. Thus, $G(m)$ is the set of functions h for which $S(f + ch(x)) - S(f)$ has a zero α of multiplicity k.

Suppose that $A(x)$ is a linear subspace of $\mathfrak{I}_1$ such that codim $A(x) \equiv$ const for every $x \in S(\mathfrak{I}_0)$. Define

(7.3) $A = \{h : h \in H \text{ and } h(x) \in A(x), \forall x \in S(\mathfrak{I}_0)\}$.

The set A contains functions from the class H whose values at x belong to $A(x)$. Clearly A is a linear subspace of H. Denote

(7.4)   codim $A \overset{\text{df}}{=}$ codim $A(x)$.

Let $A(H)$ be the class of all sets of the form (7.3). We are ready to define the mth index of S.

Definition 7.1   We shall say that index(S,m) is the mth *index of the problem S* iff

(7.5) $\text{index}(S,m) = \min_{A \subset G(m) \text{ and } A \in A(H)} \text{codim } A. \quad\blacksquare$

Theorem 7.1   Let $\mathfrak{N} : \mathfrak{I}_1 \times S(\mathfrak{I}_0) \to \mathfrak{I}_3$ be an arbitrary linear information operator such that $\mathfrak{N} \in \text{Lip}(k(m))$ and card$(\mathfrak{N}) < \text{index}(S,m)$. Then

(7.6) $p(\mathfrak{N},S) < m. \quad\blacksquare$

PROOF   Assume that there exists an information operator $\mathfrak{N}$ such that $\mathfrak{N} \in \text{Lip}(k(m))$, $n = \text{card}(\mathfrak{N}) < \text{index}(S,m)$ and $p(\mathfrak{N},S) \geq m$. Let

$$\mathfrak{N} \asymp [L_1, L_2, \dots, L_n],$$

where $L_i \in \text{Lip}(k)$, $i = 1, 2, \dots, n$. Define

(7.7) $K = \{h : h \in H \text{ and } L_i(h(x),x) = 0, i = 1, 2, \dots, n, \forall x \in S(\mathfrak{I}_0)\}$.

Since $L_1(\cdot,x), \dots, L_n(\cdot,x)$ are linearly independent for every $x \in S(\mathfrak{I}_0)$, then $K \in A(H)$ and codim $K = n$. We show that $K \subset G(m)$. Let h be an arbitrary element of K. Take any $f \in \mathfrak{I}_0$, $\alpha = S(f)$, and define $\tilde{f}(x) = f + ch(x)$ for $|c| < \delta(f)/\|h(\alpha)\|$. Then $\tilde{f}(x) \in \mathfrak{I}_0$ for x close to α and $\mathfrak{N}(\tilde{f}(x),x) = \mathfrak{N}(f,x)$ since $h(x) \in \ker \mathfrak{N}(\cdot,x)$. Thus $\tilde{f} \in V(f)$. Since $p(\mathfrak{N},S) \geq m$, Definition 2.5 yields $S(f + ch(x)) - S(f) = o(\|x - \alpha\|^{m - \eta})$ for every $\eta > 0$. This means that $h \in G(m)$ and implies $K \subset G(m)$. From (7.5), we get

index$(S,m) \leq$ codim $K = n = \text{card}(\mathfrak{N})$.

This is a contradiction. Hence $p(\mathfrak{N},S) < m$ which completes the proof. $\blacksquare$

Theorem 7.1 states that any regular linear information operator $\mathfrak{N}$ with order of information $p(\mathfrak{N},S)$ greater or equal to m has to have cardinality at least equal to the mth index of the problem S.

We now construct a linear information operator $\mathfrak{N}$, $\mathfrak{N} \in \mathrm{Lip}(k)$ and $\mathrm{card}(\mathfrak{N}) = \mathrm{index}(S,m)$, such that $p(\mathfrak{N},S) \geq m$.

Let $A^* = \{h : h \in H$ and $h(x) \in A^*(x),\ \forall x \in S(\mathfrak{I}_0)\}$ be a linear subspace of H such that the minimum in (7.5) is attained for A^*, i.e.,

(7.8) $\qquad A^* \subset G(m) \qquad$ and $\qquad \mathrm{codim}\ A^* = \mathrm{index}(S,m)$.

This means that $\mathrm{codim}\ A^*(x) = \mathrm{index}(S,m)$ for every $x \in S(\mathfrak{I}_0)$. Decompose $\mathfrak{I}_1 = A^*(x) \oplus A^*(x)^{\perp}$, i.e., for every $f \in \mathfrak{I}_1$,

(7.9) $$f = f_0 + \sum_{i=1}^{n^*} L_i^*(f,x)\xi_i(x),$$

where $n^* = \mathrm{index}(S,m)$, $f_0 \in A^*(x)$ and $L_1^*(\cdot,x), \ldots, L_{n^*}^*(\cdot,x)$ are linearly independent linear functionals for every $x \in S(\mathfrak{I}_0)$.

Let

(7.10) $\qquad\qquad \mathfrak{N}_m^*(f,x) = [L_1^*(f,x), \ldots, L_{n^*}^*(f,x)]$

be an mth *basic linear information operator*.

Definition 7.2 We shall say a *solution operator S belongs to the class* $\mathrm{Lip}(k(m))$, $S \in \mathrm{Lip}(k(m))$, iff there exists $\mathfrak{N}_m^*$ of the form (7.10) such that $\mathfrak{N}_m^* \in \mathrm{Lip}(k(m))$. ∎

Note that $\mathfrak{N}^* \in \mathrm{Lip}(k)$ means that $L_i^* \in \mathrm{Lip}(k)$ for $i = 1, 2, \ldots, n^*$. Clearly $\mathrm{card}(\mathfrak{N}_m^*) = \mathrm{index}(S,m)$. From (7.9), we have $L_i^*(\xi_j(x),x) = \delta_{ij}$ and we can assume that the functions $\xi_j \in \mathrm{Lip}(k)$.

We estimate the order $p(\mathfrak{N}_m^*,S)$ of the mth basic linear information $\mathfrak{N}_m^*$.

Theorem 7.2 Assume that $S \in \mathrm{Lip}(k(m))$. Then

(7.11) $\qquad\qquad\qquad p(\mathfrak{N}_m^*,S) \geq m$. ∎

PROOF Take any $f \in \mathfrak{I}_0$, $\alpha = S(f)$, and any $\tilde{f} \in V(f)$. Then $h(x) = \tilde{f}(x) - f$ belongs to $\ker \mathfrak{N}_m^*(\cdot,x)$, i.e., $L_i^*(h(x),x) \equiv 0$. Since $\tilde{f} \in \mathrm{Lip}(k)$, then also $h \in \mathrm{Lip}(k)$. From this and (7.9), we get that $h(x) \in A^*(x)$ for every $x \in S(\mathfrak{I}_0)$. This means $h \in A^*$. Since $A^* \subset G(m)$, $h \in G(m)$. Note that $\|h(\alpha)\| < \delta(f)$ which implies that we can put $c = 1$ in (7.2), getting

$$\|S(\tilde{f}(x)) - S(f)\| = o(\|x - \alpha\|^{m-\eta}) \qquad \forall \eta > 0.$$

This proves that $p(\mathfrak{N}_m^*,S) \geq m$. ∎

Theorems 7.1 and 7.2 state that we have to use linear information operators with cardinality at least equal to the mth index to assure the existence of an mth order iteration. If $\mathrm{index}(S,m) = +\infty$, then linear information operators with finite cardinality do not supply enough information to guarantee the existence of an mth order iteration.

As in Section 6, we prove that in some cases any linear information operator $\mathfrak{N}$ with order of information at least equal to m contains the information $\mathfrak{N}_m^*$.

Theorem 7.3   Let $G(m)$ be a linear set such that $G(m) = A^*$. Let $\mathfrak{N}$ be a linear information operator such that $\mathfrak{N} \in \mathrm{Lip}(k(m))$. Then $p(\mathfrak{N},S) \geq m$ implies $\mathfrak{N}_m^* \subset \mathfrak{N}$. ∎

PROOF   Take any $f \in \mathfrak{I}_0$, $\alpha = S(f)$. Since $p = p(\mathfrak{N},S) \geq m$, then

$$S(\tilde{f}(x)) - S(f) = o(\|x - \alpha\|^{p-\eta}) \qquad \forall \eta > 0,$$

for any $\tilde{f} \in V(f)$. Let g_0 be an arbitrary element of ker $\mathfrak{N}(\cdot,\alpha)$. From Lemma 4.4, we can find a function h such that $h \in \mathrm{Lip}(k)$, $\mathfrak{N}(h(x),x) \equiv 0$, and $h(\alpha) = g_0$. Since $S(f + ch(x)) - S(f) = o(\|x - \alpha\|^{m-\eta})$, $\forall \eta > 0$, and $|c| < \delta(f)/\|h(\alpha)\|$, then $h \in G(m) = A^*$. Thus, $h(x) \in A^*(x)$, i.e., $L_i^*(h(\alpha),\alpha) \equiv 0$. This implies that $g_0 = h(\alpha) \in$ ker $\mathfrak{N}_m^*(\cdot,\alpha)$ and ker $\mathfrak{N}(\cdot,\alpha) \subset$ ker $\mathfrak{N}_m^*(\cdot,\alpha)$. Since α is arbitrary, $\mathfrak{N}_m^* \subset \mathfrak{N}$ which completes the proof. ∎

Compare with Theorem 4.2 in Traub and Woźniakowski [76b] in which a similar problem is considered for nonlinear equations.

We illustrate the concept of the mth index by two examples.

Example 7.1   *Nonlinear Equations*   As in Example 6.3, define $S(f) = f^{-1}(0)$. Let $h \in H$. Now $h(x)$ is a function from $\mathfrak{I}_1$, i.e., $h(x) = h(x,\cdot)$. It is easy to verify that

$$G(m) = \left\{ h : \frac{\partial^j h(\alpha,\alpha)}{\partial x^j} = 0, j = 0, 1, \ldots, k(m), \forall \alpha \in S(\mathfrak{I}_0) \right\},$$

where $k = k(m)$ is defined by (7.1).

This means that $h(x,t) = O((x - t)^{k+1})$ belongs to $G(m)$. The mth index is given by

$$\mathrm{index}(S,m) = N \binom{N + k}{k},$$

where $N, N < +\infty$, is the dimension of the problem. For $N = +\infty$, $\mathrm{index}(S,m) = +\infty$. The information $\mathfrak{N}_m^*$ is now given by

$$\mathfrak{N}_m^*(f,x) = \{ f(x), f'(x), \ldots, f^{(k)}(x) \}$$

and was intensively studied by Traub and Woźniakowski [76b, 77a,b]. ∎

Example 7.2   *Linear Equations*   Consider, as in Example 7.1, $S(f) = f^{-1}(0)$ with the additional assumption that f is an affine function, i.e., $f(t) = At - b$, where A is a nonsingular $N \times N$ matrix and b is an $N \times 1$ vector. Then

$$G(m) = \left\{ h : \frac{\partial^j h(\alpha,\alpha)}{\partial x^j} = 0, j = 0,1, \ldots, k(m), \forall \alpha \in S(\mathfrak{I}_0) \right\},$$

where $h(x,t)$ is a linear function of t. It is easy to check that

$$\text{index}(S,m) = N(N + 1) \qquad \forall m > 1.$$

From this follows that every algorithm φ which uses any linear information operator $\mathfrak{N}$ with cardinality less than $N(N + 1)$ has order $p(\varphi) \leq p(\mathfrak{N},S) \leq 1$. Furthermore, every interpolatory algorithm φ^I which uses the linear information operator $\mathfrak{N}^*(f,x) = [f(x),f'(x)]$, $\text{card}(\mathfrak{N}^*) = N(N + 1)$, has order $p(\varphi^I) = p(\mathfrak{N}^*,S) = +\infty$. Since $f(x) = Ax - b$ and $f'(x) \equiv A$, f is fully determined by $\mathfrak{N}^*(f,x)$ and φ^I requires the solution of a linear system. Hence φ^I is a direct algorithm. ∎

8. COMPLEXITY INDEX FOR ITERATIVE LINEAR INFORMATION

We specify our model of computation for iterative linear information operators which is similar to the model for the linear case in Part A.

Model of Computation for Iterative Linear Information

(i) Let P be a given collection of primitives. We assume that the addition of two elements of $\mathfrak{I}_2$, $f + g$, and the multiplication of an element of $\mathfrak{I}_2$ by a scalar, cf, are primitive operations which belong to P. We also assume that every linear functional $L(\cdot,x)$ is a primitive operation which belongs to P for every x under consideration. This implies that any linear information operator $\mathfrak{N} = [L_1,L_2,\ldots,L_n]$ of finite cardinality is *permissible*, where $L_1, L_2, \ldots, L_n$ are arbitrary linear functionals.

(ii) To normalize the complexity measure, we assume that the cost of the addition of two elements of $\mathfrak{I}_2$ and the multiplication of an element of $\mathfrak{I}_2$ by a scalar is taken as unity. Assume that the complexity of evaluating a linear functional $L(\cdot,x)$ does not depend on x and let $\text{comp}(L) = \text{comp}(L(f,x))$. Let $\mathfrak{N} = [L_1,\ldots,L_n]$ be a linear information operator, $\text{card}(\mathfrak{N}) = n$. We assume that $\mathfrak{N}(f,x)$ is computed by the independent evaluation of $L_1(f,x), \ldots, L_n(f,x)$ and the information complexity of $\mathfrak{N}$ is given by

$$(8.1) \qquad \text{comp}(\mathfrak{N}) = \sum_{i=1}^{n} \text{comp}(L_i).$$

If $\text{comp}(L_i) \equiv c_1$, then $\text{comp}(\mathfrak{N}) = nc_1$ which shows how the information complexity depends on the cardinality of $\mathfrak{N}$.

(iii) Let φ be a permissible algorithm which uses $\mathfrak{N}$ and finds an ε'-approximation to $\alpha = S(f)$. Let $\text{comb}(\varphi)$ be the combinatory complexity of φ. For all problems of practical interest, φ has to use every $L_i(f,x)$, $i = 1, 2, \ldots, n$, and the current approximation x at least once and therefore $\text{comb}(\varphi) \geq n$. We rule out special problems and information operators, assuming that $\text{comb}(\varphi) \geq n$ for every algorithm under consideration. ∎

We analyze the complexity index of iterative algorithms using a linear information operator $\mathfrak{N}$. Assume that $p(\mathfrak{N},S)$ is greater than unity, where $\mathfrak{N} = [L_1, \ldots, L_n]$. Let φ be a permissible algorithm from the class $\Phi_{\text{perm}}(\mathfrak{N},S)$ and $p(\varphi) > 1$. (See Section 3.) Therefore, φ also belongs to the class $\text{IT}(\mathfrak{N},S)$ of iterative algorithms. (See Corollary 5.1.) The complexity index $z(\varphi)$ is defined (compare with (3.11)) as

$$(8.2) \qquad z(\varphi) = \frac{\text{comp}(\mathfrak{N}) + \text{comb}(\varphi)}{\log p(\varphi)},$$

where $\text{comp}(\mathfrak{N}) = \sum_{i=1}^{n} \text{comp}(L_i)$ is the information complexity and

$$\text{comb}(\varphi) = \sup_{(f,x)} \text{comp}(\varphi(x,\mathfrak{N}(f,x)))$$

is the combinatory complexity of φ. For simplicity, assume that $\text{comp}(L_i) \equiv c_1$. From Theorem 2.3, we know that $p(\varphi) \leq p(\mathfrak{N},S)$. Since $\text{comb}(\varphi) \geq n$, this yields

$$(8.3) \qquad z(\varphi) \geq \frac{nc_1 + n}{\log p(\mathfrak{N},S)}.$$

Furthermore, Theorem 2.4 guarantees the existence of algorithms whose order is equal to the order of information. Assume that one of these maximal order algorithms is permissible. Let $\overline{\text{comb}}(\mathfrak{N},S)$ denote the minimal combinatory complexity of algorithms with maximal order, i.e.,

$$(8.4) \qquad \overline{\text{comb}}(\mathfrak{N},S) = \inf_{\varphi : p(\varphi) = p(\mathfrak{N},S)} \text{comb}(\varphi).$$

Let $\underline{\text{comb}}(\mathfrak{N},S)$ denote the minimal combinatory complexity of algorithms φ from $\Phi_{\text{perm}}(\mathfrak{N},S)$ with $p(\varphi) > 1$, i.e.,

$$(8.5) \qquad \underline{\text{comb}}(\mathfrak{N},S) = \inf_{\varphi : p(\varphi) > 1} \text{comb}(\varphi).$$

Of course, $n \leq \underline{\text{comb}}(\mathfrak{N},S) \leq \overline{\text{comb}}(\mathfrak{N},S)$. From (8.4) and (8.5), we get bounds on the complexity index $z(\mathfrak{N},S)$ of the problem S with the information $\mathfrak{N}$. (See Definition 3.1.)

Lemma 8.1

$$(8.6) \qquad \frac{nc_1 + \underline{\text{comb}}(\mathfrak{N},S)}{\log p(\mathfrak{N},S)} \leq z(\mathfrak{N},S) \leq \frac{nc_1 + \overline{\text{comb}}(\mathfrak{N},S)}{\log p(\mathfrak{N},S)}. \qquad \blacksquare$$

Note that if $nc_1 \gg \overline{\text{comb}}(\mathfrak{N},S)$, then

$$(8.7) \qquad z(\mathfrak{N},S) \cong \frac{nc_1}{\log p(\mathfrak{N},S)},$$

and every maximal order algorithm φ is close to a minimal complexity index algorithm since $z(\varphi) \cong nc_1/\log p(\mathfrak{N},S)$.

The preceding discussion motivates the following problem. For fixed n, find a linear information operator $\mathfrak{N}$ with card$(\mathfrak{N}) \leq n$ and maximal order of information. Let Ψ_n be the class of all linear information operators $\mathfrak{N}$ which are sufficiently regular and card$(\mathfrak{N}) \leq n$. We assume that S is also sufficiently regular.

Definition 8.1 We shall say $p(n,S)$ is the nth *maximal order of the problem* S iff

$$(8.8) \qquad\qquad p(n,S) = \sup_{\mathfrak{N} \in \Psi_n} p(\mathfrak{N},S).$$

We shall say $\mathfrak{N}^{mo}$, $\mathfrak{N}^{mo} \in \Psi_n$, is an nth *maximal order information for the problem* S iff

$$(8.9) \qquad\qquad p(\mathfrak{N}^{mo},S) = p(n,S). \quad \blacksquare$$

The results of Sections 6 and 7 enable us to find the nth maximal order and the nth maximal order information. Recall that index(S,m) is the mth index of the problem S, $m \geq 1$. (See Definition 7.) Note that index(S,m) is a nondecreasing function of m.

Suppose first that index$(S,1) = +\infty$. Then index$(S,m) \equiv +\infty$ and Theorem 7.1 yields $p(\mathfrak{N},S) = 0$ for every $\mathfrak{N}$ with card$(\mathfrak{N}) < +\infty$. Thus, $p(n,S) \equiv 0$ and every linear information operator satisfies (8.9).

Suppose then that

$$(8.10) \qquad\qquad n_0 = \text{index}(S,1) < +\infty.$$

From Theorem 7.1, we immediately get $p(n,S) = 0$ for $n < n_0$. For $n \geq n_0$, define

$$(8.11) \qquad\qquad q(n,S) = \inf\{m : n < \text{index}(S,m)\}$$

with the convention $q(n,S) = +\infty$, if index$(S,m) \leq n$, $\forall m$. The function $q(n,S)$ is a nondecreasing function of n and $q(n_0,S) \geq 1$. We are ready to prove

Theorem 8.1 Let $n_0 = \text{index}(S,1)$. Then

$$(8.12) \qquad\qquad p(n,S) = \begin{cases} 0 & \text{for} \quad n < n_0, \\ q(n,S) & \text{for} \quad n \geq n_0. \end{cases} \quad \blacksquare$$

PROOF It suffices to prove (8.12) for $n_0 < +\infty$ and $n \geq n_0$. Let $q = q(n,S)$. Take any positive η. From (8.11) index$(S, q - \eta) \leq n$. Theorem 7.2 guarantees that there exists a linear information operator $\mathfrak{N}$ such that card$(\mathfrak{N}) = $ index$(S, q - \eta) \leq n$ and $p(\mathfrak{N},S) \geq q - \eta$. Since η is arbitrary, $p(n,S) \geq q = q(n,S)$.

From (8.11), we have $n < \text{index}(S, q + \eta)$. Theorem 7.1 states that for any $\mathfrak{N} \in \Psi_n$, $p(\mathfrak{N},S) < q + \eta$. Thus $p(n,S) \leq q$. Hence, $p(n,S) = q(n,S)$ which completes the proof. $\blacksquare$

Theorem 8.1 states the dependence of the nth maximal order $p(n,S)$ on the problem S. Since we assume S and the linear information operators $\mathfrak{N}$ are sufficiently regular, the nth maximal order $p(n,S)$ is an integer. We find an nth maximal order information for $n \geq n_0 = \text{index}(S,1)$, where $n_0 < +\infty$. Let $\eta \in (0,1)$ and let $\mathfrak{N}_n$ be a linear information operator such that $\text{card}(\mathfrak{N}_n) = \text{index}(S, p(n,S) - \eta) \leq n$ and $p(\mathfrak{N}_n,S) \geq p(n,S) - \eta > p(n,S) - 1$. The existence of $\mathfrak{N}_n$ is guaranteed by Theorem 7.2. Since $p(\mathfrak{N}_n,S)$ is an integer, we get $p(\mathfrak{N}_n,S) = p(n,S)$. This proves the following theorem.

Theorem 8.2   Let $n_0 = \text{index}(S,1) < +\infty$. For $n \geq n_0$, $\mathfrak{N}_n$ is an nth maximal order information for the problem S. ∎

We examine the complexity index $z(\mathfrak{N},S)$ of the information operators $\mathfrak{N}$ from the class Ψ_n.

Definition 8.2   We shall say $z(n,S)$ is the nth *minimal complexity index of the problem S* iff

$$(8.13) \qquad\qquad z(n,S) = \inf_{\mathfrak{N} \in \Psi_n} z(\mathfrak{N},S).$$

We shall say that $n_{\text{opt}} = n_{\text{opt}}(S)$ is the *optimal cardinality number of the problem S with respect to the complexity index* iff

$$(8.14) \qquad\qquad z(n_{\text{opt}},S) = \inf_n z(n,S).$$

We shall say a linear information operator $\mathfrak{N}^{\text{oi}}$ is an *optimal information operator for the problem S* iff

$$(8.15) \qquad \text{card}(\mathfrak{N}^{\text{oi}}) = n_{\text{opt}}(S) \qquad \text{and} \qquad p(\mathfrak{N}^{\text{oi}},S) = p(n_{\text{opt}},S). \; ∎$$

Let $\underline{\text{comb}}(n,S) = \inf_{\mathfrak{N} \in \Psi_n} \text{comb}(\mathfrak{N},S)$ and $\overline{\text{comb}}(n,S) = \inf_{\mathfrak{N} \in \Psi'_n} \text{comb}(\mathfrak{N},S)$ (where $\Psi'_n = \{\mathfrak{N} : \mathfrak{N} \in \Psi_n$ and $p(\mathfrak{N},S) = p(n,S)\}$) be the minimal combinatory complexity of $\text{comb}(\mathfrak{N},S)$ and the minimal combinatory complexity of $\text{comb}(\mathfrak{N},S)$ for the information operators with the maximal order $p(n,S)$ respectively. (See (8.4) and (8.5).) From Lemma 8.1 and Theorems 8.1 and 8.2, we get the following estimates of the nth minimal complexity index.

Lemma 8.2

$$(8.16) \qquad \frac{nc_1 + \underline{\text{comb}}(n,s)}{\log p(n,S)} \leq z(n,S) \leq \frac{nc_1 + \overline{\text{comb}}(n,S)}{\log p(n,S)}. \; ∎$$

Observe that (8.16) is simpler if $nc_1 \gg \overline{\text{comb}}(n,S)$. Then

$$(8.17) \qquad\qquad z(n,S) \cong \frac{nc_1}{\log p(n,S)}.$$

This shows that the nth minimal complexity index depends on the cardinality of information and the nth maximal order. To find the optimal cardinality number, one seeks the minimum of the function $f(n) = z(n,S)$. The value of $n_{opt}(S)$ depends only on how fast the functions $\underline{comb(n,S)}$, $\overline{comb(n,S)}$, and $p(n,S)$ tend to infinity with n. Knowing $n_{opt}(S)$, one can easily find an optimal information operator $\mathfrak{R}^{oi}$ from Theorem 8.2. We conclude this section by an example.

Example 8.1 As in Example 6.3, consider the solution of nonlinear equations, $S(f) = f^{-1}(0)$, where N is the dimension of the problem. $N < +\infty$. Traub and Woźniakowski [77b] showed that

$$n_{opt}(S) = \begin{cases} 3 & \text{for } N = 1, \text{ the scalar case,} \\ N(N+1) & \text{for } N \geq 2, \text{ the multivariate case.} \end{cases}$$

The optimal information operator $\mathfrak{R}^{oi}(f,x) = [f(x), f'(x), f''(x)]$ for $N = 1$ and $\mathfrak{R}^{oi}(f,x) = [f(x), f'(x)]$ for $N \geq 2$. ∎

9. INFORMATION OPERATOR WITH MEMORY

In this section, we briefly indicate how the concepts of the limiting diameter and order of information can be generalized for information operators with memory. As always, we want to approximate the solution $\alpha = S(f)$. Suppose that $x_0, x_1, \ldots, x_r, r \geq 1$, are known distinct approximations to α. Let

(9.1) $$\mathfrak{R}: D_{\mathfrak{R}} \subset \mathfrak{I}_1 \times \mathfrak{I}_2^{r+1} \to \mathfrak{I}_3$$

be an information operator with memory (not necessarily linear), where $(f, x_0, x_1, \ldots, x_r) \in D_{\mathfrak{R}}$ for all $f \in \mathfrak{I}_0$ and all distinct $x_0, x_1, \ldots, x_r$ sufficiently close to α, and $\mathfrak{I}_3$ is a given space. The parameter r measures the size of the memory used by $\mathfrak{R}$. In general, $\mathfrak{R}(f, x_0, x_1, \ldots, x_r)$ does not uniquely define the solution α and many different problem elements have the same information at $x_0, x_1, \ldots, x_r$ as f. We define the concept of equality with respect to $\mathfrak{R}$.

Definition 9.1 We shall say $\tilde{f} = \tilde{f}(x_0, x_1, \ldots, x_r)$ *is equal to* f *with respect to* $\mathfrak{R}$ iff

(i) $\tilde{f}: D_{\tilde{f}} \subset \mathfrak{I}_2^{r+1} \to \mathfrak{I}_0$, $\tilde{f} \in W$,

where W is a given space and there exists $\Gamma = \Gamma(f) > 0$ such that

$$[J(\Gamma)]^{r+1} \subset D_{\tilde{f}}, \text{ where } J(\Gamma) = \{x : \|x - \alpha\| \leq \Gamma\} \text{ with } \alpha = S(f),$$

(ii) $\mathfrak{R}(\tilde{f}(x_0, x_1, \ldots, x_r), x_0, x_1, \ldots, x_r) = \mathfrak{R}(f, x_0, x_1, \ldots, x_r)$

$$\forall (x_0, x_1, \ldots, x_r) \in D_{\tilde{f}}.$$

For brevity, we write $\tilde{f} \in V(f)$, where $V(f)$ is the set of all functions $\tilde{f}$ which are equal to f with respect to $\mathfrak{R}$. ∎

The space W describes the regularity of $\tilde{f}$ with respect to $x_0, x_1, \ldots, x_r$. The limiting diameter $d(\mathfrak{R}, S)$ of $\mathfrak{R}$ for the problem S is defined as follows.

Definition 9.2 We shall say $d(\mathfrak{N},S)$ is the *limiting diameter of information* $\mathfrak{N}$ *for the problem S* iff

$$(9.2) \quad d(\mathfrak{N},S) = \sup_{f \in \mathfrak{I}_0} \sup_{\tilde{f}_1,\tilde{f}_2 \in V(f)} \limsup_{\substack{x_i \to \alpha \\ i = 0,1,\ldots,r}} \left\| S(\tilde{f}_1(x_0,\ldots,x_r)) - S(\tilde{f}_2(x_0,\ldots,x_r)) \right\|.$$

We shall say the *information* $\mathfrak{N}$ *is convergent for the problem S* (or simply $\mathfrak{N}$ is convergent) iff

$$(9.3) \qquad\qquad\qquad d(\mathfrak{N},S) = 0. \quad \blacksquare$$

Let

$$(9.4) \qquad\qquad \varphi : D_\varphi \subset \mathfrak{I}_2^{r+1} \times \mathfrak{N}(D_\mathfrak{N}) \to \mathfrak{I}_2$$

be a *stationary algorithm with memory* which generates the sequence of approximations by

$$(9.5) \qquad x_{j+1} = \varphi(x_j, x_{j-1}, \ldots, x_{j-r}, \mathfrak{N}(f, x_j, x_{j-1}, \ldots, x_{j-r}))$$

for $j = r, r+1, \ldots$. Note that φ uses r previously computed approximations and the information $\mathfrak{N}$ computed at them. Usually, $\mathfrak{N}$ consists of "new information" at x_j and reuses previously computed information at $x_{j-1}, \ldots, x_{j-r}$. Let $\Phi(\mathfrak{N},S)$ be the class of *all* stationary algorithms with memory. We want to examine the convergence of the sequence $\{x_j\}$ to α. As in Section 2, we define the limiting error of φ as follows.

Definition 9.3 We shall say $e(\varphi)$ is the *limiting error of algorithm* φ iff

$$(9.6) \qquad e(\varphi) = \sup_{f \in \mathfrak{I}_0} \sup_{\tilde{f} \in V(f)} \limsup_{\substack{x_i \to \alpha \\ i = 0,1,\ldots,r}} \left\| \varphi(x_0, \ldots, x_r, \mathfrak{N}(f, x_0, \ldots, x_r)) \right.$$

$$\left. - S(\tilde{f}(x_0, \ldots, x_r)) \right\|.$$

The algorithm φ is called convergent iff $e(\varphi) = 0$. $\quad \blacksquare$

Using the same argument as in Theorem 2.1 it is easy to prove

Theorem 9.1 For any algorithm φ, $\varphi \in \Phi(\mathfrak{N},S)$,

$$(9.7) \qquad\qquad\qquad e(\varphi) \geq \tfrac{1}{2} d(\mathfrak{N},S). \quad \blacksquare$$

Thus, $\tfrac{1}{2} d(\mathfrak{N},S)$ is the inherent error of information $\mathfrak{N}$ for any algorithm φ. The diameter $d(\mathfrak{N},S)$ is an upper bound on "interpolatory algorithms" which are defined analogously as in Section 2.

Definition 9.4 We shall say φ^{I}, $\varphi^{\mathrm{I}} \in \Phi(\mathfrak{N},S)$, is an *interpolatory algorithm* iff

$$(9.8) \qquad \varphi(x_0, \ldots, x_r, \mathfrak{N}(f, x_0, \ldots, x_r)) = S(\tilde{f}(x_0, \ldots, x_r))$$

for some $\tilde{f} \in V(f)$. $\quad \blacksquare$

This means that an interpolatory algorithm φ constructs as the new approximation to α the exact solution of a problem element $\tilde{f}(x_0, \ldots, x_r)$ which shares the same information as f at $x_0, \ldots, x_r$. In practice, $\tilde{f}(x_0, \ldots, x_r)$ should be "simpler" than f. It is straightforward to prove

Theorem 9.2 For any interpolatory algorithm φ^1, $\varphi^1 \in \Phi(\mathfrak{N}, S)$,

$$(9.9) \qquad e(\varphi^1) \leq d(\mathfrak{N}, S). \quad \blacksquare$$

Thus, there exists a convergent algorithm in $\Phi(\mathfrak{N}, S)$ iff $\mathfrak{N}$ is convergent. Assume then that $\mathfrak{N}$ is a convergent information operator. We examine how fast $S(\tilde{f}(x_0, \ldots, x_r))$ tends to $S(f)$ as x_i approaches α for $i = 0, 1, \ldots, r$. We generalize the concept of order of information as follows. Let A be a set of real $(r + 1)$-tuples defined by

$$(9.10) \quad A = \left\{ (q_0, q_1, \ldots, q_r) : q_i \geq 0, \sum_{i=0}^{r} q_i \geq 1 \text{ such that} \right.$$

$$\forall f \in \mathfrak{I}_0, \alpha = S(f), \text{ and } \forall \tilde{f} \in V(f) \text{ we have}$$

$$\left. \lim_{||x_0 - \alpha|| \leq \cdots \leq ||x_r - \alpha|| \to 0} \frac{||S(\tilde{f}(x_0, \ldots, x_r)) - S(f)||}{||x_0 - \alpha||^{q_0 - \eta} \cdot \cdots \cdot ||x_r - \alpha||^{q_r - \eta}} = 0, \forall \eta > 0 \right\}.$$

Suppose for a moment that A is nonempty and define

$$(9.11) \quad t(A) = \sup\{t : t^{r+1} = q_0 t^r + q_1 t^{r-1} + \cdots + q_r \text{ for } (q_0, q_1, \ldots, q_r) \in A\}.$$

Note that the polynomial $t^{r+1} - (q_0 t^r + \cdots + q_r)$ for $q_i \geq 0$ and $\sum_{i=0}^{r} q_i \geq 1$ has a unique positive zero $t_1 \geq 1$. Therefore, $t(A) \geq 1$.

Definition 9.5 We shall say $p(\mathfrak{N}, S)$ is the *order of information $\mathfrak{N}$ for the problem S* iff

$$(9.12) \qquad p(\mathfrak{N}, S) = \begin{cases} 0 & \text{if } A \text{ is empty,} \\ t(A) & \text{otherwise.} \end{cases} \quad \blacksquare$$

For $r = 0$, i.e., for information operators without memory, the set A coincides with the set defined by (2.17) and $t(A) = \sup\{q : q \in A\}$. Therefore, Definition 9.5 for $r = 0$ coincides with Definition 2.5. See also Kung and Traub [76a] in which a similar technique for measuring the speed of convergence is proposed.

We state a theorem which often helps the calculation of the order of information $p(\mathfrak{N}, S)$.

Theorem 9.3 If there exist $p_0, p_1, \ldots, p_r$ such that $p_i \geq 0$, $\sum_{i=0}^{r} p_i \geq 1$, and

(i) for all $f \in \mathfrak{I}_0$ and all $\tilde{f} \in V(f)$,

$$(9.13) \quad ||S(\tilde{f}(x_0, \ldots, x_r)) - S(f)|| \leq c_1(f, \tilde{f})||x_0 - \alpha||^{p_0} \cdot \cdots \cdot ||x_r - \alpha||^{p_r};$$

for all $x_0, \ldots, x_r$ sufficiently close to $\alpha = S(f)$, $||x_0 - \alpha|| \leq \cdots \leq ||x_r - \alpha||$, and $c_1(f, \tilde{f}) < +\infty$, then

$$(9.14) \qquad p(\mathfrak{N}, S) \geq p,$$

where p is the unique positive zero of $t^{r+1} - (p_0 t^r + \cdots + p_r)$.

(ii)   If, additionally, there exist $f \in \mathfrak{I}_0$ and $\tilde{f} \in V(f)$ such that

(9.15) $\|S(\tilde{f}(x_0, \ldots, x_r)) - S(f)\| \geq c_2(f, \tilde{f})\|x_0 - \alpha\|^{p_0} \cdot \ldots \cdot \|x_r - \alpha\|^{p_r}$

for all $x_0, \ldots, x_r$ sufficiently close to $\alpha = S(f)$, $\|x_0 - \alpha\| \leq \cdots \leq \|x_r - \alpha\|$, and $c_2(f, \tilde{f}) > 0$, then

(9.16) $p(\mathfrak{N}, S) = p.$ ∎

PROOF   From (9.13), it follows that $(p_0, p_1, \ldots, p_r) \in A$. Thus, $p(\mathfrak{N}, S) = t(A) \geq p$ due to (9.11). This proves (9.14).

Suppose that (9.15) holds. Let $(q_0, q_1, \ldots, q_r) \in A$. Then

(9.17) $c_2(f, \tilde{f})\|x_0 - \alpha\|^{p_0} \cdot \ldots \cdot \|x_r - \alpha\|^{p_r} \leq \|S(\tilde{f}(x_0, \ldots, x_r)) - S(f))\|$

$$= o(\|x_0 - \alpha\|^{q_0 - \eta} \cdot \ldots \cdot \|x_r - \alpha\|^{q_r - \eta})$$

$$\forall \eta > 0.$$

We need the following lemma.

Lemma 9.1   Let $\omega(t) = p_0 t^r + \cdots + p_r$, $u(t) = q_0 t^r + \cdots + q_r$, where $p_i \geq 0$, $q_i \geq 0$ and $\sum_{i=0}^{r} q_i \geq 1$. Let p and q be the unique positive zeros of $t^{r+1} - \omega(t)$ and $t^{r+1} - u(t)$, respectively. Then

(9.18) $p_0 + p_1 + \cdots + p_i \geq q_0 + q_1 + \cdots + q_i$ $\forall i \in [0, r]$,

implies $p \geq q$.   ∎

PROOF   Let $\omega(t) = \omega(1) + \omega'(1)(t - 1) + \cdots + (1/r!)\omega^{(r)}(1)(t - 1)^r$. Note that $\omega^{(k)}(1) = \sum_{i=0}^{r-k} p_i c_i$, where $c_i = (r - i)!/(r - i - k)!$ for $i = 0, 1, \ldots, r - k$. Set $c_{r-k+1} = 0$. It is easy to see that $c_i > c_{i+1}$ for $i = 0, 1, \ldots, r - k$. Multiply the inequality $p_0 + p_1 + \cdots + p_j \geq q_0 + q_1 + \cdots + q_j$ by $(c_j - c_{j+1})$ and add for $j = 0, 1, \ldots, r - k$. Then

$$\omega^{(k)}(1) = \sum_{i=0}^{r-k} p_i c_i = \sum_{i=0}^{r-k} p_i \sum_{j=i}^{r-k} (c_j - c_{j+1}) = \sum_{j=0}^{r-k} (c_j - c_{j+1}) \sum_{i=0}^{j} p_i$$

$$\geq \sum_{j=0}^{r-k} (c_j - c_{j+1}) \sum_{i=0}^{j} q_i = u^{(k)}(1)$$

for $k = 0, 1, \ldots, r - 1$. This implies $\omega(t) \geq u(t)$ for $t \geq 1$. Set $t = q$. Then $\omega(q) \geq u(q) = q^r$. Thus $q^r - \omega(q) \leq 0$ which yields $p \geq q$.   ∎

Let i be any integer from $[0, r]$. Assume that $x_{i+1}, \ldots, x_r$ are fixed and let $x_1 = x_2 = \cdots = x_i$ tend to α. Then (9.17) yields $p_0 + p_1 + \cdots + p_i \geq q_0 + q_1 + \cdots + q_i$. From Lemma 9.1 and (9.11), we get $p(\mathfrak{N}, S) \leq \sup q \leq p$. Due to (9.14), $p(\mathfrak{N}, S) = p$ which proves (9.16).   ∎

Let φ be a convergent algorithm from $\Phi(\mathfrak{N}, S)$. The order of φ is defined in a similar way as the order of information. Let B be a set of real $(r + 1)$-tuple sequences defined by

(9.19) $B = \left\{ (q_0, q_1, \ldots, q_r) : q_i \geq 0; \sum_{i=0}^{r} q_i \geq 1 \text{ such that} \right.$

$\forall f \in \mathfrak{I}_0, \alpha = S(f), \text{ and } \forall \tilde{f} \in V(f), \text{ we have}$

$$\lim_{\|x_0 - \alpha\| \leq \ldots \leq \|x_r - \alpha\| \to 0} \frac{\left\| \varphi(x_0, \ldots, x_r, \mathfrak{N}(f, x_0, \ldots, x_r)) - S(\tilde{f}(x_0, \ldots, x_r)) \right\|}{\|x_0 - \alpha\|^{q_0 - \eta} \cdot \cdots \cdot \|x_r - \alpha\|^{q_r - \eta}} = 0$$

$\left. \forall \eta > 0 \right\}.$

Suppose for a moment that B is nonempty and recall that

$$t(B) = \sup\{t : t^{r+1} = q_0 t^r + q_1 t^{r-1} + \cdots + q_r \text{ for } (q_0, \ldots, q_r) \in B\}.$$

Definition 9.6 We shall say $p(\varphi)$ is the *order of algorithm* φ iff

(9.20) $$p(\varphi) = \begin{cases} 0 & \text{if } B \text{ is empty,} \\ t(B) & \text{otherwise.} \end{cases} \quad \blacksquare$$

For $r = 0$, Definition 9.6 is equivalent to Definition 2.6 since $p(\varphi) = t(B) = \sup B$ for nonempty B. Note that in (9.19) we compare the value of $\varphi(x_0, \ldots, x_r, \mathfrak{N}(f, x_0, \ldots, x_r))$ to every solution element $S(\tilde{f}(x_0, \ldots, x_r))$, where $\tilde{f} \in V(f)$. The "classical" definition of the order is based on the comparison between $\varphi(x_0, \ldots, x_r, \mathfrak{N}(f, x_0, \ldots, x_r))$ and the solution $\alpha = S(f)$. We show in the appendix that for all algorithms of practical interest the "classical" order is equal to $p(\varphi)$.

It is obvious that a result analogous to Theorem 9.3 can be stated for an algorithm φ. Namely, if

(9.21) $$\|\varphi(x_0, \ldots, x_r, \mathfrak{N}(f, x_0, \ldots, x_r)) - S(\tilde{f}(x_0, \ldots, x_r))\|$$
$$= O(\|x_0 - \alpha\|^{p_0} \cdot \cdots \cdot \|x_r - \alpha\|^{p_r})$$

for $p_i \geq 0$ and $\sum_{i=0}^{r} p_i \geq 1$, then the order $p(\varphi) \geq p$, where p is the unique positive zero of $t^{r+1} - (p_0 t^r + \cdots + p_r)$. Furthermore, if (9.21) is sharp, then $p(\varphi) = p$.

We now prove that the order of an algorithm cannot exceed the order of information.

Theorem 9.4 For any algorithm φ, $\varphi \in \Phi(\mathfrak{N}, S)$,

(9.22) $$p(\varphi) \leq p(\mathfrak{N}, S). \quad \blacksquare$$

PROOF Compare with the proof of Theorem 2.3. Without loss of generality, we can assume $B \neq \varnothing$. Let $(q_0, q_1, \ldots, q_r)$ be any element of B. Let $f \in \mathfrak{I}_0$ and

$\tilde{f} \in V(f)$. Then

$$\|S(\tilde{f}(x_0, \ldots, x_r)) - S(f)\| \le \|\varphi(x_0, \ldots, x_r, \mathfrak{N}(f, x_0, \ldots, x_r)) - S(\tilde{f}(x_0, \ldots, x_r))\|$$
$$+ \|\varphi(x_0, \ldots, x_r, \mathfrak{N}(f, x_0, \ldots, x_r)) - S(f)\|$$
$$= o(\|x_0 - \alpha\|^{q_0 - \eta} \cdot \cdots \cdot \|x_r - \alpha\|^{q_r - \eta}) \qquad \forall \eta > 0.$$

This proves that $(q_0, q_1, \ldots, q_r) \in A$ (see (9.10)) and $t(B) \le t(A)$. This yields $p(\varphi) \le p(\mathfrak{N}, S)$. ∎

The bound $p(\mathfrak{N}, S)$ is achieved by the order of any interpolatory algorithm.

Theorem 9.5   For any interpolatory algorithm φ^1, $\varphi^1 \in \Phi(\mathfrak{N}, S)$,

(9.23) $p(\varphi^1) = p(\mathfrak{N}, S)$. ∎

PROOF   The method of proof is similar to that used in Theorems 2.4 and 9.4. ∎

Theorems 9.4 and 9.5 prove that any interpolatory algorithm achieves the maximal order $p(\mathfrak{N}, S)$ in the class $\Phi(\mathfrak{N}, S)$.

Example 9.1   *Nonlinear Equations*   As in Example 6.3, define $S(f) = f^{-1}(0)$. Consider

(9.24) $\mathfrak{N}(f, x_0, \ldots, x_r) = [f(x_0), f'(x_0), \ldots, f^{(k)}(x_0), \ldots, f(x_r),$
$$f'(x_r), \ldots, f^{(k)}(x_r)]$$

for a given $k \ge 0$. Let W be the class of analytic functions. Thus, $\tilde{f} \in W$ means that $\tilde{f}(x, t)$ is analytic with respect to x and t.

CASE 1 $N = 1$.   Thus, f is a scalar function of a real or complex scalar variable. Then $\tilde{f} \in V(f)$ yields that $f(x, t) = f(t) + G(x, t) \prod_{i=0}^{r} (x - x_i)^{k+1}$ for an analytic function $G(x, t)$. It is easy to verify that the order of information $p(\mathfrak{N}, S) = p(k, r)$, where $p(k, r)$ is the unique positive zero of the polynomial $t^{r+1} - (k + 1) \sum_{j=0}^{r} t^j$ and $k + 1 \le p(k, r) < k + 2$, $\lim_{r \to +\infty} p(k, r) = k + 2$. It is well known that the interpolatory algorithm $\varphi_{r,k}^1$ is now defined as follows:

(i)   find an interpolatory polynomial ω of degree $\le (k + 1)(r + 1) - 1$ such that

$$\omega^{(i)}(x_j) = f^{(i)}(x_j), \qquad i = 0, 1, \ldots, k, \quad j = 0, 1, \ldots, r,$$

(ii)   define $x_{r+1} = \varphi_{r,k}^1(x_0, \ldots, x_r, \mathfrak{N}(f, x_0, \ldots, x_r))$ as a zero of ω with a certain criterion of its choice (for instance, the nearest zero to x_0).

For some values of k and r, we get the known iterations. For example, $k = 0$ and $r = 1$ yields the secant iteration, $k = 1$ and $r = 0$ yields the Newton iteration. For a detailed discussion, see Traub [64].

CASE 2 $N \geq 2$.   Thus, f is a multivariate or an abstract function. From Theorem 5 in Woźniakowski [74] follows

$$(9.25) \qquad p(\mathfrak{N},S) = k + 1 \qquad \forall r.$$

This means that $p(\mathfrak{N},S)$ does not depend on r and the information contained in $f^{(i)}(x_j)$ for $i = 0, 1, \ldots, k$ and $j = 1, 2, \ldots, r$ does not help to increase the order. However, if a certain position of $x_1, x_2, \ldots, x_r$ is assumed, then $p(\mathfrak{N},S)$ can be larger than $k + 1$ for $r \geq N + 1$. Examples include multivariate secant iteration ($k = 0, r = N$), and the generalization of the interpolatory iteration $\varphi_{r,k}^1$. See, for instance, Brent [72], Jankowska [75], Ortega and Rheinboldt [70], Pleshakov [77], and Woźniakowski [74]. ∎

We briefly discuss the complexity of an algorithm φ which uses an information operator $\mathfrak{N}$ with memory. Suppose that for given initial approximations $x_0, x_1, \ldots, x_r$, the algorithm φ produces the sequence $\{x_i\}$,

$$x_{i+1} = \varphi(x_i, \ldots, x_{i-r}, \mathfrak{N}(f,x_i, \ldots, x_{i-r})),$$

such that

$$(9.26) \quad e_i = G_i e_{i-1}^{p_0} e_{i-2}^{p_1} \cdots \cdot e_{i-r-1}^{p_r}, \quad e_i = \|x_i - \alpha\|, \quad i = r + 1, \ldots,$$

where $p_i \geq 0$ and $q = \sum_{i=0}^r p_i > 1$. Let p be the unique positive zero of $t^{r+1} - (p_0 t^r + \cdots + p_r)$. Assume that the constants $G_i = G_i(f)$ satisfy the relation

$$(9.27) \qquad 0 < \underline{G} \leq G_i \leq \bar{G} < +\infty.$$

To simplify the complexity analysis, we assume that there exist constants Γ, $\Gamma < 1$, C_1, and C_2 such that

$$(9.28) \qquad C_1 \Gamma^{p^i} \leq e_i \leq C_2 \Gamma^{p^i} \qquad \text{for} \quad i = 1, 2, \ldots, r,$$

$$(9.29) \qquad C_1^{1-q} < \underline{G}, \qquad C_2^{1-q} > \bar{G}.$$

Then, it is easy to verify that

$$(9.30) \qquad C_1 \Gamma^{p^i} \leq e_i \leq C_2 \Gamma^{p^i} \qquad \forall i.$$

Recall that we want to find x_k such that k is the smallest integer for which $e_k \leq \varepsilon' e_0$ for given $\varepsilon' \in (0,1)$. Let ε, $\varepsilon \leq \varepsilon'$, be such that $e_k = \varepsilon e_0$. From (9.30), we find

$$(9.31) \qquad \frac{\log(\log(C_1/\varepsilon e_0)/\log(1/\Gamma))}{\log p} \leq k \leq \frac{\log(\log(C_2/\varepsilon e_0)/\log(1/\Gamma))}{\log p}.$$

For small ε, $k \cong (\log \log 1/\varepsilon)/\log p$. Let $\text{comp}(\varphi,f)$ be the complexity of computing x_k from $x_0, x_1, \ldots, x_r$. Assume that the cost of every iterative step is equal to $c(\varphi,f)$. Then

$$(9.32) \qquad \text{comp}(\varphi,f) = kc(\varphi,f) = z(\varphi,f)g(k,\varepsilon),$$

where $g(k,\varepsilon) = k \log p = (\log \log 1/\varepsilon)(1 + O(\varepsilon))$ and

$$(9.33) \qquad\qquad\qquad z(\varphi,f) = c(\varphi,f)/\log p$$

is the *complexity index of φ for f.*

We discuss the cost of one iterative step. To perform the ith step, we have to compute $\mathfrak{N}(f,x_{i-1},x_{i-2},\ldots,x_{i-r-1})$ and next $\varphi(x_{i-1},\ldots,x_{i-r-1}, \mathfrak{N}(f,x_{i-1},\ldots,x_{i-r-1}))$. Assume that

$$(9.34) \quad \mathfrak{N}(f,x_{i-1},\ldots,x_{i-r-1}) = [\overline{\mathfrak{N}}(f,x_{i-1}),\overline{\mathfrak{N}}(f,x_{i-2}),\ldots,\overline{\mathfrak{N}}(f,x_{i-r-1})]$$

for a certain permissible information operator $\overline{\mathfrak{N}}$ without memory. Thus, the information complexity $\mathrm{comp}(\mathfrak{N}(f,x_{i-1},\ldots,x_{i-r-1})) = \mathrm{comp}(\overline{\mathfrak{N}}(f,x_{i-1}))$ since we reuse the previously computed information $\overline{\mathfrak{N}}(f,x_{i-2}),\ldots,\overline{\mathfrak{N}}(f,x_{i-r-1})$. This means that the information complexity does not depend on r, the size of memory. Let $\mathrm{comp}(\mathfrak{N}) \equiv \mathrm{comp}(\overline{\mathfrak{N}}(f,x_i))$.

Let $\mathrm{comb}(\varphi,r)$ denote the combinatory complexity of an algorithm φ. Then

$$(9.35) \qquad\qquad\qquad z(\varphi,f) = \frac{\mathrm{comp}(\mathfrak{N}) + \mathrm{comb}(\varphi,r)}{\log p}.$$

We seek an algorithm with the minimal complexity index. Since the order of an algorithm φ is no larger than the order of information $p(\mathfrak{N},S)$ and there exist algorithms φ with $p(\varphi) = p(\mathfrak{N},S)$, we get

$$(9.36) \qquad \frac{\mathrm{comp}(\mathfrak{N}) + \underline{\mathrm{comb}(\mathfrak{N},r)}}{\log p(\mathfrak{N},S)} \le \inf_{\varphi} z(\varphi,f) \le \frac{\mathrm{comp}(\mathfrak{N}) + \overline{\mathrm{comb}(\mathfrak{N},r)}}{\log p(\mathfrak{N},S)},$$

where $\underline{\mathrm{comb}(\mathfrak{N},r)}$ is the minimal combinatory complexity of algorithms which use information $\mathfrak{N}$ and have order greater than unity, $\overline{\mathrm{comb}(\mathfrak{N},r)}$ is the minimal combinatory complexity of algorithms with the maximal order $p(\mathfrak{N},S)$. If $\mathrm{comp}(\mathfrak{N}) \gg \overline{\mathrm{comb}(\mathfrak{N},r)}$, then

$$(9.37) \qquad\qquad\qquad \inf_{\varphi} z(\varphi,f) \cong \frac{\mathrm{comp}(\mathfrak{N})}{\log p(\mathfrak{N},S)}.$$

Note, however, that the inequality $\mathrm{comp}(\mathfrak{N}) \gg \overline{\mathrm{comb}(\mathfrak{N},r)}$ can usually hold only for small r or for "sufficiently hard-to-compute" f.

10.   EXTENSIONS AND OPEN PROBLEMS

We conclude Part B by a partial list of extensions and open problems which will be studied in the future.

1.   We show that if the index of a problem is infinite, the problem cannot be solved iteratively by a one-point linear information operator with finite cardinality. What is the characterization of all problems with finite index? Furthermore, what is the characterization of all problems with index(S) = s for

a given integer s? We are also interested in the same question for the mth index of a linear information operator with or without memory.

We proved that finite-dimensional nonlinear equations $S(f) = f^{-1}(0)$ can be solved by iterative algorithms of finite order using linear information operators with finite cardinality. Is that the most general form of S which can be solved iteratively? The following example shows this not to be the case.

Example 10.1 Let $f:[0,1] \to \mathbb{R}$ be a smooth scalar function. Define

$$[F(f)](t) = [Af](t) + d(t,f(t)), \qquad t \in [0,1],$$

where A is a linear operator and d a smooth function of two variables. Let $\mathfrak{J}_0$ be a class of functions f for which the equation $[F(f)](t) = 0$ has a unique and simple zero in $[0,1]$. Define the problem

(10.1) $$S(f) = g([F(f)]^{-1}(0)),$$

where g is a smooth one-to-one function and its inverse g^{-1} is a Lipschitz function. Thus, (10.1) means find the solution β of the equation $[F(f)](t) = 0$ and then compute $\alpha = g(\beta)$. Consider the one-point linear information operator

$$\mathfrak{N}(f,x) = [[Af](y),[Af]'(y),f(y),f'(y)], \qquad \text{where} \quad y = g^{-1}(x).$$

Note that $\operatorname{card}(\mathfrak{N}) \leq 4$. Knowing $\mathfrak{N}(f,x)$, we can compute $[F(f)](y) = [Af](y) + d(y,f(y))$ and $[F(f)]'(y) = [Af]'(y) + \partial_1 d(y,f(y)) + \partial_2 d(y,f(y))f'(y)$ and then apply Newton iteration to approximate β. (∂_i denotes the partial derivative with respect to the ith argument.) Consider the algorithm

$$\varphi(x,\mathfrak{N}(f,x)) = g(y - [F(f)](y)/[F(f)]'(y)).$$

Then

$$\varphi(x,\mathfrak{N}(f,x)) - \alpha = g(\beta + O((y - \beta)^2)) - g(\beta) = O((g^{-1}(x) - g^{-1}(\alpha))^2)$$
$$= O((x - \alpha)^2).$$

This proves that the order of φ and the order of information are at least equal to 2. Thus, problem (10.1) can be solved by iteration. As a particular example, set $Af = f^{(j)}$, $d \equiv 0$, and $g(x) = x$. Then $\alpha = S(f)$ is the unique and simple solution of the equation $f^{(j)}(t) = 0$. ∎

We wish to find the most general form of S which can be solved iteratively and propose the following conjecture.

Let D_0 and D_1 be open subsets of $\mathbb{C}^N$, $1 \leq N < +\infty$. Let $\mathfrak{J}_1$ denote the class of functions $f:D_0 \to D_1$.

Conjecture 10.1 If the problem S can be solved by iterative algorithms using one-point linear information operators with finite cardinality, then there exist an operator $F:D_F \subset \mathfrak{J}_1 \to \mathfrak{J}_1$ and a function $g:D_g \subset \mathbb{C}^N \to \mathfrak{J}_2$ such that

(10.2) $$S(f) = g([F(f)]^{-1}(0)) \qquad \forall f \in \mathfrak{J}_0. \qquad ∎$$

Conjecture 10.1 states that essentially only nonlinear equations can be solved by iteration. Indeed, (10.2) means that $\alpha = S(f)$ is the transformed value of β, $\alpha = g(\beta)$, where β is a solution of the transformed nonlinear equation $[F(f)](t) = 0$. Note that in Conjecture 10.1 we do not specify properties of F and g. We merely assume their existence. Example 10.1 provides an example of F and g such that the problem $g([F(f)]^{-1}(0))$ can be solved iteratively. It would be interesting to find the most general form of F and g which permits a problem to be iteratively solved.

In Part A, we showed many apparently diverse problems could be handled within the same general framework. However, if Conjecture 10.1 is true, then essentially only problems that we already knew could be solved by iteration are included within the iterative information model of this book.

2.   We discuss the classification of more general linear information operators than those considered here. These operators are also of practical and theoretical interest. We define an iterative linear information operator as follows. Let

$$(10.3) \qquad \mathfrak{N}(f,x_0,\ldots,x_r) = [L_1(f,\xi_1(x_0)),\ldots,L_n(f,\xi_n(x_0)),\ldots,$$
$$L_1(f,\xi_1(x_r)),\ldots,L_n(f,\xi_n(x_r))],$$

where

$$\xi_1(x) = x,$$
$$\xi_{j+1}(x) = \xi_{j+1}(x,L_1(f,\xi_1(x)),\ldots,L_j(f,\xi_j(x))), \qquad j = 1, 2, \ldots, n-1,$$

and $L_j(f,x)$ is a linear functional with respect to f. Thus, $L_j(f,\xi_j(x))$ depends on the previously computed information. The parameter r measures the size of "memory." For $r = 0$, $\mathfrak{N}$ is an *iterative linear information operator without memory*. For $r \geq 1$, $\mathfrak{N}$ reuses the previously computed information at $x_1, \ldots, x_r$ and is an *iterative linear information operator with memory*. For $\xi_j(x) \equiv x, j = 1, 2, \ldots, n$, we get a *one-point iterative linear information operator* which for $r = 0$ was considered in Sections 4–8. If there exists j such that $\xi_j(x) \not\equiv x$, $\mathfrak{N}$ is a *multipoint iterative linear information operator*, since the information $L_1(f,\xi_1(x)),\ldots,L_n(f,\xi_n(x))$ is computed at least at two different points. Examples of multipoint iterations for nonlinear equations may be found in Brent [76], Kacewicz [75], Kung and Traub [74], Meersman [76a,b], and Woźniakowski [76].

This classification is schematized in Figure 2.

	$r = 0$	$r \geq 1$
$\forall j, \xi_j(x) \equiv x$	One-point without memory	One-point with memory
$\exists j, \xi_j(x) \not\equiv x$	Multipoint without memory	Multipoint with memory

Figure 2

REMARK 10.1   For nonlinear iterative information operators, there is no difference between one-point and multipoint operators. For the nonlinear case, we can distinguish iterative information operators without memory $\mathfrak{N} = \mathfrak{N}(f,x_0)$ which are considered in Section 2 and iterative information operators with memory which are considered in Section 9. ∎

It would be of interest to generalize the results of Sections 4–8 to multipoint iterative linear information operators with or without memory. In particular, we are interested in the questions raised in extension 1 to multipoint iterations or iterations with memory. We would also like to extend the concept and properties of basic linear information operator (see Sections 6 and 7). That is, given a number m, find a linear information operator $\mathfrak{N}^*$ with minimal cardinality of order at least m. Does the conclusion of Theorem 7.3 continue to hold?

3.   What is the minimal number of linear functionals in (10.3) to iteratively solve the system of nonlinear equations $f(x) = 0$, $f : D \subset \mathbb{C}^N \to \mathbb{C}^N$? Newton iteration shows that $O(N^2)$ linear functionals are sufficient. It follows from Lemma 4.3 and Theorem 4.2 in Traub and Woźniakowski [76b] that any iteration based on a one-point linear information operator requires at least the evaluation of f and f'. This holds even if $N = +\infty$. This result is related to informational requirements of convergent price mechanisms in mathematical economics. (See Saari and Simon [78].)

Kacewicz [77] conjectures that $N + cN^2$ linear functionals are needed (without the restriction to one-point iteration), where c is a positive constant. Kacewicz has obtained partial results on the minimal number of linear functionals for $N = 1$ and 2.

4.   In order to derive lower bounds on complexity, we require upper bounds on the order of information for fixed information. In particular, let f be a scalar nonlinear function with a simple zero and let $S(f) = f^{-1}(0)$. Kung and Traub [74] show there exists a multipoint linear information operator using the linear functionals

(10.4) $L_j(f,x) \equiv f^{(k_j)}(x), \qquad k_j \geq 0, \qquad j = 1, 2, \ldots, n,$

such that

(10.5) $p(\mathfrak{N},S) = 2^{n-1}.$

They conjecture that

(10.6) $p(\mathfrak{N},S) \leq 2^{n-1}$

for all linear multipoint information operators which use the functionals of the form (10.4). This conjecture was established for $n = 1, 2$ by Kung and Traub [76b], for $n = 3$ by Meersman [76a,b], and for "Hermite" information with arbitrary n by Woźniakowski [76]. Wasilkowski [77] proves this conjecture holds whenever the information operator is well poised in the sense of Birkhoff complex interpolation.

We generalize the Kung–Traub conjecture.

Conjecture 10.2   Let f be any nonlinear problem with a simple zero and let $S(f) = f^{-1}(0)$. Let $L_1, \ldots, L_n$ be arbitrary linear functionals and let $\xi_1, \ldots, \xi_n$ be arbitrary functions. Then

$$p(\mathfrak{N},S) \le 2^{n-1} \qquad \text{for} \quad r = 0,$$

$$p(\mathfrak{N},S) < 2^n \qquad \text{for} \quad 0 < r < +\infty. \quad \blacksquare$$

See also Kacewicz and Woźniakowski [77] in which the maximal order of information for multipoint iterations is discussed.

11.   COMPARISON OF RESULTS FROM GENERAL AND ITERATIVE INFORMATION MODELS

For the reader's convenience we compare some of the results from the general information model of Part A and the iterative information model of Part B. See Figures 3 and 4 for a summary.

	$p(\mathfrak{N},S) = 0$	$p(\mathfrak{N},S) = 1$	$1 < p(\mathfrak{N},S) < \infty$	$p(\mathfrak{N},S) = \infty$
$d(\mathfrak{N},S)$	>0	0	0	0
$\text{comp}(\mathfrak{N},S,\varepsilon)$	Undefined	$\simeq \lg(1/\varepsilon)$	$\simeq \lg \lg(1/\varepsilon)$	Const

Figure 3.   Part B: iterative information model.

	$r(\mathfrak{N},S) = 0$	$0 < r(\mathfrak{N},S) \le \varepsilon$	$r(\mathfrak{N},S) > \varepsilon$
$p(\mathfrak{N},S)$	∞	0	0
$\text{comp}(\mathfrak{N},S,\varepsilon)$	Const	"Any" monotonically increasing function	Undefined

Figure 4.   Part A: general information model.

In both figures, we give the asymptotic dependence of $\text{comp}(\mathfrak{N},S,\varepsilon)$ as a function of ε. If the problem cannot be solved to within ε, we say the complexity is undefined. In the iterative information model, the order of information, $p(\mathfrak{N},S)$, is basic. Thus, if $p(\mathfrak{N},S) = 0$, the diameter of information is positive and the complexity is undefined. In the general information model, the radius of information, $r(\mathfrak{N},S)$, is basic. Thus, if $r(\mathfrak{N},S) = 0$, then the order is infinite and the complexity is independent of ε.

Order of information was not defined in Part A. However, the definition of Part B can be used if we recognize that in Part A, $\mathfrak{N}(f,x)$ is independent of x. Then the order of information must be either zero or infinite.

APPENDIX

We discuss the relationship between the order $p(\varphi)$ of an algorithm φ defined by (2.21) or (9.20) and the "classical" definition of order which can be stated as follows. Let $p_0, \ldots, p_r$ be real numbers such that $p_i = 0$ or $p_i \geq 1$ and $q = \sum_{i=0}^{r} p_i > 1$. Suppose that φ uses the information operator $\mathfrak{N}(f, x_0, \ldots, x_r)$ for $f \in \mathfrak{I}_0$. Assume that for every $f \in \mathfrak{I}_0$, $\alpha = S(f)$, there exist $\Gamma(f) > 0$ and $c(f) < +\infty$ such that

(A.1)
$$\|\varphi(x_0, \ldots, x_r, \mathfrak{N}(f, x_0, \ldots, x_r)) - \alpha\|$$
$$\leq c(f)\|x_0 - \alpha\|^{p_0}\|x_1 - \alpha\|^{p_1} \cdots \cdot \|x_r - \alpha\|^{p_r}$$

for all $\|x_0 - \alpha\| \leq \|x_1 - \alpha\| \leq \cdots \leq \|x_r - \alpha\| \leq \Gamma(f)$.

Assume that (A.1) is sharp, i.e., there exists $f_0 \in \mathfrak{I}_0$, $\alpha_0 = S(f_0)$, such that

(A.2)
$$\|\varphi(x_0, \ldots, x_r, \mathfrak{N}(x_0, \ldots, x_r, f_0)) - \alpha_0\|$$
$$\geq c_0(f_0)\|x_0 - \alpha_0\|^{p_0}\|x_1 - \alpha_0\|^{p_1} \cdots \cdot \|x_r - \alpha_0\|^{p_r}$$

for all $\|x_0 - \alpha_0\| \leq \cdots \leq \|x_r - \alpha_0\| \leq \Gamma(f_0)$ and $c_0(f_0) > 0$. Then we shall call the unique positive zero p of the polynomial $t^{r+1} - (p_0 t^r + \cdots + p_r)$ the "*classical*" *order* of φ. (See, among others, Traub [64].) Note that for $r = 0$, i.e., $\mathfrak{N}$ is an information operator without memory, $p = p_0$.

It is easy to verify that if

(A.3)
$$c(f)\Gamma(f)^{q-1} < 1$$

and all initial approximations $x_0, x_1, \ldots, x_r$ satisfy

(A.4)
$$e_i = \|x_i - \alpha\| \leq \Gamma(f) \qquad \text{for} \quad i = 0, 1, \ldots, r,$$

then φ generates the sequence $\{x_i\}$, $x_i = \varphi(x_i, \ldots, x_{i-r}, \mathfrak{N}(f, x_i, \ldots, x_{i-r}))$ for $i \geq r + 1$, which has the property

(A.5)
$$\lim_i x_i = \alpha,$$

(A.6)
$$e_i = O(\zeta^{p^i}),$$

for a certain $\zeta < 1$.

The number $\Gamma(f)$ can be interpreted as the radius of a ball of convergence since φ converges for all initial approximations from $J = \{x : \|x - \alpha\| \leq \Gamma(f)\}$. The constant $c(f)$ can be interpreted as the "asymptotic constant" which satisfies

(A.7)
$$\overline{\lim_i} \frac{e_{i+1}}{e_i^{p_0} \cdots \cdot e_{i-r}^{p_r}} \leq c(f).$$

We are ready to prove

Lemma A.1   Let $\mathfrak{N}$ be convergent information. Suppose that (A.1) and (A.2) hold and additionally

$$(A.8) \qquad \Gamma_0 = \lim_{x_0,\ldots,x_r \to \alpha} \Gamma(\tilde{f}(x_0,x_1,\ldots,x_r)) > 0,$$

$$(A.9) \qquad c_0 = \overline{\lim_{x_0,\ldots,x_r \to \alpha}} \, c(\tilde{f}(x_0,\ldots,x_r)) < +\infty$$

for all $f \in \mathfrak{I}_0$ and all $\tilde{f} \in V(f)$. Then $p(\varphi) = p$. ∎

PROOF   Since $\mathfrak{N}$ is convergent, $\tilde{\alpha} = S(\tilde{f}(x_0,x_1,\ldots,x_r))$ tends to $\alpha = S(f)$ as $x_0, x_1, \ldots, x_r$ approach α for every $\tilde{f} \in V(f)$. From (A.8), we get

$$\|x_i - \tilde{\alpha}\| \le \Gamma(\tilde{f}(x_0,\ldots,x_r)), \qquad i = 0, 1, \ldots, r,$$

for sufficiently small $\max_{0 \le i \le r}\|x_i - \alpha\|$. This means that $x_0, \ldots, x_r$ can be treated as approximations to α and $\tilde{\alpha}$. Let $x = \varphi(x_0,\ldots,x_r,\mathfrak{N}(f,x_0,\ldots,x_r))$. From (A.1), we have

$$\|\alpha - \tilde{\alpha}\| \le \|x - \alpha\| + \|x - \tilde{\alpha}\|$$

$$(A.10) \qquad \le c(f) \prod_{i=0}^{r} \|x_i - \alpha\|^{p_i} + c(\tilde{f}(x_0,\ldots,x_r)) \prod_{i=0}^{r} \|x_i - \tilde{\alpha}\|^{p_i}.$$

Choose j such that $p_j \ge 1$. Since $\sum_{i=0}^{r} p_i > 1$ and $c(\tilde{f}(x_0,\ldots,x_r))$ is bounded due to (A.9), then (A.10) can be rewritten

$$\|\alpha - \tilde{\alpha}\| = o(\|x_j - \alpha\|) + o(\|x_j - \alpha\| + \|\alpha - \tilde{\alpha}\|).$$

This yields $\|\alpha - \tilde{\alpha}\| = o(\|x_j - \alpha\|)$. From this and (A.10), we get

$$\|\alpha - \tilde{\alpha}\| = O\left(\prod_{i=0}^{r} \|x_i - \alpha\|^{p_i} \right).$$

This proves that $(p_0,p_1,\ldots,p_r) \in B$ (see (9.19)), and consequently $p \le p(\varphi)$. Set $\tilde{f}(x_0,\ldots,x_r) = f_0$, where f_0 satisfies (A.2). Let $(q_0,q_1,\ldots,q_r) \in B$. Then

$$\lim_{\|x_0-\alpha\| \le \cdots \le \|x_r-\alpha\| \to 0} \frac{\|\varphi(x_0,\ldots,x_r,\mathfrak{N}(f_0,x_0,\ldots,x_r)) - \alpha\|}{\|x_0 - \alpha\|^{q_0-\eta} \cdot \cdots \cdot \|x_r - \alpha\|^{q_r-\eta}} = 0 \qquad \forall \eta > 0.$$

From (A.2), it easily follows that

$$q_0 + q_1 + \cdots + q_i \le p_0 + p_1 + \cdots + p_i \qquad \text{for} \quad i = 0, 1, \ldots, r.$$

From Lemma 9.1, we get $t \in (0,p]$, where $t^{r+1} = q_0 t^r + \cdots + q_r$. Since $(q_0,\ldots,q_r)$ is any element of B, $p(\varphi) = t(B) \le p$. Thus, $p(\varphi) = p$ which completes the proof. ∎

Equations (A.8) and (A.9) state that the radius $\Gamma(\tilde{f}(x_0,\ldots,x_r))$ of the problem element $\tilde{f}(x_0,\ldots,x_r)$ is bounded from below roughly by $\Gamma_0 > 0$, and the asymptotic constant $c(\tilde{f}(x_0, \ldots, x_r))$ is bounded from above roughly by $c_0 <$ $+\infty$. Assumptions (A.8) and (A.9) hold for all algorithms of practical interest since $\Gamma(\tilde{f}(x_0, \ldots, x_r))$ and $c(\tilde{f}(x_0, \ldots, x_r))$ are continuous with respect to $x_0, \ldots, x_r$ and $\Gamma_0 = \Gamma(\tilde{f}(\alpha, \ldots, \alpha)) > 0$, $c_0 = c(\tilde{f}(\alpha, \ldots, \alpha)) < +\infty$. Therefore, for all practical cases, the "new" definition of order coincides with the "classical" one, $p(\varphi) = p$.

Bibliography

AHO, A. V., HOPCROFT, J. E., AND ULLMAN, J. D.
[74] "The Design and Analysis of Computer Algorithms." Addison-Wesley, Reading, Massachusetts, 1974.

BRENT, R. P.
[72] The computational complexity of iterative methods for systems of nonlinear equations, *in Complexity Symp.* (R. E. Miller and J. W. Thatcher, eds.), pp. 61–71. Plenum Press, New York, 1972.
[76] A class of optimal-order zero-finding methods using derivative evaluations, *in* "Analytic Computational Complexity" (J. F. Traub, ed.) pp. 59–75. Academic Press, 1976.

BRENT, R. P., WINOGRAD, S., AND WOLFE, P.
[73] Optimal iterative processes for rootfinding, *Numer. Math.* **20** (1973), 327–341.

JANKOWSKA, J.
[75] Multivariate Secant Method. Ph.D. Thesis, Univ. of Warsaw (1975). See also *SIAM J. Numer. Anal.* **16** (1979), 547–562.

KACEWICZ, B.
[75] Integrals with a Kernel in the Solution of Nonlinear Equations in N Dimensions. Dept. of Computer Science Rep., Carnegie-Mellon Univ. (1975). See also *J. Assoc. Comput. Mach.* **26**, (1979), 239–249.
[76a] An integral-interpolation iterative method for the solution of scalar equations, *Numer. Math.* **26**, (1976), 355–365.
[76b] The use of integrals in the solution of nonlinear equations in N dimensions, *in* "Analytic Computational Complexity" (J. F. Traub, ed.), pp. 127–141. Academic Press, New York, 1976.
[77] Private communication.

KACEWICZ, B., AND WOŹNIAKOWSKI, H.
[77] A survey of recent problems and results in analytic computational complexity, "Mathematical Foundations of Computer Science 77," (J. Gruska, ed.), Lecture Notes in Computer Science No. 53, pp. 93–107. Springer-Verlag, Berlin and New York, 1977.

KNUTH, D. E.
[76] Big omicron and big omega and big theta, *SIGACT News* (April 1976), 18–24.

KUNG, H. T.
[76] The complexity of obtaining starting points for solving operator equations by Newton's method, *in* "Analytic Computational Complexity" (J. F. Traub, ed.), pp. 35–57. Academic Press, New York, 1976.

KUNG, H. T., AND TRAUB, J. F.
[74] Optimal order of one-point and multipoint iterations, *J. Assoc. Comput. Mach.* **21** (1974), 643–651.
[76a] All algebraic functions can be computed fast. Dept. of Computer Science Rep., Carnegie-Mellon Univ. (1976). See also *J. Assoc. Comput. Mach.* **25** (1978), 245–260.
[76b] Optimal order and efficiency for iterations with two evaluations, *SIAM J. Numer. Anal.* **13** (1976), 84–99.

MEERSMAN, R.
[76a] Optimal use of information in certain iterative processes, *in* "Analytic Computational Complexity" (J. F. Traub, ed.), pp. 127–141. Academic Press, New York, 1976.
[76b] On maximal order of families of iterations for nonlinear equations. Doctoral Thesis, Vrije Univ. Brussels, Brussels (1976).

ORTEGA, J. M., and RHEINBOLDT, W. C.
[70] "Iterative Solution of Nonlinear Equations in Several Variables." Academic Press, New York, 1970.

PLESHAKOV, G. N.
[77] On efficiency of the multidimensional interpolation iterations (in Russian), *Zh. Vychisl. Mat. Mat. Fiz.* **17**, (1977), 1153–1160.

SAARI, D. G., AND SIMON, C. P.
[78] Effective price mechanisms, *Econometrica* **46**, (1978), 1097–1125.

TRAUB, J. F.
[61] On functional iteration and calculation of roots, *Preprints Papers, 16th Nat. ACM Conf.,* Session 5A-1, pp. 1–4. Los Angeles, California (1961).
[64] "Iterative Methods for Solution of Equations." Prentice-Hall, Englewood Cliffs, New Jersey, 1964.

TRAUB, J. F., and WOŹNIAKOWSKI, H.
[76a] Strict lower and upper bounds on iterative computational complexity, *in* "Analytic Computational Complexity" (J. F. Traub, ed.), pp. 15–34. Academic Press, New York, 1976.
[76b] Optimal linear information for the solution of nonlinear equations, *in* "Algorithms and Complexity: New Directions and Recent Results" (J. F. Traub, ed.), pp. 103–119. Academic Press, New York, 1976.
[76c] Optimal radius of convergence of interpolatory iterations for operator equations. Dept. of Computer Science Rep., Carnegie-Mellon Univ. (1976). To appear in *Aequationes Math.*
[77a] Convergence and complexity of Newton iteration for operator equations. Dept. of Computer Science Rep., Carnegie-Mellon Univ. (1977). See also *J. Assoc. Comput. Mach.* **26**, (1979), 250–258.
[77b] Convergence and complexity of interpolatory-Newton iteration in a Banach space. Dept. of Computer Science Rep., Carnegie-Mellon Univ. (1977). To appear in *Comp. and Maths. with Appls.*

WASILKOWSKI, G. W.

[77]   N-Evaluation conjecture for multipoint iterations for the solution of scalar nonlinear
       equations. Master's Thesis, Dept. of Mathematics, Univ. of Warsaw (1977). To appear
       in *J. Assoc. Comput. Mach.*

[78]   Can any stationary iteration using linear information be globally convergent? Dept.
       of Computer Science Rep., Carnegie-Mellon Univ. (1978). To appear in *J. Assoc. Comput.
       Mach.*

[79]   Any iteration for polynomial equations using linear information has infinite complexity.
       Dept. of Computer Science Rep., Carnegie-Mellon Univ. (1979).

WERSCHULZ, A. G.

[77a] Maximal order and order of information for numerical quadrature. Mathematics Research
       Rep. 77–2, Univ. of Maryland, Baltimore County (1977). See also *J. Assoc. Comput.
       Mach.* **26** (1979), 527–537.

[77b] Maximal order for approximation of derivatives. Mathematics Research Rep. 77–8. Univ.
       of Maryland Baltimore County (1977). See also *J. of Comput. System Sci.* **18** (1979). 213–217.

WOŹNIAKOWSKI, H.

[72]   On nonlinear iterative processes in numerical methods. Doctoral Thesis, Univ. of Warsaw
       (1972) (in Polish).

[74]   Maximal stationary iterative methods for the solution of operator equations, *SIAM J.
       Numer. Anal.* **11**, (1974), 934–949.

[75]   Generalized information and maximal order of information for operator equations, *SIAM
       J. Numer. Anal.* **12** (1975), 121–135.

[76]   Maximal order of multipoint iterations using *n* evaluations, *in* "Analytic Computational
       Complexity" (J. F. Traub, ed.), pp. 75–107. Academic Press, New York, 1976.

Glossary

We list basic concepts used throughout Part B. We mention a symbol, its meaning, and section reference where the symbol appears for the first time.

Symbol	Meaning	Section, Reference
S	the solution operator, sometimes called the problem, $S:\mathfrak{I}_0 \to \mathfrak{I}_2$, $\mathfrak{I}_0 \subset \mathfrak{I}_1$	2, (2.1)
$\mathfrak{I}_0$	the domain of S	2
$\mathfrak{I}_1$	linear space, $\mathfrak{I}_0 \subset \mathfrak{I}_1$	2
$\mathfrak{I}_2$	the range of S	2
α	the solution element, $\alpha = S(f)$	2
f	the problem element, $f \in \mathfrak{I}_0$	2
ε'	error parameter	2
x_0	a given initial approximation	2
$y = y(f)$	an ε'-approximation, $\|y(f) - \alpha\| \leq \varepsilon'\|x_0 - \alpha\|$	2
$\mathfrak{N}$	the iterative information operator, $\mathfrak{N}:D_{\mathfrak{N}} \to \mathfrak{I}_3$	2
$\mathfrak{I}_3$	the range of $\mathfrak{N}$	2
$\tilde{f}$	a function, $\tilde{f}:D_f \to \mathfrak{I}_0$, $f \in W$, and $\mathfrak{N}(\tilde{f}(x),x) = \mathfrak{N}(f,x)$	2, Def. 2.1
W	the regularity space	2, Def. 2.1
$V(f)$	the set of functions $\tilde{f}$	2, Def. 2.1
$d(\mathfrak{N},S)$	the limiting diameter of information $\mathfrak{N}$ for the problem S	2, Def. 2.2
φ	an algorithm, $\varphi:D_\varphi \subset \mathfrak{I}_2 \times \mathfrak{N}(D_{\mathfrak{N}}) \to \mathfrak{I}_2$	2, (2.11)
$\Phi(\mathfrak{N},S)$	the class of all algorithms using the information $\mathfrak{N}$ for the problem S.	2
$e(\varphi)$	the limiting error of algorithm φ	2, (2.13)

φ^I	an interpolatory algorithm	2, (2.15)
$p(\mathfrak{N},S)$	the order of information $\mathfrak{N}$ for the problem S	2, Def. 2.5
$p(\varphi)$	the order of algorithm φ	2, Def. 2.6
P	the set of primitives	3
$\mathrm{comp}(\mathfrak{N}(f,x))$	the information complexity of computing $\mathfrak{N}(f,x)$, where $\mathfrak{N}$ is a permissible information operator	'3
$\mathrm{comp}(\varphi(x,\mathfrak{N}(f,x)))$	the combinatory complexity of computing $\varphi(x,\mathfrak{N}(f,x))$, where φ is a permissible algorithm	3
$z(\varphi,f)$	the complexity index of φ for f	3, (3.10)
$z(\varphi)$	the complexity index of algorithm φ	3, (3.11)
$z(\mathfrak{N},S)$	the complexity index of the problem S with the information $\mathfrak{N}$	3, Def. 3.1
$\Phi_{\mathrm{perm}}(\mathfrak{N},S)$	the class of all permissible algorithms	3, Def. 3.1
φ^{mc}	a minimal complexity index algorithm	3, Def. 3.1
$\mathrm{comp}(\mathfrak{N})$	the information complexity	3, (3.17)
$\mathfrak{N}^\circ$	an optimal information operator in the class Ψ	3, Def. 3.2
$\mathfrak{N}_1 \subset \mathfrak{N}_2$	$\ker \mathfrak{N}_2(\cdot,x) \subset \ker \mathfrak{N}_1(\cdot,x),\ \forall x$	4, Def. 4.1
$\mathfrak{N}_1 \asymp \mathfrak{N}_2$	$\ker \mathfrak{N}_1(\cdot,x) = \ker \mathfrak{N}_2(\cdot,x),\ \forall x$	4, Def. 4.1
$A^\perp$	algebraic complement of A	4, (4.4)
$\mathrm{codim}\ A$	codimension of A	4, (4.5)
$\mathrm{card}(\mathfrak{N})$	the cardinality of the information $\mathfrak{N}$	4, Def. 4.2
$\mathrm{Lip}(k)$	the class of k-times differentiable functions whose kth derivatives satisfy a Lipschitz condition	4, (4.15)
$\mathfrak{N} \in \mathrm{Lip}(k)$	$\mathfrak{N}$ belongs to the class $\mathrm{Lip}(k)$	4, Def. 4.3
$\mathrm{IT}(\mathfrak{N},S)$	the class of all iterative algorithms using the information $\mathfrak{N}$ for the problem S	5, Def. 5.1
$\mathrm{index}(S)$	the index of the problem S	6, Def. 6.1
$\mathfrak{N}^*$	a basic linear information operator	6, (6.7)
$S \in \mathrm{Lip}(0)$	the solution operator S belongs to the class $\mathrm{Lip}(0)$	6, Def. 6.2
$\mathrm{index}(S,m)$	the mth index of the problem S	7, Def. 7.1
$\mathfrak{N}_m^*$	an mth basic linear information operator	7, (7.10)
$S \in \mathrm{Lip}(k(m))$	the solution operator S belongs to the class $\mathrm{Lip}(k(m))$	7, Def. 7.2
Ψ_n	the class of all regular linear information operators $\mathfrak{N}$ with $\mathrm{card}(\mathfrak{N}) \le n$	8
$p(n,S)$	the nth maximal order of the problem S	8, Def. 8.1
$\mathfrak{N}^{\mathrm{mo}}$	an nth maximal order information for the problem S	8, Def. 8.1
$z(n,S)$	the nth minimal complexity index of the problem S	8, Def. 8.2
$n_{\mathrm{opt}}(S)$	the optimal cardinality number of the problem S	8, Def. 8.2
$\mathfrak{N}^{\mathrm{oi}}$	an optimal information operator for the problem S	8, Def. 8.2

PART C

BRIEF HISTORY AND ANNOTATED BIBLIOGRAPHY

1. INTRODUCTION

The annotated bibliography of over 300 papers and books on optimal algorithms and analytic complexity covers both the eastern European and the western literature. We also include a very brief history of the subject. Each bibliographic item consists of a bibliographic reference, a set of keywords, and a short description.

To keep the bibliography of manageable size we have limited ourselves almost entirely to items central to the subject; such items are assigned the keyword *core*. We have included a few exceptionally relevant mathematical works; these are assigned the keyword *mathematics*. Even in the core, we have been selective, choosing only what we regard as the most important works, a process which has necessitated some hard choices.

We have generally included only items which study problems which can be solved only approximately and have omitted those dealing with problems which are solved approximately for reasons of efficiency. Thus, we have omitted iterative solution of large linear systems and approximate solution of hard combinatorial problems (such as NP-complete problems); each of these has its own extensive literature. Also excluded is the huge literature on analysis of a specific algorithm, since there is then no study of optimality. In particular, we have excluded the large literature on convergence, order, etc. of iterative

277

algorithms for the solution of nonlinear equations. We have omitted most items dealing with mathematical theories, even if germane to our subject. Examples include theories of approximation, differential equations, and iteration.

We have sometimes chosen to use in our descriptions unifying concepts and terminology only recently introduced rather than the terminology of the author. Thus, we use words such as information and complexity index.

Many different definitions of optimality appear in the literature. Often an optimal algorithm is defined as enjoying minimal error in some restricted class of algorithms, such as the class of linear algorithms. If the author is considering optimal algorithms in some restricted sense, then we use the phrase *optimal algorithm* in our keyword list and in our description. If the algorithm has minimal error in the class of all algorithms using the same information, then it is called an *optimal error algorithm*.

Many authors state their results in terms of *n*, which we call "cardinality of information." Although we have generally followed the author's usage, it would be routine to restate such results as a complexity function of ε.

We provide English titles for Russian and Polish papers which have not been translated into English. For translated papers, we provide references to both the original and the translation and use the title provided by the translator. If a periodical supplies a translated title for an untranslated paper, we use that title.

Since the theory of optimal algorithms and analytic complexity is a rapidly evolving subject, we shall prepare updated versions of this bibliography. The literature of this subject is large and diffuse, and since it is widely applicable, appears in many different periodicals. We solicit suggestions from our readers on important entries that we may have overlooked.

2. BRIEF HISTORY

The theory of optimal algorithms and analytic complexity has two major streams joining for the first time in this monograph. One stream, which in our terminology studies general information, started with the work of Kiefer, Sard, and Nikolskij around 1950. The other stream, which studies iterative information, began with Traub in 1961. We indicate a very few of the major achievements in each of these two streams starting with the general information case. The annotated bibliography gives a more complete history.

Kiefer [53] showed that if function evaluations are used, then Fibonacci search is optimal in searching for the maximum of a unimodal function. Professor Kiefer has informed us that this work work was done as an MIT Master's Thesis in 1948 but was only published later with the encouragement of J. Wolfowitz.

Sard [49] studied optimal algorithms for quadrature which use function evaluations at fixed points and discussed extending his results to the approximation of linear functionals. Independently, Nikolskij [50] posed the same problem and permitted the points of evaluation to be optimally chosen. Sard and Nikolskij restricted themselves to linear algorithms. In his dissertation, Smolyak [65] proved that for any linear functional defined on a balanced set and for any information operator consisting of n linear functionals, there exists a linear optimal error algorithm. Therefore, linear algorithms optimal in the sense of Sard or Nikolskij are optimal error algorithms, provided the set of elements is balanced and convex.

Golomb and Weinberger [59] performed the first systematic study of optimal error algorithms for approximation of a linear functional with information consisting of n linear functionals.

For many problems, optimal error algorithms are based on interpolatory splines. Schoenberg [64] was the first to recognize the close connection between splines and optimal algorithms in the sense of Sard.

Winograd [76] discussed using adversary arguments to obtain lower bounds in analytic complexity.

Micchelli and Rivlin [77] studied optimal algorithms for linear problems using linear information.

Traub and Woźniakowski [77] made the concept of information basic. This model permits linear and nonlinear problems and information. They introduced the concept of radius (diameter) of information in a general setting and used it to obtain a very powerful adversary principle. This initiated the study of lower bounds on complexity in a general setting. The notion of central algorithm was introduced. Some of these concepts had been implicitly used in special cases in a number of earlier papers.

One of the basic problems of analytic complexity is to find the most relevant information for a given problem. This was first studied for specific problems by Nikolskij [50] and Kiefer [53]. The idea of varying the information operator, which leads to the concept of optimal information operators, was introduced by Traub and Woźniakowski [77].

The second major stream of research in analytic complexity, which studied iterative information and iterative algorithms, had its inception in the work of Traub [61, 64]. Iterative algorithms were classified by the information they use. Theorems were obtained and conjectures proposed on maximal order of iterative algorithms for solving scalar nonlinear equations. Such maximal order results are needed to obtain lower bounds on complexity. The term "analytic computational complexity" was coined by Traub [72], although the work described in that paper is restricted to the portion of analytic complexity which would now be called iterative computational complexity.

Brent, Winograd, and Wolfe [73] used an adversary argument to obtain a maximal order theorem for the case of a class of nonstationary one-point iterations with memory using standard information to solve scalar nonlinear equations. Woźniakowski [75] introduced the concept of order of information which provides a powerful tool for establishing maximal order in an abstract space. He showed that maximal order in a class of algorithms depends only on the information used by an algorithm and not on the structure of the algorithm.

Traub and Woźniakowski [76] posed a new question: What information is relevant to the solution of a problem? A complete answer is provided for one-point iterations with linear information.

Traub and Woźniakowski [78] obtained a powerful adversary principle for establishing lower bounds on complexity in the iterative information model.

The two long reports of Traub and Woźniakowski [77, 78] for the first time bring together the two streams by including both in the same abstract setting and noting that they differ principally in whether general or iterative information is used. The material from these reports is included as part of this monograph.

Bibliography for Section 2

BRENT, R. P., WINOGRAD, S., AND WOLFE, P.
[73] Optimal iterative processes for root-finding, *Numer. Math.* **20** (1973), 327–341.

GOLOMB, M., AND WEINBERGER, H. F.
[59] Optimal approximation and error bounds, *in* "On Numerical Approximation" (R. E. Langer, ed.), pp. 117–190. Univ. of Wisconsin Press, Madison, Wisconsin, 1959.

KIEFER, J.
[53] Sequential minimax search for a maximum, *Proc. Amer. Math. Soc.* **4** (1953), 502–505.

MICCHELLI, C. A., AND RIVLIN, T. J.
[77] A survey of optimal recovery, *in* "Optimal Estimation in Approximation Theory" (C. A. Micchelli and T. J. Rivlin, eds.), pp. 1–54. Plenum Press, New York, 1977.

NIKOLSKIJ, S. M.
[50] On the problem of approximation estimate by quadrature formulae, *Usp. Mat. Nauk* **5** (1950), 165–177.

SARD, A.
[49] Best approximate integration formulas; Best approximation formulas, *Amer. J. Math.* **71** (1949), 80–91.

SCHOENBERG, I. J.
[64] Spline interpolation and best quadrature formulae, *Bull. Amer. Soc.* **70** (1964), 143–148.

SMOLYAK, S. A.
[65] On an optimal restoration of functions and functionals of them (in Russian), Candidate Dissertation, Moscow State Univ. (1965).

TRAUB, J. F.
[61] On functional iteration and calculation of roots, *Preprints of Papers 16th Nat. ACM Conf.*, Session 5A-1, pp. 1–4, Los Angeles, California, 1961.
[64] "Iterative Methods for Solution of Equations." Prentice-Hall, Englewood Cliffs, New Jersey, 1964.
[72] Computational complexity of iterative processes, *SIAM J. Comput.* **1** (1972), 167–179.

TRAUB, J. F., AND WOŹNIAKOWSKI, H.
[76] Optimal linear information for the solution of nonlinear equations, *in* "Algorithms and Complexity: New Directions and Recent Results" (J. F. Traub, ed.), pp. 103–119. Academic Press, New York, 1976.

[77]   General theory of optimal error algorithms and analytic complexity, part A: General information model. Dept. of Computer Science Rep., Carnegie-Mellon Univ. (1977).

[78]   General theory of optimal error algorithms and analytic complexity, Part B: Iterative information model. Dept. of Computer Science Rep., Carnegie-Mellon Univ. (1978).

WINOGRAD, S.

[76]   Some remarks on proof techniques in analytic complexity, *in* "Analytic Computational Complexity" (J. F. Traub, ed.), pp. 5–15. Academic Press, New York, 1976.

WOŹNIAKOWSKI, H.

[75]   Generalized information and maximal order of information for operator equations, *SIAM J. Numer. Anal.* **12** (1975), 121–135.

3. ANNOTATED BIBLIOGRAPHY

Adamski, A., Korytowski, A., and Mitkowski, W., A conception of optimality for algorithms and its application to the optimal search for a minimum, *Zastos. Mat.* **14** (1977), 499–509.

core, extremum, scalar, optimal error algorithms

Considers the concept of "strong" optimal algorithms and tests this concept on the problem of searching for the minimum of a unimodal or convex scalar function. The information is the values of f at n points. Results are related to those of Kiefer [53].

Ahlberg, J. H., and Nilson, E. N., The approximation of linear functionals, *SIAM J. Numer. Anal.* **3** (1966), 173–182.

core, approximation of linear functionals

Considers approximation of a linear functional L for a class of scalar functions f such that $f^{(n)} \in L_2$. The information is the values of f and its derivatives. Shows that the value of L on an interpolatory spline minimizes a certain type of error.

Ahlberg, J. H., Nilson, E. N., and Walsh, J. L., "The Theory of Splines and Their Applications." Academic Press, New York, 1967.

mathematics and core, splines

Considers the theory of splines and their applications for a variety of problems in numerical analysis. Optimal properties of splines are studied. Shows the relationship between splines and optimal approximation in the sense of Sard.

Aksen, M. B., and Tureckij, A. H., Best quadrature formulas for certain classes of functions (in Russian), *Dokl. Akad. Nauk SSSR*, **166** (1966), 1019–1021 [*English transl.: Soviet Math. Dokl.* **7** (1966), 203–205].

core, integration, scalar, optimal points of information, optimal linear algorithms

Considers integration for the class of scalar functions with bounded rth derivative in L_q. The information is the values of $f, f', \ldots, f^{(r-2)}$ at m points. The errors of optimal linear algorithms with optimally chosen points of information are derived for even r.

Alhimova, V. M., Best quadrature formulas with equidistant nodes (in Russian), *Dokl. Akad. Nauk SSSR* **204** (1972), 263–266 [*English transl.: Soviet Math. Dokl.* **13** (1972), 619–623].

core, integration, scalar, optimal linear algorithms

Considers integration for some classes of scalar functions with bounded rth derivative in L_q. The information is the values of f. Presents optimal quadrature formulas with equidistant points of information. The errors of such formulas are also given.

Aliev, R. M.: *See* Ibragimov, I. I.

Aphanasjev, A. Yu., On the search of minimum function with limited second derivative (in Russian), *Zh. Vychisl. Mat. Mat. Fiz.* **14** (1974), 1018–1021 [*English transl.*: Afanas'ev, A. Yu., The search for the minimum of a function with a bounded second derivative, *U.S.S.R Computational Math. and Math. Phys.* **14** (1974), 191–195].

core, extremum, scalar, optimal error algorithms

  Considers the search for the minimum in the class of scalar functions f whose second derivative belongs to $[a,b]$ with $a > 0$. The information is the values of f at two points. The interval in which a minimum of f lies is derived. Optimal points of information are obtained.

Aphanasjev, A. Yu., and Novikov, V. A., On the search of minimum of a function with the limited third derivative (in Russian), *Zh. Vychisl. Mat. Mat. Fiz.* **17** (1977), 1031–1034.

core, extremum, scalar, optimal error algorithms

  Considers the search for the minimum in the class of scalar unimodal functions whose third derivative lies in a given interval. The information is the values of f at three points. An algorithm that finds an interval at which a minimum of f lies is proposed.

Arestov, V. V., On the best approximation of differentiation operators (in Russian), *Mat. Zametki* **1** (1967), 149–154 [*English transl.*: *Math. Notes* **1** (1967), 100–103].

core, differentiation, optimal approximation by bounded linear operators

  Continuation of Stechkin [67]. Considers approximation of $f^{(k)}$ for the class of scalar functions with bounded nth derivative in L_1 or C, $0 < k < n$, by means of linear operators φ whose norm is bounded by a given constant. For small n, shows that the nearly optimal operator φ requires a few evaluations of f.

Arestov, V. V., On the best uniform approximation of differentiation operators (in Russian), *Mat. Zametki* **5** (1969), 273–284 [*English transl.*: *Math. Notes* **5** (1969), 167–173].

core, differentiation, optimal approximation by bounded linear operators

  Continuation of Arestov [67]. Studies the existence and uniqueness of optimal linear operators for the approximation of $f^{(k)}$.

Arro, V. K. *See* Levin, M. I.

Aubin, J. P., Best approximation of linear operators in Hilbert spaces, *SIAM J. Numer. Anal.* **5** (1968), 518–521.

core, approximation of linear operators

  Considers approximation of a linear operator $A: E \to F$ for the unit ball; E, F are Hilbert spaces. The information operator is a given linear operator $r_n: E \to E_n$. For a given linear operator $s_n: F \to F_n$, the operator A is approximated by a linear operator $A_n: E_n \to F_n$ such that $\|A_n r_n u - s_n A u\|_{F_n}$ for $\|u\|_E \leqslant 1$ is minimized. The solution is obtained in terms of the operator of "best interpolation."

Avriel, M., and Wilde, D. J., Optimal search for a maximum with sequences of simultaneous function evaluations, *Management Sci.* **12** (1966), 722–731.

core, extremum, scalar, optimal algorithms

  Considers the search for the maximum in a class of scalar unimodal functions. The information is the values of f. Optimal algorithms based on simultaneous function evaluations are studied.

Babenko, V. F., Asymptotically sharp bounds for the remainder for the best quadrature formulas for several classes of functions (in Russian), *Mat. Zametki* **19** (1976), 313–322 [*English transl.*: *Math. Notes* **19** (1976), 187–193].

core, integration, multivariate, asymptotic error bounds

  Considers cubature formulas for the class of scalar functions of several variables with bounded modulus of continuity. The information is the values of f at n points. The asymptotic error for the optimal cubature formula (i.e., for large n) is derived.

Babuška, I., Problems of optimization and numerical stability in computations, *Apl. Mat.* **1** (1968), 3–26.

core, integration, scalar, optimal algorithms, numerical stability

This is a paper presented at the conference "Basic problems of numerical mathematics" in Liblice 1967. Among the problems considered is integration for a class of periodic functions. The information is the values of f. Optimal points of information and optimal algorithms are studied. See also Babuška, I., Über universal optimale quadraturformeln, *Apl. Mat.* **4** (1968), 305–338; **5** (1968), 388–404.

Babuška, I., and Sobolev, S. L., Optimization of numerical methods (in Russian), *Apl. Mat.* **1** (1965), 96–130.

core, survey of optimal algorithms

Considers optimal numerical algorithms for the solution of algebraic and analytic problems. Optimal or asymptotically optimal error algorithms for the approximation of linear functionals and linear equations with compact inverse operators are studied. Relations to Kolmogorov n-widths and entropy are given. Good bibliography.

Bakhvalov, N. S., On the approximate calculation of multiple integrals (in Russian), *Vestnik MGU. Ser. Math. Mech. Astron. Phys. Chem.* **4** (1959), 3–18.

core, integration, multivariate, optimal linear algorithms, lower bounds

Considers integration for the class of scalar functions of s variables with all derivatives up to order p bounded and the pth derivative satisfying the Hölder condition of order λ. The information is the values of f. Shows that a lower bound on the error of any quadrature formula with n points is roughly $n^{-(p+\lambda)/s}$. The expected value of the error has lower bound roughly $n^{-(p+\lambda)/s-0.5}$. The bounds are shown to be sharp.

Bakhvalov, N. S., An estimate of the mean remainder in quadrature formulae (in Russian), *Zh. Vychisl. Mat. Mat. Fiz.* **1** (1961), 64–77 [*English transl.: U.S.S.R. Computational Math. and Math. Phys.* **1** (1961), 68–82].

core, integration, multivariate, average case analysis

This is a continuation of Bakhvalov [59]. For the same classes of functions, derives the expected values of the errors in quadrature formulas with f evaluated at random points.

Bakhvalov, N. S., On optimal methods of specifying information in the solution of differential Equations (in Russian), *Zh. Vychisl. Mat. Mat. Fiz.* **2** (1962), 569–592 [*English transl.: U.S.S.R. Computational Math. and Math. Phys.* **2** (1962), 608–640].

core, differential equations, entropy, n-widths

Considers the differential equation $u_t = P(t,x,u)$, where $u = u(t,x)$ and P is a given operator, for a class of functions $u(0,x)$. The information is the values of $u(0,x)$. Studies the problem of the minimal number of evaluations of $u(0,x)$ to approximate $u(t,x)$ to within ε. Shows that this problem is related to ε-entropy. Also considers the approximation of a function from its n values by means of algorithms whose range has dimension n. Shows the relation to Kolmogorov n-widths.

Bakhvalov, N. S., On the estimate of the amount of computational labor necessary in approximate solutions (in Russian), Appendix IV in the book of S. K. Godunov and W. S. Riabenki, "Theory of Difference Schemes—An Introduction," pp. 316–329. Moscow, 1962 [English translation of the book published by American Elsevier, New York, 1964, pp. 268–279].

core, integration, multivariate, differential equations, lower and upper bounds

Considers the complexity of integration for multivariate functions and the complexity of certain types of differential equations. The information is the values of f at n points. Lower and upper bounds on the complexity are derived.

Bakhvalov, N. S., Optimal properties of Adams and Gregory formulae of numerical integration (in Russian), *in* "Problems of Computational Mathematics and Computational Technique" (L. A. Ljusternik, ed.), pp. 9–26. Mashgiz, Moscow, 1963.

core, integration, scalar, differential equations, optimal algorithms

Considers integration for the class of scalar functions with bounded rth derivative. The information is the values of f and its derivatives at equidistant points. Shows that the Euler and Gregory formulas are nearly optimal. Considers also ordinary differential equations, $y' = f(x, y)$, for a class of functions f, and shows that the Adams formulas are asymptotically optimal.

Bakhvalov, N. S., On optimal bounds for the convergence of quadrature formulas and Monte-Carlo type integration methods for classes of functions (in Russian), *in* "Numerical Methods for the Solution of Differential and Integral Equations and Quadrature Formulas," pp. 5–63. Moscow, 1964.

core, integration, multivariate, asymptotically optimal algorithms, lower bounds

Considers integration for many classes of scalar functions of several variables. The information is the values of f at n points. Presents linear algorithms whose errors differ from the optimal error by at most a factor of $\ln^{\gamma} n$ for a positive γ.

Bakhvalov, N. S., On the optimal speed of integrating analytic functions (in Russian), *Zh. Vychisl. Mat. Mat. Fiz.* **7** (1967), 1011–1020 [*English transl.: U.S.S.R. Computational Math. and Math. Phys.* **7** (1967), 63–75].

core, integration, analytic functions, scalar, optimal algorithms

Considers integration for the class of analytic scalar functions bounded by a constant on the ellipse with foci ± 1 and the sum of semiaxes equal to a given c. The information is the values of f and f' at n points. Asymptotic optimality of Gauss quadrature is proven. The optimal error is roughly c^{-2n}.

Bakhvalov, N. S., On optimal methods for the solution of problems (in Russian), *Apl. Mat.* **1** (1968), 27–38.

core, survey of optimal algorithms

This is a paper presented at the conference "Basic problems of numerical mathematics" in Liblice 1967. Surveys Russian work on the optimal solution of many numerical problems.

Bakhvalov, N. S., Properties of optimal methods for the solution of problems of mathematical physics (in Russian), *Zh. Vychisl. Mat. Mat. Fiz.* **10** (1970), 555–568 [*English transl.: U.S.S.R. Computational Math. and Math. Phys.* **10** (1970), 1–20].

core, integration, multivariate, differential equations, lower bounds

Considers optimal methods for solving problems of mathematical physics. The information is the values of f at n points. The minimum number of function evaluations to solve the problem to within ε is studied for multivariate integration and for parabolic differential equations. An asymptotically optimal algorithm with linear combinatory complexity is proposed for parabolic differential equations.

Bakhvalov, N. S., On the optimality of linear methods for operator approximation in convex classes of functions (in Russian), *Zh. Vychisl. Mat. Mat. Fiz.* **11** (1971), 1014–1018 [*English transl.: U.S.S.R. Computational Math. and Math. Phys.* **11** (1971), 244–249].

core, approximation of linear functionals, optimal linear algorithms

Considers approximation of linear functionals for a balanced convex class. The information is the values of n linear functionals. Contains Smolyak's lemma which states that there exists a linear optimal error algorithm. Some extensions to the approximation of linear operators are presented.

Bakhvalov, N. S., Optimization of methods of solving ordinary differential equations with strongly oscillating solutions (in Russian), *Zh. Vychisl. Mat. Mat. Fiz.* **11** (1971), 1318–1322 [*English transl.: U.S.S.R. Computational Math. and Math. Phys.* **11** (1971), 287–292].

core, differential equations, lower bounds

Considers the equation $\mu^2 y'' + a(x)y = f(x)$ for small positive μ for the class of functions such that $a(x) \geq a_0 > 0$, $|f^{(i)}(x)| \leq A$ for $i = 0, 1, \ldots, m$, $\forall x \in [0,1]$, and $|\mu y'(0)| \leq b_0$. The information is the values of a and f. Shows that a lower bound on any algorithm using n evaluations of a has error at least roughly $\min(1, 1/(\mu n^m))$. For $m = 1$ or 2, this bound is sharp.

Bakhvalov, N. S., A lower bound for the asymptotic characteristics of classes of functions with dominating mixed derivative (in Russian), *Mat. Zametki* **12** (1972), 655–664 [*English transl.: Math. Notes* **12** (1972), 833–838].

core, integration, interpolation, periodic functions, multivariate, lower bounds

Considers integration and interpolation for the class of scalar periodic functions of several variables with bounded derivative. The information is the values of f and its derivatives. Using a new representation of the class of functions, lower bounds for the errors in integration and interpolation, and a lower bound for the ε-entropy are given.

Barnhill, R. E., Optimal quadratures in $L_\zeta^2(E)$. I and II, *SIAM J. Numer. Anal.* **4** (1967), 390–397, 534–541.

core, integration, analytic functions, scalar, optimal linear algorithms

Considers integration for the class of scalar analytic functions bounded by a constant in the ellipse E_ζ with foci ± 1, semiaxes a, b, and $\zeta = (a + b)^2$. The information is the values of f. Optimal algorithms are derived.

Barnhill, R. E., Asymptotic properties of minimum norm and optimal quadratures, *Numer. Math.* **12** (1968), 384–393.

core, integration, analytic functions, scalar, optimal linear algorithms, Gauss quadrature

Considers integration for the class of scalar analytic functions on the ellipse E_ζ with foci at ± 1 semiaxes a, b, and $\zeta = (a + b)^2$. The information is the values of f. Asymptotic properties of optimal quadrature formulas are studied. Shows that the weights and points of an optimal quadrature converge to the weights and points of Gauss quadrature as $\zeta \to +\infty$.

Barnhill, R. E., and Wixom, J. A., Quadratures with remainders of minimum norm. I and II, *Math. Comp.* **21** (1967), 66–75, 382–387.

core, integration, analytic functions, scalar, optimal points of information, optimal linear algorithms

Considers integration for a class of scalar analytic functions defined on the ellipse E_ζ with foci ± 1, semiaxes a, b, and $\zeta = (a + b)^2$. The information is the values of f. Optimal quadrature formulas for fixed and varying points of information are discussed.

Barnhill, R. E., and Wixom, J. A., An error analysis for interpolation of analytic functions, *SIAM J. Numer. Anal.* **5** (1968), 522–528.

core, interpolation, analytic functions, scalar, optimal linear algorithms

Considers interpolation for a class of scalar analytic functions defined on the ellipse E_ζ with foci ± 1, semiaxes a, b, and $\zeta = (a + b)^2$. The information is the values of f. Linear optimal algorithms are studied. Asymptotic properties of the optimal weights are considered as $\zeta \to +\infty$.

Barrar, R. B., Loeb, H. L., and Werner, M., On the existence of optimal integration formulas for analytic functions, *Numer. Math.* **23** (1974), 105–117.

core, integration, scalar, optimal linear algorithms

Considers integration for a class of analytic functions. The information is the values of f and its derivatives. The existence of weights and points of a quadrature formula which minimizes a certain error is proven.

Baudet, G. M., Asynchronous iterative methods for multiprocessors, *J. Assoc. Comput. Mach.* **25** (1978), 226–244.

core, nonlinear equations, multivariate, iterative algorithms, asynchronous

Introduces and analyzes a general class of asynchronous iterative methods. Establishes general convergence theorem and obtains complexity bounds. Presents experimental results for certain problems which show that "purely asynchronous" iterative methods are best.

Beamer, J. H., and Wilde, D. J., Time delay in minimax optimization of unimodal functions of one variable, *Management Sci.* **15** (1969), 528–538.

core, extremum, scalar, optimal algorithms

Considers the search for the maximum in a class of scalar unimodal functions. The information is the values of f. Optimal algorithms are derived for two cases. In the first, each function evaluation is performed before the preceding result is known. In the second case, each is performed before the two preceding results are known.

Beamer, J. H., and Wilde, D. J., Minimax optimization of unimodal functions by variable block search, *Management Sci.* **16** (1970), 529–541.

core, extremum, scalar, optimal algorithms

Considers the search for the maximum in a class of scalar unimodal functions. The information is the values of f. Optimal algorithms based on simultaneous function evaluations are studied.

Beamer, J. H., and Wilde, D. J., Minimax optimization of unimodal functions by variable block derivative search with time delay, *J. Comb. Theory*, **10** (1971), 160–173.

core, extremum, scalar, optimal algorithms

Considers the search for the maximum in a class of scalar unimodal functions. The information is the values of the first derivative of f. The search algorithms use a sequence of blocks of simultaneous evaluations of f'. Optimal error algorithms are derived. A method of optimizing the number of evaluations per block is given.

Bojanov, B. D., Optimal rate of integration and ε-entropy of a class of analytic functions (in Russian), *Mat. Zametki* **14** (1973), 3–10 [*English transl.: Math. Notes* **19** (1973), 551–556].

core, integration, analytic functions, scalar, lower bounds

See Bojanov [74]. The ε-entropy of a class of analytic functions is derived.

Bojanov, B. D., Best quadrature formula for a certain class of analytic functions, *Zastos. Mat.* **14** (1974), 441–447.

core, integration, analytic functions, scalar, optimal linear algorithms, lower bounds

Considers optimal quadrature formulas for the class of real functions on $[-1, 1]$ which can be analytically extended to the unit disk and whose extension is bounded by unity. The information is the values of f, f' at n points. Using Smolyak's lemma, the linear optimal error algorithm and its error are derived. For optimal points, the error is roughly $\exp(-\pi\sqrt{n/2})$.

Bojanov, B. D., Best methods of interpolation for certain classes of differentiable functions (in Russian), *Mat. Zametki* **17** (1975), 511–524 [*English transl.: Math. Notes* **17** (1975), 301–309].

core, interpolation, scalar, optimal linear algorithms

Considers the interpolation problem for the class of scalar functions with bounded rth derivative in L_q. The information is the values of $f, f', \ldots, f^{(r-1)}$ at n points. The linear optimal error algorithm is derived and shown to be a spline. The optimal points of information are proven to be equidistant.

A slightly extended version of B. D. Bojanov, Optimal methods of interpolation in $W^{(r)}L_q(M; a, b)$, in English, *Comptes Rendus Acad. Bulg. Sci.* **17** (1974), 885–888.

Bojanov, B. D., Optimal methods of integration in the class of differentiable functions, *Zastos. Mat.* **15** (1976), 105–115.

core, integration, scalar, optimal linear algorithms, lower bounds

Considers optimal quadrature formulas for the class of r times piecewise continuously differentiable functions whose rth derivative in L_q is bounded by a constant. The information is the values of $f, f', \ldots, f^{(r-1)}$ at n points. Using Smolyak's lemma, the linear optimal error algorithm and its error are derived. For optimal points, the error is obtained.

Bojanov, B. D., and Chernogorov, V. G., An optimal interpolation formula, *J. Approx. Theory* **20** (1977), 264–274.

core, interpolation, approximation, scalar, optimal linear algorithms

Considers approximation of a linear functional for a given class of functions. The information is the values of n linear functionals on f. Optimal error algorithms are studied. Linear optimal error algorithms are presented for the interpolation and approximation problems in the class of scalar functions whose second derivative is bounded in L_∞ by a constant.

Booth, R. S., Location of zeros of derivatives, *SIAM J. Appl. Math.* **15** (1967), 1496–1501.

core, nonlinear equations, scalar, optimal points of information, error bounds

Considers the search for a zero α of the kth derivative in the class of scalar functions for which $f^{(k)}$ changes sign only at α. The information is the values of f at n points. Studies the asymptotic character of the error of an optimal algorithm for optimally chosen points of information.

Booth, R. S., Location of zeros of derivatives. II, *SIAM J. Appl. Math.* **17** (1969), 409–415.

core, nonlinear equations, scalar, optimal points of information, error bounds

Continuation of Booth [67].

Borodin, A., and Munro, I., "The Computational Complexity of Algebraic and Numeric Problems." American Elsevier, New York, 1975.

core, algebraic numbers, optimal iterations, algebraic complexity, iterative complexity, maximal order

A text on algebraic complexity. Includes chapter on parallel processing in numeric computation. Of particular relevance to analytic complexity is a chapter on "The complexity of rational iterations" which covers the Paterson–Kung theory of the complexity of iterations which approximate algebraic numbers. See also Paterson [72] and Kung [72, 73].

Brent, R. P., The computational complexity of iterative methods for systems of nonlinear equations, *in* "Complexity of Computer Computations" (R. E. Miller and J. W. Thatcher, eds.), pp. 61–71. Plenum Press, New York, 1972.

core, nonlinear equations, multivariate, iterations with memory, complexity index, maximal order

Compares complexity of classes of algorithms for solving the system of nonlinear equations $f = 0$. The information is the values of f. The classes of algorithms considered include multivariate polynomial interpolatory methods as well as two new classes.

Brent, R. P., "Algorithms for Minimization without Derivatives." Prentice-Hall, Englewood Cliffs, New Jersey, 1973.

core, nonlinear equations, extremum, scalar, multivariate, order

A valuable monograph on algorithms and programs for computing zeros and extrema of scalar nonlinear functions and extrema of multivariate nonlinear functions. The information is values of the function. Contains much original material. Some discussion of optimality and complexity. Good bibliography.

Brent, R. P., Some efficient algorithms for solving systems of nonlinear equations, *SIAM J. Numer. Anal.* **10** (1973), 327–344.

core, nonlinear equations, multivariate, iterative algorithms, secant iteration, iterative complexity, complexity index, order

Considers iterative algorithms for solving a multivariate nonlinear system $f = 0$. The information is the values of f. Introduces two new classes of algorithms and establishes their local convergence. Computes a complexity index for these algorithms and compares with known methods. Poses an open problem on the optimal complexity index.

Brent, R. P., Computer Solution of Nonlinear Equations. Lecture Notes, Computer Science Dept., Stanford Univ. (1975).

core, nonlinear equations, scalar, multivariate, one-point iterations, multipoint iterations, iterations with memory, iterative complexity

A book-length set of notes. Surveys iteration algorithms and iterative complexity for both scalar and multivariate nonlinear equations. Good bibliography.

Brent, R. P., Some high-order zero-finding methods using almost orthogonal polynomials, *J. Austral. Math. Soc. Ser. B* **19** (1975), 1–29.

core, nonlinear equations, scalar, iterative information, multipoint iterations, complexity index

Considers iterative algorithms for computing a zero of a scalar nonlinear function f. The information is one evaluation of f and n evaluations of f'. The points of evaluation are determined from $n \geq 0$, and k satisfying $m + 1 \geq k > 0$, there are algorithms of order $m + 2n + 1$. To establish convergence, results are obtained on orthogonal and "almost orthogonal" polynomials. Discusses a complexity index for the iterations. Good bibliography.

Brent, R. P., A class of optimal-order zero-finding methods using derivative evaluations, *in* "Analytic Computational Complexity" (J. F. Traub, ed.), pp. 59–73. Academic Press, New York, 1976.

core, nonlinear equations, scalar, iterative information, multipoint iterations, maximal order

Considers iterative algorithms for computing a zero of a scalar nonlinear function f. The information is one evaluation of f and n evaluations of f'. The points of evaluation are determined from certain orthogonal or "almost orthogonal" polynomials. These iterations are of maximal order.

Let $x'(t) = g(x)$. The preceding results are used to obtain an explicit nonlinear Runge–Kutta method of order $2n - 1$ which uses n evaluations of g.

See also Brent [75] (Some high-order zero-finding methods using almost orthogonal polynomials).

Brent, R. P., Multiple-precision zero-finding methods and the complexity of elementary function evaluation, *in* "Analytic Computational Complexity" (J. F. Traub, ed.), pp. 151–176. Academic Press, New York, 1976.

core, nonlinear equations, formal power series, scalar, fast algorithms, multiple precision

Introduces fast algorithms and analyzes their complexity for multiprecision computation of certain numbers, arithmetic operations, and functions. Among the computations discussed are reciprocation, square roots, zero-finding, evaluation of π, evaluation of elementary transcendental functions, and the solution of scalar equations. Fast algorithms for such formal power series operations as logarithm and powering are given.

Brent, R. P., Fast multiple-precision evaluation of elementary functions, *J. Assoc. Comput. Mach.* **23** (1976), 242–251.

core, approximation, scalar, multiple precision, fast algorithms

Shows that elementary functions can be evaluated with relative error of $\Theta(2^{-n})$ in $\Theta(M(n) \log n)$ operations where $M(n)$ is the number of single-precision operations required to multiply n-bit integers. Special cases include the evaluation of constants such as π, e, and e^{π}.

Brent, R. P., The complexity of multiple-precision arithmetic, *in* "Complexity of Computational Problem Solving" (R. Anderssen and R. P. Brent, eds.), pp. 125–165. Univ. of Queensland Press, 1976.

core, approximation and nonlinear equations, scalar, multiple precision, fast algorithms, lower bounds, upper bounds, iterative complexity

Studies complexity of performing multiple-precision computations. Among the computations considered are arithmetic operations and elementary function evaluations. Upper bounds and some lower bounds are obtained. Complexities of various iterations for computing zeros of nonlinear scalar functions using variable-length multiple-precision arithmetic are also compared.

Brent, R. P., and Kung, H. T., $O((N \log N)^{3/2})$ algorithms for composition and reversion of power series, in "Analytic Computational Complexity" (J. F. Traub, ed.), pp. 217–225. Academic Press, New York, 1976.

core, composition, reversion, formal power series, fast algorithms, algebraic complexity

First announcement of the result of the title. See Brent and Kung [78] (Fast algorithms for manipulating formal power series) for proof.

Brent, R. P., and Kung, H. T., Fast algorithms for composition and reversion of multivariate power series, *Proc. Conf. Theoret. Comput. Sci.* pp. 149–158. Univ. of Waterloo, Waterloo, Canada, 1977.

core, composition, reversion, formal power series, multivariate, fast algorithms, algebraic complexity

Extends results of Brent and Kung [78] to the multivariate case. Shows that every reversion problem can be associated with a composition problem in the sense that if the composition problem can be solved fast so can the reversion problem. Presents fast algorithms for composition and reversion of power series which require substantially fewer operations than classical methods. The improvement increases as the number of variables increases.

Brent, R. P., and Kung, H. T., Fast algorithms for manipulating formal series, *J. Assoc. Comput. Mach.* **25** (1978), 581–595.

core, composition, reversion, differential equations, formal power series, fast algorithms, algebraic complexity

Gives algorithm for computing first N terms of composite of two power series in $O((N \log N)^{3/2})$ operations. Shows that the complexity of composition and reversion are asymptotically equivalent. Let MULT(N) be the minimal number of operations for computing the first N terms of the product of two polynomials. Proves that the evaluation of the reversion series truncated to N terms can be done in $O(\text{MULT}(N))$ operations. Shows that the first N terms of the power series solutions to many types of differential equations can be obtained in $O(\text{MULT}(N))$ operations.

Brent, R. P., and Traub, J. F., On the complexity of composition and generalized composition of power series, Computer Science Dept. Rep., Carnegie-Mellon Univ. (1978). To appear in *SIAM J. Comput.*

core, composition, generalized composition, formal power series, fast algorithms, algebraic complexity

Let $F^{[q]}(x)$ be the qth composite of a formal power series. Shows that $F^{[q]}(x)$ can often, but not always, be defined for general q. Gives fast algorithms and complexity bounds for computing the first N terms of $F^{[q]}(x)$ whenever it is defined. If q is an integer, the fast algorithms eliminate the complexity factor of $\log_2 q$ of the "repeated squaring" algorithm.

Brent, R. P., Winograd, S., and Wolfe, P., Optimal iterative processes for root-finding, *Numer. Math.* **20** (1973), 327–341.

core, nonlinear equations, scalar, iterative algorithms, iterations with memory, maximal order

Considers locally convergent nonstationary one-point iterations with memory for computing a zero of a scalar nonlinear function f. The information used to compute the kth iterate is the values of $f, f', \ldots, f^{(d)}$ at all previous iterates. Proves that the maximal order of any such iterate is at most $d + 2$. Settles a conjecture of Traub [64], at least for the case of iterations which use all previous information.

Busarova, T. N., Best quadrature formulae for a class of differentiable and periodic functions (in Russian), *Ukrain. Mat. Z.* **25** (1973), 291–301.

core, integration, scalar, optimal points of information, optimal linear algorithms

Considers integration for the class of scalar periodic functions with bounded third derivative in L_∞. The information is the values of f or f and f'. Shows that the optimal points of information are equidistant and the coefficients of the optimal quadrature formulas are equal. The errors of the optimal algorithms are given.

Butcher, J. C., On Runge–Kutta processes of high order, *J. Austral. Math. Soc.* **4** (1964), 179–194.

core, differential equations, Runge–Kutta methods, maximal order

Considers the solution of $y' = f(x,y)$, $y(x_0) = y_0$, by Runge–Kutta methods. The information is the values of f at n adaptively chosen points. Studies the maximal order $p(n)$ of such methods. Finds the maximal order for $n = 5$ and improves the bounds for $n = 6$.

Butcher, J. C., On the attainable order of Runge–Kutta methods, *Math. Comp.* **19** (1965), 408–417.

core, differential equations, Runge–Kutta methods, maximal order

Continuation of Butcher [64]. Finds the maximal order for $n \leq 9$ and the bound $p(n) \leq n - 2$ for $n \geq 10$.

Butcher, J. C., An order for Runge–Kutta methods, *SIAM J. Numer. Anal.* **12** (1975), 304–315.

core, differential equations, Runge–Kutta methods, maximal order

Continuation of Butcher [65]. Proves that there does not exist an explicit Runge–Kutta method which uses n evaluations of f and has order $p \geq u_k$ unless $n > p + k$, where $u_0 = 5$, $u_{n+1} = (4u_n + 2n + 3)/3$.

Casuli, V., and Trigiante, D., The convergence order for iterative multipoint procedures, *Calcolo* **14** (1977), 25–44.

core, nonlinear equations, scalar, multipoint iterations, maximal order

Considers iterative solution of scalar nonlinear equations. Restrictive assumptions are made concerning the information and algorithms used. As in Kung and Traub [74] (Optimal order of one-point and multipoint iterations), obtains maximal order for this class of iterations.

Casuli, V., and Trigiante, D., Computational complexity for a class of multipoint iterative procedures without or with internal memory, *Calcolo* **14** (1977), 225–235.

core, nonlinear equations, scalar, multipoint iterations, complexity

Considers iterative solution of scalar nonlinear equations. As in Kung and Traub [74] (Computational complexity of one-point and multipoint iteration), includes combinatory complexity in the complexity index.

Chawla, M. M., and Kaul, V., Optimal rules for numerical integration round the unit circle, *BIT* **13** (1973), 145–152.

core, integration, scalar, optimal points of information, optimal algorithms

Considers integration and approximation of a linear functional for a class of scalar analytic functions defined on a circular annulus in a Hilbert space with a reproducing kernel. The information is the values of f. Optimal weights and points of information are derived in terms of the representers of the functionals. Optimal quadrature formulas for the unit circle and the interval $[-1,1]$ are presented.

Chentsov, N. N., On quadrature formulae for functions of an infinitely large number of variables (in Russian), *Zh. Vychisl. Mat. Mat. Fiz.* **1** (1961), 418–424 [*English transl.: U.S.S.R. Computational Math. and Math. Phys.* **1** (1961), 455–464].

core, integration, abstract, lower bounds

Considers integration for a class of scalar functions of infinitely many variables. The information is the values of f. A sharp lower bound on the error of linear quadrature formulas for the class of Lipschitz functions is found.

Chernogorov, V. G.: *See* Bojanov, B. D.

Chernousko, F. L., An optimal algorithm for finding the roots of an approximately computed function (in Russian), *Zh. Vychisl. Mat. Mat. Fiz.* **8** (1968), 705–724 [*English transl.: U.S.S.R. Computational Math. and Math. Phys.* **8** (1968), 1–24].

core, nonlinear equations, scalar, optimal points of information, optimal error algorithms

Considers the search for a zero in the class of scalar functions f such that $m \le (f(x_1) - f(x_2))/(x_1 - x_2) \le M$ for $x_1, x_2 \in [a,b]$ and $m > 0$. The information is the perturbed values of f. The optimal error algorithm is derived. Optimal points of information are obtained. Assuming that the cost of evaluating f to a certain accuracy is measured by a given cost function, the optimal error algorithm with fixed cost is discussed.

Chernousko, F. L., Optimal search for extrema of unimodal functions (in Russian), *Zh. Vychisl. Mat. Mat. Fiz.* **10** (1970), 922–933 [*English transl.: U.S.S.R. Computational Math. and Math. Phys.* **10** (1970), 146–161].

core, extremum, scalar, optimal error algorithms

Considers the search for the extremum in the class of unimodal scalar functions satisfying a Lipschitz condition with a given constant. The information is the values of f at n points. The optimal error algorithms with respect to two different criteria are found. Results are related to those of Kiefer [53].

Chernousko, F. L., Optimal search for the minimum of convex functions (in Russian), *Zh. Vychisl. Mat. Mat. Fiz.* **10** (1970), 1355–1366 [*English transl.: U.S.S.R. Computational Math. and Math. Phys.* **10** (1970), 20–34].

core, extremum, scalar, convex functions, optimal error algorithms

Considers the search for the minimum in the class of convex scalar functions. The information is the values of f at n points. The optimal or nearly optimal error algorithms are presented.

Chzhan, Guan-Tsynan, On the minimum number of interpolation points in the numerical integration of the heat-conduction equation (in Russian), *Zh. Vychisl. Mat. Mat. Fiz.* **2** (1962), 80–88 [*English transl.: U.S.S.R. Computational Math. and Math. Phys.* **2** (1962), 78–87].

core, differential equations, n-widths

Considers the heat-conduction equation $u_t = u_{xx}$, where $u(x,0)$ belongs to the class H_p of functions f such that $f^{(2i)}(0) = f^{(2i)}(\pi) = 0$ for $2i < p$ and $f^{(p)} \in L_2$. The information is the values of $u(x,0)$. Studies the problem of the minimal number n of evaluations of $u(x,0)$ to approximate $u(x,t)$ to within ε by means of an algorithm whose range has dimension at most n. Derives an algorithm with $n = \Theta(\varepsilon^{-1/p})$. Since n is the largest number such that $d_n(H_p) \le \varepsilon$, where $d_n(H_p) = \Theta(\varepsilon^{-1/p})$, is the Kolmogorov n-width, asymptotic optimality of the algorithm follows.

Cohen, A. I., and Varaiya, P., Rate of convergence and optimality conditions of root finding and optimality algorithms, Dept. of Electrical Engineering and Computer Science Rep., Univ. of California, Berkeley, California (1970).

core, nonlinear equations, scalar, multivariate, maximal order, one-point iterations, iterations with memory

Considers maximal order of iterations for solving scalar and multivariate nonlinear equations, $f = 0$. The information is the values of f and its derivatives at one or more points. Following S. Winograd, shows it may be possible to encode memory in an infinite precision number and therefore change maximal order of a class of algorithms. Adds condition to definition of order which ensures that encoding does not affect order. Finds maximal order of a class of algorithms.

Coman, Gh., Monosplines and optimal quadrature formulae in L_p, *Rend. Mdt.* **5** (1972), 567–577.

core, integration, scalar, optimal points of information, optimal linear algorithms

Considers integration for the class of scalar functions whose rth derivative is bounded in L_p by a constant. The information is the values of $f, f', \ldots, f^{(r-1)}$. Based on monosplines with least deviation, the weights and points of optimal linear quadrature formulas are derived.

Coman, Gh., and Micula, Gh., Optimal cubature formulae, *Rend. Mat.* **4** (1971), 303–311.

core, integration, multivariate, optimal points of information, optimal linear algorithms

Considers integration for a class of scalar functions of two variables with bounded derivatives in L_2. The information is the values of f and its derivatives. The weights and points of optimal linear quadrature formulas are derived.

Cooper, G. J., and Verner, J. H., Some explicit Runge–Kutta methods of high order, *SIAM J. Numer. Anal.* **9** (1972), 389–405.

core, differential equations, Runge–Kutta methods

Considers the solution of $y' = f(y)$, where y and f are vectors, by explicit Runge–Kutta methods. The information is the values of f. Studies Runge–Kutta methods of high order.

Cooper, L.: *See* Krolak, P.

Danilin, Yu. M., On one algorithm efficiency estimation of absolute minimum finding (in Russian), *Zh. Vychisl. Mat. Mat. Fiz.* **11** (1971), 1026–1030 [*English transl.*: Danilin, Yu. M., Estimation of the efficiency of an absolute-minimum-finding algorithm, *U.S.S.R. Computational Math. and Math. Phys.* **11** (1971), 261–267].

core, extremum, scalar, optimal error algorithms

Considers the search for the minimum in the class of scalar functions satisfying a Lipschitz condition with a given constant. The information is the values of f. Proves that Piavsky's algorithm (see Piavsky [72]) requires at most three times as many function evaluations as an optimal algorithm which solves the problem to within error ε.

de Boor, C., Computational aspects of optimal recovery, *in* "Optimal Estimation in Approximation Theory" (C. A. Micchelli and T. J. Rivlin, eds.), pp. 69–91. Plenum Press, New York, 1977.

core, approximation, perfect splines

Considers approximation for the class of scalar functions whose kth derivative is bounded in L_∞. The information is the values of f and its derivatives. Presents a Fortran subroutine for the construction of a perfect interpolatory spline which is the optimal error algorithm. Related to Micchelli, Rivlin, and Winograd [76] and Gaffney and Powell [76].

den Heijer, C., On the local convergence of Newton's method, Dept. of Numerical Mathematics Rep., Mathematisch Centrum, Amsterdam (1976).

core, nonlinear equations, abstract, Newton iteration, optimal convergence

Establishes the optimal radius of the ball of convergence for Newton iteration for the zero of a nonlinear operator. See Traub and Woźniakowski [77] (Convergence and complexity of Newton iteration for operator equations).

Eckhardt, U., Einige eigenschaften wilfscher quadraturformeln, *Numer. Math.* **12** (1968), 1–7.

core, integration, scalar, optimal points of information, optimal algorithms

Considers integration for a class of scalar analytic functions on the unit disk in a Hilbert space. The information is the values of f. The existence and properties of optimal weights and points of quadrature formulas in the sense of Wilf [64] are presented.

Ehrmann, H., Konstruktion und durchführung von iterationsverfahren höherer ordnung, *Arch. Rational Mech. Anal.* **4** (1959), 65–88.

core, polynomials, Newton iteration

In the complexity model of this paper, the author discusses what order iteration is best for computing a zero of a polynomial as a function of degree.

Eichhorn, B. H., On sequential search, "Selected Statistical Papers," Vol. 1, pp. 81–95. Math. Centrum, Amsterdam, 1968.

core, extremum, nonlinear equations, scalar, optimal algorithms, average case analysis

Considers the search for the maximum or zero in the class of scalar unimodal functions or in the class of monotone nonincreasing functions with one zero, respectively. The information is the values of f at adaptively chosen points. Optimal algorithms for the worst and average cases are discussed.

Elhay, S., Optimal quadrature, *Bull. Austral. Math. Soc.* **1** (1969), 81–108.

core, integration, scalar, optimal points of information, optimal linear algorithms

Considers integration for a class of smooth scalar functions for which $\sum_{j=0}^{r} a_j^2 \|f^{(j)}\|_2^2$ is bounded, a_j a real constant, $a_0 \neq 0$, $a_r \neq 0$. The information is the values of $f, f', \ldots, f^{(r-1)}$. Optimal quadrature formulas with optimal points of information are derived for $r = 1$ and 2. Some properties of optimal quadrature formulas are found for any r.

Emelyanov, K. V., and Ilin, A. M., Number of arithmetical operations necessary for the approximate solution of Fredholm integral equations of the second kind (in Russian), *Zh. Vychisl. Mat. Mat. Fiz.* **7** (1967), 905–910 [*English transl.: U.S.S.R. Computational Math. and Math. Phys.* **7** (1967), 259–266].

core, integral equations, lower and upper bounds

Considers the integral equation $y(P) = \int_D K(P,Q)y(Q)\,dQ + f(P)$, where $D \subset \mathbb{R}^m$, for the class of functions K and f whose rth derivatives are bounded. The information is the values of K and f. Shows that a lower bound on the error of any algorithm using n evaluations of K is at least $n^{-r/(2m)}$. The algorithm whose error achieves this bound is derived.

Feldstein, A., Bounds on order and Ostrowski efficiency for interpolatory iteration algorithms, UCRL-72238 Rep., Lawrence Livermore Lab. (1969).

core, nonlinear equations, scalar, interpolatory iterations, order, complexity index, parallel algorithms

As in Feldstein and Firestone [67], studies Hermite interpolatory functions (HIFs) for the calculation of a zero of a nonlinear equation. Establishes bounds and limit theorems on the order of HIFs. Uses these results to establish bounds and limit theorems on a complexity index. Obtains such results for simple parallel iterations.

Feldstein, A., and Firestone, R. M., Hermite interpolatory iteration theory and parallel numerical analysis, Division of Applied Mathematics Rep., Brown Univ. (1967).

core, nonlinear equations, scalar, interpolatory iterations, order, complexity index, parallel algorithms

Studies Hermite interpolatory iteration functions (HIFs) for a zero of a nonlinear function f. The information is the values of f and its derivatives at a number of points. Analyzes order of a HIF and studies a complexity index. Discusses application of HIFs to parallel computation and compares complexity indices of parallel algorithms.

Feldstein, A., and Traub, J. F., Order of vector recurrences with applications to nonlinear iterations, parallel algorithms, and the power method, Dept. of Computer Science Rep., Carnegie-Mellon Univ. (1974).

core, nonlinear equations, scalar, iterative complexity, iterative algorithms, parallel algorithms, order, iterations with memory, interpolatory iterations, composition

Same as Feldstein and Traub [77] with many examples of applications.

Feldstein, A., and Traub, J. F., Asymptotic behavior of vector recurrences with applications, *Math. Comp.* **31** (1977), 180–192.

core, nonlinear equations, scalar, iterative complexity, iterative algorithms, parallel algorithms, order, iterations with memory, interpolatory iterations, composition, order

Proves under very weak assumptions that the root and quotient orders of the vector recurrence $\mathbf{y}_{n+1} = \mathbf{M}\mathbf{y}_n + \mathbf{w}_{n+1}$ is the spectral radius of M. Continues work of Rice [71] on assigning a matrix representation to iterations for solving nonlinear equations. Applies this result to the analysis of parallel iteration algorithms and to the order and complexity of composite iterations. Shows that a composite iteration may have order less than, equal to, or greater than the products of the individual iterations.

Fine, T., Optimum search for the location of the maximum of a unimodal function, *IEEE Trans. Inform. Theory* **IT-12** (1966), 103–111.

core, extremum, scalar, optimal algorithms, average case analysis

Considers the search for the maximum in the class of unimodal functions assuming that the maximum points are uniformly distributed. The information is the values of f at adaptively chosen points. Optimal algorithms, i.e., algorithms which minimize the expected cost, are studied. See the referee's report by J. Kiefer in *Math. Rev.* **34**, No. 7260, where the model of the paper is discussed.

Firestone, R. M.: *See* Feldstein, A.

Forst, W., Zur optimalität interpolatorischer quadraturformeln periodischer funktionen, *Numer. Math.* **25** (1975), 15–21.

core, integration, scalar, optimal linear algorithms

Considers integration for the Favard class of scalar periodic functions whose rth derivatives satisfy a Lipschitz condition with unity. The information is the values of f at n equidistant points. Shows that the trapezoidal rule is the unique optimal quadrature formula and that its error is K_{r+1}/n^{r+1}, where K_{r+1} is the Favard constant.

Forst, W., Optimale hermite-interpolation differenzierbarer periodischer funktionen, *J. Approx. Theory* **20** (1977), 333–347.

core, interpolation, scalar, optimal algorithms, splines

Considers interpolation for the Favard class of 2π-periodic scalar functions whose rth derivative is bounded in L_∞ by unity. The information is the values of f and its derivatives. Optimal algorithms are derived in terms of periodic splines.

Gaffney, P. W., Optimal interpolation, D. Phil. Thesis, Oxford Univ. (1976).

core, interpolation, scalar, optimal error algorithms, splines

This is a Ph.D. thesis. Some of the results are published in Gaffney [77] (The range of possible values of $f(x)$), Gaffney [77] (To compute the optimal interpolation formula), and Gaffney and Powell [76].

Gaffney, P. W., The range of possible values of $f(x)$, Computer Science and Systems Division Rep., AERE, Harwell, Oxfordshire (1977).

core, interpolation, scalar, optimal error algorithms, perfect splines, lower bounds

Considers the interpolation problem for the class of scalar functions with bounded kth derivative in L_∞. The information is the values of f at n points. The range of possible values of $f(x)$ is bounded by interpolatory perfect splines of degree k. The computation of these splines at x is considered.

Gaffney, P. W., To compute the optimal interpolation formula, Computer Science and Systems Division Rep., AERE, Harwell, Oxfordshire (1977).

core, interpolation, scalar, optimal error algorithms, splines

Considers the interpolation problem for the class of scalar functions with bounded kth derivative in L_∞. The information is the values of f at n points. The optimal error algorithm is defined by an interpolatory spline Ω of degree $k-1$ with exactly $n-k$ knots. The error is expressed by a perfect spline B of degree k with the same $n - k$ knots as Ω. The computation of knots and coefficients of Ω and B is studied.

Gaffney, P. W., and Powell, M. J. D., Optimal interpolation, *in* "Numerical Analysis" (G. A. Watson, ed.), Lecture Notes in Math., Vol. 506, pp. 90–100. Springer Verlag, Berlin and New York, 1976.

core, interpolation, scalar, optimal error algorithms, splines

Considers interpolation for the class of scalar functions with bounded kth derivative in L_∞. The information is the values of f. The range of possible values $f(x)$ for functions with the same information is derived in terms of perfect splines. The optimal error algorithm and its error are presented.

Gaisarian, S. S., An optimal algorithm for the approximate computation of quadratures, (in Russian), *Zh. Vychisl. Mat. Mat. Zat. Fiz.* **9** (1969), 1015–1023 [*English transl.: U.S.S.R. Computational Math. and Math. Phys.* **9** (1969), 42–53].

core, integration, scalar, optimal points of information

Considers quadrature formulas of degree s. The information is the values of f. Minimizing the dominant error term, the optimal points of information are derived in terms of the sth derivative of f.

Gaisarian, S. S., The choice of optimal networks for the numerical solution of the Cauchy problem for a set of ordinary differential equations, *Zh. Vychisl. Mat. Mat. Fiz.* **10** (1970), 465–474 [*English transl.: U.S.S.R. Computational Math. and Math. Phys.* **10** (1970), 253–267].

core, ordinary differential equations, optimal points of information

Considers the system of ordinary differential equation $y' = f(x,y)$, $y(x_0) = y_0$. The information is the values of f. The problem is solved by a one-step method on the net $\{x_i\}$. The points $\{x_i\}$ which minimize the dominant error term are studied. Asymptotically optimal nets are found.

Gaisarian, S. S.: *See also* Tikhonov, A. N.

Gal, S., Multidimensional minimax search for a maximum, *SIAM J. Appl. Math.* **23** (1972), 513–526.

core, extremum, multivariate, optimal algorithms

Considers the search for the maximum in the class of linearly unimodal scalar functions of several variables. The information is the values of f at adaptively chosen points. Optimal algorithms are defined as algorithms which minimize the measure of the set of maxima of all functions satisfying the computed information. Proves that after n evaluations of f, the optimal error is at most roughly $(3/4)^n$. Also considers the same problem for the subclass of spherical symmetric functions.

Gal, S., and Micchelli, C. A., Optimal sequential and non-sequential procedures for evaluating a functional, Univ. of Wisconsin-Madison Rep. 1871 (1978). To appear in *Appl. Anal.*

core, approximation of linear functionals and operators, optimal adaptive and nonadaptive information

Considers approximation of a linear functional (or operator) in a given class. The information is the values, possibly perturbed, of n linear functionals. Optimal deterministic, random, and adaptive sequential information are studied. Shows that for many problems the optimal deterministic and adaptive sequential information yield the same error.

Ganshin, G. S., Calculation of the greatest value of function (in Russian), *Zh. Vychisl. Mat. Mat. Fiz.* **16** (1976), 30 J 39 [*English transl.:* Ganshin, G. S., Function maximization, *U.S.S.R. Computational Math. and Math. Phys.* **16** (1976), 26–36].

core, extremum, multivariate

Considers the search for the maximum in the class of scalar functions of several variables with a certain derivative bounded. The information is the values of f. The total number of evaluations of f needed to solve the problem to within ε is estimated.

Ganshin, G. S., Optimal algorithms of calculation of the function highest value (in Russian), *Zh. Vychisl. Mat. Mat. Fiz.* **17** (1977), 562–572.
core, extremum, scalar, optimal points of information, optimal error algorithms

Considers the search for the maximum in three classes of scalar functions. The information is the values of f at n points. The optimal error algorithms are derived by finding the optimal net of n points of the given interval. The minimum value of n for which the error is at most ε is also derived.

Garey, M. R., and Johnson, D. S., Approximation algorithms for combinatorial problems: An annotated bibliography, *in* "Algorithms and Complexity: New Directions and Recent Results" (J. F. Traub, ed.), pp. 41–52. Academic Press, New York, 1976.
survey, combinatorial complexity

Gives annotated bibliography on papers with polynomial time approximation algorithms for combinatorial problems which have no known polynomial time algorithm.

Garey, M. R., and Johnson, D. S., "Computers and Intractability." Freeman, San Francisco, California, 1979.
survey, combinatorial complexity

A text on NP-complete problem. Contains material on approximate solution of hard problems. Good bibliography.

Gentleman, W. M., Measures of efficiency, unpublished letter to J. F. Traub (1970).
core, iterative complexity

Proves that any efficiency index which is invariant under self-composition must be of a certain form. Although never published, the result has been widely quoted.

Germeier, Ju. B., "Introduction to the Theory of Operations Research." Nauka, Moscow, 1971 (in Russian).
mathematics and core, operations research, extremum, scalar, optimal algorithms

Considers among other problems the search for the maximum in the class of scalar functions satisfying a Lipschitz condition. The information is the values of f. An optimal algorithm is given.

Giršovič, Ju. M.: *See* Levin, M. I.

Golomb, M., Interpolation operators as optimal recovery schemes for classes of analytic functions, *in* "Optimal Estimation in Approximation Theory" (C. A. Micchelli and T. J. Rivlin, eds.), pp. 93–137. Plenum Press, New York, 1977.
core, approximation, optimal linear algorithms, splines, n-widths

Considers approximation for a class of complex-valued functions which have a reproducing kernel in a Hilbert space. The information is the values of f. The linear optimal error algorithm is shown to be an interpolatory spline. The optimal points of information and relation to n-widths are considered. For particular spaces of analytic functions, interpolatory splines and optimal errors are explicitly found.

Golomb, M., and Weinberger, H. F., Optimal approximation and error bounds, *in* "On Numerical Approximation" (R. E. Langer, ed.), pp. 117–190. Univ. of Wisconsin Press, Madison, Wisconsin, 1959.
core, approximation of linear functionals, optimal error algorithms

This is one of the first papers dealing with approximation of a linear functional for a given class of elements. The information is the values of n functionals. Different conditions assuring the existence of optimal error algorithms with finite error are presented. The optimal error algorithms are extensively studied for a number of absorbing classes. The linear optimal error algorithms are derived for a Hilbert case. Although the word "spline" is not used, these algorithms are primarily based on splines. Many valuable examples illustrate the paper.

Grebennikov, A. I., On optimal approximation of nonlinear operators (in Russian), *Zh. Vychisl. Mat. Mat. Fiz.* **18** (1978), 762–768.

core, approximation of nonlinear operators, splines

Considers approximation of $B(u)$, where B is a nonlinear operator from a Hilbert space H into a linear normed space, and $\|Lu\| \le R$, where L is a linear operator from H into a Hilbert space; R is a given constant. The information is the values of n linear functionals $A_n = [L_1(u)_1, \ldots, L_n(u)]$. The algorithm $\varphi(A_n) = B(u_f)$, where u_f is an interpolatory spline is introduced. Shows that φ is optimal within a factor of two. Conditions assuring optimality of φ are mentioned.

Grebennikov, A. I., and Morozov, V. A., On optimal approximation of operators (in Russian), *Zh. Vychisl. Mat. Mat. Fiz.* **17** (1977), 3–15.

core, approximation of linear operators, optimal linear algorithms

Considers primarily the approximation of a linear operator in a Hilbert space. The information is a linear operator. The existence of a linear optimal error algorithm is proven. Perturbed information is also studied. The paper is related to the concepts of central and spline algorithms.

Gross, O., and Johnson, S. M., Sequential minimax search for a zero of a convex function, *MTAC* (now *Math. Comp.*) **13** (1959), 44–51.

core, nonlinear equations, scalar, convex functions, optimal points of information, optimal algorithms

Considers the solution of nonlinear scalar equations for a class of convex continuous functions f such that $f(a) > 0$ and $f(b) < 0$. The information is the values of f at n adaptively chosen points. Studies optimal points of information, optimal algorithms, and their errors.

Haber, S., The error in numerical integration of analytic functions, *Quart. Appl. Math.* **29** (1971), 411–420.

core, integration, analytic functions, scalar, optimal linear algorithms, upper bounds

Considers integration for two classes of scalar analytic functions. The information is the values of f at n points. Shows upper bounds on the minimal error of quadrature formulas.

Herzberger, J., Über matrixdarstellungen für iterationsverfahren bei nichtlinearen gleichungen, *Computing* **12** (1974), 215–222.

core, nonlinear equations, scalar, iterative algorithms, parallel algorithms, order, interpolatory iterations, composition

Shows that the order of convergence of certain iterations for the solution of nonlinear equations is the spectral radius of a certain matrix. See also Feldstein and Traub [77].

Hindmarsh, A. C., Optimality in a class of rootfinding algorithms, *SIAM J. Numer. Anal.* **9** (1972), 205–214.

core, nonlinear equations, scalar, iterative complexity, iterative algorithms, order, interpolatory iterations

Studies Hermite interpolatory iteration functions (HIFs) for solving nonlinear equations. Shows how the order of composition of HIFs may be computed and that the order of a composite HIF is not the product of the orders. Studies the complexity of composite HIFs and discusses optimal complexity.

Hyafil, L., Optimal search for the zero of the $(n-1)$st derivative, IRIA/LABORIA Rep. No. 247 (1977).

core, nonlinear equations, scalar, optimal points of information, optimal error algorithm

Considers the search for a zero of the $(n-1)$st derivative in the class of scalar functions f for which $f^{(n-1)}$ is continuous in $[a,b]$, changes sign only once, and $f^{(n-1)}(a) < 0$, $f^{(n-1)}(b) > 0$. The information is the values of f at n points. Optimal error algorithms with optimally chosen points of information are derived for odd n and $n = 2$ or $n = 4$. The proofs are based on the zero-finding problem for an asynchronous multiprocessor.

Ibragimov, I. I., and Aliev, R. M., Best quadrature formulas for certain classes of functions (in Russian), *Dokl. Akad. Nauk SSSR* **162** (1965), 23–25 [*English transl.: Soviet Math. Dokl.* **6** (1965), 621–623].

core, integration, scalar, optimal points of information, optimal linear algorithms

Considers integration for the class of scalar functions with bounded rth derivative in L_q for $q = 1, 2,$ or $+\infty$. The information is the values of $f, f', \ldots, f^{(r-2)}$ at m points. Optimal linear algorithms with optimal points of information and their errors are derived for even r.

Ilin, A. M.: *See* Emelyanov, K. V.

Ivanov, V. V., On optimal algorithms for function minimization on certain classes of functions (in Russian), *Kibernetika* **4** (1972), 81–94.

core, extremum, multivariate, optimal algorithms, error bounds

Considers the search for the minimum in the class of scalar functions of m variables with rth derivative satisfying a Lipschitz condition with a given constant. The information is the values of f and its derivatives at n points. Shows that the error of an optimal algorithm is $\Theta(n^{-(r+1)/m})$. Also presents asymptotically optimal algorithms for the classes of analytic and entire functions.

Ivanov, V. V., On optimal algorithms for the calculation of singular integrals (in Russian), *Dokl. Akad. Nauk SSSR* **204** (1972), 21–24 [*English transl.: Soviet Math. Dokl.* **13** (1972), 576–580].

core, integration, scalar, optimal points of information, optimal algorithms

Considers a certain singular integration for a class of scalar functions. The information is the values of f. The optimal algorithms, their errors, and optimal points of information are given for several classes of functions.

Ivanov, V. V., On optimal in accuracy algorithms for approximate solution of operator equations of the first kind (in Russian), *Zh. Vychisl. Mat. Mat. Fiz.* **15** (1975), 3–12 [*English transl.: Ivanov, V. V., Algorithms of optimal accuracy for the approximate solution of operator equations of the first kind, U.S.S.R. Computational Math. and Math. Phys.* **15** (1975), 1–9].

core, operator equations, optimal error algorithms

Considers approximation of an element $R = O(I)$ for I belonging to a given set where O is an operator. The information is the value of $\Psi(I)$ for an operator Ψ. The optimal error algorithms are found. This idea is illustrated by the solution of $K\Phi = f$, where K is a linear compact operator for the class of f such that $f = KLu$; L is a given linear compact operator, $\|u\| \leq 1$. The information is $\Psi(K,f) = (K_\varepsilon, f_\varepsilon)$, where $\|K - K_\varepsilon\| \leq \varepsilon, \|f - f_\varepsilon\| \leq \varepsilon$. The asymptotically optimal error algorithms are derived.

Ivanov, V. V., On optimal algorithms approximating functions for some classes (in Russian), in "Theory of Approximation of Functions," pp. 195–200. Nauka, Moscow, 1977.

core, approximation, optimal error algorithms

This is a paper presented at the international conference in Kaluga, USSR in 1975. Surveys optimality results for the approximation problem in some classes of functions.

Jankowska, J., Multivariate secant method, Ph.D. thesis, Univ. of Warsaw, first part of thesis (1975). See also *SIAM J. Numer. Anal.* **16** (1979), 547–562.

core, nonlinear equations, multivariate, secant iteration, order of information

Considers the iterative solution of a system of nonlinear equations $f = 0$. The information is the values of f. Studies optimal properties of multivariate secant iteration. Shows how the position of points at which f is evaluated influences the order of information.

Jankowski, M., and Woźniakowski, H., Computational complexity in numerical mathematics (in Polish), *Mat. Stosow.* **5** (1975), 5–27.
core, algebraic and analytic complexity
Surveys recent problems in algebraic and analytic complexity for some numerical problems.

Jarratt, P., Some efficient fourth order multipoint methods for solving equations, *BIT* **9** (1969), 119–124.
core, nonlinear equations, scalar, multipoint iterations, iterative complexity, order
Derives a fourth-order method for solving scalar nonlinear equations that uses one function and two first derivative evaluations. Discusses the complexity index.

Jarratt, P., A review of methods for solving nonlinear algebraic equations in one variable, *in* "Numerical Methods for Nonlinear Equations" (P. Rabinowitz, ed.), pp. 1–26. Gordon and Breach, New York, 1970.
core, nonlinear equations, scalar, one-point iterations with memory, multipoint iterations, order
Surveys iterative algorithms for computing a zero of a scalar nonlinear function f. The information is the values of f or the values of f and f'. Also contains some new material on comparing the computational efficiencies of various iterations.

Jeeves, T. A., Secant modification of Newton's method, *Comm. ACM* **1** (8) (1958), 9–10.
core, nonlinear equations, scalar, Newton iteration, secant iteration

Compares complexity of secant and Newton iteration for solving a scalar nonlinear equation $f = 0$. First to observe that if the cost of evaluating f' is greater than 0.43 times the cost of evaluating f, then the secant iteration has lower complexity than Newton's iteration.

Jetter, K., Optimale quadraturformeln mit semidefiniten Peano-kernen, *Numer. Math.* **25** (1976), 239–249.
core, integration, scalar, optimal algorithms
Considers integration for a class of m-times differentiable scalar functions. The information is the values of f. Expresses the error of a quadrature formula as $cf^{(m)}(\xi)$ and studies the minimization of $|c|$.

Johnson, L. W., and Riess, R. D., Minimal quadratures for functions of low-order continuity, *Math. Comp.* **25** (1971), 831–835.
core, integration, scalar, optimal linear algorithms, upper bounds
Considers integration for a class of scalar functions of low-order continuity and for a class of scalar analytic functions on the unit disk. The information is the values of f at n points. Establishes the existence of quadrature formulas with minimal error. For the class of analytic functions, shows that the error of these formulas is at most $O(n^{-1})$.

Johnson, S. M., Best exploration for maximum is Fibonaccian, RAND Corp. Rep. P-856 (1956).
core, extremum, scalar, optimal points of information, optimal error algorithms
Considers the search for the maximum in the class of scalar unimodal functions. The information is the values of f at n adaptively chosen points. Optimality of Fibonaccian search is proven. See Kiefer [53].

Johnson, S. M.: *See also* Gross, O.

Judin, D. B., and Nemirovsky, A. S., A bound of information complexity for mathematical programming problems (in Russian), *Ekonom. Mat. Metody* **12** (1976), 128–142.
core, extremum, multivariate, optimal points of information, optimal algorithms, error bounds
Considers mathematical programming problems in terms of minimization of scalar functions of several variables. The information is the values of f and its derivatives. Shows how many

evaluations of f are necessary to find an ε-approximation for the class of functions whose $(k-1)$st derivative satisfies a Lipschitz condition, and for a class of convex functions. Asymptotically optimal algorithms are given.

Judin, D. B., and Nemirovsky, A. S., Information complexity and effective methods for the solution of convex extremal problems (in Russian), *Ekonom. Mat. Metody* **12** (1976), 357–369.

core, extremum, multivariate, optimal points of information, optimal algorithms, error bonds

Continuation of Judin and Nemirovsky [76]. Studies practical aspects of asymptotically optimal algorithms. Shows that for a class of convex problems in a Hilbert space, $\Theta(\varepsilon^{-2})$ function evaluations are necessary and sufficient to find an ε-approximation.

Judin, D. B., and Nemirovsky, A. S., Information complexity for strict convex programming (in Russian), *Ekonom. Mat. Metody* **13** (1977), 550–559.

core, extremum, multivariate, optimal points of information, optimal algorithms, error bounds

Continuation of Judin and Nemirovsky [76] (Information complexity and effective methods for the solution of convex extremal problems). Shows that for a class of strictly convex functions, $\Theta(\ln \varepsilon^{-1})$ function evaluations are necessary and sufficient to find an ε-approximation.

Kacewicz, B., Integrals with a kernel in the solution of nonlinear equations in N dimensions, Computer Science Dept. Rep., Carnegie-Mellon Univ. (1975). See also *J. Assoc. Comput. Mach.* **26** (1979), 239–249.

core, nonlinear equations, abstract, multivariate, iterative information, order of information

Considers iterations for the solution of the nonlinear operator equation $f = 0$. Defines a maximal order iteration $I^q_{-1,s}$ which uses the information consisting of evaluations of the first s derivatives of f and a certain "integral with kernel g". In his complexity model, proves that for the multivariate case with large N, $I^q_{-1,1}$ has lower complexity than $I^q_{-1,s}$, $s \geq 2$, and lower complexity than any interpolatory iteration.

Kacewicz, B., An integral-interpolation iterative method for the solution of scalar equations, *Numer. Math.* **26** (1976), 355–365.

core, nonlinear equations, scalar, iterative information, order of information

Introduces the idea of using integral information for the solution of the nonlinear scalar equation $f = 0$. Defines an iteration $I_{-1,s}$ which uses the information consisting of evaluations of the first s derivatives of f and a certain integral. Establishes the maximal order of $I_{-1,s}$.

Kacewicz, B., The use of integrals in the solution of nonlinear equations in N dimensions, *in* "Analytic Computational Complexity" (J. F. Traub, ed.), pp. 127–141. Academic Press, New York, 1976.

core, nonlinear equations, multivariate, iterative information, order of information

Introduces the idea of using integral information for the solution of the nonlinear N-dimensional equation $f = 0$. Defines a maximal order iteration $I_{-1,s}$ which uses the information consisting of evaluations of the first s derivatives of f and a certain integral. Proves in his complexity model that $I_{-1,1}$ has lower complexity than $I_{-1,s}$, $s \geq 2$. Proves that $I_{-1,1}$ has lower complexity than any interpolatory iteration using standard information, for large N.

Kacewicz, B., and Woźniakowski, H., A survey of recent problems and results in analytic computational complexity, "Mathematical Foundations of Computer Science 77" (J. Gruska, ed.), Lecture Notes in Computer Science No. 53, pp. 93–107. Springer-Verlag, Berlin and New York, 1977.

core, nonlinear equations, abstract, iterative complexity

Surveys recent research in iterative complexity. Good bibliography.

Karlin, S., Best quadrature formulas and interpolation by splines satisfying boundary conditions, and The fundamental theorem of algebra for monosplines satisfying certain boundary conditions and applications to optimal quadrature formulae, *in* "Approximations with Special

Emphasis on Spline Functions" (I. J. Schoenberg, ed.), pp. 447–466, 467–484. Academic Press, New York, 1969.

core, integration, scalar, optimal points of information, optimal linear algorithms

Considers, among other problems, integration for the class of scalar functions with bounded rth derivative in L_2. The information is the values of f at n points and the values of its derivatives at the endpoints. Optimal algorithms in the sense of Sard with fixed and varying points of information, and their relation to monosplines are studied.

Karlin, S., Best quadrature formulas and splines, *J. Approx. Theory* **4** (1971), 59–90.

core, integration, scalar, optimal algorithms

Considers integration for the class of scalar functions whose nth derivative is bounded in L_2 by unity. The information is the values of f and its derivatives at the endpoints of an integration interval. Optimal algorithms in the sense of Sard are studied. Shows the correspondence to splines satisfying boundary conditions.

Karp, R. M., and Miranker, W. L., Parallel minimax search for a maximum, *J. Combin. Theory* **4** (1968), 19–35.

core, extremum, scalar, optimal error algorithms, parallel algorithms

Considers the search for the maximum in a class of scalar unimodal functions. The information is the values of f. Parallel optimal search algorithms are studied. Results related to those of Kiefer [53] are established.

Kaul, V.: *See* Chawla, M. M.

Kautsky, J., Optimal quadrature formulae and minimal monosplines in L_q, *J. Austral. Mat. Soc.* **11** (1970), 48–56.

core, integration, scalar, optimal points of information, optimal linear algorithms

Considers integration for the class of scalar functions whose rth derivative is bounded in L_q. The information is the values of $f, f', \ldots, f^{(r-1)}$. The optimal quadrature formula with optimal points of information and its error are derived. The proof is based on monosplines of minimal L_q norm.

Keast, P., Optimal parameters for multidimensional integration, *SIAM J. Numer. Anal.* **10** (1973), 831–838.

core, integration, multivariate, optimal algorithms

Considers integration for a class of scalar periodic functions of several variables with bounded Fourier coefficients. The information is the values of f. Optimal quadrature formulas with equidistant weights are considered. Related to Korobov [63].

Kiefer, J., Sequential minimax search for a maximum, *Proc. Amer. Math. Soc.* **4** (1953), 502–505.

core, extremum, scalar, optimal points of information, optimal error algorithms

Considers the search for the maximum in the class of scalar unimodal functions. The information is the values of f at n points. Proves the classic result that Fibonaccian search is the optimal error algorithm. The optimal error is $1/F_{n+1}$, where F_{n+1} is the $(n+1)$th Fibonacci number.

Kiefer has informed us that this work was done as an MIT Master's Thesis in 1948, but was only published later with the encouragement of J. Wolfowitz.

Kiefer, J., Optimum sequential search and approximation methods under minimum regularity assumptions, *J. Soc. Indust. Appl. Math.* **5** (1957), 105–136.

core, approximation of nonlinear functionals, optimal points of information, optimal error algorithms

Considers approximation of a nonlinear functional for a class of functions. The information is the values of f. Optimal error algorithms are studied for the search of the zero of a monotonic real function, the search of the point at which the maximum of a unimodal function is attained, and the integration of a real function which is nondecreasing or satisfies a Lipschitz condition with a given constant.

Knauff, W., and Kress, R., Optimale approximation linearer funktionale auf periodischen funktionen, *Numer. Math.* **22** (1974), 187–205.
core, approximation of linear functionals, optimal algorithms
Considers approximation of a linear bounded functional for a class of periodic scalar functions. The information is the values of f at n equidistant points. Linear optimal algorithms and their errors are derived.

Knauff, W., and Kress, R., Optimale approximation mit nebenbedingungen an lineare funktionale auf periodischen funktionen, *Numer. Math.* **25** (1976), 149–159.
core, approximation of linear functionals, optimal algorithms
Considers approximation of a linear functional for a class of scalar periodic complex functions. The information is the values of f at n equidistant points. Linear optimal algorithms which satisfy given linear constraints are derived.

Kornejčuk, N. P., Best cubature formulas for some classes of functions of many variables (in Russian), *Mat. Zametki* **3** (1968), 565–576 [*English transl.*: *Math. Notes* **3** (1968), 360–367].
core, integration, multivariate, optimal points of information, optimal linear algorithms
Considers integration for the class of scalar functions of several variables with bounded modulus of continuity. The information is the values of f. Optimal points of information and optimal weights for linear cubature formulas are found. The error for optimal points is derived.

Kornejčuk, N. P., New results on extremal problems of the theory of quadratures (in Russian), appendix to the second edition of S. M. Nikolskij, "Quadrature Formulae," pp. 136–223. Nauka, Moscow, 1974.
core, integration, scalar, optimal linear algorithms
Surveys the results obtained since the first edition of Nikolskij's book in 1958. Considers the integration problem for the class of scalar functions with bounded rth derivative in L_p. The information is the values of $f, f', \ldots, f^{(\zeta)}$. Shows that the problem of optimal points of information and optimal weights of linear quadrature formulas corresponds to the minimal norm spline with dominant term t^r in the space L_q, $1/p + 1/q = 1$. The solution is obtained for $\zeta = r - 1$, and $\zeta = r - 2$ for odd r.

Kornejčuk, N. P., "Extremal Problems of Approximation Theory." Nauka, Moscow, 1976 (in Russian).
mathematics, approximation, error bounds, n-widths
Surveys sharp error bounds on the best approximation for several classes of functions (mostly periodic functions with bounded derivative or bounded modulus of continuity). The Kolmogorov and linear n-widths and extremal subspaces are considered. Good Russian bibliography.

Kornejčuk, N. P., and Lušpaj, N. E., Best quadrature formulas for classes of differentiable functions and piecewise-polynomial approximation (in Russian), *Izv. Akad. Nauk SSSR Ser. Mat.* **33** (1969), 1416–1437 [*English transl.*: *ath. U.S.S.R. Izv.* **3** (1969), 1335–1355].
core, integration, scalar, optimal points of information, optimal linear algorithms
Considers the integration problem for the class of scalar functions with bounded rth derivative in L_p. The information is the values of $f, f', \ldots, f^{(\zeta)}$. Shows that the problem of optimal points of information and optimal weights of linear quadrature formulas corresponds to the minimal norm spline with leading term t^r in the space L_q, $1/p + 1/q = 1$. The solution is obtained for $\zeta = r - 2$, and $\zeta = r - 3$ for odd r.

Korobov, N. M., "Number Theory Methods in Approximation Analysis." Fizmatgiz, Moscow, 1963 (in Russian).

core, integration, multivariate, interpolation, integral equations

Considers integration for the class of periodic scalar functions of several variables with bounded Fourier coefficients. The information is the values of f. Optimal points of information for quadrature formulas with equal weights are considered. Computation of nearly optimal points is studied using some results from number theory. The results are applied to the interpolation and integral equation problems.

Korotkov, V. B., A lower error bound on cubature formulas (in Russian), *Sibirsk. Mat. Z.* **17** (1977), 1188–1191.

core, integration, multivariate, lower bounds

Considers the integration problem for a class E of scalar functions of several variables. The information is the values of f. Shows that a lower bound on the error for this problem depends on the imbedding operator from E to the space L_1. Shows a connection between the integration and approximation problems.

Korytowski, A.: *See* Adamski, A.

Kress, R.: *See* Knauff, W.

Krolak, P., and Cooper, L., An extension of Fibonaccian search to several variables, *Comm. ACM* **6**(1963), 639–641.

core, extremum, multivariate

Considers the search for the maximum in a class of continuous unimodal scalar functions of several variables. The information is the values of f. Presents an algorithm which is a generalization of a one-dimensional Fibonaccian search algorithm. Partial optimality of this algorithm is given in Krolak, P., Property of the Krolak–Cooper extension of Fibonaccian search, *SIAM Rev.* **8** (1966), 510–517, and Krolak, P., Further extensions of Fibonaccian search to nonlinear programming problems, *SIAM J. Control.* **6** (1968), 258–265.

Krylov, V. I., "Approximate Calculation of Integrals," Chapter 8, pp. 133–149. Macmillan, New York, 1962.

core, integration, scalar, optimal points of information, optimal linear algorithms

Considers integration for the class of scalar functions whose rth derivative is bounded in L_q. The information is the values of f. Optimal weights and points of quadrature formulas are studied. The solution is derived for small r.

Kukarkin, A. B.: *See* Zhileikin, Ya. M.

Kung, H. T., A bound on the multiplicative efficiency of iteration, *Proc. 4th Ann. ACM Symp. Theory Comput.* (1972), 102–107. Revised paper in *JCSS* **7** (1973), 334–342.

core, nonlinear equations, algebraic numbers, scalar, optimal iterations, multipoint iterations, iterative complexity, maximal order

Proves that the multiplicative efficiency index for an iteration approximating an algebraic number is bounded from above by unity. The proof uses degree growth arguments which permit the removal of restrictions in Paterson [72]. Proves that the maximal order of any sequence generated by an iteration with M multiplications or divisions is bounded by 2^M.

Kung, H. T., The computational complexity of algebraic numbers, *Proc. 5th Ann. ACM Symp. Theory Comput.* pp. 152–159 (1973). Revised paper in *SIAM J. Numer. Anal.* **12** (1975), 89–96.

core, nonlinear equations, algebraic numbers, scalar, optimal iterations, multipoint iterations, iterative complexity

Defines two multiplicative efficiency indices E and $\bar{E}$. Kung [73] proves that both indices are bounded by unity for an iteration approximating an algebraic number α. Proves if $E = 1$, then α is rational, while if $\bar{E} = 1$, then α is rational or quadratic irrational.

Kung, H. T., On computing reciprocals of power series, *Numer. Math.* **22** (1974), 341–348.

core, Newton iteration, formal power series, fast algorithms, algebraic complexity

Introduces idea of using Newton iteration on formal power series to compute reciprocal of a power series fast. Shows complexity of computing first N terms of reciprocal is no greater than multiplying two Nth-degree polynomials.

Kung, H. T., Complexity of numerical computation, *Proc. Internat. Comput. Symp.* (E. Gelenbe and D. Potier, eds.), pp. 247–252. North-Holland Publ., Amsterdam, 1975.

core, survey, algebraic complexity, iterative complexity

Surveys recent research and problems in algebraic and iterative complexity. Good bibliography.

Kung, H. T., The complexity of obtaining starting points for solving operator equations by Newton's method, *in* "Analytic Computational Complexity" (J. F. Traub, ed.), pp. 35–57. Academic Press, New York, 1976.

core, nonlinear equations, abstract, global information, iterative complexity, lower and upper bounds

Includes both the complexity of searching for a starting iterate and the complexity of iteration. Discusses when is the optimal time for changing from the search phase to the iterative phase. Gives new procedure for obtaining starting points for Newton iteration.

Kung, H. T., Synchronized and asynchronous parallel algorithms for multiprocessors, *in* "Algorithms and Complexity: New Directions and Recent Results" (J. F. Traub, ed.), pp. 153–200. Academic Press, New York, 1976.

core, nonlinear equations, scalar, iterative algorithms, asynchronous, parallel algorithms, average case analysis

Classifies parallel algorithms for multiprocessors as synchronized or asynchronous algorithms. Identifies and discusses important concepts involved in the design and analysis of the two types of algorithms. Applies concepts to three examples: search for zeros, iterative algorithms for solving linear systems and nonlinear scalar equations, and adaptive asynchronous algorithms. Considers what is the optimal number of processes. Good bibliography.

Kung, H. T., and Traub, J. F., Optimal order of one-point and multipoint iterations, *J. Assoc. Comput. Mach.* **21** (1974), 643–651.

core, nonlinear equations, scalar, multipoint iterations, maximal order

Studies iterations for computing zeros of nonlinear scalar functions f. The information is the values of f and its derivatives. Constructs multipoint iteration of order 2^{n-1} which uses n evaluations. Conjectures that any multipoint iteration without memory which uses this information can be of order at most 2^{n-1}.

Kung, H. T., and Traub, J. F., Computational complexity of one-point and multipoint iteration, *in* "Complexity of Computation" (R. M. Karp, ed.) *SIAM–AMS Proc.* Vol. 7, pp. 149–160 American Mathematical Society, 1974.

core, nonlinear equations, scalar, iterative information, one-point iterations, multipoint iterations, iterative complexity, lower bounds

Introduces the inclusion of combinatory complexity as well as information complexity in the complexity measure. Defines optimal complexity over a class of algorithms. Lower and upper bounds on optimal complexity are obtained for certain families of iterations.

Kung, H. T., and Traub, J. F., Optimal order and efficiency for iterations with two evaluations, *SIAM J. Numer. Anal.* **13** (1976), 84–99.

core, nonlinear equations, scalar, iterative complexity, maximal order

Considers rational iterations without memory for solving the scalar nonlinear equation $f = 0$. The information used is two evaluations of f or its derivatives. Proves maximal order is 2 which settles, for $n = 2$, a conjecture of Kung and Traub that an iteration using n evaluations without memory is of order at most 2^{n-1}. Shows that any rational two-evaluation maximal order iteration without memory must use either two evaluations of f or one evaluation of f and one of f'. Determines minimal combinatory complexity for both these cases.

Kung, H. T., and Traub, J. F., All algebraic functions can be computed fast, *J. Assoc. Comput. Mach.* **25** (1978), 245–260.

core, formal power series, Newton iteration, Newton polygon, fast algorithms, algebraic complexity

Shows that using the Newton polygon and Newton iteration the first N terms of the expansion of any algebraic function can be computed with complexity no greater than multiplying two Nth-degree polynomials.

Kung, H. T.: *See also* Brent, R.P.

Kuzovkin, A. I., and Tikhomirov, V. M., On the number of operations for finding the minimum of convex functions (in Russian), *Ekonom. i Mat. Metody* **3** (1967), 95–103.

core, extremum, multivariate, optimal points of information, optimal algorithms, error bounds

Considers the search for the minimum in a class of scalar convex functions of several variables. The information is the values of f. Optimal algorithms are studied. Shows that $\Theta(\ln \varepsilon^{-1})$ function evaluations and arithmetic operations are necessary and sufficient to find an ε-approximation.

Larkin, F. M., Optimal approximation in Hilbert spaces with reproducing kernel functions, *Math. Comp.* **24** (1970), 911–921.

core, approximation of linear functionals, optimal algorithms

Considers approximation of a linear functional L for a class of scalar functions in a Hilbert space with a reproducing kernel. The information is the values of f. Optimal algorithms are derived in terms of the representers of information. Shows for which functions f, $L(f)$ is exactly approximated by an optimal algorithm. Proves that the Gaussian quadrature formula is nearly optimal for the integration problem in the class of analytic functions within the region $|x| < r$ for large r.

Lee, J. W., Best quadrature formulas and splines, *J. Approx. Theory* **20**, (1977), 348–384.

core, integration, scalar, optimal linear algorithms

Considers integration for a class of scalar functions. The information is the values of f and its derivatives. Optimal quadrature formulas in the sense of Sard are studied.

Levin, A. Ju., On an algorithm for the minimization of convex functions (in Russian), *Dokl. Akad. Nauk SSSR* **160** (1965), 1244–1247 [*English transl.: Soviet Math. Dokl.* **6** (1965), 286–289].

core, extremum, multivariate, convex functions, upper bounds

Considers the search for the minimum in the class of scalar convex functions of several variables which satisfies a Lipschitz condition. The information is the values of f and its gradient. Presents an algorithm which finds an ε-approximation and requires $O(\ln \varepsilon^{-1})$ evaluations of f and its gradient.

Levin, M. I., and Giršovič, Ju. M., Extremal problems for cubature formulas (in Russian), *Dokl. Akad. Nauk SSSR* **236** (1977), 1303–1306 [*English transl.: Soviet Math. Dokl.* **18** (1977), 1355–1358].

core, integration, multivariate, optimal points of information, optimal linear algorithms

Considers integration for some classes of functions of two variables with bounded derivatives in L_q. The information is the values of f and its derivatives. Optimal linear algorithms with fixed and optimal points of information and their errors are derived.

Levin, M. I., Giršovič, Ju. M., and Arro, V. K., Best quadrature formulas on sets of functions (in Russian), *Dokl. Akad. Nauk SSSR* **226** (1976), 51–54 [*English transl.: Soviet Math. Dokl.* **17** (1976), 46–50].

core, integration, scalar, optimal points of information, optimal linear algorithms

Considers integration for some classes of scalar functions with bounded rth derivative in L_q. The information is the values of f, and its derivatives at the end points. Optimal linear algorithms with optimal points of information and their errors are presented.

Ligun, A. A., Exact inequalities for splines and best quadrature formulas for certain classes of functions (in Russian), *Mat. Zametki* **19** (1976), 913–926 [*English transl.: Math. Notes* **19** (1976), 533–541].

core, integration, scalar, optimal points of information, optimal linear algorithms, splines

Considers the integration problem for the class of periodic functions with bounded rth derivative. The information is the values of f. Shows that the rectangle quadrature formula is optimal and that equidistant points of information are optimal. The error for the optimal points is derived. Proofs are based on exact estimates of the kth derivative of periodic splines in Orlicz spaces with uniform norm.

Ligun, A. A.: *See also* Zaliznyak, N. F.

Lipow, P. R., Spline functions and intermediate best quadrature formulas, *SIAM J. Numer. Anal.* **10** (1973), 127–136.

core, integration, scalar, optimal algorithms, splines

Considers integration of a class of scalar functions. The information is the values of f and its derivatives. Optimal quadrature formulas in the sense of Sard are studied. Optimal formulas are based on cardinal Hermite splines.

Lipson, J., Newton's method: A great algebraic algorithm, *Proc. Symp. Symbolic and Algebraic Computation* (R. D. Jenks, ed.), pp. 260–270. Assoc. Comput. Mach., New York, 1976.

core, formal power series, Newton iteration, fast algorithms, algebraic complexity

Shows that for a "regular" problem Newton iteration may be used to compute the first N terms of the expansion of any algebraic function with complexity no greater than multiplying two Nth-degree polynomials. See also Kung and Traub [78].

Loeb, H., and Werner, M., Optimal numerical quadrature in H_p space, *Math. Z.* **138** (1974), 111–117.

core, integration, analytic functions, scalar, optimal linear algorithms, upper bounds

Considers integration for the class H_p of scalar analytic functions on the unit disk. The information is the values of f and its derivatives at n points. Proves that the upper bound on the minimal error of quadrature formulas is roughly $\exp(-cn)$, where $c > 0$.

Loeb, H. L.: *See also* Barrar, R. B.

Lušpaj, N. E., Best quadrature formulae for some classes of functions (in Russian), *Proc. Internat. Conf. Young Res. Math.* pp. 58–62. Charkov, 1966.

See Lušpaj [69].

Lušpaj, N. E., Best quadrature formulas on classes of differentiable periodic functions (in Russian), *Mat. Zametki* **6** (1969), 475–482 [*English transl.: Math. Notes* **6** (1969), 740–744].

core, integration, scalar, optimal points of information, optimal linear algorithms

Considers integration for the class of 2π periodic functions with bounded rth derivative in L_q. The information is the values of $f, f', \ldots, f^{(\zeta)}$. The optimal points and weights of linear quadrature formulas are found for $\zeta = r - 1$, $r = 1, 2, \ldots$. The error of this linear optimal algorithm is also derived.

Lušpaj, N. E., Best quadrature formula on the class $W^r_* L_2$ of periodic functions (in Russian), *Mat. Zametki* **16** (1974), 193–204 [*English transl.: Math. Notes* **16** (1974). 701–708].

core, integration, scalar, optimal linear algorithms, error bounds

Considers the integration problem for the class of periodic functions whose rth derivative is bounded by unity in L_2. The information is the values of $f, f', \ldots, f^{(\zeta)}$ at n points. The linear optimal quadrature formula is found for $\zeta = r - 2$ and $r - 3$. The error bounds are derived.

Lušpaj, N. E.: *See also* Kornejčuk, N. P.

Majstrovskij, G. D., On the optimality of Newton's method (in Russian), *Dokl. Akad. Nauk SSSR* **204** (1972), 1313–1315 [*English transl.: Soviet Math. Dokl.* **13** (1972), 838–840].

core, nonlinear equations, multivariate, Newton iteration

Considers the solution of nonlinear equations $f = 0$ for the class of functions f from $\mathbb{R}^n$ into $\mathbb{R}^n$ with a bounded Lipschitz constant. The information is the values of f and f'. Suppose that x_0 is a sufficiently close approximation to a solution. Then it is shown that any algorithm using n values of f and f' at any points has error essentially not less than the error of Newton's method after n steps.

Mangasarian, O. L., and Schumaker, L. L., Best summation formulae and discrete splines, *SIAM J. Numer. Anal.* **10** (1973), 448–459.

core, approximation of linear functionals, optimal algorithms, splines

Considers approximation of a linear functional defined on a finite-dimensional function space. The information is the values of f. Optimal algorithms in the sense of Sard are studied. The problem is reduced to a solvable linear or quadratic programming problem. Shows discrete monosplines are related to optimal algorithms.

Mansfield, L. E., On the optimal approximation of linear functionals in spaces of bivariate functions, *SIAM J. Numer. Anal.* **8** (1971), 115–126.

core, approximation of linear functionals, optimal algorithms

Considers approximation of a linear functional for a class of bivariate functions in a Hilbert space. The information is the values of n linear functionals. Optimal algorithms are derived in terms of the representers of the appropriate functionals. Application to integration is considered.

Mansfield, L. E., Optimal approximation and error bounds in spaces of bivariate functions, *J. Approx. Theory* **5** (1972), 77–96.

core, approximation of linear functionals, optimal algorithms

Considers approximation of a linear functional for a class of scalar functions of two variables in a Hilbert space with a reproducing kernel. The information is the values of n linear functionals. Optimal algorithms and error bounds are found in terms of the representers of the functionals. Application to the integration problem is considered. Related to bivariate splines.

Marchuk, A. G., and Osipenko, K. Yu., Best approximation of functions specified with an error at a finite number of points (in Russian), *Mat. Zametki* **17** (1975), 359–368 [*English transl.: Math. Notes* **17** (1975), 207–212].

core, approximation of linear functionals, optimal linear algorithms

Considers the approximation of a linear functional for a given class of problems. The information is the perturbed values of n linear functionals. A linear optimal error algorithm is constructed. This is a generalization of Smolyak's lemma.

Maung, Čžo Njun, and Sharygin, I. F., Optimal cubature formulae on the classes $D_2^{1,c}$ and D_s^{1,L_1}, *in* "Problems of Numerical and Applied Mathematics," Vol. 5, pp. 22–27. Tashkent, 1975 (in Russian).

core, integration, multivariate, optimal linear algorithms

Considers integration for two classes of scalar functions of two and s variables, respectively, with bounded first derivative. The information is the values of f. Presents asymptotically optimal linear algorithms.

Meersman, R., Optimal use of information in certain iterative processes, *in* "Analytic Computational Complexity" (J. F., Traub, ed.), pp. 109–125. Academic Press, New York, 1976.

core, nonlinear equations, scalar, iterative information, order of information, multipoint iterations

Studies maximal order of multipoint iteration for solution of the scalar nonlinear equation $f = 0$. Information is n evaluations of f or its derivatives. Establishes the Kung and Traub conjecture that the maximal order is bounded by 2^{n-1} for the case $n = 3$. Exhibits all arrangements of information for $n = 3$ which yield optimal information.

Meersman, R., On maximal order of families of iterations for nonlinear equations, Doctoral Thesis, Vrije Univ. Brussel, Brussels, (1976).

core, nonlinear equations, scalar, Hermitian information, order of information, multipoint iterations without memory, iterative complexity

Considers the iterative solution of nonlinear scalar equations. The information is the values of f and its derivatives. Studies the n-evaluation conjecture for multipoint iterations without memory. Proves the conjecture for $n \leq 3$ and for different classes of information operators.

Meinguet, J., Optimal approximation and error bounds in seminormed spaces, *Numer. Math.* **10** (1967), 370–388.

core, approximation of linear functionals

Considers approximation of a linear functional for a given class. The information is the values of n linear functionals. The range of possible values of the functional for elements satisfying the information is derived.

Melkman, A. A., n-Widths and optimal interpolation of time- and band-limited functions, *in* "Optimal Estimation in Approximation Theory" (C. A. Micchelli and T. J. Rivlin, eds.), pp. 55–68. Plenum Press, New York, 1977.

core, interpolation, optimal points of information, n-widths

Considers the optimal recovery of functions from a subset of the Paley–Wiener class of entire functions. The information is the values of f. The linear optimal error algorithm is derived. The optimal points of information are studied. The error for optimal points is the n-width of the subset. The n-widths are found for a more general set of time- and band-limited functions.

Melkman, A. A., and Micchelli, C. A., Optimal Estimation of Linear Operators in Hilbert Spaces from Inaccurate Data, Univ. Bonn Rep. (1977). See also *SIAM J. Numer. Anal.* **16** (1979), 87–105.

core, approximation of linear operators, optimal linear algorithms

Considers approximation of a linear operator for a class in a Hilbert space. The information is the value of a perturbed information operator. Linear optimal error algorithms are derived.

Meyers, L. F., and Sard, A., Best approximate integration formulas, *J. Math. Phys.* **28** (1950), 118–123.

core, integration, scalar, optimal algorithms

Considers integration for a class of scalar m-times differentiable functions. The information is the values of f at n points. Optimal quadrature formulas in the sense of Sard are derived for small m and some n.

Meyers, L. F., and Sard, A., Best interpolation formulas, *J. Math. Phys.* **29** (1950), 198–206.

core, interpolation, scalar, optimal algorithms

Considers interpolation for a class of scalar m-times differentiable functions. The information is the values of f at n points. Optimal algorithms in the sense of Sard are derived for small n and m.

Micchelli, C. A., Optimal Estimation of Linear Functionals, IBM Research Rep. 5729 (1975).

core, approximation of linear functionals and linear operators, optimal linear algorithms

Considers the approximation of a linear functional or operator for a balanced convex set. The information is the value of a perturbed linear operator. The existence of linear optimal error algorithms is studied.

Micchelli, C. A., On an optimal method for the numerical differentiation of smooth functions, *J. Approx. Theory* **18** (1976), 189–204.

core, differentiation, optimal linear algorithms

Considers the approximation of $f'(0)$ for the class of functions for which $\|Lf\| \le \gamma$, where L is a polynomial differential operator and γ is a constant. The information is the values of a perturbed f at infinitely many points. A linear optimal error algorithm is derived.

Micchelli, C. A., and Miranker, W. L., High order search methods for finding roots, *J. Assoc. Comput. Mach.* **22** (1975), 51–60.

core, nonlinear equations, scalar, globally convergent iterations, order

Defines and analyzes higher order globally convergent search methods for solving the nonlinear scalar equation $f = 0$. The information is the values of f and global bounds on certain derivatives. The order of convergence of these methods is obtained.

Micchelli, C. A., and Pinkus, A., On a best estimator for the class M^r using only function values, *Indiana Univ. Math. J.* **26** (1977), 751–759.

core, approximation, optimal points of information, optimal linear algorithms, splines, n-widths

Considers approximation for the class of functions $f(x) = P(x) + ((r - 1)!)^{-1} \int_0^1 (x - t)_+^{r-1} \, d\lambda_f(t)$, where P is a polynomial of degree $\le r$ and the total variation $\|\lambda_f\| \le 1$. The information is the values of f. The optimal error algorithm in L_1 is shown to be a linear interpolatory spline. The error for the optimal points of information is the Kolmogorov n-width of the class.

Micchelli, C. A., and Rivlin, T. J., "Optimal Estimation in Approximation Theory." Plenum Press, New York, 1977.

core, approximation

This is the proceedings of the International Symposium on Optimal Estimation in Approximation Theory in 1976. Contains many papers of interest. See Micchelli and Rivlin [77], Melkman [77], de Boor [77], and Golomb [77].

Micchelli, C. A., and Rivlin, T. J., A survey of optimal recovery, *in* "Optimal Estimation in Approximation Theory" (C. A. Micchelli and T. J. Rivlin, eds.), pp. 1–54. Plenum Press, New York, 1977.

core, approximation of linear operators, optimal error algorithms, optimal linear algorithms

Considers the optimal approximation of a linear operator U for a balanced convex class K. The information is the value of a perturbed linear operator I. Optimal error algorithms are studied. Assuming that U is a functional, the existence and properties of the linear optimal error algorithms are extensively studied. (It is not assumed that I is finite dimensional.) The ideas are illustrated by many valuable examples. In some cases, splines are shown to be a basic tool for optimal recovery.

Micchelli, C. A., and Rivlin, T. J., Optimal recovery of best approximations, IBM T. J. Watson Research Center Rep. 7071 (1978). To appear in *Resultate der Mat.*

core, approximation, optimal error algorithms

Considers approximation of the best uniform polynomial approximation of a function from the class of continuous functions bounded in the sup norm by unity. The information is the perturbed values of f. The optimal error algorithm and its error are derived.

Micchelli, C. A., Rivlin, T. J., and Winograd, S., The optimal recovery of smooth functions, *Numer. Math.* **26** (1976), 191–200.

core, approximation, optimal error algorithms, splines

Considers approximation for the class of scalar functions with bounded kth derivative in L_∞. The information is the values of f and its derivatives. The linear optimal error algorithm is shown to be essentially an interpolatory spline. Estimates of the optimal error are given.

Micchelli, C. A.: *See also* Gal, S.; Melkman, A. A.

Micula, Gh.: *See* Coman, Gh.

Miranker, W. L., A Survey of Parallelism in Numerical Analysis, *SIAM Rev.* **12** (1971), 524–547.

core, survey, nonlinear equations and extremum, scalar, parallel algorithms

Surveys parallel algorithms for numerical problems. Among the areas covered are the solution of nonlinear equations and optimization. Good bibliography.

Miranker, W. L.: *See also* Karp, R.; Micchelli, C. A.

Mitkowski, W.: *See* Adamski, A.

Mockus, I. B., Bayesian methods for extremum search (in Russian), *Avtomat. Vyčisl. Techn.* **3** (1972), 53–62.

core, extremum, scalar, optimal Bayesian algorithms

Considers the search for the minimum in a class of scalar functions. Optimal Bayesian algorithms are presented.

Morozov, V. A.: *See* Grebennikov, A. I.

Motornyj, V. P., On the best quadrature formula of the form $\sum_{k=1}^{n} p_k f(x_k)$ for some classes of periodic differentiable functions (in Russian), *Dokl. Akad. Nauk SSSR* **211** (1973), 1060–1062 [*English transl.: Soviet Math. Dokl.* **14** (1973), 1180–1183].

core, integration, scalar, optimal points of information, optimal linear algorithms

This is an announcement of the results proven by Motornyj [74].

Motornyj, V. P., On the best quadrature formulae of the form $\sum_{k=1}^{n} p_k f(x_k)$ for some classes of periodic differentiable functions (in Russian), *Dokl. Akad. Nauk SSSR Ser. Math.* **38** (1974), 583–614.

core, integration, scalar, optimal points of information, optimal linear algorithms

Considers integration for the class of periodic scalar function whose rth derivative has bounded modulus of continuity. The information is the values of f. Optimal points of information and optimal quadrature formulas are studied. Optimality of rectangle quadrature with equidistant points is shown for three classes of functions.

Motornyj, V. P., Some extremal problems of theory of quadrature and approximation of functions (in Russian), *Mat. Zametki* **19** (1976), 299–311 [*English transl.: Math. Notes* **19** (1976), 176–183].

core, integration, approximation, scalar, periodic functions

This is the abstract of a habilitation thesis. Surveys results on quadrature and approximation for several classes of periodic differentiable functions. Good Russian bibliography.

Munro, I.: *See* Borodin, A.

Nemirovsky, A. S.: *See* Judin, D. B.

Newman, D. J., Location of the maximum on unimodal surfaces, *J. Assoc. Comput. Mach.* **12** (1965), 395–398.

core, extremum, multivariate, asymptotically optimal algorithms

Considers the search for the maximum on the lattice points in the class of scalar unimodal functions of k variables. The information is the values of f. Proves that asymptotically $\log n$ function evaluations are enough to solve the problem on the lattice with $(n + 1)^k$ points.

Nielson, G. M., Bivariate spline functions and the approximation of linear functionals, *Numer. Math.* **21** (1973), 138–160.

core, approximation of linear functionals, splines

   Considers approximation of a linear functional for a class of scalar functions of two variables in a Hilbert space. The information is the values of n linear functionals. Linear optimal algorithms are shown to be based on bivariate interpolatory splines.

Nikolskij, S. M., On the problem of approximation estimate by quadrature formulae (in Russian), *Usp. Mat. Nauk* **5** (1950), 165–177.

core, integration, scalar, optimal points of information, optimal linear algorithms

   This classic paper considers integration for the class of scalar functions whose rth derivative is bounded in L_p by a constant. The information is the values of $f, f', \ldots, f^{(\zeta)}$ at n points. Optimal linear quadrature formulas are defined. The points and weights of an optimal quadrature formula are found for $r = 1, 2$, and for arbitrary r with $\zeta = r - 2$.

Nikolskij, S. M., "Quadrature Formulae." Nauka, Moscow. 1st ed., 1958; 2nd ed., 1974 (in Russian) [*English transl.:* Hindustan Publ., Delhi, India, 1964].

core, integration, scalar, optimal points of information, optimal linear algorithms

   This classic book introduces optimal linear quadrature formulas for a class of functions. The information is the values of $f, f', \ldots, f^{(\zeta)}$. The optimal points of information are found for the class of functions with bounded rth derivative in L_∞ for small r or $\zeta = r - 2$.

Nilson, E. N.: *See* Ahlberg, J. H.

Novikov, V. A.: *See* Aphanasjev, A. Yu.

Ortega, J. M., and Rheinboldt, W. C., "Iterative Solutions of Nonlinear Equations in Several Variables." Academic Press, New York, 1970.

mathematics, nonlinear equations, multivariate, order

   This monograph surveys the basic theoretical results about nonlinear equations in n dimensions. Analyzes many iterations. Good bibliography.

Osipenko, K. Yu., Optimal interpolation of analytic functions (in Russian), *Mat. Zametki* **12** (1972), 465–476 [*English transl.: Math. Notes* **12** (1972), 712–719].

core, interpolation, analytic functions, scalar, optimal points of information, optimal linear algorithms

   Considers the interpolation problem for the class of real scalar functions on $[a,b]$ which can be analytically extended to a region G and whose extension is bounded by a constant. The information is the values of f at n points. Using Smolyak's lemma, the linear optimal error algorithm and its error are derived. Optimal points of information are considered. If G is the unit disk, the optimal points are determined by elliptic functions. If G is the ellipse with foci ± 1 and the sum of semiaxes equal to c, the optimal points are the zeros of the Chebyshev polynomial and the error is roughly c^{-n}.

Osipenko, K. Yu., Best approximation of analytic functions from information about their values at a finite number of points (in Russian), *Mat. Zametki* **19** (1976), 29–40 [*English transl.: Math. Notes* **19** (1976), 17–23].

core, approximation of linear functionals, analytic functions, scalar, optimal linear algorithms

   Generalizes Smolyak's lemma to the complex case. A linear optimal error algorithm for the approximation of a linear complex functional is proven. The approximation of analytic scalar functions is considered. Results are related to Osipenko [72].

Osipenko, K. Yu.: *See also* Marchuk, A. G.

Ostrowski, A., "Solution of Equations in Euclidean and Banach Spaces," 3rd ed. Academic Press, New York, 1973.

mathematics, nonlinear equations, scalar, multivariate, abstract, iterative algorithms, one-point iterations, iterations with memory, order, Newton iteration, secant iteration

A monograph on the numerical solution of nonlinear equations. Contains much material on convergence and order of iterations. Surveys theory of solving equations numerically.

Pallashke, D., Optimale differentiations—und integrationsformeln in $C_0[a,b]$, *Numer. Math.* **16** (1976), 201–210.

core, differentiation, integration, scalar, optimal algorithms

Considers differentiation and integration for a class of scalar continuous functions. The information is the values of f. Optimal algorithms are derived.

Parker, D. S. Jr., Studies in conjugation: Huffman tree construction, nonlinear recurrences, and permutation networks. Dept. of Computer Science Rep. UIUCDCS-R-78-930, Ph.D. Thesis, Univ. of Illinois (1978).

core, composition, formal power series, fast algorithms, parallel algorithms

Considers the evaluation of nonlinear recurrences on a parallel machine. Studies when there exists a transformation which makes this problem easy and whether this transformation process can be automated. Good bibliography. See also Brent and Traub [78].

Paszkowski, S., Optimum choice of initial approximations in interpolation methods of solving equations, *Zastos. Mat.* **12** (1971), 201–216.

core, nonlinear equations, scalar, interpolatory iterations, optimal initial approximations

Considers the iterative solution of the nonlinear scalar equation $f = 0$. The information is the values of f. Studies the optimal choice of initial approximations for interpolatory iterations. Shows that this problem is equivalent to minimization of the sup norm of the function $\prod_{i=0}^{n} |x - a_i|^{b_i}$ with respect to a_i, for $x \in [-1,1]$ and fixed positive b_i.

Paterson, M. S., Efficient iterations for algebraic numbers, *in* "Complexity of Computer Computations" (R. E. Miller and J. W. Thatcher eds.), pp. 41–52. Plenum Press, New York, 1972.

core, nonlinear equations, algebraic numbers, scalar, optimal iteration, one-point iterations, iterative complexity

Introduces question of optimal iteration for approximation of algebraic numbers. Proves that the multiplicative efficiency index (which is the reciprocal of the complexity index) must be bounded from above by unity. The proof uses results from number theory. Presents several conjectures on optimal efficiency.

Paulik, A., Zur existenz optimaler quadraturformeln mit freien knoten bei integration analytischer funktionen, *Numer. Math.* **27** (1977), 395–405.

core, integration, scalar, optimal linear algorithms

Considers integration for a class of scalar analytic functions on a circle. The information is the values of f. For fixed points of information the optimal quadrature and its error are derived. The existence of optimal points is established.

Piavsky, S. A., One algorithm for the searching of global extremum of function (in Russian), *Zh. Vychisl. Mat. Mat. Fiz.* **12** (1972), 888–896 [*English transl.: U.S.S.R. Computational Math. and Math. Phys,* **12** (1972), 57–67].

core, extremum, multivariate

Considers an algorithm finding the minimum for the class of functions $f: X \to \mathbb{R}$, where X is a compact set, such that there exists $g: X \times X \to \mathbb{R}$ and $f(x) \geq g(x,y)$, $\forall x, y \in X$. (If f satisfies a Lipschitz condition $|f(x) - f(y)| \leq L\|x - y\|$, set $g(x,y) = f(y) - M\|x - y\|$ for $M \geq L$.) The information on f is the function g. The algorithm requires finding a global minimum of $\max_{0 \leq j \leq k} g(x,y_j)$ for differnt $y_1, \ldots, y_k$. The efficiency analysis and optimality to within a factor of 3 of this algorithm is proved by Danilin [71] for the scalar case.

Pinkus, A., Asymptotic minimum norm quadrature formulae, *Numer. Math.* **24** (1975), 163–175.
core, integration, scalar, optimal linear algorithms

Considers integration for the class of scalar analytic functions whose norm is bounded by unity on a complex domain *B*. The information is the values of *f* and its derivatives. The optimal weights and points of linear quadrature formulas are studied. Shows that if *B* grows to the whole complex space, then the optimal weights and points converge to the weights and points of the Gaussian quadrature.

Pinkus, A.: *See also* Micchelli, C. A.

Pleshakov, G. N., On efficiency of the multidimensional interpolation iterations (in Russian), *Zh. Vychisl. Mat. Mat. Fiz.* **17** (1977), 1153–1160.
core, nonlinear equations, multivariate interpolatory iterations

Considers the solution of *n* nonlinear equations in *n* unknowns. The information is the values of *f*. Inverse interpolatory iterations with memory are studied. A necessary assumption on a suitable position of iteration points seems to be missing. The Ostrowski efficiency index is discussed.

Powell, M. J. D.: *See* Gaffney, P. W.

Reinsch, Ch., Two extensions of the Sard–Schoenberg theory of best approximation, *SIAM J. Numer. Anal.* **11** (1974), 45–51.
core, approximation of linear functionals, optimal algorithms, splines

Considers approximation of a linear functional for a class of functions. The information is the values of *f* and its derivatives. Optimal algorithms based on natural or periodic splines are derived. Extensions are given to a Hilbert setting and to the problem of smoothing.

Rheinboldt, W. C.: *See* Ortega, J. M.

Rice, J. R., "The Approximation of Functions," Vol. II, Chapter 10. Addison-Wesley, Reading, Massachusetts, 1969.
core and mathematics, approximation of linear functionals, optimal error algorithms

Considers, among other problems, approximation of a linear functional on the ball in a Hilbert space. The information is the values of *n* linear functionals. Optimal error algorithms are constructed using splines.

Rice, J. R., Matrix representations of nonlinear equation iterations–Application to parallel computation, *Math. Comp.* **25** (1971), 639–647.
core, nonlinear equations, scalar, iterative algorithms, parallel algorithms, order, interpolatory iterations, composition

Introduces idea of assigning a matrix representation to iterations for solving nonlinear equations. Applies his results to the analysis of iterations useful in parallel computation. See also Feldstein and Traub [77].

Rice, J. R., On the computational complexity of approximation operators, *in* "Approximation Theory" (G. G. Lorentz, ed.), pp. 449–456. Academic Press, New York, 1973.
See Rice [76].

Rice, J. R., On the computational complexity of approximation operators II, *in* "Analytic Computational Complexity" (J. F. Traub, ed.), pp. 191–204. Academic Press, New York, 1976.
core, approximation, optimal points of information, optimal algorithms

Continuation of Rice [73]. Considers approximation of scalar *p*-times differentiable or analytic functions. The information is the values of *f*. Assumes that the only operations counted are evaluations of *f*. Studies how many function evaluations are asymptotically necessary to produce an estimate with error proportional to the error of the best polynomial approximation for different classes of functions.

Richter, N., Properties of minimal integration rules, *SIAM J. Numer. Anal.* **7** (1970), 67–79.
core, integration, analytic functions, scalar, optimal algorithms

Considers integration for a class of scalar analytic functions in a Hilbert space. The information is the values of f. The existence of optimal quadrature formulas is proved. Properties of weights and points of an optimal quadrature are studied. See also Richter-Dyn, N., Properties of minimal integration rules, II, *SIAM J. Numer. Anal.* **8** (1971), 497–508.

Richter-Dyn, N., Minimal interpolation and approximation in Hilbert spaces, *SIAM J. Numer. Anal.* **8** (1971), 583–597.
core, approximation of linear functionals, optimal linear algorithms

Considers approximation of a linear functional for a class of scalar functions in a Hilbert space with a reproducing kernel. The information is the values of f. Optimal linear algorithms are studied. Application to integration is considered.

Riess, R. D.: *See* Johnson, L. W.

Rissanen, J., Maximum power feedback law, *Internat. J. Control* **14** (1971), 233–240.
core, nonlinear equations, scalar, iterative information, control theory, maximal order

Considers the iterative solution of certain scalar nonlinear equations which occur in control theory. Obtains an iterative algorithm by a process of linearization. Shows this algorithm has order about 1.55 if certain initial conditions hold. Proves that this algorithm has "maximal power" among all "smooth" algorithms using the same information.

Rissanen, J., On optimum root-finding algorithms, *J. Math. Anal. Appl.* **36** (1971), 220–225.
core, nonlinear equations, scalar, maximal order

Considers maximal order of algorithms for solving a scalar nonlinear equation. Proves that the secant method has maximal order among all algorithms using the same information as the secant method.

Ritter, K., Two-dimensional spline functions and best approximations of linear functionals, *J. Approx. Theory* **3** (1970), 352–368.
core, approximation of linear functionals, optimal algorithms, splines

Considers approximation of a linear functional for a class of scalar functions of two variables. The information is the values of f and its partial derivatives. Optimal algorithms in the sense of Sard are studied. Shows the connections with two-dimensional interpolatory splines.

Rivlin, T. J.: *See* Micchelli, C. A.

Saari, D. G., and Simon, C. P., Effective price mechanisms, *Econometrica* **46** (1978), 1097–1125.
core, economic equilibrium, nonlinear equations, multivariate, iterative information, one-point iterations

Considers how much information is required for a price mechanism to converge to an economic equilibrium. Among the results established is the following. Let U be an open subset of $\mathbb{R}^n$ and let $f: U \to \mathbb{R}^n$ be an excess demand function. Proves that any "local effective price mechanism" requires the evaluation of f and f'. This result follows from Theorem 4.2 in Traub and Woźniakowski [76] (Optimal linear information for the solution of nonlinear operator equations).

Šajdaeva, T. A., Quadrature formulae with least bound for the remainder for some classes of functions (in Russian), *Trudy Mat. Inst. Steklov. Akad. Nauk SSSR* **53** (1959), 313–341.
core, integration, scalar, optimal points of information, optimal linear algorithms

Considers integration for classes of scalar functions with bounded first, second, or third derivative. The weights and points of optimal quadrature formulas are derived.

Sard, A., Best approximate integration formulas; Best approximation formulas, *Amer. J. Math.* **71** (1949), 80–91.

core, integration, approximation of linear functionals, scalar, optimal linear algorithms

This is probably the first paper which discusses optimal algorithms. It deals with integration for the class of scalar m-times differentiable functions. The information is the values of f at n fixed points t_i. The remainder of a quadrature formula $\sum_{i=1}^{n} f(t_i)k_i$ of order $m - 1$ may be written as $\int_a^b f^{(m)}(t)k(t)\,dt$, where the kernel k depends only on the weights and points of the quadrature formula. A quadrature formula is called best (now it is called best in the sense of Sard) if $\int_a^b k^2(t)\,dt$ is minimal with respect to weights k_i. Solves this problem only for small n and m. Extension to approximation of linear functionals is discussed.

Sard, A., "Linear Approximation." American Mathematical Society, Providence, Rhode Island, 1963.

mathematics and core, approximation, integration, scalar

Considers, among other problems, integration for a class of scalar functions. The information is the values of f and its derivatives. Optimal quadrature formulas in the sense of Sard are discussed.

Sard, A., Optimal approximation, *J. Funct. Anal.* **1** (1967), 222–244.

core, approximation of linear operators, optimal algorithms

Considers approximation of a linear operator for a class of problems. The information operator is given by n linear operators. Optimal linear algorithms are studied. Generalization of Sard [49]. See also Sard, A., Optimal approximation: An addendum, *J. Funct. Anal.* **2** (1968), 368–369.

Sard, A., Approximation based on nonscalar observations, *J. Approx. Theory* **8** (1973), 315–334.

core, approximation of linear operators, splines

Considers approximation of a linear operator on a class of elements. The information is a linear operator. Studies spline approximation. This is a continuation of Sard [67].

Sard, A., and Weintraub, S., "A Book of Splines." Wiley, New York, 1971.

core, splines

Surveys the theory of spline approximation. Provides optimal approximations of a function, derivatives of a function, or its integral. The information is the values of f at regularly spaced points. A Fortran program for spline approximation is given. The cardinal splines are computed for a number of cases. Good bibliography.

Sard, A.: *See also* Meyers, L. F.

Schmeisser, G., Optimale quadraturformeln mit semidefiniten kernen, *Numer. Math.* **20** (1972), 32–53.

core, integration, scalar, optimal algorithms

Considers integration for the class of $(2k)$-times differentiable scalar functions. The information is the values of f. Assumes that the error of a quadrature formula may be expressed as $cf^{(2k)}(\xi)$ and studies the minimization of $|c|$.

Schoenberg, I. J., On best approximations of linear operators, *Nederl. Akad. Wetensch. Indag. Math.* **67** (1964), 155–163.

core, approximation of linear functionals, optimal linear algorithms, splines

Considers approximation of a linear functional L for a class of scalar m-times differentiable functions. The information is the values of f. Optimal algorithms in the sense of Sard are obtained by operating with L on the interpolatory spline.

Schoenberg, I. J., Spline interpolation and best quadrature formulae, *Bull. Amer. Math. Soc.* **70** (1964), 143–148.

core, integration, scalar, optimal linear algorithms, splines

Considers integration for a class of scalar m-times differentiable functions. The information is the values of f. Optimal quadrature formulas in the sense of Sard are shown to be the integrals of the interpolatory splines.

Schoenberg, I. J., On monosplines of least deviation and best quadrature formulae, *SIAM J. Numer. Anal. Ser. B* **2** (1965), 144–170.

core, integration, scalar, optimal linear algorithms, splines

Considers integration for a class of m-times differentiable scalar functions. The information is the values of f and its derivatives. Optimal quadrature formulas in the sense of Sard are studied. Shows the relation between optimal quadrature formulas and monosplines of least deviation from zero. Proves that the classic Hermite and Euler–Maclaurin quadrature formulas are optimal. See also Schoenberg, I. J., On monosplines of least square deviation and best quadrature formulae II, *SIAM J. Numer. Anal.* **3** (1966), 321–328.

Schoenberg, I. J., Monosplines and quadrature formulae, *in* "Theory and Applications of Spline Functions" (T. N. E. Greville, ed.), pp. 157–207. Academic Press, New York, 1969.

core, integration, scalar, optimal points of information, optimal linear algorithms, splines

Considers integration for a class of scalar m-times differentiable functions. The information is the values of f and its derivatives. Defines and studies optimal quadrature formulas in the sense of Sard as formulas for which weights *and* points of information are chosen to minimize $\int_a^b k^2(t)\, dt$. See Sard [49]. The solution is derived in terms of monosplines of least L_2 norm.

Schoenberg, I. J., A second look at approximate quadrature formulae and spline interpolation, *Advances in Math.* **4** (1970), 277–300.

core, integration, scalar, optimal linear algorithms, splines

Considers integration with a weight function for a class of scalar m-times differentiable functions. The information is the values of f and its derivatives at the endpoints of the integration interval. Optimal quadrature formulas in the sense of Sard are studied. Shows that the problem reduces to the monosplines of least L_2 norm under some boundary conditions.

Schultz, M., H., The computational complexity of elliptic partial differential equations, *in* "Complexity of Computer Computations" (R. E. Miller and J. W. Thatcher, eds.). Plenum Press, New York, 1972.

core, differential equations, optimal algorithms

Considers an elliptic partial differential equation on a square domain. The information is the values of functions which determine the elliptic equation. Presents a fourth-order algorithm on the square root mesh which yields a second-order result on the original mesh. This algorithm has combinatory complexity proportional to the number of grid points and therefore is asymptotically optimal.

Schultz, M. H., The complexity of linear approximation algorithms, *in* "Complexity of Computation" (R. M. Karp, ed.), pp. 135–148. American Mathematical Society, 1974.

core, approximation, optimal algorithms

Considers approximation for a class of scalar functions. Studies linear algorithms whose range is finite dimensional. Shows that for the class of functions f such that $\|f\| \leq 1$, any linear algorithm has error one. The Kolmogorov n-width serves as a tool to find "good" linear algorithms for the class of absolutely continuous functions whose first derivative is bounded by one in L_∞.

Schultz, M. H., Complexity and differential equations, *in* "Analytic Computational Complexity" (J. F. Traub, ed.), pp. 143–149. Academic Press, New York, 1976.

core, approximation, linear and nonlinear equations, multivariate, sparse

Summary of results on three topics: complexity of an approximation theory problem; storage complexity of algorithms for certain classes of sparse linear problems; complexity of two algorithms for solving certain sparse nonlinear systems.

Schumaker, L. L.: *See* Mangasarian, O. L.

Secrest, D., Numerical integration of arbitrarily spaced data and estimation of errors, *SIAM J. Numer. Anal. Ser. B* **2** (1964), 52–68.

core, integration, scalar, optimal linear algorithms

Considers integration for the class of scalar functions whose rth derivative is bounded in L_p by a constant. The information is the values of f. Optimal error quadrature formulas are derived.

Secrest, D., Error bounds for interpolation and differentiation by the use of spline functions, *SIAM J. Numer. Anal. Ser. B* **2** (1965), 440–447.

core, approximation of linear functionals, optimal error algorithms, error bounds

Considers approximation of a linear functional for the class of scalar functions whose nth derivative is bounded in L_2 by a constant. The information is the values of f. Optimal error algorithms are derived in terms of natural splines. Optimal error bounds for interpolation and differentiation are presented.

Secrest, D., Best approximate integration formulas and best error bounds, *Math. Comp.* **19** (1965), 79–83.

core, integration, scalar, optimal algorithms

Considers integration for the class of scalar functions whose nth derivative is bounded in L_2 by a constant. The information is the values of f. Optimal algorithms and their errors are derived in terms of natural splines.

Shamanskii, V. E., A modification of Newton's method (in Russian), *Ukr. Mat. Z.* **19** (1967), 133–138.

core, nonlinear equations, multivariate, iterative complexity

Considers a class of algorithms for the solution of a multivariate nonlinear system $f = 0$. Information used are the evaluations of f. A discrete approximation of the Jacobian in Newton iteration is held constant for m_k iterations. The m_k are chosen so as to minimize a complexity index.

These algorithms are a discretized form of algorithms introduced by Traub (see Traub [62] (The theory of multipoint iteration functions) or Traub [64]).

Sharygin, I. F., A lower estimate for the error of quadrature formulae for certain classes of functions (in Russian), *Zh. Vychisl. Mat. Mat. Fiz.* **3** (1963), 370–376 [*English transl.: U.S.S.R. Computational Math. and Math. Phys.* **3** (1963), 489–497].

core, integration, multivariate, lower bounds

Considers integration for three classes of scalar functions of several variables with bounded Fourier coefficients. The information is the values of f. Lower bounds on the error of linear quadrature formulas are given.

Sharygin, I. F., A lower bound for the error of a formula for approximation summation in the class $E_{s,p}(C)$ (in Russian), *Mat. Zametki* **21** (1977), 371–375 [*English transl.: Math. Notes* **21** (1977), 207–210].

core, integration, multivariate, optimal points of information, lower bounds

Considers the integration problem for the class $E_{s,p}(C)$ of scalar functions of s variables. The information is the values of f at p points. It is shown that for any points of information every algorithm has the error at least $\Theta(p^{-1} \ln{}^s p)$. This bound is achievable for the optimal algorithm using p equidistant points of information.

Sharygin, I. F.: *See also* Maung, Čžo Njun

Sieveking, M., An algorithm for division of power series, *Computing* **10** (1972), 153–156.

core, formal power series, fast algorithms

Gives a fast algorithm for computing the quotient of two formal power series. See also Kung [74].

Simon, C. P.: *See* Saari, D. G.

Smolyak, S. A., Interpolation and quadrature formulas for the classes W_s^z and E_s^z (in Russian), *Dokl. Akad. Nauk SSSR* **131** (1960), 1028–1031 [*English transl.: Soviet Math. Dokl.* **1** (1960), 384–387].

core, interpolation, integration, multivariate, lower bounds

Considers the interpolation and integration problems for two classes W_s^z and E_s^z of complex periodic functions of s variables with bounded Fourier coefficients. The information is the values of f at n points. Lower and upper bounds on the error for optimal points of information are derived. The upper bound is sharp for prime n.

Smolyak, S. A., On optimal restoration of functions and functionals of them (in Russian). Candidate Dissertation, Moscow State Univ. (1965).

core, approximation of linear functionals, optimal linear algorithms

This is a Ph.D. thesis not available to us which does not seem to have been published. Considers the approximation of a linear functional for a balanced convex class of functions. The information is the values of n linear functionals. Shows that there exists a linear optimal error algorithm. See Bakhvalov [71] (On the optimality of linear methods for operator approximation in convex classes of functions) for a proof of this result.

Sobol, I. M., "Multivariate Quadrature Formulas and Haar Functions." Nauka, Moscow, 1969 (in Russian).

core, integration, multivariate

Considers the integration problem for scalar functions of several variables belonging to the Fourier–Haar class of functions with bounded coefficients. The information is the values of f. Optimal or close to optimal quadrature formulas are studied. The error is derived.

Sobolev, S. L., On the order of convergence of cubature formulas (in Russian), *Dokl. Akad. Nauk SSSR* **162** (1965), 1005–1008 [*English transl.: Soviet Math. Dokl.* **6** (1965), 808–812].

core, integration, multivariate, optimal points of information, optimal linear algorithms, lower, upper bounds

Considers integration for a class of scalar functions of s variables with bounded derivatives up to order r in L_2. The information is the values of f at n points. Proves that the optimal cubature formulas has error roughly $n^{m/s}$.

Sobolev, S. L., "Introduction to the Theory of Cubature Formulas." Nauka, Moscow, 1974 (in Russian).

core and mathematics, integration, multivariate, optimal linear algorithms

Considers, among other problems, integration for the class of scalar functions of several variables with bounded rth derivative in L_2. The information is the values of f. Derives a lower bound for the error of the integration problem. Optimal points of information are studied.

Sobolev, S. L.: *See also* Babuška, I.

Sonnevend, G., On optimization of algorithms for function minimization (in English), *Zh. Vychisl. Mat. Mat. Fiz.* **17** (1977), 591–609.

core, extremum, nonlinear equations, multivariate, optimal points of information, optimal error algorithms

Considers the search for the minimum of scalar functions f of several variables and the search for the zero of systems of nonlinear equations grad $f(x) = 0$ for the class of functions f such that (grad $f(x) -$ grad $f(y), x - y)/\|x - y\|^2 \in [m,M]$, where $m \geq 0$. The information is the values of f and its derivative. The optimal points of information and optimal error algorithms with respect to different error criteria are defined. Shows a connection between these problems and dynamic programming and optimal control problems.

Stechkin, S. B., Best approximation of linear operators (in Russian), *Mat. Zametki* **1** (1967), 137–148 [*English transl.: Math. Notes* **1** (1967), 91–99].

core, linear operators, optimal approximation by bounded linear operators

In this paper, there is no concept of ε-approximation or of information and the definition of optimality is different than in our setting. Considers approximation of an unbounded linear operator U by linear operators φ whose norms do not exceed a given constant. Find optimal operators φ and their errors for several U. If U is a differentiation operator $Uf = f^{(k)}$ and k is small, the nearly optimal φ requires few evaluations of f.

Stenger, F., Optimal convergence of minimum norm approximations in H_p, *Numer. Math.* **29** (1978), 345–362.

core, integration, approximation, analytic functions, scalar, optimal linear algorithms, lower, upper bounds

Considers integration and approximation for the class H_p of scalar analytic functions on the unit disk. The information is the values of f at n points. Proves that the minimal error of linear algorithms for both problems is roughly $\exp(-c\sqrt{n})$, where $c > 0$.

Stern, M. D., Optimal quadrature formulae, *Comput. J.* **9** (1967), 396–403.

core, integration, scalar, optimal linear algorithms

Considers integration for the class of scalar functions whose second derivative is bounded in L_p by a constant. The information is the values of f. Optimal quadrature formulas and their errors are derived.

Stetter, F., On best quadrature of analytic functions, *Quart. Appl. Math.* **27** (1969), 270–272.

core, integration, analytic functions, scalar, optimal linear algorithms, upper bounds

Considers integration for a class of scalar analytic functions. The information is the values of f. Studies the optimal linear algorithm and derives upper bounds for its error.

Strongin, R. G., "Numerical Methods for Multivariate Extremal Problems." Nauka, Moscow, 1978 (in Russian).

core, extremum, scalar, optimal Bayesian algorithms

Considers the search for the minimum in the class of scalar functions satisfying a probabilistic Lipschitz-like condition. The information is the values of f. Optimal Bayesian algorithms are studied.

Sukharev, A. G., Optimal strategies of the search for an extremum (in Russian), *Zh. Vychisl. Mat. Mat. Fiz.* **11** (1971), 910–924 [*English transl.: U.S.S.R. Computational Math. and Math. Phys.* **11** (1971), 119–137].

core, extremum, multivariate, optimal points of information

Considers the search for the maximum in the class of scalar functions of N variables satisfying a Lipschitz condition with a constant L on the set K. The information is the values of f at n points. Shows that the optimal points of information form the optimal covering of K. The optimal adaptive information is shown to be nonadaptive. For $K = [0,1]^N$, $(L/\varepsilon)^N$ function evaluations are necessary to find an ε-approximation to the solution. A probabilistic choice of the points of information is also considered.

Sukharev, A. G., Best sequential search strategies for finding an extremum (in Russian), *Zh. Vychisl. Mat. Mat. Fiz.* **12** (1972), 35–50 [*English transl.: U.S.S.R. Computational Math. and Math. Phys.* **12** (1972), 39–59].

core, extremum, multivariate, optimal points of information

This is a continuation of Sukharev [71]. Assumes that the points at which a function is evaluated may depend on the previously computed information. Shows that the optimal points of information form the optimal covering of the domain of f. Related to dynamic programming.

Sukharev, A. G., "Optimal Search for Extremum." Moscow State Univ., Moscow, 1975 (in Russian).

core, extremum, nonlinear equations, scalar, optimal points of information

Considers the search for the maximum in the class of Lipschitz or unimodal scalar functions. The information is the values of f given simultaneously or adaptively. Surveys many results for this problem. Considers also the solution of nonlinear scalar equations for the class of functions f such that $f(a) \le 0$, $f(b) \ge 0$, and $(f(x) - f(y))/(x - y) \in [m, M]$ for $x \in [a,b]$ with given positive m, M.

Sukharev, A. G., Optimal search for a zero of function satisfying Lipschitz's condition (in Russian), *Zh. Vychisl. Mat. Mat. Fiz.* **16** (1976), 20–30 [*English transl.*: Sukharev, A. G., Optimal search for the roots of a function satisfying a Lipschitz conditi < n, *U.S.S.R. Computational Math. and Math. Phys.* **16** (1976), 17–26].

core, nonlinear equations, scalar, optimal points of information, optimal error algorithms

Considers the search for a zero in the class of scalar functions satisfying a Lipschitz condition with a given constant on a given interval. The information is the values of f at n points x_i. The optimal choice for the x_i is obtained for the case that f changes sign at the interval end points. If this assumption does not hold, the optimal error algorithm which minimizes the absolute value of f is obtained. In both cases, the algorithms are bisection algorithms.

Sukharev, A. G., The optimal method for constructing best uniform approximations for functions of certain class, (in Russian), *Zh. Vychisl. Mat. Mat. Fiz.* **18** (1978), 302–313 [*English transl.*: *U.S.S.R. Computational Math. and Math. Phys.* **18** (1978), 21–31].

core, approximation, optimal points of information

Considers uniform approximation for the class of scalar functions satisfying a Lipschitz condition with a given constant. The information is the values of f. The optimal points of nonadaptive and adaptive information are discussed.

Sukharev, A. G., Optimal quadrature formulas for some functional classes, Report (1978).

core, integration, multivariate, optimal points of information, optimal algorithms

Considers the integration problem for the class of generalized Lipschitz scalar functions of s variables. The information is the values of f at n points. The optimal algorithms and their errors are derived. The optimal points of information are found. The error for optimal points is roughly $n^{-1/s}$.

Taikov, L. V., Kolmogorov-type inequalities and the best formulas for numerical differentiation (in Russian), *Mat. Zametki* **4** (1968), 233–238 [*English transl.*: *Math. Notes* **4** (1968), 631–634].

core, differentiation, optimal approximation by bounded linear operators

Following Stechkin [67], considers approximation of $f^{(k)}$ for the class of scalar functions with bounded nth derivative in L_2, $0 \le k < n$, by means of linear operators whose norm is bounded by a given constant. Finds the error of such an optimal approximation.

Tanama, V. P., On the optimality of methods of solving nonlinear unstable problems, *Dokl. Akad. Nauk SSSR* **220** (1975), 1035–1037 [*English transl.*: *Soviet Math. Dokl.* **16** (1975), 213–215].

core, operator equations, optimal error algorithms

Considers a nonlinear operator equation $Ax = y$, where A is a one-to-one continuous operator and y belongs to a given set. The information is a perturbed value of Ax. Shows that the "residual principle" algorithm and the "quasi-solution" algorithm are optimal error algorithms to within a factor of two.

Tarassova, V. P., Optimal strategies of search for domain of greatest values for some classes of functions (in Russian), *Zh. Vychisl. Mat. Mat. Fiz.* **18** (1978), 886–896.

core, extremum, optimal error algorithms

Considers the search for the maximum in a class K of functions defined on $[a,b]$ with values in a linear ordered space. The information is the values of f. Assuming that for every $f \in K$ there exists a subinterval $\Delta \subset [a,b]$ of length δ such that $f(x) > f(x')$, $\forall x \in \Delta$, $x' \notin \Delta$, the optimal error algorithms are derived for $\delta = (b - a)/2$.

Tikhomirov, V. M., Best methods of approximation and interpolation of differentiable functions in the space $C[-1,1]$ (in Russian), *Mat. Sb.* **80** (122) (1969), 290–304 [*English transl.: Math. U.S.S.R. Sb.* **9** (1969), 275–289].

mathematics, n-widths

Considers approximation for the class of scalar functions whose $(r - 1)$th derivatives satisfy a Lipschitz condition with unity. The periodic case is also studied. The Kolmogorov and Gelfand n-widths are expressed in terms of the norms of perfect splines. The extremal subspaces are shown to be spanned by splines.

Tikhomirov, V. M., "Some Problems of Approximation Theory." Moscow State Univ., Moscow, 1976 (in Russian).

mathematics, approximation, n-widths

Considers general problems of approximation. Contains many classical and new results especially for different n-widths (Kolmogorov, Gelfand, linear). Good bibliography.

Tikhomirov, V. M.: *See also* Kuzovkin, A. I.

Tikhonov, A. N., and Gaisarian, S. S., The choice of optimum networks in the approximate calculation of quadratures (in Russian), *Zh. Vychisl. Mat. Mat. Fiz.* **9** (1969), 1170–1176 [*English transl. U.S.S.R. Computational Math. and Math. Phys.* **9** (1969), 252–262].

core, integration, scalar, optimal points of information

Considers quadrature formulas of degree s. The information is the values of f. Optimal points of information are considered. Shows that if $f^{(s)}$ does not change sign on an interval, then the unique optimal points exist and are determined by a three-term nonlinear recurrence. The connection with suboptimal knots which minimize the dominant error term is shown.

Todd, M. J., Optimal dissection of simplices, Department of Operations Research Rep., Cornell Univ. (1976).

core, nonlinear equations, multivariate, optimal algorithms

Considers the fixed point problem in a class of continuous functions of several variables. The information is the values of f. Gives nearly optimal triangulations for computing fixed points. Obtains bounds on the asymptotic rate at which the error decreases for optimal algorithms.

Traub, J. F., On functional iteration and the calculation of roots, *Preprints of Papers 16th Nat. ACM Conf.* Session 5A-1 pp. 1–4. Los Angeles, California (1961).

core, nonlinear equations, scalar, one-point iterations, one-point iterations with memory, iterative complexity, maximal order

Initiates research into iterative complexity. Short summary of research later published in Traub [64]. Introduces classification of one-point iteration and one-point iteration with memory (which is called multipoint iteration in this paper). Proves maximal order theorem for one-point iteration. Conjectures memory always adds less than one to order for a one-point iteration with memory.

Traub, J. F., The theory of multipoint iteration functions, *Digest of Technical Papers, ACM Nat. Conf.* Vol. 1, pp. 80–81 (1962).

core, nonlinear equations, scalar, multivariate, multipoint iterations, iterative information, iterative algorithms, order

Introduces concept of multipoint iteration functions for computing a zero of a nonlinear function f. Points out that for multipoint iterations, algorithms of order p do not require the evaluation of the first $p - 1$ derivatives of f. In particular, there exist iterations of order p using $p - 1$ values of f and one value of f', iterations of order p using one value of f and $p - 1$ values of f', and iterations of order $2(p - 1)$ using $p - 1$ values of f, one value of f' and one value of f''. See also Shamanskii [67], who has analyzed a discrete version of the first of these iterations.

Traub, J. F., Optimal m-invariant iteration functions, *Notices Amer. Math. Soc.* **9** (1962), 122.

core, nonlinear equations, scalar, one-point iterations, maximal order

Considers one-point iterations for solving a scalar nonlinear equation for a zero of known multiplicity m. Seeks a family of iterations of maximal order p for all positive integer m and p. The problem is completely solved; the maximal order iterations have coefficients depending on Stirling numbers of the first and second kind.

Traub, J. F., On the informational efficiency of iteration functions, *Abstr. Short Commun. Internat. Congr. Math., Stockholm* 202 (1962).

core, nonlinear equations, scalar, one-point iterations, one-point iterations with memory, multipoint iterations, maximal order

Considers iterations for computing a simple zero of a scalar nonlinear function. Discusses relation between the information used by an iteration and its maximal order for three classes of iterations: one-point, one-point with memory (which is called modified one-point), and multipoint.

Traub, J. F., Interpolatory iteration functions, *Nat. ACM Conf.*, (1963). (Paper abstract appears in *Comm. ACM*, **6** (1963), 357.)

core, nonlinear equations, scalar, one-point iterations, one-point iterations with memory, order

Introduces interpolatory iterations for a zero of a scalar nonlinear function. Derives two polynomial equations which determine the order of an interpolatory iteration for a simple or multiple zero.

Traub, J. F., "Iterative Methods for the Solution of Equations." Prentice-Hall, Englewood Cliffs, New Jersey, 1964.

core, nonlinear equations, scalar and multivariate, one-point and multipoint iterations, iterations with memory, interpolatory iterations, iterative information, iterative complexity, maximal order

Considers iterations for computing a simple or multiple zero of a scalar nonlinear function f. Also presents some results on multivariate nonlinear functions. The information is the values of f and its derivatives. Some of the results were announced in earlier abstracts and summaries; see papers by Traub from 1961 to 1963.

Introduces classification of iterations as one-point, one-point with memory, multipoint, multipoint with memory. Proves maximal order theorem for one-point iterations. Introduces idea of interpolatory iteration. Conjectures memory always adds less than one to order for a one-point iteration. Introduces multipoint iteration and shows the maximal order properties of multipoint iteration are very different from one-point iteration. Analyzes computational efficiency. Good bibliography.

Much of the material in this book is contained in a 140-page unpublished manuscript prepared in 1961.

Traub, J. F., Computational complexity of iterative processes, *SIAM J. Comput.* **1** (1972), 167–179.

core, survey, iterative complexity, maximal order

Surveys research in iterative complexity and gives some history. Introduces terminology *analytic computational complexity*. Good bibliography.

Traub, J. F., Optimal iterative processes: theorems and conjectures, "Information Processing," Vol. 71, pp. 1273–1277. North-Holland Publ., Amsterdam, 1972.

core, nonlinear equations, scalar, one-point iterations, one-point iterations with memory, multipoint iterations, iterative complexity, maximal order

Surveys iterative complexity as of 1972. Good bibliography.

Traub, J. F., An introduction to some current research in numerical computational complexity, *in* "The Influence of Computing on Mathematical Research and Education" (*Proc. Symp. Appl. Math.*), Vol. 20, pp. 47–55. American Mathematical Society, Providence, Rhode Island, 1974.

core, survey, algebraic complexity, analytic complexity, algebraic numbers, parallel algorithms

Surveys some research in algebraic complexity, analytic complexity, and parallel algorithms as of 1974. Good bibliography.

Traub, J. F., Parallel algorithms and parallel computational complexity, "Information Processing," Vol. 74, pp. 685–687. North-Holland Publ., Amsterdam, 1974.

core, parallel algorithms, parallel complexity

Surveys some research in parallel algorithms and parallel complexity as of 1974. Defines optimal speed-up for both algebraic and analytic problems. In the methodology of this paper, the speed-up for computing a zero of a nonlinear function is bounded by a constant for any number of processors.

Traub, J. F., Theory of optimal algorithms, *in* "Software for Numerical Mathematics" (D. J. Evans, ed.), pp. 1–13. Academic Press, New York, 1974.

core, nonlinear equations, scalar, iterative complexity, algebraic complexity

Surveys iterative complexity and some topics in algebraic complexity as of 1973.

Traub, J. F., (ed.), "Analytic Computational Complexity." Academic Press, New York, 1976.

core, iterative information, fast algorithms, iterative complexity, maximal order

Proceedings of a symposium held in 1975. Contains 13 papers on analytic computational complexity.

Traub, J. F., Introduction, *in* "Analytic Computational Complexity" (J. F. Traub, ed.), pp. 1–4. Academic Press, New York, 1976.

core, analytic complexity

Summarizes some of the reasons for studying analytic complexity. Gives brief overview of 13 invited papers.

Traub, J. F., Recent results and open problems in analytic computational complexity, *in* "Mathematical Models and Numerical Methods," Vol. 3. Banach Center Publ. PWN, Warsaw, Poland, 1978.

core, nonlinear equations, abstract, iterative complexity, maximal order

Based on a lecture presented at the Banach Center in 1975. Gives a brief survey of research and open problems as of 1975.

Traub, J. F., and Woźniakowski, H., Strict lower and upper bounds on iterative computational complexity, *in* "Analytic Computational Complexity" (J. F. Traub, ed.), pp. 15–34. Academic Press, New York, 1976.

core, nonlinear equations, iterative information, one-point iterations, iterative complexity, complexity index, lower bounds

Studies the complexity of iteration assuming that a simplified error equation holds. Introduces complexity index and shows that complexity is the product of the complexity index and the error coefficient function. Gives strict nonasymptotic lower and upper bounds on complexity. Also gives rigorous conditions for comparing the complexity of two different algorithms.

Traub, J. F., and Woźniakowski, H., Optimal linear information for the solution of nonlinear operator equations, *in* "Algorithms and Complexity: New Directions and Recent Results" (J. F. Traub, ed.), pp. 103–119. Academic Press, New York, 1976.

core, nonlinear equations, abstract, optimal linear iterative information, order of information, one-point iterations

Poses a new question: What information is relevant to the solution of a problem? Let f be a nonlinear operator. The information is any finite-dimensional linear operator on f. Proves that the maximal order of any one-point iteration using linear information is the cardinality of the information. On the other hand, any iteration of order n using linear information of cardinality n must use the standard information $f(x), f'(x), \ldots, f^{(n-1)}(x)$. That is, any even locally convergent one-point iteration must use the information $f(x), f'(x)$.

Traub, J. F., and Woźniakowski, H, Optimal radius of convergence of interpolatory iterations for operator equations, Dept. of Computer Science Rep., Carnegie-Mellon Univ. (1976). To appear in *Aequationes Math.*

core, nonlinear equations, abstract, one-point iterations, radius of convergence

Studies radius of convergence of one-point direct interpolatory iteration as a function of order. Shows for two classes of operator equations that the radius of convergence can be large for large order.

Traub, J. F., and Woźniakowski, H., Convergence and complexity of Newton iteration for operator equations, Dept. of Computer Science Rep., Carnegie-Mellon Univ. (1977). See also *J. Assoc. Comput. Mach.* **26** (1979), 250–258.

core, nonlinear equations, abstract, Newton iteration, iterative complexity, optimal convergence

Considers what conditions must be imposed to assure "good complexity" in addition to convergence. Studies Newton iteration for a zero of a nonlinear operator. Establishes optimal radius of ball of convergence with respect to a certain functional. Shows Newton may have arbitrarily high complexity when it converges and conjectures this is a general phenomenon. Establishes radius of ball of good complexity and lower bound on complexity of Newton iteration.

Traub, J. F., and Woźniakowski, H., Convergence and complexity of interpolatory-Newton iteration in a Banach space, Dept. of Computer Science Rep., Carnegie-Mellon Univ. (1977). To appear in *Comput. Math. Appl.*

core, nonlinear equations, abstract, Newton iteration, optimal iterations, iterative complexity

The class of interpolatory-Newton iterations is defined and analyzed for the computation of a simple zero of a nonlinear equation in a Banach space of finite dimension. Convergence of the class is established. Concepts of "informationally optimal class of algorithms" and "optimal algorithm" are formalized. For the multivariate case, the optimality of Newton iteration is established in the class of one-point iterations under an "equal cost assumption."

Traub, J. F., and Woźniakowski, H., General theory of optimal error algorithms and analytic complexity, part A. General information model, Dept. of Computer Science Rep., Carnegie-Mellon Univ. (1977).

core, analytic complexity, optimal linear information, optimal error algorithms, central algorithms, linear algorithms, lower bounds

This paper included in Chapters 1–3 and 5 of Part A.

Traub, J. F., and Woźniakowski, H. (Chapter 2 was written jointly with B. Kacewicz), General theory of optimal error algorithms and analytic complexity, part B. Iterative information model, Dept. of Computer Science Rep., Carnegie-Mellon Univ. (1978).

core, analytic complexity, nonlinear equations, abstract, iterative information, order of information, iterative algorithms

This paper is included in Part B.

Traub, J. F.: *See also* Brent, R. P.; Feldstein, A.; Kung, H. T.

Trigiante, D.: *See* Casuli, V.

Trojan, J. M., Tight bounds on the complexity index of one-point iterations (1979) To appear in *Comput. Math. Appl.*

core, nonlinear equations, abstract, asymptotically optimal algorithms, one-point iterations, iterative complexity

Considers maximal order one-point iterations for solving nonlinear equation $f = 0$ in a Banach space. The information used by the nth-order method is the standard information $f(x), f'(x),$ $\ldots, f^{(n-1)}(x)$. If f is finite dimensional, the combinatory complexity of these methods is linear in the number of pieces of scalar information used. This yields tight bounds on the complexity index. See also Trojan, J. M., How to decrease the combinatory complexity, *Demonstratio Math.* **11** (1978), 807–811, where the scalar case is studied.

Tureckij, A. H.: *See* Aksen, M. B.

Varaiya, P.: *See* Cohen, A. I.

Velikin, V. L., Optimal interpolation of periodic differentiable functions with bounded rth derivative (in Russian), *Math. Zametki* **22** (1977), 663–670.

core, interpolation, periodic functions, scalar, optimal points of information

Considers the interpolation problem for the class of periodic scalar functions with bounded rth derivative in L_∞. The information is the values of f, f' at n points. It is shown that the optimal points of information are equidistant and that $2n$ function evaluations are more relevant than n function and n first derivative evaluations.

Verner, J. M.: *See* Cooper, G. J.

Vitushkin, A. G., "Estimation of the Complexity of the Tabulation Problems." Fizmatgiz, Moscow, 1959 (in Russian) [*English transl.*: Vitushkin, A. G., "Theory of the Transmission and Processing Information." Pergamon, Oxford, 1961].

core, optimal coding, entropy

Considers the optimal coding for a given class F of functions defined on a domain G. Suppose that for a fixed $\varepsilon > 0$, there exist a finite set w and a function $P_k = P_k(x,y), y = (y_1, y_2, \ldots, y_p)$ which is a polynomial of degree at most k with respect to each y_i, $i = 1, 2, \ldots, p$, such that for every function $f \in F$ there exists $y(f) \in w^p$ such that $|f(x) - P_k(x, y(f))| \le \varepsilon$, $\forall x \in G$. Then $T_\varepsilon^\Phi(f) = \{y(f), P_k\}$ is called the table of f, where Φ is a space containing f and $P_k(\cdot, y)$. If $n = \text{card}(w)$ is the total number of elements of w, then n^p is the total number of different elements which can occur to code all functions from F. $P(T_\varepsilon^\Phi(F)) = \log_2 n^p$ is called the size (or complexity) of the table and measures the number of binary bits necessary to represent n^p elements. Shows that for optimal coding, $P(T_\varepsilon^\Phi(F))$ is essentially equal to the ε-entropy $H_\varepsilon(F)$ of F. For some sets F, the basic sharp inequality $p \log_2(k + 1)/\varepsilon) \ge c(F)H_\varepsilon(F)$ is proven ($c(F)$ is a positive constant). Finds ε-entropy for many important classes of functions.

Wacker, H. J., Minimierung des rechenaufwandes des globalisierungen spezieller iterationsverfahren von typ minimales residuum, *Computing* **18** (1974), 209–224.

core, nonlinear equations, abstract

Considers the solution of a nonlinear equation $T(x) = 0$ is a Hilbert space. The continuation method is employed, and $T(x, s_i) = 0$ with $0 = s_0 < s_1 < \cdots < s_b = 1$, where $T(x, 1) \equiv T(x)$ is solved by a locally convergent iteration starting with the computed approximation x_{i-1} to the solution of $T(x, s_{i-1}) = 0$. The optimal equidistant points s_i are considered.

Walsh, J. L.: *See* Ahlberg, J. H.

Wasilkowski, G. W., *N*-Evaluation conjecture for multipoint iterations for the solution of scalar nonlinear equations, Master's thesis, Dept. of Mathematics, Univ. of Warsaw (1977). To appear in *J. Assoc. Comput. Mach.*

core, nonlinear equations, scalar, order of information, multipoint iterations without memory

Considers multipoint iterations without memory for the solution of nonlinear scalar equations. The information is N values of f and its derivatives generated by an incidence matrix. Proves that the order of such iterations is no higher than 2^{N-1} whenever the corresponding Birkhoff interpolation problem is well poised in the complex case.

Wasilkowski, G. W., Can any stationary iteration using linear information be globally convergent? Dept. of Computer Science Rep., Carnegie-Mellon Univ. (1978). To appear in *J. Assoc. Comput. Mach.*

core, nonlinear equations, scalar, iterative information, one point iterations, multipoint iterations, global convergence

Proves that a stationary iteration which uses linear information cannot be globally convergent. Shows this result holds even for as simple a class of problems as the set of all analytic complex functions having only simple zeros. Conjectures the result holds even for the class of all real polynomials with real simple zeros.

Wasilkowski, G. W., The strength of nonstationary iteration, Dept. of Computer Science Rep., Carnegie-Mellon Univ. (1979).

core, nonlinear equations, abstract, linear information, global convergence

Considers the iterative solution of nonlinear equations in a Banach space. The information is linear and adaptive. Probes the existence of globally convergent nonstationary iterations for the class of analytic operators with simple zeros.

Wasilkowski, G. W., Any iteration for polynomial equations using linear information has infinite complexity, Dept., of Computer Science Rep., Carnegie-Mellon Univ. (1979).

core, nonlinear equations, scalar, linear information, complexity

Considers the iterative solution of polynomial equations with simple zeros. The information is linear and adaptive. Proves that for any iteration φ and any number k, there exists a polynomial f with all simple zeros such that the first k approximations produced by φ do not approximate a zero α of f better than a starting approximation x_0. This holds even if $|x_0 - \alpha|$ is arbitrarily small. This result implies that complexity of any iteration is infinite for the class of polynomial equations with simple zeros.

Wasilkowski, G. W., and Woźniakowski, H., Optimality of spline algorithms, Dept. of Computer Science Rep., Carnegie-Mellon Univ. (1978).

core, approximation of linear operators, spline algorithms, optimal linear algorithms, central algorithms

This paper is included in Part A as Chapter 4.

Weinberger, H. F., Optimal approximation for functions prescribed at equally spaced points, *J. Res. Nat. Bur. Std. Sect. B* **65** (1961), 99–104.

core, approximation of linear functionals, optimal error algorithms

Considers approximation of a linear functional for the class of scalar functions whose kth derivative is bounded in L_2 by a constant. The information is the values of f at n equidistant points. Optimal error algorithms and their errors are derived. The computation of the optimal algorithm involves the inversion of a matrix of size $k - 1$.

Weinberger, H. F., On optimal numerical solution of partial differential equations, *SIAM J. Numer. Anal.* **9** (1972), 182–198.

core, approximation of linear operators, differential equations, optimal algorithms

Considers approximation of a linear operator $S:B \rightarrow \Sigma$ for the unit ball in B. Information is the value of a linear operator $\mathfrak{R}:B \rightarrow \mathbb{R}^n$. For a given linear operator $M: \mathbb{R}^m \rightarrow \Sigma$, the operator S is approximated by MQN, where Q is a $n \times m$ matrix. For fixed n and m, the optimal choice of Q, M, and N is studied. The optimal error is expressed in terms of the norms of S and S^*. The results are illustrated for parabolic and elliptic differential equations.

Weinberger, H. F.: *See also* Golomb, M.

Weintraub, S.: *See* Sard, A.

Werner, M.: *See* Barrar, R. B.; Loeb, H. L.

Werschulz, A. G., Optimal order and minimal complexity of one-step methods for initial value problems, Dept. of Computer Science Rep., Carnegie-Mellon Univ. (1976).

core, ordinary differential equations, optimal order.

This is a part of a Ph.D. Thesis. Considers the solution of an initial value problem in a class of ordinary differential equations. The information is the values of the right-hand side function. Finds an optimal order of one-step methods which minimizes the complexity of obtaining an ε-approximation. Shows that under reasonable hypotheses the optimal order tends to infinity as ε goes to zero.

Werschulz, A. G., Computational complexity of one-step methods for the numerical solution of initial value problems, Dept. of Computer Science Rep., Carnegie-Mellon Univ. (1976). To appear in *Computing*.

core, ordinary differential equation, optimal order, optimal step-size

This is a part of a Ph.D. Thesis. Considers the solution of a class of ordinary differential equations. The information is the values of the right-hand side function. Studies the optimal order and optimal step-size of one-step methods which minimize the complexity of obtaining an ε-approximation. Shows that the optimal order increases monotonically and tends to infinity as ε tends to zero.

Werschulz, A. G., Computational complexity of one-step methods for systems of differential equations, Dept. of Computer Science Rep., Carnegie-Mellon Univ. (1976). To appear in *Math. Comp.*

core, ordinary differential equations, optimal order

This is a part of Ph.D. Thesis. Considers the solution of an initial value problem for a system of N ordinary differential equations. The information is the values of the right-hand side function and its derivatives. Exhibits an algorithm for the pth-order Taylor method with $\Theta(p^N \log p)$ combinatory complexity. Finds an optimal order which minimizes the complexity bounds of obtaining an ε-approximation and shows that under reasonable hypotheses the order tends to infinity as ε goes to zero.

Werschulz, A. G., Maximal order and order of information for numerical quadrature, Mathematics Research Rep. 77-2, Univ. of Maryland, Baltimore County (1977). See also *J. Assoc. Comput. Mach.* **26** (1979), 527–537.

core, integration, scalar, order of information, maximal order

Considers integration for the class of analytic functions. The information is the values of f and its derivatives. Defines the concepts of local order and order of information. Studies methods which use fixed information and have maximal order. Proves that the maximal order is equal to the order of information. Finds the order of information for "equally weighted" Hermitian information.

Werschulz, A. G., Maximal order for quadratures using n evaluations, Mathematics Research Rep. 77-7, Univ. of Maryland, Baltimore County (1977). To appear in *Aequationes Math.*

core, integration, scalar, maximal order.

Considers integration for the class of analytic functions. The information is the values of n functionals L_i of f, where $L_i(f) = f^{(j_i)}(x_i)$, $i = 1, 2, \ldots, n$, for some j_i. Conjectures that the order of information (see Werschulz [77, Rep. 77-2]) is at most $2n + 1$. Proves this conjecture for Hermitian information.

Werschulz, A. G., Maximal order for approximation of derivatives, Mathematics Research Rep. 77-8, Univ. of Maryland, Baltimore County (1977). See also *J. Comput. System Sci.* **18** (1979), 213–217.

core, differentiation, order of information, maximal order

Considers differentiation for the class of smooth scalar functions. The information is the values of f. Defines the concept of order of a method and order of information. Shows that the maximal order of methods using fixed information is equal to the order of information. Proves that the central difference formula has maximal order.

Werschulz, A. G., Maximal order and order of information for local and global numerical problems, *Proc. Conf. Informat. Sci. Systems* John Hopkins Univ., Baltimore, Maryland (1978).

core, discretized information, order of information, optimal algorithms

Considers approximation of $S(f; h)$ for f from a given set when a real h tends to zero. The information is an operator $\Re(f, h)$. Defines the order of information and proves that this is the sharp upper bound on the orders of algorithms which use $\Re(f, h)$. Several examples illustrate the paper.

Werschulz, A. G., Multipoint methods with memory using Hermitian information, *Proc. Conf. Informat. Sci. Systems* Johns Hopkins Univ., Baltimore, Maryland (1979).

core, nonlinear equations, scalar, Hermitian information, memory, maximal order

Considers multipoint iterations with memory for the solution of nonlinear scalar equations. The information is the values of f and its derivatives. Studies maximal order of such iterations and shows that 2^n is the sharp bound on the maximal order.

Wilde, D. J., "Optimum Seeking Methods." Prentice-Hall, Englewood Cliffs, New Jersey, 1964.

core, extremum, scalar, optimal algorithms

A text on the search for the maximum of a function. The information is the values of f. Considers, among others, Fibonaccian search for the class of unimodal functions.

Wilde, D. J.: *See also* Avriel, M.; Beamer, J. H.

Wilf, H. S., Exactness conditions in numerical quadrature, *Numer. Math.* **6** (1964), 315–319.

core, integration, scalar, optimal points of information, optimal algorithms

Considers integration for a class of scalar analytic functions on the unit disk in a Hilbert space. The information is the values of f. Optimal weights and points of quadrature formulas are considered.

Winograd, S., Parallel iteration methods, *in* "Complexity of Computer Computations" (R. E. Miller and J. W. Thatcher, eds.), pp. 53–60. Plenum Press, New York, 1972.

core, nonlinear equations, scalar, iterations with memory, parallel complexity, iterative complexity, maximal order

Considers solution of a scalar nonlinear equation $f = 0$ on a parallel computer. Considers a class of iterations for which the information is f and its derivatives. Shows that for the complexity model of this paper, the speed-up is logarithmic in the number of processors.

Winograd, S., Some remarks on proof techniques in analytic complexity, *in* "Analytic Computational Complexity" (J. F. Traub, ed.), pp. 5–14. Academic Press, New York, 1976.

core, lower bounds

Discusses "fooling" or "adversary" argument to obtain lower bounds in analytic complexity. For illustration, shows how this argument is used to obtain lower bounds on algorithms for the maximum of a unimodal function, zero of a scalar function, solution of an integral equation, and approximation of a scalar function.

Winograd, S.: *See also* Brent, R. P.; Micchelli, C. A.

Wixom, J. A.: *See* Barnhill, R. E.

Wolfe, P.: *See* Brent, R. P.

Woźniakowski, H., On nonlinear iterative processes in numerical methods (in Polish), Ph.D. Thesis, Univ. of Warsaw (1972).

core, nonlinear equations, abstract, one-point iterations, iterations with memory, interpolatory iterations, maximal order

Considers iterative solution of nonlinear equations. The information is the values of f and its derivatives. Studies maximal order one-point iterations without and with memory. Generalizes interpolatory iterations with memory to multivariate cases. See also Woźniakowski [74].

Woźniakowski, H., Maximal stationary iterative methods for the solution of operator equations, *SIAM J. Numer. Anal.* **11** (1974), 934–949.

core, nonlinear equations, abstract, one-point iterations, iterations with memory, interpolatory iterations, maximal order

Generalizes problem of maximal order to infinite-dimensional problems. Establishes maximal order of interpolatory algorithms in scalar case. Shows that memory does not in general increase order in multivariate case.

Woźniakowski, H., Generalized information and maximal order of iteration for operator equations, *SIAM J. Numer. Anal.* **12** (1975), 121–135.

core, nonlinear equations, abstract, order of information, one-point iterations, iterations with memory, interpolatory iterations

Introduces concept of order of information which provides general tool for establishing maximal order of an algorithm. Shows maximal order depends only on information used by an algorithm and not on the structure of the algorithm. Proves that any generalized interpolatory algorithm has maximal order.

Woźniakowski, H., Properties of maximal order methods for the solution of nonlinear equations, *Z. Angew. Math. Mech.* **55** (1975), 268–271.

core, nonlinear equations, abstract, Hermitian information, order of information, multipoint iterations for scalar problems, iterative complexity

Studies properties of maximal order iterations. Announces that the Kung and Traub conjecture holds for $n \leq 3$ and for Hermitian information.

Woźniakowski, H., Maximal order of multipoint iterations using n evaluations, *in* "Analytic Computational Complexity" (J. F. Traub, ed.), pp. 75–107. Academic Press, New York, 1976.

core, nonlinear equations, scalar, Hermitian information, iterative information, order of information, multipoint iterations

Studies maximal order of multipoint iteration for solution of the scalar equation $f^{(m)} = 0, m \geq 0$. Information is n evaluations of f or its derivatives. Let $p_n(m)$ denote the maximal order. Establishes for Hermitian information the Kung and Traub conjecture the $p_n(0) = 2^{n-1}$. Conjectures that $p_n(m) = 2^{n-1}$. Shows relation between the problems of maximal order and Birkhoff interpolation.

Woźniakowski, H.: *See also* Jankowski, M.; Kacewicz, B.; Traub, J. F.; Wasilkowski, G. W.

Yun, D. Y. Y., Hensel meets Newton—Algebraic constructions in an analytic setting, *in* "Analytic Computational Complexity" (J. F. Traub, ed.), pp. 205–215. Academic Press, New York, 1976.

core, Newton iteration, algebraic complexity

Discusses use of Newton iteration (which is usually considered an analytic technique) for the solution of algebraic problems such as p-adic approximation.

Zaliznyak, N. F., and Ligun, A. A., On optimum strategy in search of global maximum of function (in Russian), *Zh. Vychisl. Mat. Mat. Fiz.* **18** (1978), 314–321 [*English transl.: U.S.S.R. Computational Math. and Math. Phys.* **18** (1978), 31–38].

core, extremum, scalar, optimal linear information, optimal error algorithms

Considers the search for the maximum in a class which is the algebraic sum of a convex, compact, and balanced set and a finite-dimensional linear space. The information is the values of n linear functionals. The optimal error algorithms are derived. The optimal adaptive information is shown to be nonadaptive. For the class of 2π periodic functions whose rth derivative is bounded in L_∞ by unity, the optimal information is the values of f at n equidistant points, the optimal error algorithms are related to splines, and the error is the n-widths of the problem, which is equal to K_r/n^r, where K_r is the Favard constant.

Žensybkaev, A. A., On the best quadrature formula on the class $W^r L_p$ (in Russian), *Dokl. Akad. Nauk SSSR* **227** (1976), 277–279 [*English transl.: Soviet Math. Dokl.* **17** (1976), 377–380].

core, integration, periodic functions, scalar, optimal points of information, optimal linear algorithms

Considers integration for the class of scalar periodic functions with bounded rth derivative in L_p. The information is the values of f. Proves that the rectangle quadrature formula with equidistant points is an optimal linear algorithm with optimally chosen points of information. Generalizes Motornyi [73] where this result was proven for $p = \infty$.

Žensykbaev, A. A., Best quadrature formulas for some classes of nonperiodic functions (in Russian), *Dokl. Akad. Nauk SSSR* **236** (1977), 531–534 [*English transl.: Soviet Math. Dokl.* **18** (1977), 1222–1226].

core, integration, scalar, optimal points of information, optimal linear algorithms

Considers integration for the class of scalar functions with bounded rth derivative in L_p. The information is the values of f. Optimal linear algorithms with optimal points of information and their errors are studied.

Žensykbaev, A. A., On a property of the best quadrature formulae (in Russian), *Mat. Zametki* **23** (1978), 551–562.

core, integration, scalar, optimal linear algorithms

Considers integration for the class of scalar functions whose rth derivative is bounded in L_q. The information is the values of $f, f', \ldots, f^{(\zeta)}$. Shows that the optimal quadrature formulas coincide for $\zeta = 2m$ and $\zeta = 2m + 1$. This also holds for the periodic case.

Zhileikin, Ya. M., and Kukarkin, A. B., On the optimal evaluation of integrals with strongly oscillating integrand (in Russian), *Zh. Vychisl. Mat. Mat. Fiz.* **18** (1978), 294–301

core, integration, scalar, optimal error algorithms

Considers the approximation of $\int_0^1 e^{iwg(x)} f(x)\, dx$ for a fixed function g, large $|w|$, and for the class of functions f with bounded rth derivative. The information is the values of f. The optimal error algorithms and their errors are presented for $g \in C^{r+1}$.

Zhilinskas, A. G., One-step Bayesian method for searching for the extremum of functions of one variable (in Russian), *Cybernetics* **1** (1975), 139–144.

core, extremum, scalar, optimal Bayesian algorithms

Considers the search for the minimum in a class of scalar functions related to the Wiener process. Presents an optimal one-step Bayesian algorithm.

Author Index

Numbers in italics refer to the pages on which the complete references are listed.

A

Adamski, A., 16, 165, *203, 281*, 303, 310
Ahlberg, J. H., 16, 60, 69, *203, 281*, 311, 326
Aho, A. V., 17, *203*, 234, *272*
Aksen, M. B., 108, 112, 115, *203, 281*, 325
Alhimova, V. M., 59, 108, 115, *203, 281*
Aliev, R. M., 108, 112, 115, *208, 281, 298*
Anselone, P. M., 71, *203*
Aphanasjev, A. Yu, 16, 165, *203, 282*, 311
Arestov, V. V., 96, *204*, 282, *282*
Arro, V. K., 59, 108, *211, 282, 305*
Atteia, M., 71, *204*
Aubin, J. P., *282*
Avriel, M., 16, 165, *204, 282*

B

Babenko, V. F., 108, *204, 282*
Babuška, I., 16, 64, 108, 118, 162, *204, 283*, 318
Bakhvalov, N. S., 16, 55, 64, 99, 108, 132, 162, *204*, 283, *283, 284, 285*, 318
Barnhill, R. E., 99, 108, 117, *205, 285*, 329
Barrar, R. B., 108, *205, 285*, 306, 327
Baudet, G. M., *286*
Beamer, J. H., 16, 165, *205, 286*, 328

B

Bojanov, B. D., 59, 60, 69, 92, 94, 99, 108, 115, 121, 147, *205, 286, 287*, 291
Booth, R. S., 16, 164, *205*, 287, *287*
Borodin, A., 87, *205, 287*, 310
Brent, R. P., 16, 17, 23, *205*, 234, 263, 266, *272*, 279, *280, 287*, 288, *288*, 289, *289*, 312, 329
Busarova, T. N., 108, 118, *205, 289*
Butcher, J. C., 290, *290*

C

Casuli, V., *290*, 325
Chawla, M. M., 60, 108, *206, 290*, 301
Chentsov, N. N., 108, *206, 290*
Chernogorov, V. G., 60, 99, 121, *205, 287*, 291
Chernousko, F. L., 16, 165, *206, 291*
Chzhan, Guan-Tszynan, 64, 132, *206, 291*
Cohen, A. I., *291*, 325
Coman, Gh., 69, 108, 115, *206, 291, 292*, 310
Cooper, G. J., *292*, 325
Cooper, L., 16, 165, *210*, 292, *303*

D

Danilin, Yu., M., 16, 165, *206, 292*, 312
de Boor, C., 69, 121, 131, *206, 292*, 309
den Heijer, C., *292*

E

Eckardt, V., 108, *206, 292*
Edwards, R. E., 27, *206*
Ehrmann, H., *292*
Eichhorn, B. H., 16, 164, 165, *206, 293*
Elhay, S., 108, *206, 293*
Emelyanov, K. V., 132, *206, 293, 298*

F

Feldstein, A., 293, *293,* 294, 297, 313, 325
Fine, T., 16, 165, *207, 294*
Firestone, R. M., *293,* 294
Forst, W., 69, 94, 99, 108, *207, 294*

G

Gaffney, P. W., 69, 99, 100, 101, *207,* 292, 294, *294, 295,* 313
Gaisarian, S. S., 108, 132, *207,* 216, *295*
Gal, S., 16, 49, 60, 165, *207, 295*
Ganshin, G. S., 16, 165, *207, 295, 296*
Garey, M. R., *207, 296*
Gentleman, W. M., *296*
Germeier, Ju. B., *296*
Giršovič, Ju. M., 59, 108, *211,* 296, *305*
Golomb, M., 16, 60, 64, 69, 99, 104, 106, 121, *207,* 279, *280,* 296, *309,* 327
Grebennikov, A. I., 16, 69, *208, 297,* 310
Gross, O., 16, 164, *208, 297,* 299

H

Haber, S., 108, *208, 297*
Herzberger, J., *297*
Hindmarsh, A. C., *297*
Holmes, R., 71, *208*
Hopcroft, J. E., 17, *203, 272*
Hyafil, L., 16, 164, *208, 297*

I

Ibragimov, I. I., 108, 112, 115, *208,* 281, *298*
Ilin, A. M., 132, *293, 298*
Ismagilov, R. S., 65, *208*
Ivanov, V. V., 16, 108, 121, 165, *208, 298*

J

Jankowska, J., 263, *272, 298*
Jankowski, M., *299,* 329
Jarratt, P., *299*
Jeeves, T. A., *299*

Jetter, K., 108, *208, 299*
Johnson, D. S., *207, 296*
Johnson, L. W., 108, *208, 299,* 314
Johnson, S. M., 16, 165, *208, 297,* 299
Judin, D. B., 16, 165, *208, 299,* 300, *300,* 310

K

Kacewicz, B., 16, *209,* 234, 266, 267, 268, *272, 273, 300,* 329
Karlin, S., 58, 69, 108, 115, *209, 300, 301*
Karp, R. M., 16, 165, *209, 301,* 310
Kaul, V., 60, 108, *206, 290,* 301
Kautsky, J., 108, 112, 115, *209, 301*
Keast, P., 108, 118, *209, 301*
Kiefer, J., 16, 48, 49, 60, 108, 163, 164, 165, 179, 183, *209,* 278, 279, *280,* 281, 291, *294,* 299, *301*
Knauff, W., 16, 60, *209, 302,* 303
Knuth, D. E., 91, *209,* 237, *273*
Kolmogorov, A. N., 157, 159, 162, *209*
Kornejčuk, N. P., 59, 65, 69, 108, 110, 111, 115, 118, 121, 127, 128, 129, *210, 302,* 307
Korobov, N. M., 99, 108, 118, *210,* 301, *302*
Korotkov, V. B., 47, *210, 303*
Korytowski, A., 16, 165, *203, 281,* 303
Kress, R., 16, 60, *209, 302,* 303
Krolak, P., 16, 165, *210,* 292, *303*
Krylov, V. I., 59, 108, 115, *210, 303*
Kukarkin, A. B., 108, *217,* 303, *330*
Kung, H. T., 16, 23, 170, *205, 210,* 236, 243, 259, 266, 267, *273,* 287, 289, *289,* 290, 303, *303, 304,* 306, 318
Kuzovkin, A. I., 16, 165, *210, 305,* 321

L

Larkin, F. M., 16, 60, 108, 117, *210, 305*
Laurent, P. J., 71, *203*
Lee, J. W., 58, 108, 115, *210, 305*
Levin, A. Ju., 16, 165, *210,* 305
Levin, M. I., 59, 108, *211,* 282, 296, *305*
Ligun, A. A., 16, 49, 59, 69, 108, 118, 165, *211, 217,* 306, *306, 330*
Lipow, P. R., 58, 69, 108, 115, *211, 306*
Lipson, J., *306*
Loeb, H. L., 108, *205, 211, 285, 306,* 327
Lorentz, G. G., 157, 159, 162, *211*
Lušpaj, N. E., 59, 69, 108, 112, 115, 118, *210, 211, 302,* 306, *306,* 307

M

Majstrovskij, G. D., 16, 164, *211, 307*
Mangasarian, O. L., 16, 58, 69, *211, 307*, 317
Mansfield, L. E., 16, 58, 60, 69, 108, *211, 307*
Marchuk, A. G., 16, 56, *211, 307*, 311
Maung Čžo Njun, 108, *211, 307*, 317
Meersman, R., 16, *211*, 234, 266, 267, *273, 308*
Meinguet, J., 16, 60, *212, 308*
Melkman, A. A., 16, 60, 64, 69, 99, 121, *212, 308, 309*
Meyers, L. F., 16, 58, 99, 108, 115, *212, 308,* 315
Michelli, C. A., 11, 12, 14, 16, 45, 49, 52, 56, 60, 64, 69, 76, 96, 101, 120, 121, 130, 131, 164, 170, *207, 212*, 279, *280*, 292, *295, 308, 309, 309*, 310, 313, 314, 329
Micula, Gh., 108, *206, 292*, 310
Miranker, W. L., 16, 164, 165, 170, *209, 212, 301, 309*, 310, *310*
Mitkowski, W., 16, 165, *203, 281*, 310
Mockus, I. B., 16, 165, *212, 310*
Morozov, V. A., 16, 69, *208, 297*, 310
Motornyj, V. P., 59, 108, 116, 118, 121, *212,* 310, *310*
Munro, I., *287*, 310

N

Nemirovsky, A. S., 16, 165, 208, 299, 300, 310
Newman, D. J., 16, 165, *212, 310*
Nielson, G. M., 16, 60, 69, *212, 311*
Nikolskij, S. M., 16, 52, 58, 59, 108, 115, *213*, 279, *280, 311*
Nilson, E. N., 16, 60, 69, *203, 281*, 311
Novikov, V. A., 16, 165, *203, 282, 311*

O

Ortega, J. M., 263, *273, 311*, 313
Osipenko, K. Yu, 16, 56, 94, 99, 103, 121, 161, *211, 213, 307, 311*
Ostrowski, A., *311*

P

Pallashke, D., 96, 108, *213, 312*
Pan, V., 19, 21, *213*
Parker, D. S., Jr., *312*
Paszkowski, S., *312*
Paterson, M. S., 287, 303, *312*
Paulik, A., 108, *213, 312*

R

Piavsky, S. A., 16, 165, *213*, 292, *312*
Pinkus, A., 16, 45, 64, 69, 108, 117, 121, *212, 213, 309*, 313, *313*
Pleshakov, G. N., 263, *273, 313*
Powell, M. J. D., 69, 99, 100, 101, *207*, 292, 294, *295*, 313

R

Reinsch, Ch., 16, 60, 69, *213, 313*
Rheinboldt, W. C., 263, *273, 311*, 313
Rice, J. R., 16, 121, *213*, 294, *313*
Richter, N., 108, *213, 314*
Richter-Dyn, N., 16, 60, 99, 108, *213, 314*
Riess, R. D., 108, *208, 299*, 314
Rissanen, J., *314*
Ritter, K., 16, 58, 69, *213, 314*
Rivlin, T. J., 11, 12, 14, 16, 56, 60, 64, 69, 76, 101, 120, 121, 130, 131, *212*, 279, *280*, 292, 309, *309*, 314

S

Saari, D. G., 267, *273, 314*, 318
Šajdaeva, T. A., 59, 108, 115, *213, 314*
Sard, A., 16, 52, 56, 58, 99, 108, 115, *212, 214*, 278, *280, 308*, 315, *315*, 316, 327
Schmeisser, G., 108, *214, 315*
Schoenberg, I. J., 16, 58, 69, 108, 110, 115, *214*, 279, *280*, 315, 316
Schultz, M. H., 16, 35, 64, 121, *214, 316*
Schumaker, L. L., 16, 58, 69, *211, 307*, 317
Secrest, D., 60, 69, 96, 99, 108, 115, *214, 317*
Shamanskii, V. E., *317*, 322
Sharygin, I. F., 108, *211, 214, 307*, 317, *317*
Sieveking, M., *318*
Simon, C. P., 267, *273, 314*, 318
Smolyak, S. A., 16, 52, 54, 57, 58, 99, 108, 109, 118, *214*, 279, *280, 318*
Sobol, I. M., 108, *214, 318*
Sobolev, S. L., 16, 64, 77, 108, 162, *204, 215, 283*, 318, *318*
Sonnevend, G., 16, 164, 165, *215, 318*
Stechkin, S. B., 96, *215*, 282, *319*, 320
Stenger, F., 108, *215, 319*
Stern, M. D., 108, 115, *215, 319*
Stetter, F., 108, *215, 319*
Strongin, R. G., 16, 165, *215, 319*
Sukharev, A. G., 16, 49, 108, 121, 164, 165, *215, 319*, 320, *320*

T

Taikov, L. V., 96, *215, 320*
Tanama, V. P., *320*
Tarassova, V. P., 16, 165, *215, 321*
Tikhomirov, V. M., 16, 42, 45, 65, 67, 120,
    121, 127, 129, 162, 165, *209, 210, 215,*
    305, 321, *321*
Tikhonov, A. N., 108, *216,* 295, *321*
Todd, M. J., 16, 164, *216, 321*
Traub, J. F., 16, 175, 176, *210, 216,* 223, 225,
    227, 230, 231, 232, 234, 235, 237, 239,
    244, 252, 259, 262, 266, 267, *273,* 279,
    280, *280,* 289, *289,* 290, 292, 293, *293,*
    297, *304, 305,* 306, 312, 313, 314, 317,
    321, *321,* 322, *322, 323, 324,* 325, 329
Trigiante, D., *290,* 325
Trojan, J. M., 23, *216, 325*
Tureckij, A. H., 108, 112, 115, *203, 281,* 325

U

Ullman, J. D., 17, *203, 272*

V

Varaiya, P., *291,* 325
Velikin, V. L., 99, *216, 325*
Verner, J. M., *292,* 325
Vitushkin, A. G., 161, 162, *216, 325*

W

Wacker, H. J., *325*
Walsh, J. L., *281,* 326

Wasilkowski, G. W., 68, 190, *216,* 225, 234,
    268, *274, 326,* 329
Weinberger, H. F., 16, 60, 69, 132, 137, 140,
    142, *207, 216,* 279, *280, 296, 326,* 327
Weintraub, S., *315,* 327
Werner, M., 108, *205, 211, 285, 306,* 327
Werschulz, A. G., 16, 95, *217,* 234, *274, 327,*
    328
Wilansky, A., 33, *217*
Wilde, D. J., 16, 165, *204, 205, 217, 282, 286,*
    328, *328*
Wilf, H. S., 108, *217,* 292, *328*
Winograd, S., 16, 17, 69, 101, 120, 121, 130,
    131, *205, 212, 217, 272,* 279, *280, 281,*
    289, 292, *309, 328,* 329
Wixom, J. A., 99, 108, *205, 285,* 329
Wolfe, P., 16, 17, *205, 272,* 279, *280, 289,* 329
Woźniakowski, H., 16, 17, 68, 175, 176, *216,*
    217, 225, 231, 232, 234, 235, 237, 239,
    244, 252, 263, 266, 267, 268, *273, 274,*
    279, 280, *280, 281,* 292, *299, 300,* 314,
    323, 324, 326, 329, *329*

Y

Yun, D. Y. Y., *329*

Z

Zaliznyak, N. F., 16, 49, 165, *217,* 306, *330*
Žensykbaev, A. A., 59, 108, 118, *217, 330*
Zhileikin, Ya. M., 108, *217,* 303, *330*
Zhilinskas, A. G., 16, 165, *217,* 330

Subject Index

A

Absolute error, *see* Error, absolute
Adaptive information, *see* Information, adaptive
Adversary principle, 6, 8–10, 17, 23, 279, 280
Algebraic complexity, *see* Complexity, algebraic
Algorithm(s), 8, 9, 11, 158, 225, 229, 275
 adaptive, 4
 asymptotically convergent sequence of, 200, 201
 asymptotically optimal complexity, 164, 165, 168, 172, 173, 179, 221
 bisection, 164, 169, 190
 central, 8, 9, 14, 52, 60, 64, 68, 70, 74–78, 80, 91, 100–102, 105, 110, 122, 124, 164–169, 171–173, 177–179, 182, 183, 190, 219, 279
 convergent, 190, 230–235, 247, 258–260
 deviation of, 6, 68–70, 73, 74, 76, 79, 80, 220
 direct, 233, 253
 error of, 6, 12, 17, 64, 150, 157, 218
 homogeneous, 53, 54, 69, 72, 73, 74, 78
 interpolatory, 8, 9, 12–14, 52, 64, 72–74, 78, 80, 101, 102, 110, 128, 131, 164–169,
 172, 173, 177, 179, 182, 183, 190, 218, 225, 226, 230–234, 244, 253, 258, 259, 262, 276
 iterative, 4, 22, 163, 223–226, 230, 234, 242, 243, 245, 247, 249, 254, 265, 279
 limiting error of, 230, 258
 linear, 3, 6, 9, 51, 52, 55, 58, 60, 61, 64–70, 74, 76, 79, 80, 82, 84–86, 129, 131, 159, 168, 279
 linear central interpolatory, 52, 53, 64, 92, 93, 95, 96, 99
 linear central spline, 69, 123, 125–127, 132, 135–138, 141, 142, 145
 linear optimal error, 6, 52, 54, 56–61, 63, 67, 87, 94, 101–104, 109, 112, 115–117, 121, 122, 124, 128, 130–132, 160, 164, 279
 maximal order, 23, 224, 234, 238, 254, 279
 minimal complexity index, 23, 224, 225, 237, 254, 276
 nearly optimal complexity, 82, 84–86, 91, 93, 95, 96, 99, 102, 104, 107, 114, 115, 117, 123–128, 132, 136, 138, 141, 142, 145, 146, 220, 224
 nonadaptive, 4
 nonlinear, 61
 nonstationary, 230

Algorithm(s) *(cont.)*
  optimal, 1, 2, 4–8, 23, 24, 52, 69, 70, 77, 90,
      149, 164, 277–279
    in sense of Nikolskij, 6, 52, 54, 56, 58–60,
        279
    in sense of Sard, 6, 52–60, 69, 110, 279
  optimal complexity, 1, 3, 5, 6, 9, 18–20, 23,
      51, 84, 93, 164, 169, 219, 224
  optimal error, 3, 5, 6, 8, 9, 13, 14, 19, 20,
      23, 35, 48, 51–59, 64, 68, 69, 77–80, 85,
      86, 89, 93, 94, 98, 99, 102, 104, 107,
      110, 111, 114, 116, 117, 124, 126–135,
      138, 140, 141, 143, 145, 146, 164, 168,
      173, 175, 176, 187, 219, 224, 278, 279
  optimal linear, 6, 41, 52, 64, 80, 101
  order of, 225, 227, 231, 232, 233, 269
  permissible, 9, 11, 17–20, 22, 23, 83, 84,
      187, 219, 225, 235, 237, 238, 253, 254,
      276
  rectangle, *see* Quadrature formula,
      rectangle
  spline, 60, 64, 69–80, 105, 106, 110, 122,
      124, 135, 220
  stability of, 199
  stationary
    with memory, 230, 258
    without memory, 229
Analytic complexity, *see* Complexity, analytic
Applications to
  approximation of linear functional, 6, 89,
      91, 283, 284, 295, 296, 301, 305,
      307–309, 311, 313–318, 326
  approximation of linear operator, 136, 281,
      282, 295–297, 308, 309, 315, 326, 327
  approximation problem, 6, 13, 15, 25, 47,
      50, 69, 89, 108, 118, 119, 121, 123, 146,
      147, 248, 287, 288, 292, 298, 302, 309,
      310, 313, 315–317, 319–321
    for general Hilbert space, 121
    for W_2^r in L_2, 123, 125, 126, 187, 221
    for $\tilde{W}_2^r$ in L_2, 125, 187, 221
    for W_∞^r in C, 101, 102, 127, 129, 132, 187,
        188, 221
    for $\tilde{W}_\infty^r$ in $\tilde{C}$, 128, 187, 188, 221
  differential equations, 6, 25, 132, 146, 248,
      283–285, 290–292, 295, 316, 327
  differentiation problem, 16, 94, 96, 146,
      186, 188, 221, 234, 282, 309, 312, 320,
      328
  elliptic differential equation problem, 89,
      132, 143–145, 149, 187, 221
  hyperbolic differential equation problem,
      89, 132, 143–145, 149, 187, 221
  integral equations, 293
  integration problem, 6, 13, 16, 25, 46–49,
      52, 56, 58, 59, 69, 89, 96, 107, 108, 114,
      116, 118, 119, 120, 146, 147, 190,
      281–286, 289–321, 327, 328, 330
    for W_p^r, 98, 109, 115, 187, 188, 221
    for $\tilde{W}_p^r$, 187, 188, 221
    for $\tilde{W}_\infty^r$, 115, 117
  interpolation problem, 6, 13, 25, 69, 89, 93,
      99, 146, 188, 221, 285, 287, 294, 295,
      308, 311, 318, 325
    for analytic bounded functions, 103,
        104, 147, 187, 221
    for Hilbert space with reproducing
        kernel, 104, 107, 187, 221
    for W_∞^r, 102, 120, 186, 187, 221
  nonlinear equation problem, 13, 16, 17,
      163–168, 170, 173, 184, 187, 188, 221,
      225, 226, 229, 231–235, 248, 249, 252,
      257, 262, 265, 266, 279, 286–300, 303,
      304, 307, 308–329
  operator equations, 17, 164, 298, 320
  parabolic differential equation problem, 89,
      132, 136–138, 145, 149, 187, 221
  search for maximum (minimum) problem,
      7, 16, 47, 49, 163–165, 176, 179, 183,
      187, 188, 221, 278, 281, 282, 286, 287,
      291–305, 310, 312, 318–321, 328, 330
Approximation, 121, 287, *see also*
    Applications, approximation
  ϵ- (or ϵ'-), 9, 10, 14, 15, 18, 22, 26, 30, 35,
      37, 83, 93, 95, 96, 98, 99, 102, 104, 106,
      107, 139, 162, 169, 172, 178, 182, 186,
      190, 201, 218, 227, 253, 275
Asymptotically convergent sequence of
    algorithms, *see* Algorithms,
    asymptotically convergent sequence of
Asymptotically convergent sequence of
    information operators, *see* Information,
    asymptotically convergent sequence of
Asymptotically optimal complexity algorithm,
    see Algorithm, asymptotically optimal
    complexity
Asymptotically optimal sequence of
    information operators, *see* Information,
    asymptotically optimal sequence of
Asymptotic model, *see* Model, asymptotic
Asymptotic speed of convergence, 200, 201
Average case model, *see* Model, average case

B

Basic linear information operator, *see*
  Information, basic linear
Birkhoff complex interpolation, 268

C

Cardinality, 26, 29, 121, 123, 125, 129, 139,
    142, 150–156, 166, 179
  ϵ- number, 85–87, 114, 115, 117, 122–129,
    131, 132, 135, 136, 138, 139, 141, 142,
    145, 146, 147, 167, 168, 178, 179, 202,
    220
  of information, *see* Information, cardinality
    of
  optimal, number, 256, 276
Center of set, 11, 14, 74, 77, 78, 166
Central algorithm, *see* Algorithm, central
Chebyshev polynomials, 15, 98, 112
Class of algorithms
  IT $(\mathfrak{N},S)$, 243–246, 249, 254, 276
  Φ_n, 157
  Φ^s, 72, 220
  $\Phi(\epsilon)$, 18
  $\Phi_L(n)$, 64, 220
  $\Phi_n(\mathfrak{N})$, 157, 221
  $\Phi(\mathfrak{N},S)$, 11, 56, 218, 230, 238, 258, 275
  $\Phi_{perm}(\mathfrak{N},S)$, 237, 254, 276
Class of information operators
  essentially more efficient, 193
  Lip(k), 241, 246, 276
  Ψ, 20, 178, 187, 219
  Ψ^a, 164
  Ψ_f^a, 183, 189, 221
  Ψ_f^{non}, 188, 189, 221
  Ψ_L^a, 192, 221
  Ψ_L^{non}, 190, 191, 221
  Ψ_n, 35, 41, 166, 220, 221, 255, 256
  Ψ_{NON}, 192, 221
  Ψ_n^a, 49, 170
  Ψ_n^p, 107, 221
  Ψ_n^0, 117
  Ψ_U, 86, 122, 220
  Θ-equivalent, 193
Combinatorial complexity, *see* Complexity,
    combinatorial
Combinatory complexity, *see* Complexity,
    combinatory
Complete information, *see* Information,
    complete

Complexity
  algebraic, 2, 3, 9, 21, 23, 287, 289, 299,
    304–306, 323, 329
  of algorithm, 17, 18, 172, 176, 178, 182,
    219, 263
  analytic, 1, 6, 7, 9, 16, 22, 23, 26, 41, 44,
    48, 52, 64, 67, 121, 149, 157, 195, 199,
    200, 223, 277, 278, 279, 299, 323, 324
  combinatorial, 9, 21, 195, 296
  combinatory, 18, 19, 20, 23, 51, 68, 70, 72,
    80, 83, 84, 117, 164, 165, 167, 168, 172,
    178, 179, 219, 224, 235, 236, 238, 253,
    254, 256, 264, 276
  ϵ-, 19, 20, 85, 93–99, 102, 104, 107, 114, 115,
    122–129, 132, 136, 138, 141, 142, 145,
    146, 168, 169, 170, 172, 178, 179, 182,
    183, 185, 186, 188, 189, 190, 193, 219
  of problem S in class Ψ, 20, 21, 85, 86,
    117, 168
  of problem S with information $\mathfrak{N}$, 18, 84,
    85, 93, 173
  essentially greater, 186, 189
  greater, 186, 188
  index, 122, 123, 134, 225, 234, 236, 238,
    253, 256
  index of algorithm, 224, 236, 237, 238, 254,
    264, 276
  index of problem, 226, 237, 254, 256, 276
  information, 20, 169, 172, 187, 224, 236,
    238, 264
  iterative, 223, 279, 296, 322
  minimal combinatory, 254
  model, *see* Model, complexity
  problem, 1–3, 6, 17, 18, 21, 23, 24, 82, 185,
    224
  space, 2, 6, 51, 82, 90
  Θ-, 186, 188, 190, 192, 193
  time, 2, 51, 82, 90
Complexity hierarchy, 185, 186, 187, 192,
    193
  for class Ψ_f^a, 189
  for class Ψ_f^{non}, 187, 189
  for class Ψ_L^a, 191
  for class Ψ_L^{non}, 190
  for class Ψ_{NON}, 192
  for fixed problem, 192, 194
Computability, *see* Problem, ϵ- computable
Computation, *see* Model, of computation
Condition number, 199
Covering, 160
  ϵ-, 159

D

Deviation of algorithm, *see* Algorithm,
    deviation of
Diameter, 159, 174, 258
    of information, 9, 11, 25, 27, 28, 30, 31, 33,
        34, 48, 49, 119, 149–151, 174, 218, 219,
        224, 225, 227–229, 243, 245, 246, 258,
        268, 275, 279
    limiting, 225–229, 243, 245, 246, 257, 258,
        275
    nth, 138, 140, 142, 143, 146
    nth minimal, 6, 26, 35, 41, 42, 49, 67, 119,
        133, 150, 153, 154, 219, 220, 221
    problem error, 38, 220
    set, 10, 11, 228
Differential equations, *see* Applications,
    differential equations
Differentiation problem, *see* Applications,
    differentiation problem

E

Element
    problem, 10, 14, 25, 231
    solution, 196
Elliptic differential equation problem, *see*
    Applications, elliptic differential equation
    problem
Entropy, 157, 158, 161
    ϵ-, 7, 41, 150, 159, 160, 162, 221
Error
    absolute, 196–198
    algorithm, 41, 64, 218
    limiting, 258, 275
    local, 6, 14, 68, 70, 80, 101, 172, 220
    n-dimensional algorithm, 7, 41, 65, 150,
        157, 158
    nth minimal linear, 52, 65, 220
    optimal, 3, 6, 13, 176, 219
    relative, 196–198
Extremal subspace, 120, 127
    nth, in sense of Gelfand, 44–47, 59, 119,
        120, 125, 126, 130, 220
Extremum, *see* Applications, search for
    maximum (minimum) problem

F

Favard constants, 102, 116, 127, 130
Finite-complexity problem, *see* Problem,
    finite-complexity
Floating point arithmetic, 198

G

Gauss–Chebyshev quadrature formula, *see*
    Quadrature, Gauss–Chebyshev
Gauss quadrature formula, *see* Quadrature,
    Gauss
Gelfand n-width, *see* Width, Gelfand n-

H

Hermite information, *see* Information,
    Hermite
Hermite interpolatory polynomial, *see*
    Interpolatory polynomial
Hierarchy, *see* Complexity hierarchy,
    Problem, hierarchy
Homogeneous algorithm, *see* Algorithm,
    homogeneous
Hyperbolic differential equation problem, *see*
    Applications hyperbolic differential
    equation problem

I

Index, 122, 123, 134
    mth, of problem, 226, 250, 252, 255, 276
    nth minimal complexity, of problem, 226,
        256, 276
    of problem, 26, 28, 31, 35, 219, 224, 226,
        245, 247, 249, 264, 276
Infinite-complexity problem, *see* Problem,
    infinite-complexity
Information
    adaptive, 6, 26, 47, 48, 49, 50, 164, 165,
        168, 170, 171, 179, 189, 191, 220, 225
    adaptive linear, 7, 26, 47, 48, 163–165, 168,
        170, 179, 183, 196
    basic linear, 226, 246–249, 267, 276
    cardinality of, 7, 20, 25–28, 31, 32, 35, 41,
        48, 49, 52, 56, 64, 70, 77, 82–86,
        151–153, 156, 159–161, 164, 167, 190,
        192, 219–221, 225, 226, 238, 239, 244,
        246–253, 256, 264, 265, 267, 276, 278
    complete, 23, 24
    complexity, 17, 83, 187, 219, 234, 238, 253,
        254, 264, 276
    contained, 27, 151, 239
    convergent, 228, 229, 231, 232, 235, 244,
        247, 249, 258, 259, 270
    diameter of, *see* Diameter, of information
    divergent, 228–232
    equivalent, 151, 152, 239
    of f, 10

general, 278, 280
Hermite, 267
iterative, 4, 16, 157, 163, 185, 223–225, 227, 253, 266–268, 275, 278–280
linear, 4–6, 48, 52, 54, 64, 82–86, 113, 116, 129, 151, 153, 157, 159–168, 176, 177, 179, 190, 191, 196, 200, 201, 220, 223–228, 239–241, 244, 246–256, 266, 267, 276, 280
more relevant, 238
mth basic linear, 226, 251, 276
multipoint, 225, 266, 267
  with memory, 267
nonadaptive, 6, 7, 26, 47–50, 163–169, 176, 177, 179, 182, 187–189, 191, 196, 225
nonlinear, 7, 150, 151, 153, 156, 157, 159, 160, 163, 164, 170, 192, 267
nth maximal order, 226, 255, 256, 276
nth optimal, 26, 34–37, 40, 41, 45, 46, 67, 108, 111, 112, 115–117, 119–142, 145, 147, 149, 153, 164, 166, 168, 170, 178, 179, 195, 219, 221
one-point, 225, 226, 264–267
operator, 1, 4, 8–11, 14, 15, 17, 23, 25–30, 32–41, 43, 45–48, 54, 56, 58, 59, 66, 67, 70, 75, 77, 82, 83, 86, 91, 94, 96, 103, 105, 111, 113, 115, 116, 118–121, 123–131, 137, 140, 141, 145–147, 150–161, 164, 168–170, 174, 177, 178, 184–189, 191, 192, 194, 196, 198–201, 218, 219, 223–229, 232, 234, 235, 238–244, 246–253, 255–257, 259, 264–269, 276, 279
  asymptotically convergent sequence of, 200, 201
  asymptotically optimal sequence of, 201
  with memory, 257, 263, 266, 267
  without memory, 259, 264, 266, 267, 269
optimal, 2, 3, 5, 6, 26, 40, 41, 44–47, 50, 67, 111, 119, 127, 135, 140, 148–150, 176, 220, 238, 256, 257, 276, 279
  in class, 238
  linear, 4, 5, 33, 165
partial, 23, 24
permissible, 1, 9, 14, 15, 17, 18, 20, 22, 24, 45, 83, 84, 86, 150, 156
perturbed, 56
radius of, see Radius, of information
standard, 29, 232, 234, 279
stationary, with memory, 230
stationary, without memory, 227
unperturbed, 198

Integral equations, see Applications, integral equations
Integration problem, see Applications, integration problem
Interpolation problem, see Applications, interpolation problem
Interpolatory algorithm, 92–94, 96, 262
Interpolatory iteration, 263
Interpolatory polynomial, 91, 262
Interpolatory Taylor formula, 92–94
Iteration, 164
  interpolatory, 175, 263
  nonstationary, 17, 225, 234, 279
  stationary, 17, 224–226
Iterative algorithm, see Algorithm, iterative
Iterative complexity, see Complexity, iterative
Iterative information, see Information, iterative

K

Kolmogorov n-width, see Width, Kolmogorov n-
Kung–Traub conjecture, 268

L

Linear algorithm, see Algorithm, linear
Linear information, see Information, linear
Local error, see Error, local

M

Mantissa, 198
Maximal order, see Order, maximal
Memory, 226
Minimal norm, mth, of linear operator, 35, 220
Minimal subspace, 38, 39
  mth, of linear operator, 36–40, 220
Model
  asymptotic, 7, 24, 195, 199, 201, 202
  average case, 7, 24, 195, 196
  complexity, 201
  of computation, 1–3, 6, 8, 9, 15, 21, 26, 82–84, 156, 224–226, 234, 253
  perturbed, 7, 195, 198
  relative, 7, 195, 196, 198
  worst case, 24, 195
mth basic linear information, see Information, basic linear
mth index of problem, see Index, mth, of problem

*m*th minimal norm of linear operator, *see*
    Minimal norm, *m*th, of linear operator
*m*th minimal subspace of linear operator, *see*
    Minimal subspace, *m*th of linear operator

N

Newton iteration, 101, 176, 262, 265, 267
Noncomputable, *see* Problem, ϵ-noncomputable
Nonlinear equation problem, *see* Applications,
    nonlinear equation problem
Notation
    o, 91
    O, 91
    Θ, 91
*n*th extremal subspace in sense of Gelfand,
    see Extremal subspace, *n*th, in sense of
    Gelfand
*n*th maximal order information for problem,
    see Information, *n*th maximal order
*n*th maximal order of problem, *see* Order, *n*th
    maximal
*n*th minimal diameter of information, *see*
    Diameter, *n*th minimal
*n*th minimal linear error, *see* Error, *n*th
    minimal linear
*n*th minimal radius of information, *see*
    Radius, *n*th minimal
*n*th optimal information, *see* Information, *n*th
    optimal
n-width, *see* Width

O

Operation, *see* Primitive operation
Operator
    equations, *see* Applications, operator
        equations
    information, *see* Information, operator
    linear *m*th minimal norm of, *see* Minimal
        norm, *m*th, of linear operator
    *m*th minimal subspace of, *see* Minimal
        subspace, *m*th, of linear operator
    restriction, *see* Restriction operator
    solution, *see* Solution operator
Optimal algorithm, *see* Algorithm, optimal
Optimal cardinality number, *see* Cardinality,
    optimal, number
Optimal complexity algorithm, *see* Algorithm,
    optimal complexity
Optimal error, *see* Error, optimal

Optimal error algorithm, *see* Algorithm,
    optimal error
Optimal information, *see* Information,
    optimal in class; Information, optimal,
    linear
Order
    of algorithm, 232, 238, 261, 262, 264, 269,
        276
    classical, 261, 269, 270
    of exactness, 96, 98, 117
    of information, 17, 224–227, 231–233, 235,
        238, 244, 248–252, 254, 255, 257, 259,
        260–262, 264–269, 276, 279
    maximal, 4, 16, 176, 224, 226, 231, 234,
        254–256, 262, 264, 268, 279, 280
    new, 270
    *n*th maximal, 226, 255, 256, 276

P

Parabolic differential equation problem, *see*
    Applications, parabolic differential
    equation problem
Partial information, *see* Information, partial
Penalty, ϵ-, 139, 142
Permissable aglorithm, *see* Algorithm,
    permissible
Permissible information, *see* Information,
    permissible
Perturbed model, *see* Model, perturbed
Primitive operation, 9, 17, 83, 84, 156, 225,
    234, 235, 238, 247, 253
Primitives, 22
Problem, *see also,* Solution, operator
    complexity, *see* Complexity, problem
    convergent, 38, 39, 220
    element, 2, 3, 68, 218, 225, 227, 243, 257,
        270, 275
    ϵ-computable, 18, 219
    ϵ-noncomputable, 18, 26, 38, 123, 125, 219,
        220
    finite-complexity, 2, 22–24
    hierarchy, 2
    infinite-complexity, 22–24
    linear, 3–6, 15, 20, 26, 35, 38, 45, 48, 52,
        60, 82–84, 86, 87, 89, 163, 164, 189,
        195, 196, 198, 224, 237, 279
    nonlinear, 5, 7, 9, 16, 26, 47, 49, 163, 164,
        168, 179, 225, 279
    strongly noncomputable, 38, 220

Q

Quadrature formula, 89, 117, 201, 234, 278
  best in sense of Sard, 57, 58
  Gauss, 98, 99, 117
  Gauss–Chebyshev, 98
  optimal error, 117
  optimal in sense of Nikolskij, 58
  optimal in sense of Sard, 58
  rectangle, 116, 117
  trapezoidal, 199, 200

R

Radius, 12, 138, 141, 149, 159, 164
  of information, 7–9, 31, 33, 59–61, 70, 76,
    98, 101, 102, 105, 108, 111, 118, 119,
    131, 145, 146, 148, 150, 151, 163, 165,
    170–173, 177, 181, 183 ,184, 198, 199,
    201, 218, 224, 229, 268, 279
  nth minimal, 108, 111, 118, 119, 169, 182
  set, 10, 11
Random access machine, 17, 234
Relative error, *see* Error, relative
Relative model, *see* Model, relative
Restriction operator, 25, 26, 29, 32, 41, 60,
    70, 146, 199, 200, 219

S

Search for maximum problem, *see*
    Applications, search for maximum
    (minimum) problem
Secant iteration, 262, 263
Smolyak's theorem, 52, 54, 56–59, 94, 101,
    102, 109
Solution
  element, 10, 15, 48, 184, 218, 227, 230, 245,
    261, 275
  operator, 8–10, 25, 26, 30, 32, 33, 45, 59,
    91, 121, 136, 137, 140, 144, 146, 149,

    154, 159, 162, 164, 165, 170, 176, 185,
    186, 188–192, 196, 199, 200, 218, 220,
    225, 227, 242, 246, 248, 251, 275, 276
  linear, 41, 153, 159, 189
  nonlinear, 153, 159, 165, 188, 189, 191
Space
  $L_p(X)$, 90
  $W_p^r(X)$, 90
  $\tilde{W}_p^r(X)$, 90
Spline, 101, 105, 107, 110, 111, 131, 135, 220,
    279, 281, 315
  algorithm, *see* Algorithm, spline
  B-, 100
  interpolatory, 279
  natural, 57, 110
  perfect, 100, 101, 102, 120, 129, 130, 131
  unique, 131
Standard information, *see* Information,
    standard
Strongly noncomputable problem, *see*
    Problem, strongly noncomputable

T

Taylor interpolatory formula, *see*
    Interpolatory Taylor formula

U

Unimodal function, 4, 7, 16, 47, 163, 164,
    176, 178, 179, 183, 278

W

Width
  Gelfand n-, 6, 26, 35, 41, 42, 44–46, 59,
    65, 67, 108, 118, 119, 123, 129, 136,
    149, 220
  Kolmogorov n-, 7, 41, 64, 65, 149, 150, 157,
    158, 220, 221
  linear Kolmogorov n-, 6, 52, 64, 65, 67,
    123, 136, 220

ACM MONOGRAPH SERIES

Published under the auspices of the Association for Computing Machinery Inc.

Editor THOMAS A. STANDISH *University of California at Irvine*
Former Editors
Richard W. Hamming, Herbert B. Keller, Robert L. Ashenhurst

A. FINERMAN (Ed.) University Education in Computing Science, 1968

A. GINZBURG Algebraic Theory of Automata, 1968

E. F. CODD Cellular Automata, 1968

G. ERNST AND A. NEWELL GPS: A Case Study in Generality and Problem Solving, 1969

M. A. GAVRILOV AND A. D. ZAKREVSKII (Eds.) LYaPAS: A Programming Language for Logic and Coding Algorithms, 1969

THEODOR D. STERLING, EDGAR A. BERING, JR., SEYMOUR V. POLLACK, AND HERBERT VAUGHAN, JR. (Eds.) Visual Prosthesis: The Interdisciplinary Dialogue, 1971

JOHN R. RICE (Ed.) Mathematical Software, 1971

ELLIOTT I. ORGANICK Computer System Organization: The B5700/B6700 Series, 1973

NEIL D. JONES Computability Theory: An Introduction, 1973

ARTO SALOMAA Formal Languages, 1973

HARVEY ABRAMSON Theory and Application of a Bottom-Up Syntax-Directed Translator, 1973

GLEN G. LANGDON, JR. Logic Design: A Review of Theory and Practice, 1974

MONROE NEWBORN Computer Chess, 1975

ASHOK K. AGRAWALA AND TOMLINSON G. RAUSCHER Foundations of Microprogramming: Architecture, Software, and Applications, 1975

P. J. COURTOIS Decomposability: Queueing and Computer System Applications, 1977

JOHN R. METZNER AND BRUCE H. BARNES Decision Table Languages and Systems, 1977

ANITA K. JONES (Ed.) Perspectives on Computer Science: From the 10th Anniversary Symposium at the Computer Science Department, Carnegie-Mellon University, 1978

DAVID K. HSIAO, DOUGLAS S. KERR, AND STUART E. MADNICK Computer Security, 1979

ROBERT S. BOYER AND J STROTHER MOORE   A Computational Logic, 1979

J. F. TRAUB AND H. WOŹNIAKOWSKI   A General Theory of Optimal Algorithms

In preparation

R. L. WEXELBLAT   History of Programming Languages